Process Design for Cryogenics

Process Design for Cryogenics

Alexander Alekseev

WILEY-VCH

Author
Dr.-Ing. Alexander Alekseev
Linde Group
Dr.-Carl-von-Linde-Str. 6
Höllriegelskreuth
Germany, 82049

Cover Image: © Carter Hurd/
Shutterstock

Library of Congress Card No.: applied for

British Library Cataloguing-in-Publication Data
A catalogue record for this book is available from the British Library.

Bibliographic information published by the Deutsche Nationalbibliothek
The Deutsche Nationalbibliothek lists this publication in the Deutsche Nationalbibliografie; detailed bibliographic data are available on the Internet at <http://dnb.d-nb.de>.

Print ISBN: 978-3-527-34422-2
ePDF ISBN: 978-3-527-81560-9
ePub ISBN: 978-3-527-81562-3
oBook ISBN: 978-3-527-81563-0

Typesetting Straive, Chennai, India

Printed and bound by CPI Group (UK) Ltd, Croydon, CR0 4YY

C9783527344222_110724

To my wife Ekaterina, for her love and encouragement in keeping me healthy. She never ceased to convince me that the life beyond cryogenics is lovely, a notion she is absolutely correct about.

Contents

Preface *xix*
Acknowledgments *xxi*
Symbols, Signs, and Abbreviations *xxiii*

1 Introduction *1*
1.1 Historical Background *3*
1.2 Cryogenic Applications *5*
1.2.1 Natural Gas Liquefaction *6*
1.2.2 Cryogenic Air Separation *6*
1.2.3 Helium Plants *7*
1.2.4 Aerospace *7*
1.2.5 Physics *7*
1.2.6 Medical Applications *8*
1.2.7 Liquid Hydrogen for Hydrogen Economy *8*
1.2.8 Devices Based on High-temperature Superconductors for Electrical Engineering *8*
References *9*

2 Basics of Process Design *11*
2.1 Design Procedure *12*
2.2 Process Simulators and Fluid Properties *15*
2.3 Optimization *15*
2.4 Control Concept *17*
References *18*

3 Cryogenic Fluids *19*
3.1 Air *19*
3.2 Nitrogen *22*
3.3 Oxygen *23*
3.4 Argon *24*
3.5 Methane *25*
3.6 Neon *26*

3.7 Hydrogen 27
3.7.1 Spin Isomers of Hydrogen (Ortho- and Para-Hydrogen) 28
3.7.2 Consequences for Liquefaction Processes 30
3.8 Helium 31
3.8.1 Phase Diagram (Pressure–Temperature Diagram) and Lambda Line 32
3.8.2 Superfluidity 32
3.8.3 High Thermal Conductivity of Superfluid 33
3.9 Temperature–Enthalpy Diagram 34
3.10 Temperature–Entropy Diagram 36
References 38

4 Reversibility Concept 39
4.1 Reversibility and Irreversibility 39
4.1.1 Irreversibility 39
4.1.2 Reversibility 40
4.2 Carnot Engine 40
4.3 Reversible Process as Benchmark 42
4.4 Entropy 42
4.4.1 Entropy According to Rudolf Clausius 42
4.4.1.1 Extension of the Clausius Definition 43
4.4.2 Entropy According to Boltzmann 44
4.4.3 Entropy for Thermodynamic Analysis 44
4.5 Exergy Concept 45
4.5.1 Exergy of Mechanical and Electrical Energy 45
4.5.2 Exergy of Heat/Heat Flow 46
4.5.3 Exergy of Cold/Cooling Power 47
4.5.4 Exergy of Material/Material Stream 47
4.5.5 Reference States 48
4.5.6 Exergy for Multicomponent Materials/Material Flows 49
4.6 Application of Exergy Concept for Thermodynamic Analysis, Exergy Loss 49

5 Unit Operations for Cryogenic Processes 51
5.1 Isenthalpic Expansion 51
5.1.1 Joule–Thomson Effect 51
5.1.2 Expansion Valve 53
5.1.3 Joule–Thomson Coefficient 53
5.1.4 Isenthalpic Expansion in Temperature–Enthalpy Diagram 54
5.1.5 Isenthalpic Expansion in Temperature–Entropy Diagram 54
5.1.5.1 Learnings About the Joule–Thomson Coefficient from *T,s*-Diagram 54
5.1.6 Loss in an Isenthalpic Expansion Process 57
5.1.6.1 The Loss is Independent of Expansion Device Design 60
5.1.7 Joule–Thomson Expansion and Isenthalpic Expansion 60
5.1.8 Flashgas 61
5.1.9 Summary of Isenthalpic Expansion 61
5.2 Expansion in Expander 62

5.2.1 Energy Conservation Law for Expander *63*
5.2.2 Reversible Expansion *64*
5.2.3 Nonreversible Expansion (Real Turboexpander) *66*
5.2.3.1 Expander Efficiency in Terms of Enthalpy *66*
5.2.3.2 Impact of the Inlet Temperature *67*
5.2.3.3 Impact of Kappa *67*
5.2.3.4 Impact of the Isentropic Efficiency *68*
5.2.4 Exergy Loss in Expander *68*
5.2.4.1 The Loss in Expander Depends on Mechanical Design of the Expander *69*
5.2.5 Generated Power, Loading Devices *69*
5.2.6 Comparison, Isenthalpic Expansion Versus Expansion in Expander *70*
5.3 Heat Exchanger *72*
5.3.1 Double-Tube (Multitube) Heat Exchanger *72*
5.3.2 Definition: LTD and MTD *74*
5.3.2.1 Logarithmic Temperature Difference LTD *74*
5.3.2.2 Mean Temperature Difference MTD *74*
5.3.3 Definition: *Q/MTD* and *UA* *75*
5.3.3.1 What is the Q/MTD Good For? *76*
5.3.4 Multi-Stream Heat Exchanger *77*
5.3.4.1 Simplified Calculation Procedure *77*
5.3.5 Exergy Loss in Heat Exchangers *78*
5.3.5.1 General Consideration *78*
5.3.5.2 Exergy Loss in Heat Exchanger, Deep Dive *79*
5.3.6 Efficiency of Heat Exchanger *81*
5.4 Single Adiabatic Compression Stage *81*
5.4.1 Energy Conservation Law for Adiabatic Compression Stage *82*
5.4.2 Reversible Adiabatic Compression Stage *83*
5.4.3 Nonreversible Adiabatic Compression, Isentropic Efficiency *83*
5.4.3.1 Impact of the Inlet Temperature *84*
5.4.3.2 Impact of the Isentropic Efficiency *84*
5.4.3.3 Impact of the Isentropic Exponent *84*
5.4.4 Exergy Loss, Adiabatic Compressor Stage *85*
5.5 Multi-Stage Compression, Isothermal Compression *86*
5.5.1.1 Special Case: Main Air Compressor *86*
5.5.2 Reversible Isothermal Compression *88*
5.5.3 Nonreversible Isothermal Compression, Isothermal Efficiency *89*
5.5.3.1 Typical Values *90*
5.5.3.2 Connection to Adiabatic Single Compression Stage *90*
5.5.3.3 Rigorous Calculation of Multi-Stage Compressor *90*
5.5.4 Rule of Thumb for Isothermal Compression *91*
5.5.5 Energy Conservation Law For Multi-Stage Compressor *92*
5.5.6 Second Law for Multi-Stage Compressor *92*
5.5.7 Exergy Loss, Isothermal Compression *93*
5.5.8 Comparison, Adiabatic Versus Isothermal Compression *94*
References *94*

6 Key Hardware Components *95*
6.1 Thermal Insulation/Coldbox *95*
6.1.1 Low-performance Insulation *97*
6.1.2 High-performance Insulation *97*
6.1.3 Dewar Vessel *99*
6.2 Heat Exchanger *101*
6.2.1 Aluminum Plate-fin Heat Exchanger *102*
6.2.1.1 Mechanical Design *102*
6.2.1.2 Thermal Design, Hydraulic Design *104*
6.2.1.3 Attributes *105*
6.2.1.4 Restrictions and Limitations *107*
6.2.2 Spiral-wound Heat Exchanger *108*
6.2.2.1 Mechanical Design *108*
6.2.2.2 Thermal and Hydraulic Design *110*
6.2.2.3 Restrictions and Limitations Relevant for Use and Operation *114*
6.2.2.4 Summary: Spiral-wound Heat Exchanger *115*
6.3 Expanders *115*
6.3.1 General Issues *116*
6.3.1.1 Expansion Process in Detail *116*
6.3.1.2 Isentropic Efficiency *118*
6.3.1.3 Bearing System *118*
6.3.2 Requirements for Design of a Process with Expander *121*
6.3.2.1 Limited Pressure Ratio *121*
6.3.2.2 No Phase State Change *122*
6.3.3 Large Turboexpanders with Oil Bearings *123*
6.3.3.1 Mechanical Design *124*
6.3.3.2 Bearings for Nitrogen Liquefiers, Labyrinth Sealing, and Seal Gas *125*
6.3.3.3 Control and Instrumentation *126*
6.3.4 Helium Expander *127*
6.3.4.1 Mechanical Design *128*
6.3.4.2 Bearing System *129*
6.3.4.3 Utilization of Expander Power *132*
6.3.4.4 Control and Instrumentation *133*
6.3.4.5 Assessment of Helium Expander *133*
6.3.5 Joule–Thomson Expander *133*
6.3.5.1 Capacity *135*
6.3.5.2 Techno-economics *135*
6.3.5.3 Mechanical Design *136*
6.4 Expansion Valve *136*
6.5 Compressors *136*
6.5.1 General Issues *136*
6.5.1.1 Energy Transformation Chain for Compression *137*
6.5.1.2 Compressor Lubrication *138*

6.5.2 Centrifugal Compressor *139*
6.5.2.1 Mechanical Design and Principle of Compression in a Centrifugal Compressor Stage *139*
6.5.2.2 Impeller Design *141*
6.5.2.3 Multi-staging *142*
6.5.2.4 Thermodynamics of Centrifugal Compressors *143*
6.5.2.5 Control of Centrifugal Compressors *146*
6.5.2.6 Driver for Compressor *149*
6.5.3 Screw Compressor *149*
6.5.3.1 Screw Compressor, Design, and Working Principle *150*
6.5.3.2 Tuning of Available Screw Compressors *152*
6.5.3.3 Process Parameters *152*
6.5.3.4 Oil Removal System (ORS) *153*
6.5.3.5 Compressor Control *153*
6.5.3.6 Assessment of Screw Compressors *153*
6.5.4 Reciprocating Compressor *154*
6.5.4.1 Design and Working Principle *154*
6.5.4.2 Process-relevant Aspects *159*
6.5.4.3 Mechanical Design *161*
6.5.4.4 Load Range *162*
6.5.4.5 Control *165*
6.5.4.6 Pulsations and Pulsation Control *165*
6.5.4.7 Short Assessment of Reciprocating Compressors *166*
References *167*

7 Cryogenic Refrigeration *169*
7.1 Principle of Cryogenic Refrigeration *172*
7.1.1 Refrigerator and Definitions Around It *172*
7.1.2 Energy Conservation Law for Refrigerator *173*
7.1.2.1 Energy Conservation Law for Refrigerator *173*
7.1.2.2 Three Interfaces *174*
7.1.2.3 The Pump Analogy *174*
7.1.3 Parasitic Heat Flow Through Thermal Insulation *176*
7.1.4 Exercise: Energy Conservation for Refrigerator *177*
7.1.5 Reversible Refrigerator and Carnot Equation *177*
7.1.5.1 Reversible Refrigerator *177*
7.1.5.2 Carnot Equation *178*
7.1.5.3 Extension of Carnot Equation to Irreversible Refrigerator *178*
7.1.5.4 What Carnot Equation Expresses? *178*
7.1.6 The Second Law of Thermodynamics for Refrigerator *179*
7.1.7 Minimum Power Requirement for Typical Cooling Temperatures *179*
7.1.8 Key Performance Indicators for Refrigerators *180*
7.1.8.1 Coefficient of Performance *180*

7.1.8.2 Carnot Efficiency *181*
7.1.8.3 Specific Power *181*
7.2 Joule–Thomson Process (Nitrogen Joule–Thomson Process) *181*
7.2.1 Main Elements *182*
7.2.1.1 Thermal Insulation/Coldbox *182*
7.2.1.2 Heat Exchanger *182*
7.2.1.3 Compressor *183*
7.2.1.4 Expansion Valve *183*
7.2.1.5 Evaporator *183*
7.2.1.6 Joule–Thomson Refrigerator Assembled *184*
7.2.2 Process and Process Features *185*
7.2.2.1 Cool-Down Procedure *185*
7.2.2.2 Joule–Thomson Process Description *187*
7.2.2.3 Cooling Capacity of the Joule–Thomson Process *187*
7.2.2.4 Typical Process Pressure Levels *192*
7.2.3 Example Process Calculation, Nitrogen-Joule–Thomson Process *194*
7.2.3.1 Task Definition *194*
7.2.3.2 Process and Hardware Concept *194*
7.2.3.3 Process Simulation *196*
7.2.3.4 Process Units *199*
7.2.3.5 Overall Performance of the Designed Process *202*
7.2.3.6 Process Analysis *202*
7.2.4 Process Optimization *208*
7.2.4.1 Variables *208*
7.2.4.2 Example of Process Optimization *209*
7.2.4.3 In Summary *210*
7.2.5 Control of a Joule–Thomson Refrigerator *211*
7.2.6 Assessment of the Joule–Thomson Process *213*
7.3 Brayton Process *213*
7.3.1 Main Elements *214*
7.3.1.1 Comment About Compression *215*
7.3.2 Process and Process Features *215*
7.3.2.1 Process Description *215*
7.3.2.2 Phase State Conditions *216*
7.3.2.3 Working Fluids *216*
7.3.2.4 Cooling Temperature *217*
7.3.2.5 Typical Process Pressure Levels *217*
7.3.2.6 Cooling Capacity *217*
7.3.2.7 Reversible Brayton Process *219*
7.3.3 Example, Process Calculation, Helium Brayton Process *220*
7.3.3.1 Task Definition *220*
7.3.3.2 Properties Data *220*
7.3.3.3 Process and Hardware Concept *220*
7.3.3.4 Process Simulation *223*
7.3.3.5 Process Units *224*

7.3.3.6 Overall Performance of the Process 227
7.3.3.7 Process Analysis 227
7.3.4 Process Optimization 230
7.3.4.1 Variables 230
7.3.4.2 Constraints 230
7.3.4.3 Example of Process Optimization 230
7.3.4.4 In Summary 232
7.3.5 Control of a Brayton Refrigerator 233
7.3.6 Hardware Components 233
7.3.6.1 Heat Exchanger 233
7.3.6.2 Compressor 234
7.3.6.3 Expander 234
7.3.7 Applications 235
7.3.7.1 Air Separation 235
7.3.7.2 Hydrogen Liquefaction 235
7.3.7.3 Turbo-Brayton Refrigerator 235
7.3.8 Assessment of the Brayton Process 235
7.4 Claude Process 236
7.4.1 Basic Claude Process, Main Elements 236
7.4.1.1 Morphology 237
7.4.1.2 Special Feature 238
7.4.1.3 Extended Definition of the Claude Process 238
7.4.2 Process and Process Features 238
7.4.2.1 Main Idea of the Claude Process 238
7.4.2.2 Basic Claude Process, Process Description 239
7.4.2.3 Typical Process Pressure Levels 240
7.4.2.4 Cooling Capacity 241
7.4.3 Nitrogen Claude Process 243
7.4.3.1 Process Description 243
7.4.4 Example, Process Calculation, Nitrogen Claude Process 245
7.4.4.1 Task Definition 245
7.4.4.2 Process and Hardware Concept 245
7.4.4.3 Properties Data 247
7.4.4.4 Process Simulation 247
7.4.4.5 Process Units 249
7.4.4.6 Overall Performance 254
7.4.4.7 Process Analysis 254
7.4.5 Helium Claude Process 257
7.4.5.1 Process Description 259
7.4.5.2 Special Feature 260
7.4.5.3 Staging 260
7.4.6 Example, Process Calculation, Helium Claude Process 261
7.4.6.1 Task Definition 261
7.4.6.2 Process and Hardware Concept 261
7.4.6.3 Properties Data 264

7.4.6.4 Process Simulation *264*
7.4.6.5 Process Units *264*
7.4.6.6 Overall Performance *270*
7.4.6.7 Process Analysis *270*
7.4.6.8 Review of Simulation Procedure *275*
7.4.7 Process Optimization *276*
7.4.8 Control of a Claude Refrigerator *277*
7.4.9 Hardware Components *277*
7.4.9.1 Compressor *277*
7.4.9.2 Heat Exchanger *277*
7.4.9.3 Expander *278*
7.4.9.4 Commercial Availability *278*
7.4.10 Applications *279*
7.4.11 Assessment of the Claude Process *279*
7.5 Mixed-Fluid Joule–Thomson Process *279*
7.5.1 Main Elements *280*
7.5.2 Mixed Fluid as Refrigerant, Properties *280*
7.5.2.1 Main Components of Mixed Fluid *281*
7.5.2.2 Thermodynamic Fundamentals for Mixed Fluids *281*
7.5.2.3 T,s-Diagram for Typical Mixed Fluid *282*
7.5.2.4 T,h-Diagram for a Typical Mixed Fluid *286*
7.5.2.5 Isobaric Cooling/Heating Of Mixed Fluid *289*
7.5.2.6 Comparison Mixed Fluid Versus Nitrogen *291*
7.5.3 Process Features *294*
7.5.3.1 Phase State Conditions *294*
7.5.3.2 Difference Between the Mixed-Fluid-Based and Nitrogen-Based Joule–Thomson Process *297*
7.5.3.3 Temperature Profile in Heat Exchanger *297*
7.5.3.4 Location of Pinch Point(S) *300*
7.5.3.5 Cooling Capacity of a Mixed-Fluid Joule–Thomson Process *301*
7.5.3.6 Typical Process Pressure Levels *303*
7.5.4 Example Process Calculation, Mixed-Fluid Joule–Thomson Process *304*
7.5.4.1 Task Definition *305*
7.5.4.2 Properties Data *305*
7.5.4.3 Process and Hardware Concept *305*
7.5.4.4 Process Simulation *307*
7.5.4.5 Process Units *308*
7.5.4.6 Overall Performance of the Designed Process *311*
7.5.4.7 Process Analysis *311*
7.5.5 Design of the Mixed Fluid *315*
7.5.5.1 Mixed-Fluid Composition *315*
7.5.5.2 Ballast Gas Addition *319*
7.5.5.3 Solidification *320*
7.5.6 Optimization of a Mixed-Fluid Joule–Thomson Process *323*

7.5.7 Control of a Mixed-Fluid Joule–Thomson Refrigerator *323*
7.5.7.1 A New Option *324*
7.5.8 Hardware Components *324*
7.5.8.1 Compressor for Mixed-Fluid Refrigerator *325*
7.5.8.2 Heat Exchanger for Mixed-Fluid Joule–Thomson System *327*
7.5.9 Assessment of the Mixed-Fluid Joule–Thomson Process *330*
7.5.9.1 Challenging Issues *330*
7.5.9.2 Applications *330*
7.6 Cryogenic Refrigeration Processes, Life Cycle *330*
7.6.1 Joule–Thomson Process *330*
7.6.2 Brayton Process *331*
7.6.3 Claude Process *331*
7.6.4 Mixed-Fluid Joule–Thomson Process *331*
References *331*

8 Liquefaction of Cryogenic Gases *333*
8.1 Thermodynamic Basics for Liquefaction *333*
8.1.1 Definitions for Liquefaction Processes *333*
8.1.1.1 Isobaric Liquefaction Process in T,s-Diagram *333*
8.1.1.2 Liquefier *334*
8.1.1.3 Liquefaction Capacity *335*
8.1.1.4 Connection to Refrigeration Processes *335*
8.1.2 Minimum Liquefaction Power/Work *335*
8.1.2.1 Minimum Liquefaction Power via Carnot Equation *335*
8.1.2.2 Minimum Liquefaction Power via Exergy *336*
8.1.2.3 Minimum Liquefaction Work for Typical Cryogenic Gases *337*
8.1.2.4 Key Performance Indicators for Liquefiers *337*
8.1.3 Specifics of Process Design for Liquefaction *338*
8.1.3.1 Early Liquefaction Processes *338*
8.1.3.2 Process Design Approaches for Recent Liquefaction Plants *343*
8.1.4 Classification of Liquefaction Processes *348*
8.1.5 Liquefaction with External Cooling *348*
8.1.5.1 Principle of External Cooling *348*
8.1.5.2 Management of Flashgas *353*
8.1.5.3 Internal Versus External Cooling *354*
8.2 Nitrogen Liquefaction *355*
8.2.1 General Setup, Feed Gas, and Products *355*
8.2.2 Power Requirements *356*
8.2.3 Main Elements *356*
8.2.3.1 Compressor *356*
8.2.3.2 Heat Exchanger *357*
8.2.3.3 Large Expander *357*
8.2.3.4 Impact of Heat Exchanger Design on Process Design *357*
8.2.3.5 Impact of Expander Power on Process Design *359*

8.2.4 Process Design for Nitrogen Liquefaction *359*
8.2.4.1 Guidelines/Principles of Process Design for Nitrogen Liquefaction *360*
8.2.4.2 Core Process for Nitrogen Liquefaction *361*
8.2.4.3 Process Extension for Pressureless LIN As Product *364*
8.2.4.4 Process Extension for Pressureless Nitrogen as Feed *366*
8.2.4.5 Example of a Dual-Expander Claude Process for Nitrogen Liquefaction *366*
8.2.5 Example of an Industrial Plant *369*
8.2.5.1 Process Design *370*
8.2.5.2 Mechanical Design *372*
8.2.5.3 Brief Assessment of This Process *372*
8.2.6 Operation of Nitrogen Liquefier in Combination with ASU *373*
8.2.6.1 Liquid Nitrogen as Cold Carrier for Other Liquid Products *373*
8.2.6.2 Flexible Operation for Optimization of Power Bill *373*
8.2.7 Status Quo Nitrogen Liquefaction *374*
8.3 Helium Liquefaction *374*
8.3.1 Overall Helium Liquefier Setup, Feed Gas, and Product *376*
8.3.2 Performance Considerations *378*
8.3.2.1 Power Requirements *378*
8.3.2.2 Definition of Liquefaction Capacity for Helium Liquefiers *380*
8.3.2.3 Performance Comparison *380*
8.3.3 Main Elements and Impact on Process Design *381*
8.3.3.1 Heat Exchanger *381*
8.3.3.2 Compressor *382*
8.3.3.3 Helium Expander *383*
8.3.3.4 Impact of Expander Power on Process Design *383*
8.3.4 Process Design *385*
8.3.4.1 Typical Features of Helium Liquefaction Process *385*
8.3.4.2 Core Process for Helium Liquefaction *386*
8.3.4.3 Process Extension, Helium Purification System *392*
8.3.4.4 Example of a Claude Process For Helium Liquefaction *393*
8.3.5 Example of a Commercial Helium Liquefier *395*
8.3.5.1 Process Design *395*
8.3.5.2 Compressor System *396*
8.3.5.3 Cold Box, Mechanical Design *397*
8.3.5.4 Cold Box, Thermal Insulation *397*
8.3.6 Status Quo, Helium Liquefaction *398*
8.4 Natural Gas Liquefaction *400*
8.4.1 Natural Gas Composition *401*
8.4.2 LNG Infrastructure and Process Chain *402*
8.4.2.1 Infrastructure Process Chain *403*
8.4.2.2 Feed Gas and Product *404*
8.4.3 Power Requirements *404*
8.4.4 Key Hardware Components *406*
8.4.5 Process Design for Natural Gas Liquefaction *406*

8.4.5.1 Basic Mixed-Fluid Joule–Thomson Process *408*
8.4.5.2 Impact of Maldistribution on Design of a Mixed-Fluid-Based Process *418*
8.4.5.3 Mixed-Fluid Joule–Thomson Process with Single Phase-Separator *421*
8.4.5.4 Mixed-Fluid Joule–Thomson Process with Two-/Multiple-Phase Separators *427*
8.4.5.5 Mixed-Fluid Joule–Thomson Process with External Pre-Cooling Stage *427*
8.4.6 Examples of Industrial Process for Natural Gas Liquefaction *432*
8.4.6.1 Process Example with Propane PreCooling *432*
8.4.6.2 Process Example with Aluminum Plate-Fin Heat Exchanger *435*
8.4.7 Outlook, Natural Gas Liquefaction *437*
References *438*

9 Supplements *439*
9.1 Supplements for Cryogenic Refrigeration *439*
9.1.1 Example of Techno-economic Optimization *439*
9.1.2 Process Design Approaches for Improvement of Joule–Thomson Process *441*
9.1.2.1 Relocate the Process to a Higher Pressure Level *441*
9.1.2.2 Implementation of a Precooling Stage *444*
9.1.2.3 A Combination of These Approaches *445*
9.1.2.4 Approaches for Improvement of Joule–Thomson Process, Summary *445*
9.2 Supplements for Liquefaction *445*
9.2.1 Original Claude Process *445*
9.2.2 Kapitsa Process *447*
9.2.3 Example of a Nitrogen Liquefaction Process *452*
9.2.3.1 Task Definition *452*
9.2.3.2 Process and Hardware Concept *452*
9.2.3.3 Process Design Calculation, Cold Part of Process *454*
9.2.3.4 Heat Exchanger *457*
9.2.3.5 Compressor *458*
9.2.3.6 Expander-booster Units *459*
9.2.3.7 Overall Process *461*
9.2.3.8 Losses *461*
9.2.3.9 Evaluation *462*
9.2.4 Helium Liquefaction Process with Integrated Liquid Nitrogen Precooling *462*
9.2.4.1 Procedure for "Process-to-Hardware" Design" *462*
9.2.4.2 Formal Task Definition *463*
9.2.4.3 Compressor *465*
9.2.4.4 Expanders *467*
9.2.4.5 Expansion Valve *469*
9.2.4.6 Liquid Nitrogen *469*

9.2.4.7 Overall Performance and Losses *469*
9.2.5 Basic Mixed-Fluid Joule–Thomson Process for Natural Gas Liquefaction *470*
9.2.5.1 Task Definition *471*
9.2.5.2 Process and Hardware Concept *472*
9.2.5.3 Thermodynamic Properties *474*
9.2.5.4 Simulation Results *474*
9.2.5.5 Heat Exchanger *475*
9.2.5.6 Compressor *478*
9.2.5.7 Overall Performance and Losses *478*
9.2.6 Mixed Fluid Joule–Thomson Process with Single-phase Separator *479*
9.2.6.1 Task Definition *479*
9.2.6.2 Simulation Results *479*
9.2.6.3 Heat Exchanger *479*
9.2.6.4 Compressor *483*
9.2.6.5 Overall Performance and Losses *483*
9.2.7 Mixed Fluid Joule–Thomson Process with Two or More Phase Separators *483*
9.2.7.1 Primary Concept *484*
9.2.7.2 Mixed-fluid Composition *486*
9.2.7.3 Process Variations *486*
9.2.7.4 Brief Assessment *486*
References *487*

Index *489*

Preface

This book is an outgrowth of courses in "Cryogenics and Industrial Cryogenic Plants" taught at the Technical University of Munich over the past 12 years. This course emphasizes process design and simulation of cryogenic processes. Consequently, the book specifically addresses topics and aspects related to process design, rather than covering cryogenic matter in its entirety. This relatively narrow specialization of the book reflects the increasing professional differentiation within the engineering community. This trend is also observable in the technical literature about cryogenics: there are books tailored for physicists by physicists (such as [1–3]), for mechanical engineers by mechanical engineers (such as [4–8]), and for material property experts by their counterparts. However, there is a scarcity of literature on low-temperature technology specifically tailored for chemical engineers by process engineers, with only Isalski's book from 1989 being mentioned. With this book, I endeavor to bridge this gap.

The main features of this book are as follows:

1. Focus on Four Basic Processes:

   This book exclusively focuses on four fundamental cryogenic processes: Joule–Thomson process, mixed fluid Joule–Thomson process, Brayton process, and Claude process. The decision to limit the study material to these four processes was made deliberately.

   These four processes are comprehensively examined in Chapter 7 about cryogenic refrigeration. They are then revisited within the context of their application in Chapter 8 about liquefaction processes. The approach and methodology employed are universally applicable and can be extended to all other cryogenic and non-cryogenic processes.
2. Process Examples:

   The book includes several examples of process calculations that illustrate the typical relationships.
3. Thermodynamics:

   Approximately one-quarter of the book (Chapters 4 and 5) is devoted to thermodynamic fundamentals. While these concepts are commonly covered in other texts, I have meticulously curated the most essential and valuable content to optimize its application in process analysis, particularly in Chapters 7, 8, and 9.

4. Hardware Components:
   The detailed design (sizing/dimensioning) of key hardware components (compressors, expanders, heat exchangers, etc.) is not extensively covered in this book. But since these hardware components play a pivotal role in determining process design, they are briefly outlined in Chapter 6 to provide an overview.

The book contains some sections addressing specific aspects or topics, which possess a broad applicability across all cryogenic processes (such as the drawbacks of subatmospheric operating conditions or methods for comparing two processes). However, these text blocks are embedded in chapters for individual basic processes, where they fit best in terms of content. Creating a separate chapter for it would lead to a complicated structure of the book. I expect that the attentive reader will recognize and correctly assign these contents.

Not covered within the scope of the book are cascade processes (as well as cascading as a process design method), processes based on regenerators as heat exchange device, and electromagnetic (electrocaloric, magnetocaloric, or optical) methods. Topics such as the thermodynamic description of material properties (equations of state, etc.), measurement technology, and instrumentation are also not addressed in this book; there are many good books available on these subjects.

The time constraints and other limitations imposed by my full-time professional commitments may have led to the text not being as polished as I would have liked. However, I am eager to move forward and publish the book in its current state. I welcome feedback, corrections, and suggestions for improvement. Please send them directly to me.

References

1 Pobell, F. (2007). *Matter and Methods at Low Temperatures*. Berlin, Heidelberg: Springer-Verlag. ISBN: 978-3-540-46356-6; eBook ISBN: 978-3-540-46360-3. https://doi.org/10.1007/978-3-540-46360-3.

2 Ventura, G. and Risegari, L. (2010). *The Art of Cryogenics: Low-Temperature Experimental Techniques*. Elsevier.

3 Van Sciver, S.W. (2012). *Helium Cryogenics*, 470. New York: Springer Science. http://dx.doi.org/10.1007/978-1-4419-9979-5.

4 Jungnickel, H., Agsten, R., and Kraus, W.E. (1990). *Grundlagen der Kältetechnik*. Berlin: Verlag Technik GmbH.

5 Flynn, T.M. (1997). *Cryogenic Engineering*. CRC Press.

6 Barron, R.F. (1985). *Cryogenic Systems*. Monographs on Cryogenics.

7 Timmerhaus, K.D. and Flynn, T.M. (2013). *Cryogenic Process Engineering*. New York: Springer. ISBN: 978-1-4684-8758-9; eBook ISBN: 978-1-4684-8756-5. https://doi.org/10.1007/978-1-4684-8756-5.

8 Hausen, H. and Linde, H. (1985). *Tieftemperaturtechnik*. Berlin, Heidelberg, New York, Tokyo: Springer Verlag.

Acknowledgments

I learned to apply the exergy concept as a tool for thermodynamic analysis from Prof. Viktor Brodianski in Moscow, where Vladimir Mogorychny also made significant contributions to my development.

During my doctoral studies at the Dresden University of Technology, I gleaned the fundamental principles of process engineering from Prof. Hans Quack. Our most engaging conversations often occurred on the way to lunch in the cafeteria. I am confident he will recognize his foundational ideas and arguments, and I hope he finds satisfaction in their application.

Later, I started to work as a process engineer at Linde AG, where I benefited immensely from the expertise of Horst Corduan and Dr. Dirk Schwenk. Furthermore, my initial department head, Gerhard Pompl, provided steadfast support for all my process engineering investigations. Six additional names from the Linde Group deserve special mention – Prof. Hans Kistenmacher, Prof. Aldo Belloni, Dr. Andreas Opfermann, Dr. Harald Ranke, Dr. Krish Krishnamurthy, and Prof. Harald Klein, who is now Chair of the Department for Plant and Process Technology at the Technical University of Munich. These colleagues have been instrumental in supporting and inspiring me to establish the lecture "Cryogenics and Industrial Cryogenic Plants" at the Technical University of Munich, which serves as the foundation for the content of this book. In addition, I am grateful for the unwavering support provided by the current management, Dr. Amitabh Gupta and Dr. Minish Shah. My heartfelt thanks go out to all the aforementioned personalities.

I express my most sincere appreciation to several people who contributed to the success of this project: Dr. Wolfgang Eberhard Kraus reviewed and provided feedback on the contents of the first five chapters. While I was unable to address all his comments, I am immensely grateful to him for his efforts. Paul Glatt and Elisabeth Mae Wilbur have corrected my English and contributed to making the text more comprehensible. Many thanks to the doctoral students who helped me with lectures: Johanna Hemaurer, Marc Xia, Umberto Cardella, Christian Wolf, and Chengyuan Wu, to Senior Engineer Sebastian Rehfeldt, and Secretary Melanie Laubenbacher, as well as to the students who prepared the German version of the book: Martin Kibili, Manuel Mayr, Alexander Vogt, Sebastian Obermeier, Patrick Brandl, Lola Mercier, Ahmed Elkhoshet, Giovanni Turi da Fonte Dias, Mauricio Platteau, and Philipp Schwab.

Symbols, Signs, and Abbreviations

Latin Symbols

Symbol	Description	Example of units
COP	coefficient of performance	
C_p	heat capacity at constant pressure	J/K
c_p	mass [or molar] specific heat capacity at constant pressure	J/(kg K) J/(mol K)
C_v	heat capacity at constant volume	J/K J/K
c_v	mass [or molar] specific heat capacity at constant volume	J/(kg K) J/(mol K)
D	exergy loss	J, kWh
$\dot{D}$	exergy loss flow	W, kW
E	exergy	J
$\dot{E}$	exergy flow	W
e	mass [or molar] specific exergy	J/kg J/mol
H	enthalpy	J, kWh
$\dot{H}$	enthalpy flow	W, kW
h	mass [or molar] specific enthalpy	J/kg J/mol
h_b	mass [or molar] specific enthalpy at the boiling point	J/kg J/mol
h_d	mass [or molar] specific enthalpy at the dew point	J/kg J/mol
m	mass	kg
$\dot{m}$	mass flow or mass flowrate	kg/s
M or M_r	molecular weight	g/mol
MTD	mean temperature difference in heat exchanger	K

Symbol	Description	Example of units
LTD	logarithmic temperature difference in heat exchanger	K
LF	liquid fraction = molar [or mass] fraction of liquid in a two-phase fluid/material stream	
N	molar amount of substance	mol
$\dot{N}$	molar flow or molar flowrate	mol/s Nm^3/h
p	pressure	bar
p_s	boiling pressure or saturation pressure or vapor pressure	bar
P	power as mechanical or electric power	W, kW
P_{EXP}	mechanical power generated by an expander	W, kW
P_{COMP}	mechanical power required for driving a compressor	W, kW
Q	heat	J, kWh
$\dot{Q}$	heat flow	W, kW
$\dot{Q}_o$	cooling capacity (or cooling power) of a refrigerator	W, kW
$\dot{Q}_{amb}$	waste heat rejected into environment/ambient (to cooling air or cooling water)	W, kW
$\dot{Q}^{HEX}$	heat flow, transferred from warm side/fluid to cold side/fluid in a heat exchanger = heat exchanger duty	W, kW
$\dot{Q}_{is}$	heat flow through non-ideal thermal insulation	W, kW
q	mass [or molar] specific heat	J/kg
$\dot{q}$	mass [or molar] specific heat flow	W/kg
R_m	universal gas constant $R = 8.314$ J/(mol K)	J/(mol K)
S	entropy	J/K
$\dot{S}$	entropy flow	W/K
s	mass [or molar] specific entropy	J/(kg K) J/(mol K)
s_b	mass [or molar] specific entropy at the boiling point	J/(kg K) J/(mol K)
s_d	mass [or molar] specific entropy at the dew point	J/(kg K) J/(mol K)
T	temperature	K
T_s	boiling temperature or saturation temperature	K
T_o	cooling temperature – temperature of the object being cooled	K
T_{amb}	ambient temperature = environment temperature or room temperature	K
T_c	critical temperature	K
u	fluid velocity	m/s
VF	vapor fraction = molar [or mass] fraction of liquid in a two-phase fluid/material stream	mol/mol or kg/kg

Symbol	Description	Example of units
W	mechanical work	J, kWh
x_i	molar [or mass] fraction of a component in liquid mixture	mol/mol kg/kg
y_i	molar [or mass] fraction of a component in gaseous/vapor mixture	mol/mol kg/kg
z_i	molar [or mass] fraction of a component in a mixture	mol/mol kg/kg

Greek Symbols

γ	gamma = heat capacity ratio c_p/c_v = isentropic exponent for an ideal gas	–
ε	synonymous to coefficient of performance COP	–
$\kappa = \kappa$	kappa = isentropic exponent for a real gas	–
μ	Joule–Thomson coefficient	K/bar
μ	in chapter 4.4.1 it is used for chemical potential	J/mol
η	efficiency	
η_e	exergy efficiency = Carnot efficiency	
η_s	isentropic efficiency of a compressor stage or expander stage	
η_T	isothermic efficiency of a multistage compressor	
η_{LIQ}	efficiency of a liquefier	
π	pressure ratio in an expander or a compressor	
ρ	density of fluid	kg/m^3
ϑ	temperature in °C	°C
λ	thermal heat conductivity	W/(m K)
Σ	sum	

Mixed Symbols

Δh_v	molar [or mass] enthalpy of vaporization	J/kg J/mol
$\Delta \dot{H}^{WE}$	enthalpy [flow] difference at the warm end of a heat exchanger	kW
ΔS	entropy rise	J/K
Δs	specific entropy rise	J/(mol K)
ΔT	temperature difference	K
ΔT_{min}	minimal temperature difference in a heat exchanger	K
ΔT_{max}	maximal temperature difference in a heat exchanger	K
ΔT^{CE}	temperature difference at the cold end of a heat exchanger	K
ΔT^{WE}	temperature difference at the warm end of a heat exchanger	K
T_λ	temperature at λ – point (for liquid helium)	K

Subscripts

amb	used for ambient/environment conditions
c	used for cold objects (cold material streams, cold expanders and similar)
c	sometimes relates to critical states, for example, critical temperature T_c or critical pressure p_c
Carnot	relates to Carnot process or generally to a reversible process
COMP	relates to compressor
EXP	relates to expander
g	used for gaseous phase
HP	relates to High Pressure level
i	used either for a mixture component i or for heat exchanger zone i
in	used for inlet streams (material streams, energy streams and other streams)
JT	used for Joule–Thomson (Joule–Thomson process, Joule–Thomson expansion and similar)
L or *l*	used for liquid phase
LP	relates to Low Pressure level
out	used for outlet streams (material streams, energy streams and other streams)
V or *v*	used for vapor phase
w	used for warm objects (warm material streams, warm expanders and similar)
1, 2, 3, 4, 5, etc.	relate to corresponding state in process flow diagram
2phase	relates to two-phase state

Superscripts

COMP	relates to compressor
EXP	relates to expander
HEX	relates to heat exchanger
in	used for inlet streams (material streams, energy streams and other streams)
JT	relates to Joule Thomson valve
out	used for outlet streams (material streams, energy streams and other streams)
REF or *o*	relates to reference state

Further Abbreviation

AIR	air
CNG	compressed natural gas
GAN	gaseous nitrogen
GAP	gaseous air (pressurized)
GOX	gaseous oxygen
LAP	liquid air (pressurized)
LAR	liquid argon
LH2	liquid hydrogen
LHe or LHE	liquid helium
LIN	liquid nitrogen
LNG	liquefied natural gas
LOX	liquid oxygen
PGAN	pressurized gaseous nitrogen

Symbols Used for Process Flow Diagrams

	gaseous material stream
	liquid material stream
	two-phase fluid
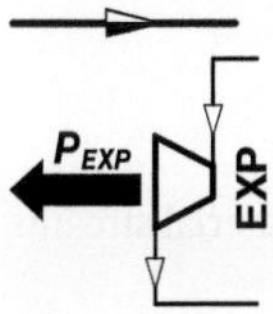	expander or turboexpander
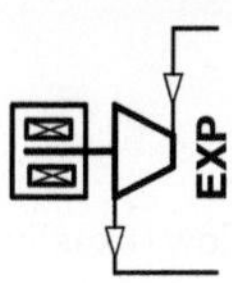	expander with generator brake
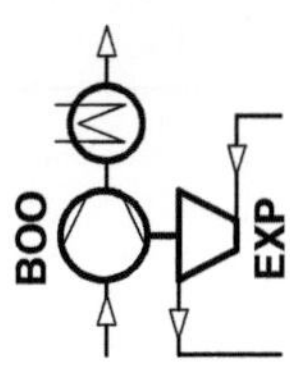	expander-booster-combination
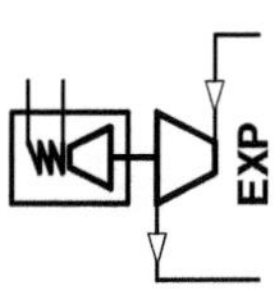	expander with oil brake or gas brake

expansion valve or
Joule-Thomson-valve or
control valve

multistage quasi-isothermic compressor unit

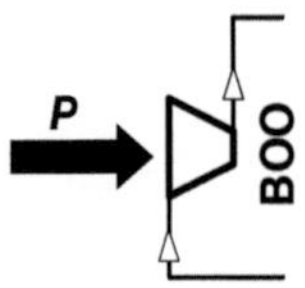

compressor (turbocompressor) stage or booster stage

water cooler (usually after a compressor unit)

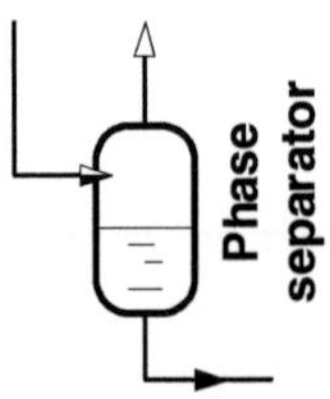

phase separator

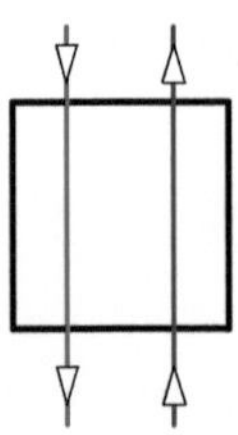

heat exchanger with a warm (red) and cold (blue) material streams

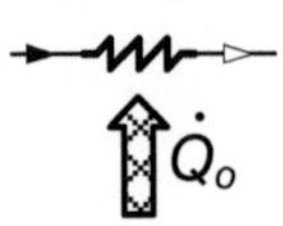

evaporator [or cold gas heat exchanger], where a heat flow (usually from object being cooled) is applied to a cooling fluid

1
Introduction

Generally, refrigeration technology is the engineering science of how to cool a corresponding cooling object to a temperature below the ambient temperature and/or how to keep it cold continuously at this low temperature for the required time. Usually, the cooling is done by means of a special device – a refrigerator. Typical examples of refrigerators known from our everyday life are household refrigerators, chest freezers, and air conditioners.

Cryotechnology (from the ancient Greek κρύος (*kryos*) – cold, ice) is a part of refrigeration technology covering the range of very low temperatures (see Figure 1.1). In terms of numbers, probably, the most successful commercial cryosystem is the superconducting magnet at the core of an magnetic resonance imaging (MRI) system, as found in most large hospitals today. Other typical examples of cryosystems are helium liquefier, air separation plant, or liquid hydrogen storage.

Principally, the cryotechnology consists of two large areas: (i) technologies for use/application of very low-temperature cold and (ii) that for generation of cryogenic cold (see Figure 1.1). The latter is called cryogenics (also from ancient Greek κρύος (*kryos*) – cold, ice and γενεά (*geneá*) – generation, or γένεσις (*génesis*) – origin), which is literally translated as “deep cold generation.” A typical example of a **cryogenic** system is a helium liquefier. In contrast, a superconducting magnet at the core of MRI systems, as mentioned above, is a typical example of a cryo application.

In English-speaking community, the terms “cryogenic” and “cryogenics” are used often as synonym for “cryo” and “cryo technology”: a cryo application can be referred to as a “cryogenic application,” although no active cold generation takes place within such systems.

In cryogenic engineering, it is common to specify and define the temperature $\boldsymbol{T}$ in degrees Kelvin (K) as absolute temperature instead of relative temperature Θ usually measured in degrees Celsius (°C). The conversion is simple and is performed as follows:

$$\mathrm{T}\,(\mathrm{K}) = \Theta\,(^\circ\mathrm{C}) + 273.15$$

$$\Theta\,(^\circ\mathrm{C}) = T\,(\mathrm{K}) - 273.15$$

Process Design for Cryogenics, First Edition. Alexander Alekseev.

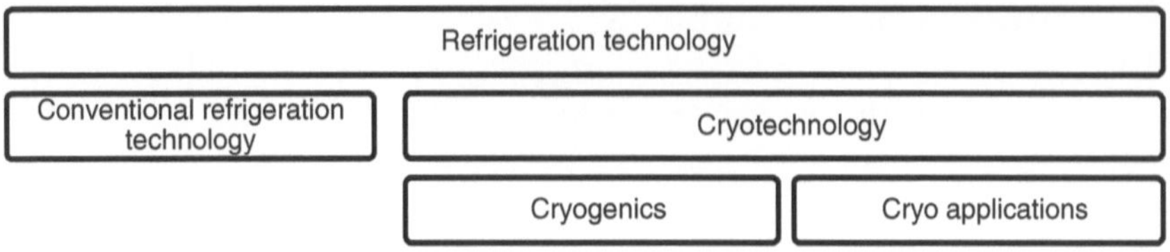

Figure 1.1 Refrigeration technology.

The use of degrees Kelvin has the advantage of eliminating negative temperature values. The zero point is the absolute zero point. "That frees us from annoying minus signs and subtractions," as Kurt Mendelssohn writes in [1].

According to the definition established during the 13th Congress of the International Institute of Refrigeration in Paris (1971), the formal boundary between conventional refrigeration and cryotechnology is a temperature of 120 K (approx. −153 °C), which corresponds to the liquefaction temperature of natural gas. This definition is to be considered in historical context – at that time, the first large natural gas liquefaction plants were being built, the business around liquefied natural gas (LNG) was booming, and technologies for this temperature level were emerging.

Although this 120 K-level definition was formally established, the discussion about the borderline between the cryo and the conventional refrigeration is still ongoing. The discussion as such does not generate any useful, applicable knowledge, but it provides an indication of what is the most emerging cryogenic application at the moment. For example, in the 1990s, when the first applications based on the so-called high-temperature superconductors (HTSCs) (see Section 1.2) were intensively developed, a lively debate revolved around the question of whether to reset the boundary temperature to the critical temperature of the best high-temperature superconductive materials.

Recently, the gas industry prefers to consider the temperature of 217 K (approx. −56 °C) as a practical boundary between cryotechnology and conventional refrigeration technology. This temperature corresponds to the triple point of carbon dioxide (CO_2) and is attributed to increasing commercial value of this gas. The technology and the logistics chain for CO_2 supply on an industrial scale closely resemble those of liquid oxygen or liquid nitrogen, and therefore liquid CO_2 is often considered a cryogenic gas.

From a process design perspective (since this book is about process design), it would make sense to define the differentiation criteria based on thermodynamic process used in a plant. For example, a system based on a Joule–Thomson process or a mixed fluid Joule–Thomson process, Brayton process, Claude process, or cascade with more than two stages would be allocated to cryogenics, whereby a system based on the Rankine process or its derivates would be considered as conventional refrigeration technology.

This book is focused on cryogenics and cryogenic processes for temperatures below 120 K, according to the formal definition of cryogenics.

1.1 Historical Background

The era of cryogenics began with the liquefaction of oxygen, which was reported by two scientists – Louis Paul Cailletet and Raoul Pierre Pictet – in December 1877, independently from each other.[1] It was a classical example of a disruptive development causing a considerable acceleration in corresponding science, in this case in cryogenics.

The oxygen liquefaction realized by Pictet and Cailletet was an exceptional achievement and must be highly honored. Science historians have pointed out the phenomenon that "at any given time there is a consensus among scientists of a given field regarding the fundamental hypotheses of their discipline" (see [2]). The "standard" theory at that time postulated that all known gaseous substances could be categorized into two groups:

- gases which can be liquefied by compression and cooling, such as chlorine, ammonia, sulfur dioxide, hydrogen sulfide and
- gases which cannot be liquefied at all, regardless of pressure and temperature conditions – the so-called "permanent" gases such as nitrogen, oxygen, or hydrogen.

This thermodynamic rule was accepted by most engineers/scientists without doubt. Only a minority – scientists like Pictet and Cailletet – dared to disbelieve and question the established theory. The exceptional achievement of Pictet and Cailletet was not only about the oxygen liquefaction, but it was more about breaking the rules.

After the announcement of the successful liquefaction of oxygen, the theory of permanent gases was suddenly (literally overnight) no longer valid. From this moment on, it was evident that *all* gases could be liquefied, without exception. The race for the liquefaction of permanent gases and thus "the quest for absolute zero" was opened at this moment. Only five years later, in April 1883, Karol Olszewski and Zygmunt Wroblewski (Cracow, Poland) succeeded in producing liquid air and later liquid nitrogen. This operation required a temperature of 77 K (−196 °C), 14° lower than the temperature of liquid oxygen (91 K ≈ −182 °C). It took another 15 years until hydrogen, with a boiling temperature of 20.1 K (−253 °C), was liquefied by James Dewar in London (1898). Ten years later (1908), Heike Kammerlingh Onnes was able to produce liquid helium in Leiden (Netherlands (NL)). The boiling temperature of helium is 4.2 K (≈−269 °C). This opened the domain of extremely low temperatures to scientists.

Further developments in cryogenics in the following years were dominated primarily by two communities:

- Physicists and chemists, who made amazing and groundbreaking discoveries such as superconductivity or superfluidity (see Table 1.1) and
- Innovators like Carl Linde, George Claude, or Samuel Collins, who launched commercial products, mostly cryogenic systems or cryogenic hardware components, and created the demand and market for cryogenic applications. Several examples are listed in Table 1.2.

---

1 This event is described in detail in [1].

Table 1.1 Milestones in cryogenic + science.

Year	Discovery/event	Temperature (K)	Scientist(s)
1911	Superconductivity	4.2	Heike Kammerlingh Onnes
1933	Meißner–Ochsenfeld effect	4.2	Walther Meißner, Robert Ochsenfeld
1938	Superfluidity	<2.7	Peter Kapitsa (original: Capița)
1957	Superconductivity (Bardeen-Cooper-Schrieffer (BCS)) theory	—	Bardeen, Cooper, Schrieffer
1972	Superfluidity in ^{3}He	0.002	David Lee, Robert Richardson, Douglas Osheroff
1986	High-temperature superconductivity	>35	Johannes Bednorz, Alexander Müller
2003	Nobel prize for physics	—	Vitaly Ginzburgh, Alexei Abrikosov, Anthony Leggett
2017	Nobel prize (chemistry) for cryo-electron microscopy	>4.2	Joachim Frank, Jacques Dubochet, Richard Henderson

Table 1.2 Milestones in cryogenic engineering.

Year	Innovation	Inventor(s)
1893	Dewar flask	James Dewar
1895	Industrial air liquefaction	Carl Linde
1902	Low-temperature piston expansion machine and Claude cycle	George Claude
1905	Industrial air separation	Carl Linde
1938	Centrifugal expander for air separation	Peter Kapitsa
1947	Industrial helium liquefaction plant	Samuel Collins
1952	Stirling cooler	J.W.L. Köhler
1957	Atlas rocket with liquid hydrogen as fuel	NASA, I am not sure about the name
1958	Superinsulation	I am not sure about the name
1958	Mixed gas cycle for natural gas liquefaction	A.P. Klimenko
1963	Gifford–McMahon cooler, pulse tube cooler	E. Gifford, R. Longsworth
1990s	Joule–Thomson expansion machine	I am not sure about the name
1990s	Cold compressor	I am not sure about the name
1990s	Structured packing for air separation plants, especially for cryogenic argon separation	Linde AG, I am not sure about the name

The discussion about which group – scientists with their discoveries, or innovators with their inventions – created a greater contribution to our culture is a confusing one. In reality, both these groups were linked so closely to each other that it is almost impossible to distinguish between them: the physicist James Dewar was also a great innovator. The same applies to Heike Kammerlingh Onnes or Peter Kapitsa, who not only discovered superfluidity, but also built the first industrial air separation plant with a turbo expander. Carl Linde, who founded the air separation industry, invested an immense amount of time in the development of an equation of state (similar to van der Waal's equation). From the author's point of view, both groups – scientists and innovators – emanated unbelievable creative impulses, and both groups were extremely "cool" (in the cryogenic sense).

1.2 Cryogenic Applications

Cryogenics is normally not of prime concern to its users. Instead, it is a sort of utility only for a primary application (for example, steelmaking). Two typical functions delivered by cryogenics are shown in Figure 1.2: either it provides a cooling capacity or it supplies the main application with the required industrial gases such as nitrogen, argon, oxygen, and hydrogen.

Use case cooling capacity: The cooling capacity can be provided directly by a cryogenic refrigerator or indirectly in form of a liquefied cryogen (refrigerant), such as liquid helium or liquid nitrogen, which is filled into the corresponding device and usually evaporates during the operation of the application.

Use case gas supply: there are two options available:

1. **Remote cryogenic plant**: The required gas is produced and liquefied in a corresponding remote cryogenic production plant or facility, and the customer is supplied with compressed or liquefied gas (liquid oxygen, liquid nitrogen, etc.) by pipelines, trucks, or other transport methods. The delivered liquefied gas is evaporated on the customer's site to provide the gas to the main application. This method is used for small and medium gas quantities (up to 20 t/d),
2. **On-site cryogenic plant**: an appropriate cryogenic production plant is installed on the customer's site directly. It supplies the application with the required amount of gas. This option is economically more feasible for large amounts of gas.

Almost all industrial gases can be produced by a non-cryogenic method as well. The decision for or against the cryogenic option is always based on techno-economic considerations according to the given requirements and conditions. The cryogenic system is usually a cost-intensive option. It is competitive only

- if the alternative technical solution is not able to deliver the required product quality (for example, the gas purity) or
- if high gas amount/flow produced by means of a large-scale cryogenic plant is required. Since the "economy of scale" works well in this case, a very competitive specific product cost can be easily achieved.

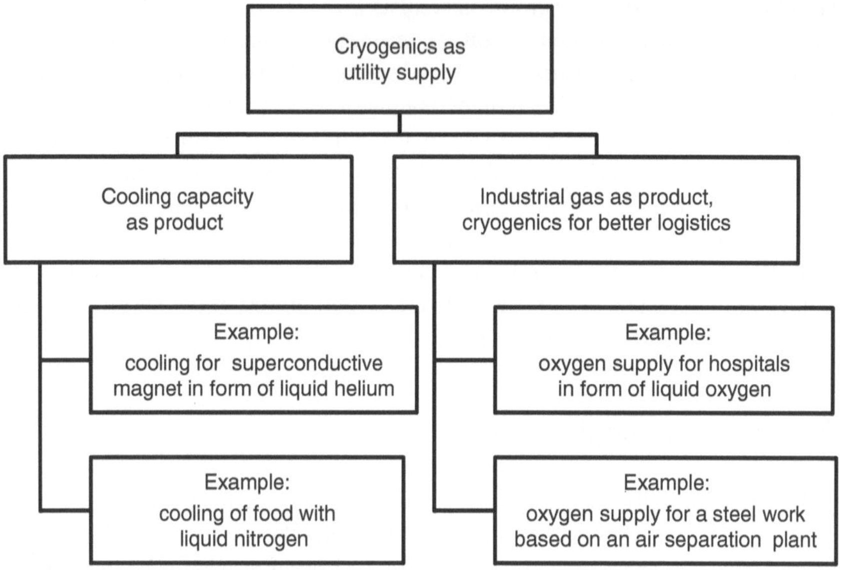

Figure 1.2 Cryogenics as utility.

Several typical cryogenic applications are briefly described in this chapter as a rough overview. For a deep dive into this topic, the book [3] is recommended.

1.2.1 Natural Gas Liquefaction

The largest cryogenic systems in the world are the so-called "base-load natural gas liquefaction plants." The natural gas is mainly liquefied for logistics reasons: the density of the LNG is roughly 600 times higher than the density of natural gas under normal conditions; therefore, the transport of natural gas in form of LNG over longer distances provides a reasonable economic option in comparison to other transport methods, for example by means of a pipeline. A further valuable advantage of LNG transport is its flexibility in terms of geography: it is possible to supply customers at different destinations in different countries using the same ship. The pipeline-based logistics (the alternative option) does not provide a comparable flexibility.

In former times, the LNG industry was based primarily on individual, more or less self-contained projects that required immense capital investment (in the range of double-digit billions of US dollars) and which therefore were fraught with high risk. In the meantime, "the industry has proven its reliability and stability under different market and economic conditions. Currently, demand growth and high energy prices, coupled with advances in technology, are driving more planned and proposed LNG projects than at any point in history," according to [4].

1.2.2 Cryogenic Air Separation

The second-largest cryogenic systems are air separation plants. With a double-digit number of plants installed or replaced annually, the air separation industry represents the largest market for cryogenic systems. Medium-size systems (10 000–90 000 m^3/h oxygen) dominate this market. These systems are used for production of oxygen, nitrogen, argon, neon, and xenon, at high purity and in large quantities. Modern metallurgy, especially steel production processes, cannot be realized without oxygen, nitrogen, and argon, and oxygen production is an integral part of every steel plant. In similar fashion, the chemical industry in its present form is inconceivable without oxygen, hydrogen, and nitrogen supplies. These gases are essential for ammonia production and for bleaching of pulp, gas-to-liquid processes, and general partial oxidation methods. Large quantities of high-purity nitrogen are used in the electronics industry and chip production, where it is not only applied as a process and carrier gas, but also used for purging purposes. Air separation plants produce not only gaseous products, but also cryogenic liquids like liquid nitrogen and liquid oxygen for numerous applications in the food and beverage industry, welding, chemistry, and so forth.

1.2.3 Helium Plants

40–50 years ago, the availability of liquid helium was very limited. However, recently it has become a sort of commodity: today every large university owns a helium liquefier and maintains its own local liquid helium infrastructure for research purposes. The global helium supply infrastructure is well developed: large amounts of helium are produced, liquefied, and shipped successfully from one continent to another.

1.2.4 Aerospace

Space launch vehicles require large quantities of liquid oxygen as an oxidizer for almost all types of fuel. Therefore, cryogenics plays an essential role in the space industry. Another cryogenic substance – liquid hydrogen – is still used as fuel for heavy rockets, such as the European Ariane, US-based Atlas V, and Chinese Long Marche. However, the newest commercial space launchers do not use liquid hydrogen. This fuel is being replaced with more common substances like kerosene due to better economics, availability, and easier logistics. The development of the reusable "Falcon" or "Antares" rocket families confirms this trend. The use of liquid (or solid) methane (cryo-option) as rocket fuel is discussed as an attractive option for future rockets.

1.2.5 Physics

Cryogenics plays a considerable role in high-energy physics. Large particle accelerators and fusion research systems, especially, include large cryo components such as:

- superconducting magnets and cavities,
- cryovacuum pumps for generation of high vacuum,
- detectors, targets, and further scientific instruments cooled to cryo temperature.

The most prominent projects of the last decade were the massive Large Hadron Collider (LHC) particle accelerator at CERN in Geneva and the Wendelstein 7X stellarator in Greifswald. The International Thermonuclear Experimental Reactor (ITER) fusion reactor in France will be the largest scientific machine upon its completion. It consists of a significant cryogenic system for cooling of gigantic magnets and vacuum pumps.

Moreover, there are many medium- and small-size scientific cryo-based machines (accelerators, cold neutron sources, neutrino detectors) available in almost every major urban area in the world.

1.2.6 Medical Applications

The most successful cryogenic commercial application ever realized is the MRI tomograph found in many medical institutions. The core of this instrument is a superconducting magnet submerged in liquid helium. The modern version of the MRI device includes an additional integrated cryocooler, which helps to reduce the evaporation rate of liquid helium and therefore extend the recharging interval of the magnet with liquid helium. When inside the MRI instrument, this cooling option can be noticed as an audible pulse approximately once per second.

Medical oxygen is used for treatment of patients with respiratory problems (including asthma or COVID).

Blood plasma and other biological samples are usually stored at low temperatures (cryopreservation) in special cryo storage devices (vessels or cabinets).

A relatively new application is the production of extremely sensitive bio-based pharmaceuticals (such as special bacteria, probiotica, lipid-rich substances, and similar) via cryo-based freeze–drying technology. It ensures very gentle handling of substances with minimized destruction of these living objects and guarantees acceptable shelf life.

1.2.7 Liquid Hydrogen for Hydrogen Economy

Liquid hydrogen is seen as an important part of the logistics chain in the future hydrogen-based economy. From today's perspective, it is still a great vision. It is still difficult to predict when this vision will become reality. The hydrogen-based economy will require a completely new infrastructure and a considerable investment (billions and billions) in its development. Therefore, it is considered to be a long-term project.

1.2.8 Devices Based on High-temperature Superconductors for Electrical Engineering

Devices based on HTSC – ceramic materials without electrical resistance at temperatures below 100 K – are regarded as an interesting option for some special tasks in electrical engineering.

- HTSC-based motors or generators are not only extremely efficient and compact, but they also have a broader load range and the fastest ramping capability. These machines can be used in the maritime sector (or other bulk transport) in the future due to extremely high specific power density.

- Another unique device is an electrical cable based on HTSC for power transmission and distribution. It is a compact system, which allows a virtually invisible power supply without power limitations, without thermal losses, and without electromagnetic emissions. This could be an attractive solution for power supply in densely populated areas like cities, where bulky overhead lines are no longer imaginable today and the installation of conventional copper underground cables is too expensive due to the large footprint required.
- The next interesting HTSC-based device is the fault current limiter developed for safe disconnection (and protection) of the grids and grid segments in case of an electrical fault. The HTSC-based design is an extremely fast-acting system. It allows safe and reliable operation of electrical grids at close to maximum load. It also helps to reduce the overdesign margin (the standard option today) and creates a more sustainable and economical grid solution.

The commercial success of all these applications is still unclear due to the high competitiveness of existing technology, at least in terms of cost. But increasing urbanization and electrification of cities (e-mobility, digitalization, air conditioners, etc.) create more and more opportunities for HTSC applications.

References

1 Mendelssohn, K. and von Pape, G. (1966). *Die Suche nach dem absoluten Nullpunkt*, Kindlers Universitäts Bibliothek. München: Kindler.

2 Kahnemann, D. Thinking, fast and slow. Farrar, Straus and Giroux, 2011.

3 Haering, H.-W. (ed.) (2008). *Industrial Gases Processing*. Wiley-VCH Verlag GmbH & Co KGaA. ISBN: 978-3-527-31685-4.

4 Tusiani, M.D. (2007). *LNG: A Non-Technical Guide*. PennWell. ISBN: 978-0-87814-885-1.

2

Basics of Process Design

The classic book of Peters et al. [1] on engineering of process plants covers almost all aspects from process design to economics and ethics. A further recommendable text is the book of Blass [2] on systematics of process design, including task definition, process synthesis, process analysis, and process assessment, as well as procedures for decision taking based on technical and economic feasibility and ecological risk assessment.

The design procedure introduced in this book is shown in Figure 2.1. Development of the process idea (visualized by means of a basic process flow diagram) and development of the corresponding hardware concept are performed simultaneously. It is slightly different compared to the conventional design procedure (described for example in the book [1], which presumes a step-by-step approach, starting with process idea, followed by process simulation, and development of the hardware components concept based on results of the simulation.

Since development of both concepts (process concept and hardware components concept) is performed simultaneously, the engineer must be able to think at least two-dimensionally.[1] At first glance, it may appear that the 2D-thinking is more complex and requires more effort. However, "according to the latest theory on the function of associative memory, the majority of associative thinking effort occurs silently, below the threshold of consciousness," see Kahneman [3]. Therefore, in the end, it is considered the most natural and efficient way of thinking.

The process design requires a lot of creativity, but it also includes a lot of routine. In order to understand a process, it is recommended to invest as much time as possible into the process simulation and process calculation. The advantages and disadvantages of a process become clearly visible as you observe it under high-resolution microscope.

1 In reality, process development often includes more than two dimensions, for example, related aspects such as the control concept or transportation concept typically have to be considered as well.

Process Design for Cryogenics, First Edition. Alexander Alekseev.

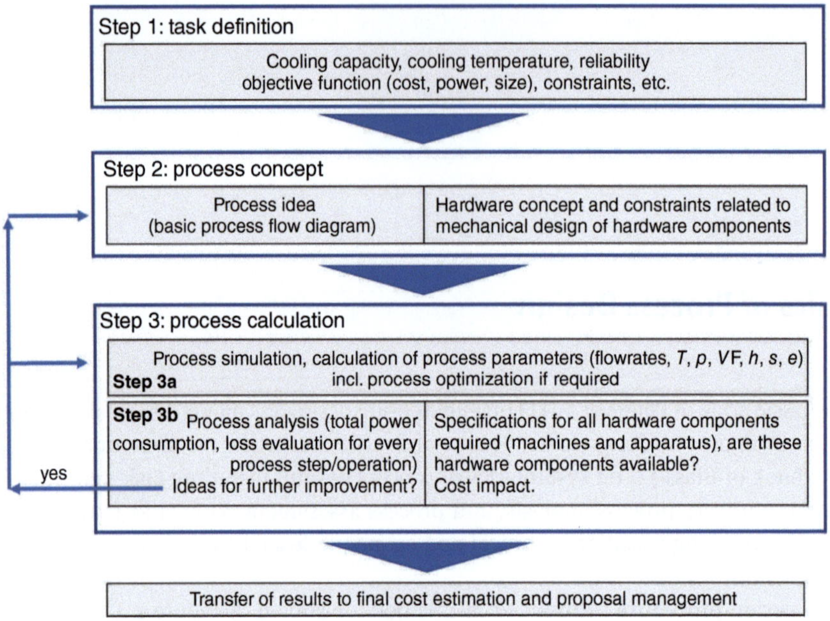

Figure 2.1 Process design, typical procedure.

2.1 Design Procedure

The typical design procedure (see Figure 2.1) includes three steps:

1. Task definition
2. Development of the process concept
3. Process calculation

before results will be transferred to the department(s) responsible for proposal management.

Step 1: ***Task definition*** **(sometimes referred to as BoD – Basis of Design):**
During this step, key information must be collected and fixed in a set of documents. Usually, this set includes:

1. main technical requirements, consisting of:
   - basic requirements on (i) cooling temperature, (ii) cooling capacity, and (iii) reliability (mean time between maintenance or similar) of refrigerator, and
   - special requirements

   e.g. fast cool-down time or even more exotic specifications such as max level of mechanical vibrations;
2. ambient conditions and utilities:
   ambient conditions are air temperature, ambient pressure (which depends on elevation), and humidity. Utilities are cooling water, power supply, instrumentation air, etc.;

3. objective function or targets for power consumption and total plant cost including a set of corresponding parameters (power cost, amortization time, etc.). Sometimes these numbers are not clearly specified, but at least a rough direction (power-optimized system or cost-optimized system) must be given;
4. all constraints (transport limitations, space available) and other information important for process design.

The more accurately these specifications are provided, the higher the probability that the customer will be satisfied by the performance of the considered system.

Step 2: ***Development of a process concept*** is the second step. It includes development of

- process idea in the form of a basic process flow diagram (BPFD),
- hardware components concept (type of compressor, kind of heat exchanger, thermal insulation, etc.).

The point is that an industrial refrigeration system may be realized only on the basis of commercially **available** hardware components with corresponding references, guarantees, etc. The hardware concept impacts the process idea, cost, and process parameters because it sets constraints for the process design. For example, every type of compressor has a typical compressor efficiency curve, or pressure loss in heat exchanger depends on the kind of heat exchanger used. These constraints must be considered during the following process calculation. Therefore, the development of the hardware concept must be performed simultaneously with the development of the process idea.

Step 3 ***Process calculation*** is the next step. The target is to obtain some performance indicators for the techno-economic assessment (in comparison to similar systems) and to carry out an assessment of the proposed process based on these numbers. Two of the most important performance indicators are:

(a) total power consumption of the plant,
(b) total plant cost.

Methodologically, the process calculation phase consists of two steps:

- Step 3a: process simulation (for a given process idea and proposed hardware concept)
- Step 3b: process analysis, specification of hardware components

---

Step 3a **The target of the *process simulation step*** is to find out

1. process parameters (flow rate, temperature, pressure, phase state, composition, enthalpy, entropy, and exergy) at all relevant process points;
2. values required for the design of hardware components, for example, power consumption and volumetric flow rate at the inlet of a compressor, mean temperature difference and heat duty in a heat exchanger, etc.

required for estimation of total power consumption as well as total plant cost.

How to find a solution?

Some cryogenic processes can be calculated manually or by means of simple tools like EXCEL. But normally, the process simulation is performed by means of an appropriate software – the so-called process simulators (see Section 2.2).

The calculation of process parameters at all relevant process points is a procedure similar to the solving of a puzzle or a logic game (such as the well-known Sudoku game), in which some of the numbers (information) are given, and the remaining information must be "discovered" according to certain rules (e.g. energy balances and material balances).

Challenges: The real process simulations tasks are characterized either by the lack or by the excess of information.

- In case of missing, fragmented, or ambiguous information, it is sometimes not possible to achieve a clear solution by means of process simulation directly. Therefore, the engineer must complete the missing numbers on his own. For this reason, corresponding assumptions (as realistic as possible) must be made.
- The case of excess information is challenging, too.
  Usually, it means that the task has multiple correct solutions. The engineer must find out the best one. Therefore, an **optimization** (see Section 2.3) must be carried out in order to find it. The optimization procedure is challenging and time-consuming.

Step 3b ***Process analysis and specification of hardware components:*** The process simulation (Step 3a) delivers process parameters and other important figures. During Step 3b, these numbers are analyzed and converted into the performance indicators (total power consumption and total cost).

The total power consumption results from the process simulation directly as sum of power[s] for all power-consuming and power-generating devices (compressors, expanders, pumps, heaters, etc.). The total power consumption can be reduced by avoidance or reduction of irreversibilities in the process. The systematic process analysis (analysis of losses in process units) makes these irreversibilities visible and tangible, and delivers valuable input for further improvement of the process design. The checking and implementation of these new ideas require further process simulations. Therefore, the process calculation itself is an iterative procedure consisting of numerous loops following a sequence:

(a) process simulation,
(b) process analysis,
(c) development of new ideas,
(d) implementation of these ideas in process simulation, i.e. going back to step (a) process simulation.

Estimation of the total plant cost is a challenging job. Usually, a process design engineer is not able to estimate the full cost of the system. His responsibility is limited to estimates for the most important (and expensive) hardware components. These are typically being formally specified, inquired, and discussed with original equipment manufacturers. The target is to get a confirmation of availability and cost.

In the early phase (where the process design is not finalized yet) the whole cost information is not really required. The cost considerations and decisions may be done on the basis of differences (cost deltas) from a known process design (the so-called "reference design").

Example:

- the new heat exchanger is larger and is, therefore, more expensive by US$50 000,
- the main compressor is compact and therefore cheaper by US$100 000,
- therefore, the total system is cheaper by US$50 000.

Other Aspects Related to Process Design

A common requirement is that the cryogenic plant must perform not only in a single operating point but in other operation modes (operating cases), too, for example in underload or overload conditions. Therefore, the whole process design sequence (including optimization) is usually executed for all specified operation cases: normal operation case, underload case, etc.

2.2 Process Simulators and Fluid Properties

The most common method to perform a process simulation is by means of process simulators. These software packages include integrated fluid properties data as well as ready-for-use operation units, which may be simply connected to complex process chains. Some process simulators include optimization algorithms. The process simulators simplify and accelerate simulation procedures considerably. The most known commercial packages are ASPEN, PROII, HYSIS, UNISIM, and gPROMS. There are also some suitable "open source" process simulators available, for example, DWSIM.

The properties of liquids and gases are required for process simulation (Step 3a). The best case would be to have these properties data in the form of an appropriate software package which delivers all necessary fluid properties (density, heat conductivity, heat capacity, specific enthalpy, specific exergy, etc.) for a set of input parameters, e.g. pressure and specific enthalpy. There are some commercial software packages available for this purpose. I suspect, one of the most well-known is the software "REFPROP" or "miniREFPROP" from NIST, the latter often being used for academic studies, because it is provided free of charge.

2.3 Optimization

According to Wikipedia, "optimization is the selection of a best element (with reference to certain criterion) from a set of available alternatives. In the simplest case, an optimization problem consists of maximizing or minimizing a real function (called the 'objective function') by systematically choosing input values from within an allowed set and computing the value of the function." The engineering optimization problems are characterized by:

a. multiple objective functions,
b. variables are continuous or discrete,
c. variables are often specified by a set of constraints – conditions that the solution (set of variables) must satisfy.

Examples of Constraints: An example of a typical constraint relevant for process simulation in cryogenics is the requirement of minimum temperature difference in the main heat exchanger. Other examples are maximum height or diameter of separation columns, maximum volume or size, or weight of cold boxes. These last restrictions are typically being derived from transport limitations or conditions on site.

Examples of Variables: The minimum driving power of the process of a plant can be achieved by changing process parameters such as pressures, temperatures, composition of the working fluid, and flow rates. The costs can be positively influenced, e.g. by specifying larger temperature differences and pressure losses in heat exchangers, which leads to more compact and therefore cheaper equipment, or by using "pre-engineered," standardized, and therefore cost-efficient hardware components.

Examples of Objective Function: Typical examples of objective functions are maximal argon yield or maximal liquid production (in an air separation process). Two examples of more exotic objective functions are maximal load range and max load ramping rate of a system. But the most known and most popular objective functions used in engineering are:

1. minimal power consumption and
2. minimal costs of the system.

An important point about these two functions is that neither min power nor min cost is used as a single objective function. There are different concepts for which reasonable objective function to use, but at the end it is always a formula considering two cost blocks – capital expenditures – "CapEx" – (which represent plant costs) and operational expenditures – "OpEx" – (which are broadly related to the costs of power consumption). In the easiest case, it is simply CapEx + OpEx. The optimization dilemma is the following: the power consumption (and therefore OpEx) can be reduced by using more efficient but more expensive (higher CapEx) hardware components, such as more efficient compressors, heat exchangers with larger heat transfer surface, and separation columns with higher number of theoretical trays The optimization process is about how to achieve min CapEx at reasonable OpEx.

I would like to emphasize the fact that **process optimization is only a part of the overriding investment optimization process.** Industrial companies are business-driven units. Every system/plant is considered primarily as a type of investment project, which must generate profit. Therefore, the optimization must be performed for the overall business case. The target is

(a) maximization of profit and
(b) minimization of the so-called total cost of ownership "TCO"[2]
(c) within the framework of a long-term (10, 15, or 20 years) project, typically being financed (at least partially) by credit.

The calculation of these three items is extremely challenging because of various uncertainties: e.g. the profit often depends on sales development, turnover, product price level, competition, etc.; the credit conditions (interest rate, foreign currency exchange rate) may change during the project time; OpEx depends on power tariffs, which may vary considerably. Theoretically, this optimization problem could be simplified by making viable assumptions for interest rates and business growth. Unfortunately, the economists are often not able to derive these assumptions

2 TCO includes total cost of acquisition and operating costs, as well as costs related to replacement, upgrades during the whole life cycle including de-installation of the system at the end. More information you will find until: https://en.wikipedia.org/wiki/Total_cost_of_ownership.

with appropriate accuracy, and thus the task remains complex and has no clear mathematical solution. In this situation, the targets cannot be clearly defined, and some decisions will be driven more by intuition (mainly based on experience). The process engineers are important members of the project team. Usually, they are well informed about what is more important for the customer – low power consumption or low costs. Accordingly, the system may be developed either as "cost-optimized" or as "energy-optimized" solution. Surprisingly, engineers can manage this sort of task successfully, despite the aforementioned uncertainties.

2.4 Control Concept

According to Wikipedia, "a **control system** manages, commands, directs, or regulates the behavior of other devices or systems using control loops." An example of an automatic control device is the so-called feedback controller: "the control system compares the value or status of the process variable being controlled with the desired value or setpoint and applies the difference as a control signal to bring the process variable output of the plant to the same value as the setpoint."

Control for Cryogenic Refrigeration

A cryogenic refrigeration system must also be controlled. The typical target is to keep the cooled object at the specified constant cooling temperature. For this purpose, the cooling capacity of a refrigerator must be equivalent to the heat flow generated by the cooling object. If the cooling capacity is smaller than required, the temperature of the cooling object will rise, and the functionality of the cooled device cannot be guaranteed. In the opposite case – if the cooling capacity is higher than the cooling demand – the functionality of the cooled application is maintained, but the higher power consumption of the refrigerator becomes visible in the form of a high-volume billing for electricity at the end of the month.

The typical control loop for a refrigerator looks as follows: the control system

- monitors the temperature of the object being cooled,
- compares the temperature value with specified setpoint value,
- increases the cooling capacity of the refrigerator if the cooled object becomes warmer than the setpoint, or
- reduces the cooling capacity of refrigerator if the cooled object becomes colder than the setpoint.

Process Design Engineer and Engineer Responsible for Control/Operation of Refrigerator – Different Ways of Thinking

The special feature of every refrigerator with an active control system is that such a plant never operates in a steady-state condition (as calculated and desired by process design engineer). In every single moment:

- either it is generating more cold than required, and the control loop is reducing the cooling capacity,
- or it is generating less cold than required and the control loop is trying to increase the cold production.

The engineers responsible for operation and control consider every plant as a dynamic system; they monitor the dynamic trajectory of process variables and derive control measures from this information. From this perspective, the steady-state condition is the most unpleasant state that may occur because the control system cannot really recognize what happens actually, what will occur in next moment, and what kind of action it should be prepared for. The steady-state operation is rigorously avoided by control engineers. The dynamic development – slow or fast – must be present.

From a process design engineer's perspective, every dynamic is an enormous challenge: the simulation effort for dynamic processes is at least one order of magnitude higher than for steady-state processes, it requires considerably more information on the mechanical design of hardware components, and the mathematical models are significantly more complex and consume more computing time. Therefore, process design engineers actively try to avoid dynamic simulations and perform them only if absolutely necessary.

This difference in the way of thinking may be considered as a potential conflict situation. However, in real life, the computational challenges are well understood by both sides (control engineers and process design engineers), so there is no room for conflict.

References

1 Peters, M.S., Timmerhaus, K.D., and West, R.E. (2003). *Plant Design and Economics for Chemical Engineers*. McGraw-Hill Higher Education.

2 Blass Eckhart, *Entwicklung verfahrenstechnischer Prozesse: Methoden, Zielsuche, Lösungssuche, Lösungsauswahl* Chemische Technik Verfahrenstechnik (German Edition) Taschenbuch1997, ISBN-10: 9783540618232, ISBN-13: 978-3540618232.

3 Kahneman, D. (2011). *Thinking, Fast and Slow*. Farrar, Straus and Giroux, ISBN: 978-0-141-03357-0.

3

Cryogenic Fluids

Typical cryogenic working fluids are briefly described in current section. The data properties of cryogenic gases are shown in Table 3.1. Since thermodynamic diagrams are very useful for visualization of thermodynamic properties, description of temperature–enthalpy diagram and temperature–entropy diagram is given in Sections 3.9 and 3.10.

Only a few fluids remain liquid or gaseous at cryogenic temperatures (<120 K):

- Main components of air such as nitrogen, oxygen,
- Light noble gases such as helium, neon, and argon,
- Hydrogen,
- Light hydrocarbons such as methane, ethane, ethylene, propane, propylene, as well as the halogenated versions of these substances,
- Isotopes of substances described above, mixtures of substances described above, chemical compounds (for example NO) of substances described above,
- Fluorine and carbon monoxide.

The handling of fluorine and carbon monoxide is challenging since fluorine is a chemically highly reactive substance and carbon monoxide is toxic. Neither of these gases is used in cryogenics, because other gases such as nitrogen or argon have very similar thermodynamic properties (boiling temperature, evaporation enthalpy, etc.) but do not require special safety measures.

The other cryogenic gases are harmless for humans. Most of them are odorless and tasteless. Cryogenic gases and liquids are colorless, except for liquid oxygen, which is slightly blue. Most of these substances are components of ambient air or natural gas (Figure 3.1).

3.1 Air

Ambient air is the natural source for production of oxygen, nitrogen, and noble/rare gases such as argon, neon, krypton, xenon. The large-scale production of oxygen and nitrogen is, of course, the main concern.

The composition of atmospheric air remains nearly constant up to an altitude of 90 km according to Ref. [2]. The data in Table 3.2 refer to dry air without impurities

Process Design for Cryogenics, First Edition. Alexander Alekseev.

Table 3.1 Data properties for cryogenic gases, acc. to Ref. [1].

			Normal boiling point (pressure = 1.013 bar)			Critical point			Triple point		Gas at pressure 1 bar and temperature 290 K	
		Molar mass (g/mol)	Temperature (K)	Liquid density (kg/m^3)	Evaporation heat (kJ/kg)	Temperature (K)	Pressure (bar)	Density (kg/m^3)	Temperature (K)	Pressure (bar)	Density (kg/m^3)	Isobaric heat capacity (kJ/[kg K])
Helium	He	4.003	4.22	124.67	20.56	5.20	2.28	69.58			0.166	5.193
Normal hydrogen	n-H_2	2.016	20.37	70.85	448.71	33.15	12.96	31.26	13.96	0.0736	0.084	14.275
Para-hydrogen	p-H_2	2.016	20.27	70.83	446.07	32.94	12.86	31.32	13.80	0.0704	0.084	14.485
Neon	Ne	20.179	27.10	1207.00	85.75	44.49	26.79	481.91	24.56	0.4346	0.837	1.030
Nitrogen	N_2	28.013	77.36	806.08	199.18	126.19	33.96	669.02	63.15	0.1252	1.162	1.042
Air	AIR	28.965	78.903 - 81.72	875.21	204.81	132.63	37.85	691.75	59.75	0.0526	1.202	1.006
Argon	Ar	39.948	87.30	1395.40	161.14	150.69	48.63	535.60	83.81	0.6889	1.658	0.522
Oxygen	O_2	31.999	90.19	1141.20	213.06	154.58	50.43	436.14	54.36	0.0015	1.328	0.919
Carbon monoxide	CO	28.010	81.64	793.21	214.68	132.86	34.94	303.91	68.16	0.1545	1.162	1.042
Fluor	F	37.997	85.04	1501.80	174.37	144.41	51.72	592.86	53.48	0.0024	1.577	0.821
Methane	CH_4	16.04	111.67	422.36	510.83	190.56	45.992	162.66	90.69	0.1170	0.667	2.214

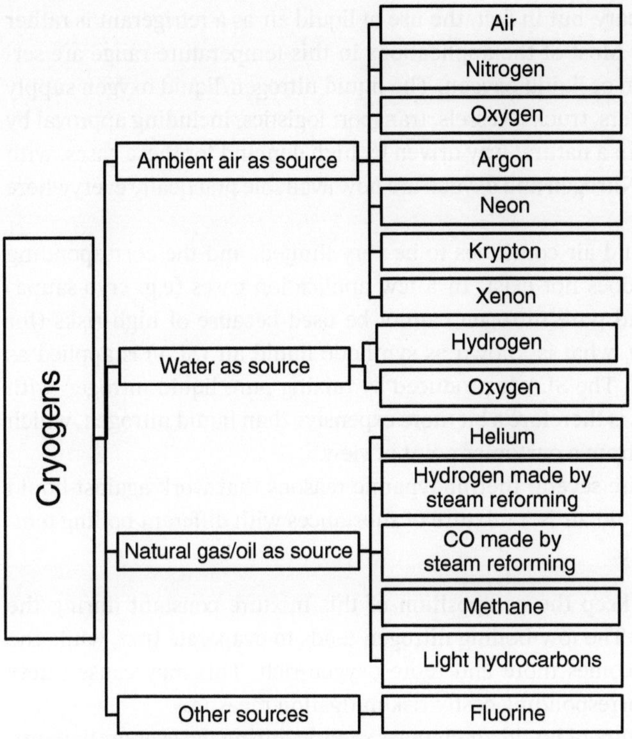

Figure 3.1 Cryogenic fluids, sorted by source.

Table 3.2 Composition of dry air.

Substance	Concentration	Substance	Concentration
N_2	78.118 mol%	CO_2	Approx. 400 ppm
O_2	20.95 mol%	H_2	0.4 ppm
Ar	0.932 mol%	CO	0.1 ppm
Ne	18.18 ppm	N_2O	0.3 ppm
He	5.24 ppm	CH_4	2 ppm
Kr	1.14 ppm	C_2H_2	<1 ppb
Xe	87 ppb	Other C_nH_m	<10 ppb

1 ppm (parts per million) = 0.000 001 mol/mol.
1 ppb (parts per billion) = 0.000 000 001 mol/mol.

(dust, etc.). The humidity depends on weather conditions, with an average value in Europe of less than 1 mol%; the maximum value amounts to 4 mol%. The CO_2 content has been on the rise for decades. Ten years ago, it was lower than 350 ppm, now it is slightly higher than 400 ppm. In general, it is expected to reach or even surpass 500 ppm over the next decade. This also depends on the season (summer versus winter) due to vegetation cycles.

It may sound a bit scary, but in fact, the use of liquid air as a refrigerant is rather an uncommon option. Most of the applications in this temperature range are serviced by liquid nitrogen or liquid oxygen. The liquid nitrogen/liquid oxygen supply infrastructure (containers, trucks, vessels, transport logistics, including approval by authorities, etc.) grew in a natural way driven by high demand for these gases, with numerous customers. Nitrogen and oxygen are now available practically everywhere at reasonably low cost.

The demand for liquid air continues to be very limited, and the corresponding infrastructure simply does not exist. In a few application cases (e.g. cryo-sauna) where pure oxygen and pure nitrogen cannot be used because of high risks (for example asphyxiation), what is known as synthetic liquid air (SLA) is applied as an alternative solution. The SLA is produced by mixing pure liquid nitrogen with pure liquid oxygen and is therefore a bit more expensive than liquid nitrogen, which can be a bit confusing from a customer point of view.

Additionally, there are several thermodynamic reasons that work against liquid air as a refrigerant. Liquid air is a mixture of substances with different boiling temperatures, and therefore:

- It is challenging to keep the composition of this mixture constant during the evaporation process. The low-boiling nitrogen tends to evaporate first, while the remaining liquid becomes more and more oxygen-rich. This may cause safety issues and require corresponding costly risk mitigation measures.
- The boiling temperature of liquid air changes slightly during the evaporation process as well. This issue may be critical for some applications that require a stable cooling temperature.

The thermodynamic properties of air are dominated by the properties of nitrogen (see Section 3.2). For example, 1 m^3 of air weighs 1.29 kg under standard conditions (1.25 kg for nitrogen).

3.2 Nitrogen

Nitrogen is not distinguishable by normal human senses. Even the high nitrogen concentration of 78.12 mol% in the ambient air is harmless for humans.

Liquid nitrogen boils at 77.4 K at 1.013 bar, which is a bit lower than the boiling temperature of oxygen (90.2 K). This difference in properties of nitrogen and oxygen is small, but it is large enough to exploit for air separation (separation of air into pure oxygen and pure nitrogen) by means of cryogenic rectification.

The triple point of nitrogen is at 63.2 K, which places it between the triple points of oxygen and argon.

Typical applications for liquid nitrogen are listed below:

- Food freezing (frozen pizza, dough, etc.),
- Storage of biological and medical probes and samples, as well as freeze-drying of biological material,

- Cooling of infrared detectors or germanium detectors for radiation control,
- Soil freezing in civil engineering in order to solidify water-containing soil layers,
- In mechanical engineering, liquid nitrogen is used, for example, to shrink some components, such as gear shafts prior to fitting of gears,
- Recycling of composite materials such as tires and cables: the plastics and rubber-based materials become brittle at low temperatures; these materials can be destructed or crushed in appropriate milling devices and separated into fractions of valuable materials. This makes economic recycling of such complex materials feasible.

The following product abbreviations are used in the gas industry:

- GAN – for gaseous nitrogen
- LIN or LN2 – for liquid nitrogen

3.3 Oxygen

At standard conditions ($T = 0\,°C$ and $p = 1.013$ bar), 1 m^3 of oxygen weighs 1.43 kg (compared to 1.29 kg for air). In the liquefied state, oxygen's density is 1.14 kg/l; hence, liquid oxygen is slightly heavier than water. Liquid oxygen boils at 90.2 K at 1.013 bar; the liquid is light blue in color.

A special feature of oxygen is the anomalously low triple point temperature of 54.4 K, which is lower than the triple point of nitrogen (63.16 K) and argon (83.86 K), although the normal boiling temperatures of these gases are lower than those of oxygen.

Oxygen is paramagnetic.[1] Therefore, measurement of oxygen concentration in a gaseous mixture is a relatively easy task if the accuracy of measurement methods based on magnetic properties meets the requirements. The concentration measurement for other cryogenic gases (nitrogen, argon, helium, neon, and hydrogen) in a mixture is more complex because other (expensive and discontinuous) methods such as gas chromatography and spectrometry are required for this purpose.

Oxygen is a nonflammable gas, but it is considered a dangerous substance since oxygen or oxygen-enriched air accelerates the combustion of all burnable materials considerably. Even metals like aluminum or steel burn in an oxygen-rich atmosphere, which is unusual in ambient air at normal conditions. Organic materials burn fast and vigorously in oxygen. Hydrocarbon-based materials such as oils and greases can even detonate unexpectedly upon contact with oxygen. Therefore, systems for storage and handling of liquid and gaseous oxygen must be kept completely free of hydrocarbons and hydrocarbon-containing materials.

1 Paramagnetism is one form of magnetism: Paramagnets are only magnetic as long as they are in an external magnetic field, in contrast to the permanent magnetization of ferromagnets. Paramagnetic materials are attracted by other magnets.

The largest consumer of oxygen is metallurgy, including steelmaking, followed by the chemical industry (coal gasification, paper production, partial oxidation, etc.). Oxygen is used in aerospace as an oxidizing agent for rocket propulsion systems. The most common applications of O_2 are found in welding and respiratory medicine.

The following abbreviations for oxygen products are often used in the gas industry:

- GOX – for Gaseous Oxygen
- LOX – for Liquid Oxygen

3.4 Argon

Formally, argon was discovered by W. Ramsay and Lord Rayleigh in 1894. They recognized that the density of nitrogen gas generated as a product of a chemical reaction differs from the density of nitrogen that remains in the air after the removal of oxygen. A reasonable explanation of this fact was that the air contains another gas (in addition to nitrogen and oxygen), which must be heavier than nitrogen.

Argon is indeed considerably heavier than nitrogen: at standard conditions, 1 m^3 of argon weighs 1.66 kg (by comparison, air density is 1.29 kg/m^3). In the liquid state, argon has a density of 1.4 kg/l; the density of liquid argon is therefore higher than that of water. The argon concentration in the air is relatively high: 0.932 mol%. Liquid argon boils at 87.3 K at ambient pressure. This temperature is a bit higher than the normal boiling temperature of nitrogen but lower than that of oxygen.

A special feature of argon is that the triple point temperature of argon (83.8 K) is very close to the normal boiling temperature (87.3 K), the difference being only 3.4 K. That means that a relatively minor subcooling by 4 K can lead to solidification. This special property – a small difference between the normal boiling temperature and the triple temperature – is typical for all rare gases such as neon, krypton, xenon, and helium.[2] This phenomenon shows that atoms/molecules of rare gases in the liquid phase are already extremely well ordered in terms of atomic/molecular structure. This liquid structure (and density) does not differ much from the crystalline structure and density of the solid phase. Therefore, the transition (from liquid to solid) requires only a relatively low degree of subcooling.

Argon[3] is an inert gas. This property is very valuable for some processes, for example, as a shielding gas for welding purposes. Argon does not react with other substances, even at high temperature (and high energy density) at the welding point. In steel production, relatively large quantities of argon are injected directly into the molten metal without risk of chemical reaction inside the molt.

Further applications of argon are as:

- Extinguishing agent, mainly for protection of electrical systems. The extinguishing effect is achieved by oxygen displacement. Pure argon or a mixture of argon and nitrogen is used for this application.

---

2 Helium is an exception, the cooling of liquid helium below the normal boiling point leads to the formation of another liquid phase – the so-called suprafluid liquid – instead of a solid phase.

3 The origin for the name "argon" is the ancient Greek word ἀργός *argos*, which means "inert."

- Carrier gas for gas chromatography in analytics.
- Conventional electric bulbs are filled with a mixture of argon and nitrogen, as this gas reduces the sublimation of the metal filament. Argon has a low thermal conductivity in comparison to nitrogen and is cheaper in comparison to krypton or xenon (which could be an alternative solution because of its lower heat conductivity). An advantage of the lower thermal conductivity is a higher possible filament temperature and, therefore, higher efficiency.
- Filling gas for high-performance insulating glass (IG), more commonly known as double glazing (or double-pane and increasingly triple glazing/pane), which consists of two or three glass window panes separated by a gas-filled space to reduce heat transfer across a part of the building envelope.
- Fluorescent gas with a typical violet color in cold cathode fluorescent lamps (like neon lamps). With the addition of mercury, the color changes to blue.
- Laser gas for argon-ion lasers.
- Food additive (E938) as protective gas in food packaging and wine production.

The following product abbreviations are used in the gas industry:

- GAR – for gaseous argon (gaseous argon),
- LAR – for liquid argon.

3.5 Methane

Methane (CH_4) is the main component of conventional natural gas, or what is known as "associated gas" in the production of crude oil. Shale gas produced by fracking (a controversially discussed but cost-efficient method) also consists primarily of methane. Hard coal includes some methane gas – also known as "mine gas." Biogas (which contains roughly 60% methane) is considered a viable source of methane, too. However, biogas contains many other impurities, like carbon dioxide and hydrogen sulfide, and therefore requires a complex treatment procedure. In the universe, methane is present in large quantities on planets, comets, and moons. Methane is also contained in the earth's atmosphere, albeit at a very low concentration of 1.75 ppm. In air separation systems, methane is considered to be a critical contamination that should be removed (at least partially) prior to the main rectification process.

The thermodynamic properties are shown in Table 3.1: methane boils at 111.7 K at 1.013 bar. The triple point is at 90.7 K and 0.12 bar, while the critical point is at 190.6 K and 46 bar.

From a thermodynamic perspective, methane is a relatively simple substance. In the liquid or vapor phase, it has no anomalies or effects/phases/states that require special processing steps during process design.

Methane is used mainly as a medium for heat generation and as a fuel for combustion engines. The caloric value of methane can easily be memorized because it is a round figure of 50 MJ/kg. Methane is an important feedstock to produce hydrogen and synthesis gas (mixture of CO and H_2), which is required for the large-scale production of basic chemicals like ammonia and methanol. Chemically, methane is a very stable compound that is difficult to break down into its constituents C, CO, and H. The corresponding process requires certain special conditions like high temperatures and pressures, and the presence of a catalyst.

Methane is a highly flammable gas, with the flash point of −188 °C and ignition temperature of 600 °C. It burns in the air with a bluish, non-sooty flame, as is well known from conventional gas stoves in the kitchen. Especially mixtures of 4.4–16.5 vol% of methane in air are dangerous because they are explosive. Therefore, cylinders or vessels for storage of methane should be placed only in well-ventilated locations or in explosion-protected areas.

No specific abbreviation for methane is used in the gas industry. However, there are two special terms – "LNG" and "CNG," which are used for natural gas and mainly consists of methane:

- LNG: liquefied natural gas (liquid natural gas) and
- CNG: compressed natural gas.

In some countries, especially in Scandinavia, LNG is used as fuel similar to gasoline or diesel. It can be purchased as a conventional commercial product and is delivered by specialized companies. CNG is a commercial product as well. It can be purchased at some petrol stations as conventional car fuel.

3.6 Neon

Neon was discovered by W. Ramsay and M.W. Travers by boiling liquid air. It was discovered relatively late, in 1898. The concentration of neon in ambient air is extremely low (18.18 ppm); therefore, a significant amount of air must be processed to obtain a small amount of neon. The first air liquefier developed by Carl Linde did become commercially available in 1896; therefore, the time frame of neon discovery is more or less self-explanatory.

The first application of the newly discovered gas was the neon lamp developed in 1910 by Georges Claude. Filled into a glass tube, neon gas glows if high voltage is applied. Today, neon is mainly used in plasma TV sets and monitors.

Recently, liquid neon has been considered the preferred refrigerant for cooling high-voltage applications based on high-temperature superconductors like motors, generators, and transformers (see Section 1.4). These devices are still in the pilot phase or demonstration phase and are not commercially available yet.

Neon is a relatively light gas, with a density of 0.9 kg/m^3 at standard conditions.

Triple Point–Boiling Point

At atmospheric pressure, neon boils at the relatively low temperature of 27.1 K. The triple point at 24.6 K is only 2.5 K lower than the normal boiling point (this feature of rare gases was discussed already in the section regarding argon). Therefore, a relatively low amount of subcooling, by only 3 K, will induce transition to the solid phase.

No Liquid Between 44.5 and 54.4 K

The critical point of neon is about 44.5 K; at higher temperatures neon is gaseous. Helium and hydrogen are gaseous at this temperature as well. Oxygen is the fluid with the lowest boiling point of all remaining pure substances; it stays liquid at 54.4 K (triple point of oxygen). Therefore, there are no pure substances[4] available in nature that remain liquid within the temperature range between 44.5 and 54.4 K.

Another "no-liquid-range" can be found at temperatures below the triple point of hydrogen.

Neon Isotopes

Several neon isotopes between ^{16}Ne and ^{34}Ne are known. Three of them – ^{20}Ne, ^{21}Ne, and ^{22}Ne – are stable isotopes. Their natural abundance is given as:

- ^{20}Ne: 90.48%,
- ^{21}Ne: 0.27%
- ^{22}Ne: 9.25%.

This natural isotope distribution (^{20}Ne/^{22}Ne: 90.48%/9.25%) is an exceptional property of neon. A typical cryogenic gas consists of a "main" isotope with a fraction higher than 99%; correspondingly, the fraction of all remaining isotopes amounts to less than 1%. In the case of neon, the distribution of isotopes is very unusual.

The isotopes of a chemical element must be considered thermodynamically as different physical substances. They have very similar but distinctly different densities, boiling temperatures, critical parameters (pressure and temperature), triple point parameters, and other properties. Therefore, from a thermodynamic point of view, a mixture of isotopes should be considered and treated as a mixture of different physical substances. It requires a more complex thermodynamic model and additional effort for calculation of properties. With other elements, these methods are not applied because the isotopic composition of most substances is dominated by a single "main" isotope. However, in the case of neon, the isotopic composition must be taken into account if high accuracy is required. In this case, natural neon can be considered as a mixture of two isotopes (^{20}Ne/^{22}Ne).

3.7 Hydrogen

Three hydrogen isotopes are described in the literature. In this section, only the most common isotope ^{1}H (protium) is considered.

Hydrogen is the most common element in the universe. It is a very small molecule, consisting of two atoms, each one with a single proton in the nucleus. Therefore, hydrogen is the substance with the lowest density in both gaseous and liquid states. The viscosity of hydrogen is very low.

Hydrogen boils at the very low temperature of 20.4 K at atmospheric pressure. The specific power required to achieve and generate cold at this temperature is extremely high; therefore, liquefied hydrogen is a very valuable material. From an

4 a mixture of nitrogen (22.5 mol%) and oxygen remains liquid at 50.1 K, according to M.Ruhemann. Therefore the "no-liquid-range" is a bit smaller: only from 44.5 to 50.1 K.

economic perspective, it is difficult to establish successful applications based on liquefied hydrogen if they may work with less expensive gaseous hydrogen; hence, only a few liquid hydrogen applications are known today. The most famous one is the use as fuel for heavy-load space launchers. For the same reason, the automotive sector is focused more on compressed hydrogen, and vehicles with liquid hydrogen on board are still very rare. Potentially, in the future, heavy-duty vehicles, trains, ships, and airplanes will use LH2 as a fuel, due to its better volumetric and gravimetric density in comparison to compressed hydrogen.

In former times, liquid hydrogen was used as a refrigerant for some physical experiments. Recently, liquid hydrogen has been completely replaced by liquid helium for this purpose due to the well-developed liquid helium infrastructure.

Hydrogen is an explosive and flammable gas (ignition temperature: 580 °C, ignition or explosion range in mixture with air at ambient conditions: 5–75%). The management of corresponding risks requires special equipment and is therefore expensive.

The abbreviation LH2 is used for liquid hydrogen in the industrial gas community.

3.7.1 Spin Isomers of Hydrogen (Ortho- and Para-Hydrogen)

Under standard conditions, hydrogen H_2 is a mixture of two sorts of molecules, differing from each other by the (rotation) direction of their nuclear spin.[5] These two isomers are called ortho- and para-hydrogen (o- and p-hydrogen). In an ortho-hydrogen molecule, the spins of both nuclei have the same (parallel) direction, whereas the para-hydrogen molecule consists of two atoms with opposite (antiparallel) spin directions, as shown in Figure 3.2.

Most of the physical properties of hydrogen, such as vapor pressure, liquid density, triple point temperature, and pressures, depend slightly upon the ortho–para composition.

The equilibrium ortho–para composition (abbreviation: e-H_2) is temperature-dependent (Figure 3.3). At room temperature, it corresponds to approximately 75% ortho-hydrogen and 25% para-hydrogen. This mixture has its own name, "normal

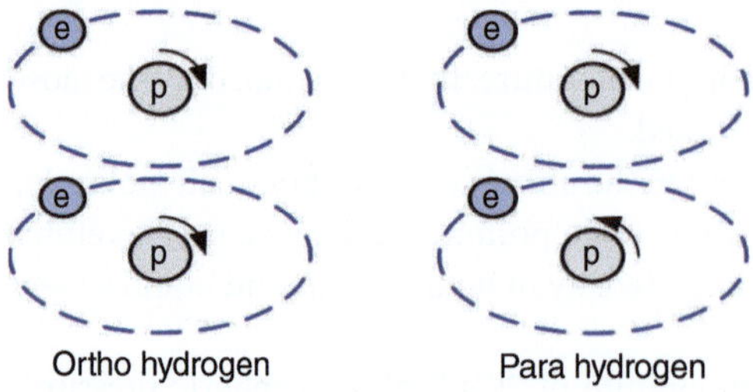

Figure 3.2 Ortho-hydrogen (left) and para-hydrogen (right).

5 The nuclear spin is a quantum mechanical property of particles. For visualization purposes, this property is described as the total angular momentum of an atomic nucleus around its gravity center. Usually, the impact of nuclear spin on macroscopic properties of gases or liquids is extremely small and therefore is neglected. But in case of liquid hydrogen, the nuclear spin has a visible impact on macroscopic properties and has to be taken into account.

$$\frac{y_{para}}{y_{ortho}} = \frac{1}{3} \cdot \frac{1 + 5 \cdot \exp\left(-6\frac{B}{T}\right)}{3 \cdot \exp\left(-2\frac{B}{T}\right) + 7 \cdot \exp\left(-12\frac{B}{T}\right)}$$

where

T – temperature, K

y – molar fraction, mol/mol

$$B = \frac{h_P^2}{8 \cdot \pi^2 \cdot I_T \cdot k_B} = 86.2\ K, \text{ since}$$

$$h_p = 6.626 \cdot 10^{-34}\ J \cdot s$$

$$k_B = 1.3806 \cdot 10^{-23}\ J/K$$

$$I_T = 4.64 \cdot 10^{-48}\ \text{kg} \cdot \text{m}^2$$

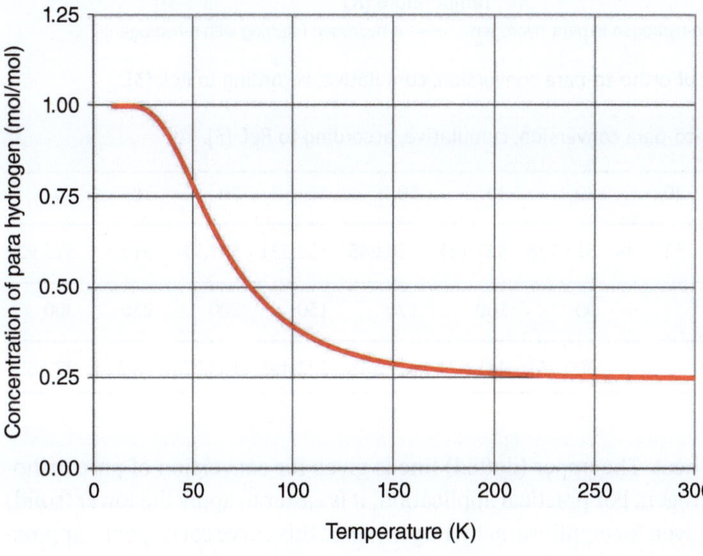

Figure 3.3 Para-hydrogen fraction in equilibrium hydrogen, depending on temperature.

hydrogen" (abbreviation: n-H_2). At the boiling point, the ortho–para equilibrium shifts significantly to para-hydrogen and amounts to 0.21% o-H_2 and 99.79% p-H_2, respectively. Due to the considerable heat of conversion, the calculation of the ortho-para-conversion becomes an important aspect in hydrogen liquefaction. Therefore, the quality of these data defines the accuracy of the whole process design. Although collected and published more than 50 years ago, the data available from National Institute of Standards (NIST, former NBS) are still unique, process design engineers worldwide use these data as main reference. The corresponding publication [4] can be downloaded from the official NIST site. The following figure is based on numbers from this publication. A good estimation for the ortho-para equilibrium is given by following equation using Boltzmann distribution [5]. Figure 3.3 shows the calculated profile according to this equation.

Figure 3.4 and Table 3.3 show the heat of ortho-to-para conversion from normal to para hydrogen according to Ref. [3], which, at the end, is based on Ref. [4]. It is

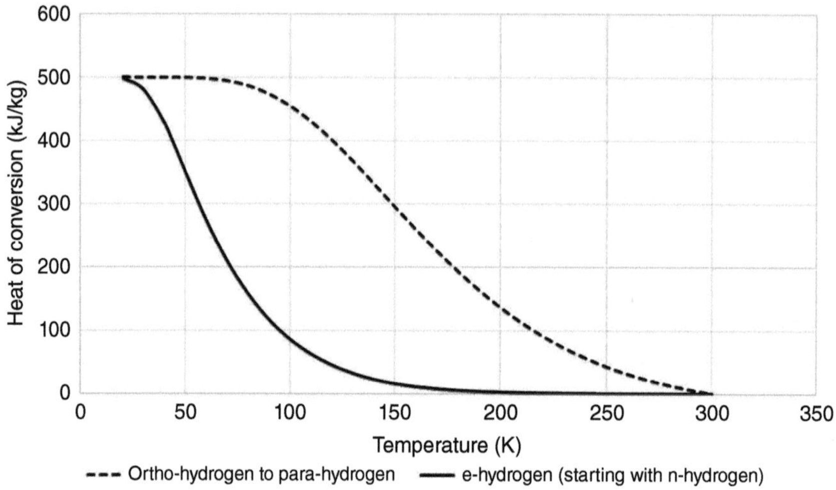

Figure 3.4 Heat of ortho-to-para conversion, cumulative, according to Ref. [3].

Table 3.3 Heat of ortho-to-para conversion, cumulative, according to Ref. [3].

Temperature	K	20	30	40	50	60	70	78	80
Conversion heat	kJ/kg	527.138	527.138	527.117	526.845	525.531	521.77	515.5	513.932
Temp	K		90	100	120	150	200	250	300
Conversion heat	kJ/kg		500.757	481.671	427.248	322.495	163.774	70.524	27.562

the cumulative heat. The upper (dotted) line is given for conversion of pure ortho- to pure parahydrogen. For practical application, it is easier to apply the lower (solid) curve, which is given for equilibrium hydrogen, since this curve corresponds approximately to the process in real liquefiers. The total heat of conversion from normal hydrogen to parahydrogen is roughly 500 kJ/kg if starting from 300 K to liquid at 20 K. But the curve is radically non-linear: the heat of conversion from 300 down to 80 K (liquid nitrogen temperature level) is roughly 155 kJ/kg or roughly ⅓ of the total heat. The further cooling from 80 to 20 K requires 345 kJ/kg (⅔ of the total value).

The specific heat of conversion for equilibrium hydrogen is shown in Figure 3.5. This curve has a maximum at approx. 50 K. According to this curve, the heat of ortho-to-para conversion rises from 0.5 kJ/(kg K) at 150 K to more than 6 kJ/(kg K) at 35–65 K. Below this temperature, the specific conversion heat becomes lower and lower.

3.7.2 Consequences for Liquefaction Processes

From the liquefaction perspective, the most important fact is that the internal energy content of ortho-hydrogen is higher in comparison to para-hydrogen. It means that the conversion from ortho-to-para-hydrogen leads to heat generation.

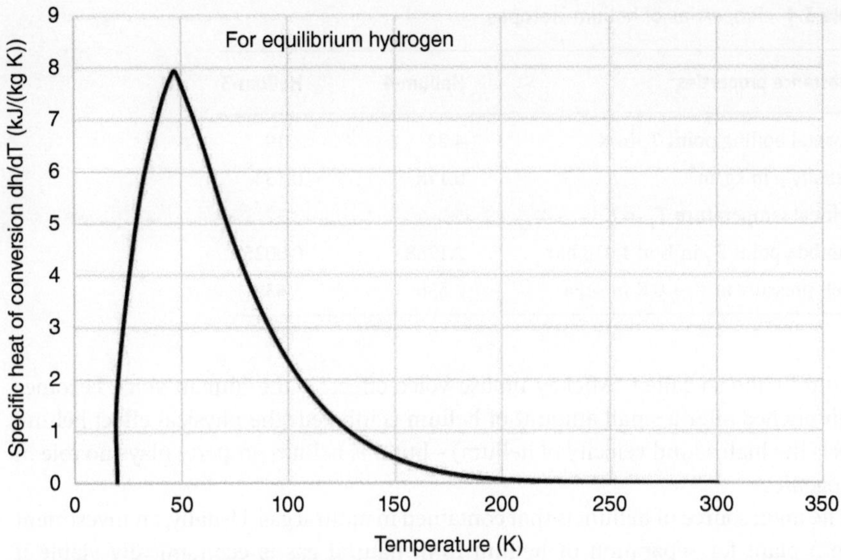

Figure 3.5 Specific heat of ortho-to-para conversion.

The interaction between the atomic nucleus and the electrons in hydrogen molecules is extremely weak. Therefore, natural (not catalyzed) ortho-to-para conversion is a very slow process and can take several days or weeks. However, the ortho–para conversion process can be accelerated considerably by means of appropriate catalysts (e.g. iron oxide or chromium oxide).

In a hydrogen liquefier, the hydrogen gas can be cooled down and liquefied very quickly (in a few seconds). In this case, the ortho–para composition of the resulting liquid hydrogen will correspond to the normal composition at room temperature, since virtually no natural ortho-to-para conversion may occur in such a short period of time. Unfortunately, the liquid produced this way is not suitable for medium- & long-term storage, since the natural ortho-to-para conversion will appear in the background anyway, and the generation of the corresponding conversion heat (which is comparable with the evaporation heat of the liquid hydrogen) will lead to continuous evaporation of the stored liquid.

For the production of storable liquid hydrogen, the ortho–para conversion must be integrated into the liquefaction process. Both conversion processes – liquefaction and conversion–must be executed simultaneously in the liquefier. The target is to achieve equilibrium composition (mainly para-hydrogen) at the final temperature of the liquid hydrogen. The corresponding approaches are discussed in Section 8.5 about hydrogen liquefaction.

3.8 Helium

Helium is the most valuable cryogenic gas. It is the material with the lowest boiling point – 4.22 K at atmospheric pressure. One of the most well-known properties of

Table 3.4 Properties of helium isotopes.

Substance properties	Helium-4	Helium-3
Normal boiling point T_s in K	4.22	3.19
Density ρ in kg/m^3	0.178	0.134
Critical temperature T_c in K	5.2	3.32
Lambda point T_λ in K at 1.013 bar	2.1768	0.0025
Melt pressure at T → 0 K in MPa	2.536	3.439

helium is the so-called "Mickey mouse voice effect" – the human voice becomes high-pitched after a small amount of helium is inhaled (the physical effect behind this is the high sound velocity of helium) – but this helium property plays no role in cryogenics.

The main source of helium is that contained in natural gas. Usually, an investment into a plant for separation of helium from natural gas is economically viable if the helium concentration is relatively high (>0.1%). Suitable wells are distributed throughout the globe. Currently, the USA, Qatar, Algiers, Russia, Australia, and Poland are the main suppliers. Theoretically, helium can be extracted from ambient air as well, but the very low concentration of helium in air (roughly 6 ppm) makes this source uncompetitive with helium extracted from natural gas.

In the gas industry, the abbreviation LHe (liquid helium) is used for liquid helium.

Chemically, helium is a very inert substance. The helium atom is the second-smallest among all chemical elements, after hydrogen.

Generally, eight isotopes of helium are known, but only two of them – helium-3 (^{3}He) and helium-4 (^{4}He) – are stable. The physical properties of helium-4 and helium-3 are very different (see Table 3.4). Therefore, these gases must be considered as two different physical substances.

In this section, only the most common isotope ^{4}H is considered. Helium-4 is a very special gas and has some surprising and outstanding properties.

3.8.1 Phase Diagram (Pressure–Temperature Diagram) and Lambda Line

A special feature of Helium-4 is that it has no classical triple point, where the solid, the liquid, and the vapor phase coexists. The solid helium phase occurs at elevated pressures above 25 bar only, and it can coexist with two liquid helium phases (normal liquid helium and superfluid liquid helium), but not with the vapor phase, see Figure 3.6. The vapor phase can coexist only with both liquid phases mentioned above.

3.8.2 Superfluidity

The cooling of liquid helium to temperatures lower than 2.2 K (the so-called lambda temperature) leads to the formation of a second liquid phase called "superfluid liquid" or helium-II (He-II) instead of a solid phase as could be normally expected.

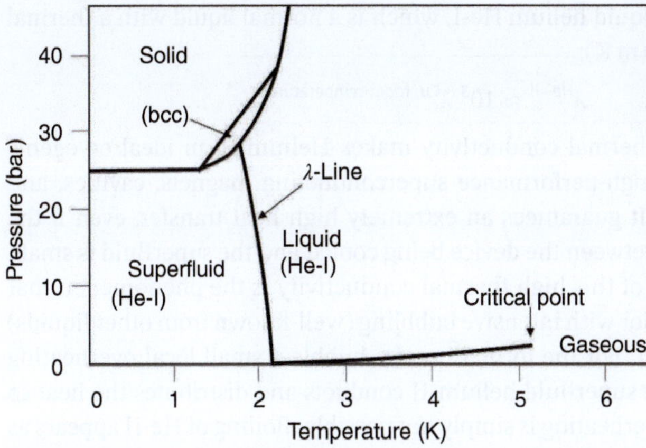

Figure 3.6 Phase diagram of ^{4}He.

Figure 3.7 Lambda curve of ^{4}He.

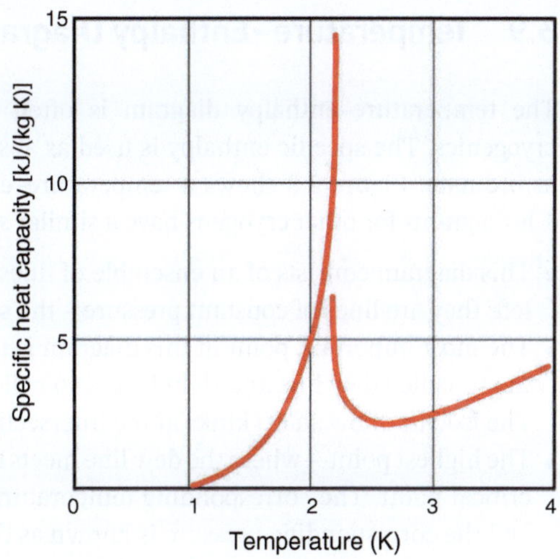

Conventional liquid helium above the lambda point is called helium-I (He-I). The term "lambda point" originates from the very distinctive graph of heat capacity versus temperature (see Figure 3.7), which illustrates the transition from He-I to He-II. This curve resembles the Greek letter lambda. The "lambda temperature" itself depends on pressure, and in a ***p,T***-phase diagram it forms a corresponding line called the "lambda" line.

The He-II is called suprafluid because of its outstanding high fluidity or – conversely – low viscosity. For example, it can flow through extremely narrow capillaries ($10^{-8}\cdots10^{-7}$ m diameter) without any (at least no measurable) viscosity-related hydraulic resistance.

3.8.3 High Thermal Conductivity of Superfluid

The low viscosity of superfluid helium is linked to an extremely high effective thermal conductivity, which is roughly 1000-times higher than the heat conductivity of pure copper at room temperature or 1 000 000-times higher than the heat conductivity of normal liquid helium He-I, which is a normal liquid with a thermal conductivity of $\lambda = 2\,\text{W/(m K)}$:

$$\lambda^{\text{He-II}} \approx 10^6\,\lambda^{\text{He-I}}, \qquad \lambda^{\text{He-II}} \approx 10^3\,\lambda^{\text{Cu, room temperature}}$$

This extremely high thermal conductivity makes Helium-II an ideal cryogenic refrigerant for cooling high-performance superconducting magnets, cavities, and other sensitive devices. It guarantees an extremely high heat transfer, even if the temperature difference between the device being cooled and the superfluid is small.

Another consequence of this high thermal conductivity is the phenomenon that the typical boiling behavior with intensive bubbling (well-known from other liquids) does not exist with He-II. For the formation of a bubble a small local overheating of liquid is required. But superfluid helium-II conducts and distributes the heat so successfully that local overheating is simply not possible. Boiling of He-II appears as an exhalation from the liquid surface.

3.9 Temperature–Enthalpy Diagram

The temperature–enthalpy diagram is often used for process visualization in cryogenics. The specific enthalpy is used as abscissa and the absolute temperature as ordinate. Figure 3.8 shows a temperature–enthalpy diagram for nitrogen. The T,h-diagrams for other cryogens have a similar structure. Note the following points:

- This diagram consists of an ensemble of lines going from top right to the bottom left: they are lines of constant pressure – the so-called isobars.
- The most important point in this diagram is the two-phase area, bordered left by the so-called dew line and right by the so-called boiling line.
  The isobars show sharp kinks at the intersection of the dew line and boiling line.
- The highest point—where the dew line meets the boiling line—is referred to as the critical point. The corresponding temperature is called the critical temperature, and the corresponding pressure is known as the critical pressure; this is the point where the liquid and vapor phases become indistinguishable.
- The area below the critical temperature is divided into three spaces:
  - liquid state left to the boiling line,
  - vapor state right to the dew line and
  - two-phase area (vapor–liquid) in-between.
- In two-phase area, the definition of a state by specification of pressure and temperature (as it is used outside of the two-phase area) is insufficient. A new parameter – the molar vapor fraction *VF*[6] (or molar liquid fraction *LF*) – is required to define the state there. This parameter (vapor fraction) is used to calculate specific enthalpy in the two-phase area h_{2phase} by Eq. (4.3). Some data property packages cannot deliver this value directly, but rather generate two other specific enthalpies: the specific enthalpy at the dew point (saturated vapor) h_d and the specific enthalpy at the boiling point (saturated liquid) h_b.

$$h_{2phase} = (1 - VF) \cdot h_d + VF \cdot h_b \tag{3.1}$$

---

6 molar vapor fraction is defined as $VF = \frac{\dot{m}^V}{\dot{m}^V + \dot{m}^L}$, where $\dot{m}^V$ – molar flowrate of vapor phase, $\dot{m}^L$ – molar flowrate of liquid phase

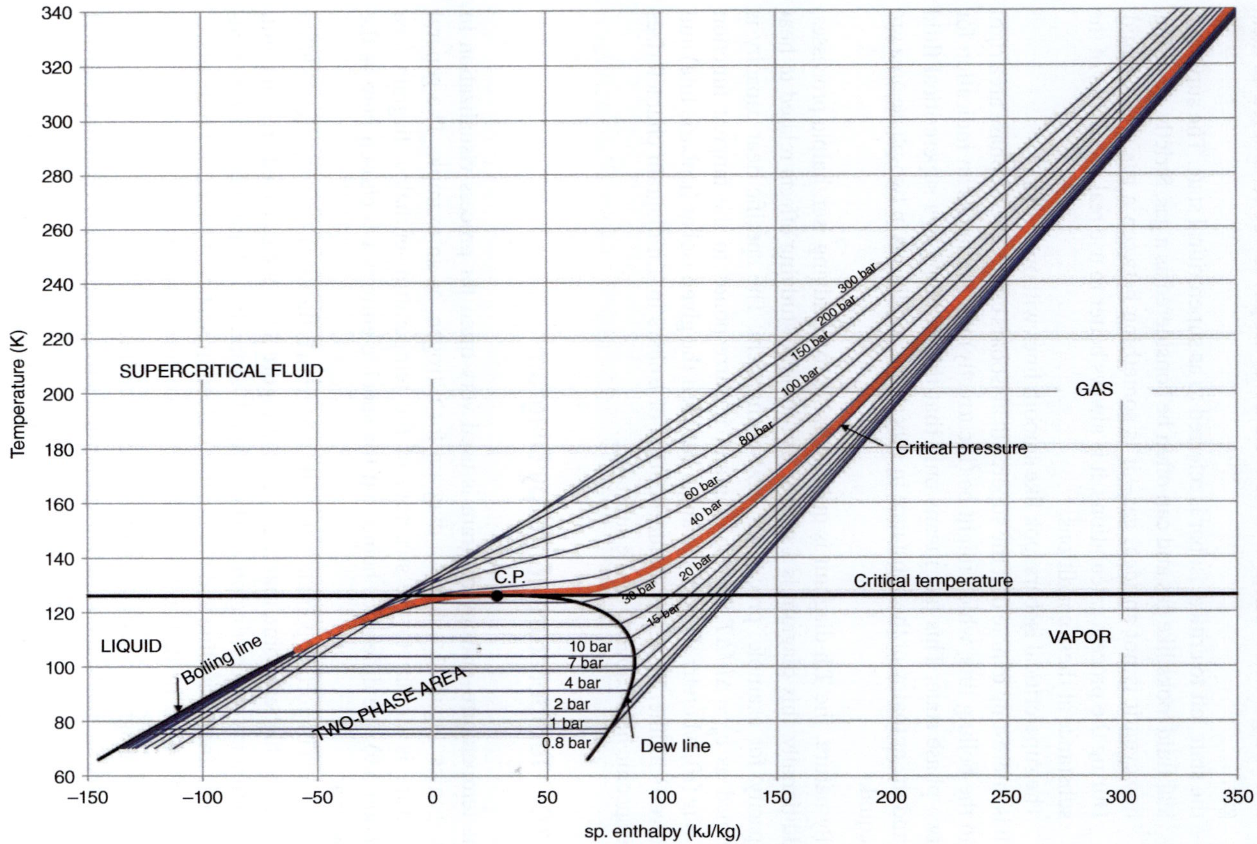

Figure 3.8 Temperature–enthalpy diagram for nitrogen.

- The area above the critical temperature is divided into two spaces by the critical isobar (isobar going through the critical point):
  - the state right to the critical isobar is referred to as gaseous. The isobars look like straight lines. Often, at these conditions, the gas can be considered an ideal gas,
  - the state left to critical isobar is referred to as supercritical state. The supercritical fluid looks like gas and can often be considered as a gas. Strictly speaking, however, it is not an ideal gas—it is something between a gas and a liquid. During the process calculations, it is always better to use real properties of the substance at these conditions.

    The supercritical isobars look like smooth lines without kinks.
- It is interesting to note that the supercritical isobars cross the two-phase area right to the boiling line, which might be (incorrectly) interpreted as an indication for two-phase state. This fact means only that the enthalpy of a supercritical fluid and the enthalpy of the substance in subcritical conditions in two-phase area are equal.

By nature, the T,h-diagram is quite suitable for studying isenthalpic processes. Additionally, this diagram is a convenient tool for studying effects related to heat capacity, for example, process in heat exchangers. The specific heat capacity is defined as $c_p = \Delta h/\Delta T$. In T,h-diagram, it corresponds to the tangent function $c_p = tg\,(\alpha)$, whereby the α is the angle between the given isobar (isobaric line) and the vertical line. It means the declination of isobars in T,h-diagram characterizes the specific heat capacity of the substance.

3.10 Temperature–Entropy Diagram

The temperature–entropy diagram is used very often for process visualization in cryogenics. Figure 3.9 shows a diagram for nitrogen, as an example. The general structure is similar to the structure of the temperature–enthalpy diagram (see Section 3.9), the difference being that the specific entropy $\boldsymbol{s}$ is chosen here as the abscissa.

The most important point to note in this diagram is the two-phase area, resembling a hut. The highest point, marked as C.P. in Figure 3.9 is the so-called critical point. All comments concerning phase state (liquid, vapor, two-phase, gas, and supercritical) and lines (dew line and boiling line), given in Section 3.9 for the T,h-diagram, are valid for the T,s-diagram, too. There are two deviations to note:

- The T,s-diagram consists of **two** ensembles of lines
  - isobars going from top right to the bottom left, known from the T,h-diagram and
  - isenthalps (isenthalpic lines) extending from the top left to the bottom right in the lower part of the diagram. In the upper right, isenthalps are more like horizontal lines.
- The supercritical isobars do not cross the two-phase area but are instead always left to the boiling line.
- Equation (3.2) is valid for calculation of specific entropy in two-phase area s_{2phase} since most of the data property packages cannot deliver this value, they generate

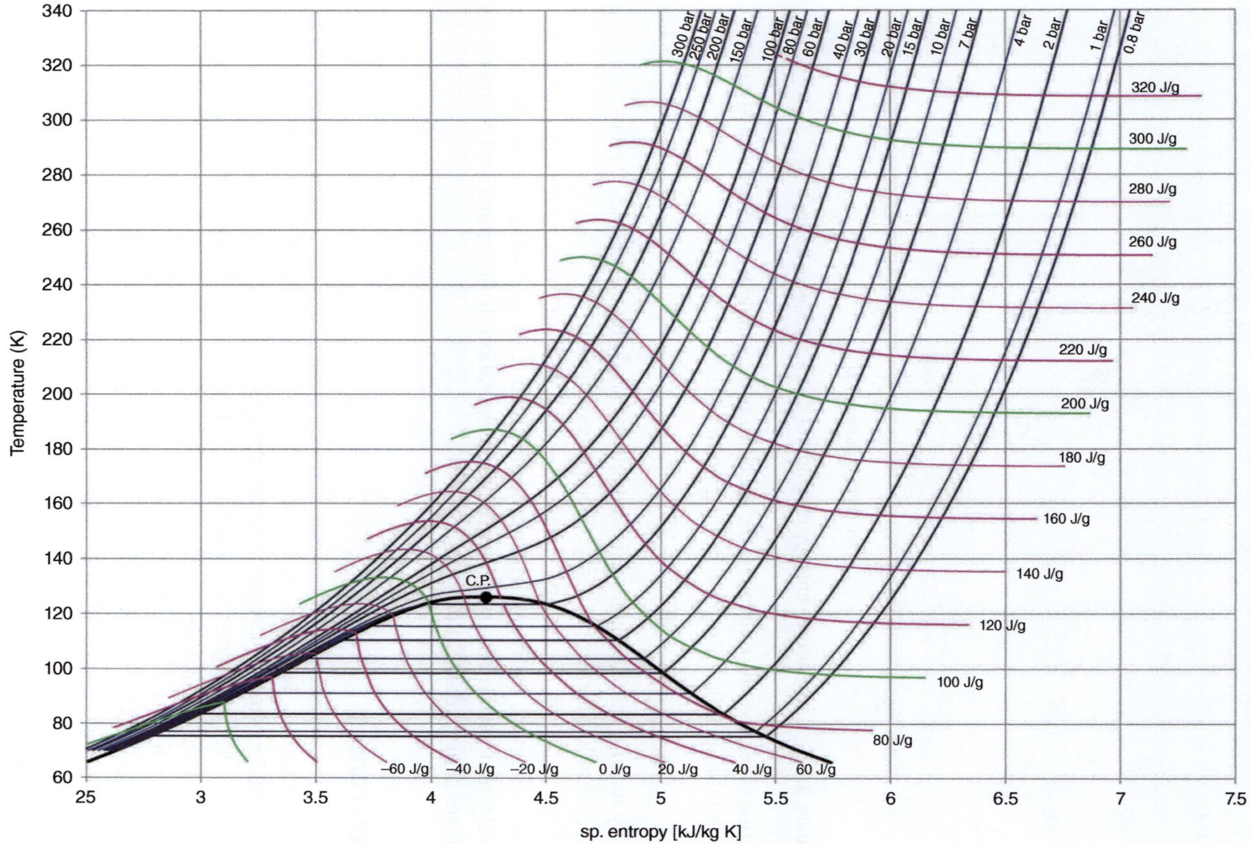

Figure 3.9 temperature–entropy diagram for nitrogen.

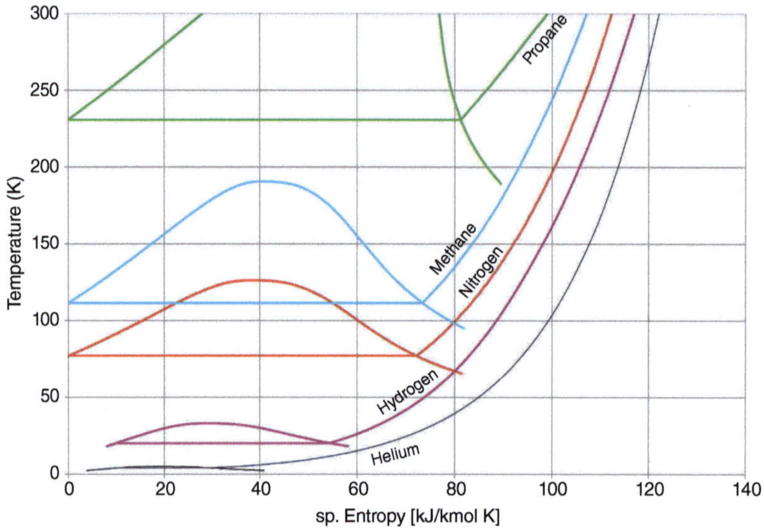

Figure 3.10 temperature–entropy diagram for some cryogenic fluids.

two other specific entropies: the specific entropy in dew point (saturated vapor) s_d and the specific entropy in boiling point (saturated liquid) s_b.

$$s_{2phase} = (1 - VF)\, s_d + VF\, s_b \tag{3.2}$$

By nature, the T,s-diagram is suitable for studying isentropic or near-isentropic processes, like isentropic compression or expansion.

Figure 3.10 shows two-phase areas as well as 1 bar-isobars for classic cryogenic fluids (helium, hydrogen, nitrogen, and methane) in comparison to propane as refrigerant used often in conventional refrigeration. It illustrates well temperature ranges of cryogenic fluids.

References

1 Leachman, J., Jacobsen, R., Lemmon, E., and Penoncello, S. (2017). *Thermodynamic Properties of Cryogenic Fluids*. Springer.

2 Jungnickel, H., Agsten, R., and Kraus, W.E. (1990). *Grundlagen der Kältetechnik*. Berlin: Verlag Technik GmbH.

3 McIntosh, G.E. (2015). Applications of ortho-para hydrogen catalyst. *IOP Conference Series: Materials Science and Engineering, Volume 101, Advances in Cryogenic Engineering: Proceedings of the Cryogenic Engineering Conference (CEC) 2015* (28 June to 2 July 2015), Tucson, AZ, USA. https://doi.org/10.1088/1757-899X/101/1/012079.

4 McCarty, R.D., Hord, J., and Roder, H.M. (1981). *Selected Properties of Hydrogen (Engineering Design Data)*. National Bureau of Standards.

5 Kinard, G.E. (1998). The commercial use of liquid hydrogen over the last 40 years. *Proceedings of the 17th International Cryogenic Engineering Conference*, Bournemouth.

4
Reversibility Concept

It is assumed that the reader has some basic knowledge and understanding of thermodynamics already, at least for topics such as open/closed thermodynamic system, enthalpy, and isobaric/isochoric heat capacity. Nevertheless, this section starts with some basic terms and definitions. The goal is to refresh one's knowledge by using typical language and logic of thermodynamics. If the reader feels uncomfortable with contents of this section, we would recommend to revitalize his knowledge by reading a good textbook about basics of thermodynamics.

Reversibility concept is a powerful concept for analysis and design of cryogenic processes. In this chapter, a compressed introduction of this concept is given. In first step, the key terms "reversibility" and "irreversibility" are introduced as well as basic considerations are given, as to why the reversible process is the best process in terms of efficiency; in second step, the entropy as a measure for irreversibility is introduced; in next step, two important terms – (i) term "exergy" as an extension of term "entropy" and (ii) term "exergy loss" as a clear quantitative measure for process loss – are introduced; in last step, the application of exergy concept for analysis of thermodynamic processes is explained.

4.1 Reversibility and Irreversibility

4.1.1 Irreversibility

If you imagine a video sequence showing two glass jars filled with a colored gas, and then when a valve between the vessels is actuated, the gas collects in one of these two vessels, you immediately know that this movie is running backward. Viewers can normally recognize[1] when a process is shown going backward because most processes can definitely go in one direction only. Some examples of such processes are as follows:

– a vehicle deaccelerating due to friction when the engine is turned off,

1 This is not an intuitive decision, where one person may decide for one option and another person for another option. A human (more precisely, the human cognitive system) is indeed able to identify a "wrong direction" process easily and rapidly, based on clear built-in criteria such as feasibility and probability.

Process Design for Cryogenics, First Edition. Alexander Alekseev.

- an apple falling from a tree,
- sugar dissolving in tea, and
- gas escaping from a pressurized cylinder when the valve is opened.

Such processes are considered ***not reversible*** or ***irreversible***.

4.1.2 Reversibility

But sometimes we are confronted with other types of processes where we are not able to discern the proper direction of the movie: right or wrong? We can imagine that either direction is possible. An example of such a process is the conversion of mechanical energy into electrical energy by means of an electric generator. The reverse process – the transformation of electrical energy into mechanical energy by means of an electric motor – is quite realistic as well. Such processes are called ***reversible processes***.

Two popular examples (described in every thermodynamics textbook) are the isothermal and the adiabatic processes[2] of compression and expansion:

- In the former, the compression process is carried out extremely slowly in order to avoid any increase in fluid temperature caused by compression. This requires efficient heat exchange with the environment. The reverse process – slow expansion to ambient conditions in combination with heat exchange (isothermal expansion) – returns the gas to its original state before compression.
- The idea of adiabatic compression is the opposite of isothermal compression: the goal is to compress the gas as quickly as possible, so that no heat exchange may occur between the gas and the environment. The reverse process – a quick adiabatic expansion (no heat exchange with environment) returns the gas to its original state.

For both processes – isothermal as well as adiabatic compression/expansion – the mechanical energy required for compression is equal to the mechanical energy released during the expansion process. The mechanical energy is not converted into another form of energy.

Of course, the sharp-sighted observer will recognize that ***all processes are basically irreversible***. An appropriate high-resolution monitoring device would indeed deliver evidence that every fast-running process and every slow-running process suffer minor (nonzero) losses. Hence, a 100% reversible process is a virtual process.

4.2 Carnot Engine

The classical reversible process/engine is the Carnot[3] process/engine described in every thermodynamic textbook. Usually, the Carnot engine is explained as a device

2 The easiest way is to consider a compression process is a compression/expansion in a reciprocating compressor/expander.

3 The Carnot process is the most popular reversible process, but it is not the only reversible process available. Other examples are Stirling cycle (https://en.wikipedia.org/wiki/Stirling_cycle) and Brayton process (https://en.wikipedia.org/wiki/Brayton_cycle).

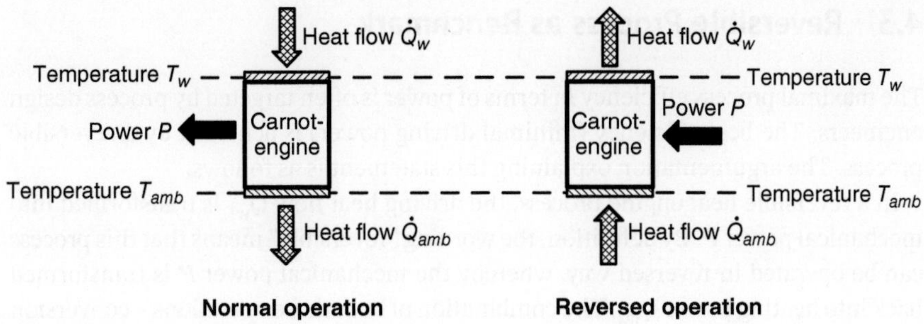

Figure 4.1 Carnot engine, main functionalities.

for converting heat $\boldsymbol{Q}$ in mechanical work $\boldsymbol{W}$. Principally, if this device operates continuously, it is able to convert a continuous heat flow $\dot{\boldsymbol{Q}}$ into mechanical power $\boldsymbol{P}$. Such a device is shown in Figure 4.1 left.

For normal operation, this device is

(a) heated on one side/surface by driving heat flow $\dot{\boldsymbol{Q}}_w$. The temperature at the heated side (surface) of the engine achieves the temperature $\boldsymbol{T}_w$,
(b) and simultaneously cooled on the other side/surface, which means that a heat flow is rejected from the engine. Usually, this heat flow is absorbed by environment, whereby the cooled side/surface of the engine is being kept at the ambient temperature $\boldsymbol{T}_{amb}$. Therefore, this heat flow is often referred to as $\dot{\boldsymbol{Q}}_{amb}$.

In normal operation, the Carnot engine converts heat flow $\dot{\boldsymbol{Q}}_w$ into mechanical power $\boldsymbol{P}$.[4]

The special feature of Carnot engine is that the process inside this engine (Carnot process) is a reversible process. By definition, the wording "reversible" means that the Carnot engine is able to operate in reversed way, whereby all energy flows (power and heat flows) run in opposite direction, as shown in Figure 4.1 right.

In reversed operation mode, the mechanical power $\boldsymbol{P}$ is applied to this device. This power is used for generation of heat flow $\dot{\boldsymbol{Q}}_w$, which can be used for heating purposes. During the reversed operation

(a) one side (surface) of the engine becomes hot, and the temperature there achieves the value $\boldsymbol{T}_w$. The heat flow $\dot{\boldsymbol{Q}}_w$ will be removed from this surface,
(b) Simultaneously, the other side (surface) of the Carnot engine becomes colder. For normal operation, this side (surface) is being kept at the ambient temperature $\boldsymbol{T}_{amb}$ by means of closer thermal contact with environment. It means that this side (surface) of Carnot engine is continuously warmed by a heat flow $\dot{\boldsymbol{Q}}_{amb}$ from the environment.

A thorough description of Carnot engine and the so-called "reversed Carnot cycle" (a version of Carnot cycle for a refrigeration system) is available at: https://en.wikipedia.org/wiki/Carnot_heat_engine, https://en.wikipedia.org/wiki/Carnot_cycle.

4 Like a conventional power station, where the coal or natural gas is combusted and the generated heat power is converted into mechanical and later electrical power.

4.3 Reversible Process as Benchmark

The maximal process efficiency in terms of power is often targeted by process design engineers. The best efficiency (minimal driving power) is achieved by a reversible process. The argumentation explaining this statement is as follows:

In a reversible heat engine process, the driving heat flow $\dot{Q}_{w1}$ is transformed into mechanical power $\boldsymbol{P}$. By definition, the wording "reversible" means that this process can be operated in reversed way, whereby the mechanical power $\boldsymbol{P}$ is transformed back into heating power $\dot{Q}_{w2}$. The combination of both these operations – conversion of $\dot{Q}_{w1}$ into $\boldsymbol{P}$ and thereafter back $\boldsymbol{P}$ into $\dot{Q}_{w2}$ – in such a way that $\dot{Q}_{w2} = \dot{Q}_{w1}$ – is feasible only if loss[es] (of power, energy, and heat) are avoided during all these transformations. Therefore, every reversible process is considered as loss-free process (see Section 4.5 for definition of losses), the driving power $\boldsymbol{P}$ is minimal, and efficiency of this process is maximal.

From these considerations, three takeaways can be derived:

1. The best efficiency is achieved by means of a reversible process.
2. The reversible process may be used as a benchmark for characterization of thermodynamic processes.
3. Every real (nonreversible) process can be improved by making it more reversible, or, in other words, by avoiding every kind of irreversibility in this process.

4.4 Entropy

It is hard to explain the thermodynamic term entropy. Entropy is not a really tangible property – it cannot be measured directly like temperature, pressure, or flow rate. It does not follow conservation rules like the conservation of energy, mass, particle number, or electrical charge. Entropy has a touch of mystery.

4.4.1 Entropy According to Rudolf Clausius

Rudolf Clausius (1822–1888) established the theory of conversion of heat into power, by restating the Carnot process. It is not possible to reconstruct his way of thinking, because thermodynamics in its current form did not even exist back then. But at the end, R. Clausius was able to introduce an equation (today referred to as Carnot equation), which is valid for every reversible process or engine:

$W = Q_w (T_w - T_{amb})/T_w$ (for conversion of heat Q into mechanical work W)	$P = \dot{Q}_w (T_w - T_{amb})/T_w$ (for conversion of heat flow $\dot{Q}$ into mechanical power P)

In combination with the energy conservation law for the engine,

$Q_w = W + Q_{amb}$ (for conversion of heat Q into mechanical work W)	$\dot{Q}_w = P + \dot{Q}_{amb}$ (for conversion of heat flow $\dot{Q}$ into mechanical power P)

the following equation can be obtained

$Q_{amb}/T_{amb} = Q_w/T_w$ (for conversion of heat Q into mechanical work W)	$\dot{Q}_{amb}/T_{amb} = \dot{Q}_w/T_w$ (for conversion of heat flow $\dot{Q}$ into mechanical power P)

This equation expresses that the quotient $\boldsymbol{Q/T}$ (respectively $\boldsymbol{\dot{Q}/T}$) in a reversible heat engine/process remains constant.

The exceptional achievement of Rudolf Clausius was that he was able to recognize this fact. To emphasize its importance, he introduced the new term – entropy[5] – as follows:

$S = Q/T$ (entropy)	$\dot{S} = \dot{Q}/T$ (entropy flow)

According to this original definition, the term "entropy" is linked to the term "heat" or "heat flow." Entropy represents a kind of characteristic/property of the heat. It considers not only the absolute value of heat (in Joules) but also the temperature level (in K), which corresponds to that heat.

4.4.1.1 Extension of the Clausius Definition

Heat (or heat flow) is not the only option in processes of energy conversion or transfer in a thermodynamic system. There are several other options available. Using the first law of thermodynamics, which states that the change in internal energy is the result of mechanical work and the heat applied to the system:

$$dU = \delta W + \delta Q,$$

as well as the knowledge that mechanical work may be impacted by volume change (by the means of compression), composition change, etc. according to the equation:

$$\delta W = -p\,dV + \mu\,dN + \ldots$$

the next equation for the change of entropy as a function of the thermodynamic variables can be obtained:

$$dS = (dU + p\,dV - \mu\,dN - \ldots)/T \tag{4.1}$$

This way, the term entropy becomes applicable to all process steps or units related to the transformation of energy from heat transfer to composition change of a multicomponent flow. It can be used for characterization of processes in compressors, pumps, heat exchangers, mixers, flashes, separation columns, etc.

Additionally, this equation is a sort of definition for the entropy of fluids and material streams. This is very useful since most thermodynamic processes are realized by means of working fluids. The entropy of a fluid stream is a characteristic of the energy (thermal, mechanical, chemical, etc.) contained in it.

5 The word entropy itself is an artificial word created by R. Clausius by combining the Ancient Greek words energia (activity, operation, and work) with "trop" (transformation).

4.4.2 Entropy According to Boltzmann

The most general interpretation of entropy as a measure of the uncertainty about a system is based on the idea originally created by Ludwig Boltzmann (1844–1906). According to this concept,[6] heat, entropy, or other macroscopic quantities result from movement of individual molecules. For calculation of these functions, Boltzmann developed a two-step approach:

(1) in the first step, the mechanics of every individual molecule (velocity, momentum, and energy) is described/modeled by a set of equations,
(2) in the second step, the properties of a substance/continuum are created by means of integration through all of the individual molecules.

Boltzmann showed that his definition of entropy was equivalent to the entropy according to Clausius's definition if multiplied by a factor known as Boltzmann's constant now. In summary, the Clausius's definition of entropy provides the experimental definition of entropy, whereas the statistical definition of entropy extends the concept, providing a sort of tangibleness for this function. More information about entropy can be found here:

https://en.wikipedia.org/wiki/Entropy.

4.4.3 Entropy for Thermodynamic Analysis

Thermodynamic analysis is an important step during the development of new processes or the further development of existing processes. The target is always to develop the most reversible process and therefore minimize the power consumption of the system.

For this purpose, entropy is an extremely useful function: in reversible processes, entropy remains constant; in irreversible processes, entropy increases. Therefore, every irreversibility can be immediately recognized by calculating the entropy increase. Additionally, the degree of irreversibility can easily be quantified: a small entropy increase means that the process step approximates a reversible process, whereas an irreversible process is characterized by a large increase in entropy.

Surprisingly, the entropy was not widely used by engineers in their daily routines for process assessment. The reason for this is unclear. Perhaps they lie in the fact that an entropy increase, usually given in kW/K or kJ/K, does not per se indicate the impact of this value on the performance of the entire system; instead, an additional calculation procedure is required for this purpose. In this context, engineers set forth their search for a more convenient method of process evaluation. There are different approaches used. The most universal one is based on **exergy [concept]**, see Section 4.5.

Generally, the exergy concept is a further development of entropy-based analysis. An advantage of the exergy-based analysis concept is that the corresponding procedure is well-structured. It is universally applicable to every process unit and may be simply algorithmized. The results are transparent and "ready for discussion."

---

6 Today this approach is referred to as the statistical thermodynamics.

4.5 Exergy Concept

The term EXERGY was introduced in the 1950s by Zoran Rant and developed further in the 1960s and 1970s by many other scientists. The initial question, investigated by Rant, was the following:

> The term "energy" itself is universal: it is used for different kinds of energy (electrical, mechanical, thermal, magnetic, etc.), which are united by common units of measure (Joule or Watt). But all these kinds of energy are nonequivalent in terms of transformation. For example, one kilowatt of electrical power can always be converted in one kilowatt of heat by means of a simple electric heater. But the reverse transformation (from heat to electrical power) does not work in the same way. According to the Second Law of Thermodynamics it is not possible to generate one kilowatt electric power from one kilowatt of heat power alone. That means one kilowatt of heat power is less valuable than one kilowatt of electricity.

The idea of Zoran Rant was to create a new universal characteristic for energy and energy streams – a sort of thermodynamic currency that allows a fair comparison of all kinds of energies in terms of transformability. For this purpose, the term EXERGY, and a corresponding method for using this term were introduced.

4.5.1 Exergy of Mechanical and Electrical Energy

The most valuable kind of energy is the mechanical or electrical (electromagnetic) energy. In terms of exergy, mechanical and electromagnetic energy are equivalent. The background is that electrical energy can be converted into mechanical energy in a reversible way by means of a reversible electric motor (or generator for the reverse process). These devices are, of course, sort of virtual devices only, but the real machines perform this transformation in an extremely efficient way, too.

For these reasons, mechanical and electrical energy (power) were chosen as a benchmark for all other types of energy. By definition, the exergy of electrical energy (power) is equivalent to the absolute value of this energy: the same is valid for the mechanical energy (power):

$$E_W = W \quad \text{(for conversion of heat } Q \text{ in mechanical work } W)$$
$$\dot{E}_P = P \quad \text{(for conversion of heat flow } \dot{Q} \text{ in mechanical power } P) \tag{4.2}$$

One consequence of this decision/definition relates to unit of measure for exergy. Namely, the exergy and exergy streams are measured in the same units (J, kJ, W, kW, etc.) as energy and energy streams. That is a bit confusing since the term "exergy" in broader sense is not equivalent to energy (in terms of reversibility this function is closer to term "entropy" as it will be shown in Sections 4.5.3 and 4.5.4). Principally, it could make sense to introduce a separate unit of measure for exergy – a sort of exergetic Joule or exergetic Watt – to avoid any mistakes. But a general consensus for an appropriate adjustment is difficult to achieve.

4.5.2 Exergy of Heat/Heat Flow

The exergy of thermal energy or thermal power depends on the temperature T_w at which this heat is available. The background is that every temperature difference between the ambient temperature T_{amb} and a higher temperature T_w can be used to drive a heat engine, converting the heat into mechanical energy. Theoretically, if a reversible heat engine (for example, a Carnot engine) is used, the heat can be converted into mechanical work W_{Carnot} in reversible (best possible) way. The produced mechanical power W_{Carnot} is equivalent to the exergy of the given heat Q. The Carnot equation defines the exergy of the heat E_Q as follows:

$$
\begin{aligned}
E_Q &= W_{Carnot} = Q(T_w - T_{amb})/T_w \quad \text{(for conversion of heat } Q \text{ in mechanical work } W\text{)}\\
\dot{E}_{\dot{Q}} &= P_{Carnot} = \dot{Q}(T_w - T_{amb})/T_w \quad \text{(for conversion of heat flow } \dot{Q} \text{ in mechanical power } P\text{)}
\end{aligned} \tag{4.3}
$$

If the temperature T_w is high, the exergy of this thermal energy is high because the mechanical work (power) that can be gained from the available temperature difference ($T_w - T_{amb}$) is high. If the temperature T_w is close to the ambient temperature, the exergy of this heat is low.

Example 4.1 ***Exergy of Cooling Water***

Some devices, for example, compressors, are cooled by means of cooling water. Formally, this way, the heat from the device is transferred to the cooling water. The value of the heat contained in the cooling water can be very high in terms of enthalpy difference (between the inlet and the outlet). But it is considered close to zero from exergy perspective because it is available at the ambient temperature only:

$$T_w \to T_{amb} \qquad \to \qquad \dot{E}_{\dot{Q}} = P_{Carnot} = \dot{Q}\,(T_w - T_{amb})/T_w \to 0$$

Example 4.2 ***Water Kettle***

A person puts a kettle on to boil water but then leaves the kitchen to do something else. An hour later, the person tries to make tea with the water but finds that the water is cold. Question: What went on in the kitchen during that hour from thermodynamic perspective?

Answer: The heat of the boiled water has been transferred to the ambient air – as a result, the air in the kitchen has become slightly but not discernibly warmer. The energy has not been lost, but rather the heat of the boiled water (at the relatively high-temperature level of 100 °C) was converted into the heat at room temperature level (20 °C). This latter heat has no value (in terms of power/exergy) – it does not make the room appreciably warmer, and we cannot use it to make tea. The high-value thermal energy of the boiled water has been transformed into the zero-value thermal energy of the ambient air.

Question: How to calculate the exergy loss?

Answer: (a) calculate the exergy of the freshly boiled water by means of Eq. (4.3); (b) the exergy of the water at room temperature will be zero. The difference between these two values – the exergy loss – quantifies how many kWh of power were wasted during this process.

4.5.3 Exergy of Cold/Cooling Power

The same procedure is applicable for considerations of exergy of cold/cooling power. The temperature difference between the ambient temperature T_{amb} and the cooling temperature T_o can be used to drive a heat engine and convert the heat from environment into mechanical energy. If a reversible engine is used for this conversion, the ambient heat can be converted into the power in reversible (best possible) way, without entropy generation. The produced mechanical work/power corresponds to the exergy of the given cold/cooling power. The Carnot equation defines this value $\boldsymbol{E_{Qo}}$ as:

$$E_{Qo} = W_{Carnot} = Q_o(T_{amb}-T_o)/T_o \quad \text{(for conversion of heat } Q_o \text{ in mechanical work } W\text{)}$$
$$\dot{E}_{Qo} = P_{Carnot} = \dot{Q}(T_{amb}-T_o)/T_o \quad \text{(for conversion of heat flow } \dot{Q}_o \text{ in mechanical power } P\text{)} \tag{4.4}$$

If the temperature T_o is low, then this cold/cooling power is very valuable, because the amount of mechanical work/power generated from the temperature difference $(T_{amb} - T_o)$ is high. An example illustrating the value of the cold/cooling capacity is shown in Section 7.1.7.

4.5.4 Exergy of Material/Material Stream

The thermodynamic function enthalpy describes the thermomechanical energy of a material or material stream. Analogously, a function can be introduced for definition of the exergy of a material/material stream. Here we use the symbols $\boldsymbol{E}$, $\dot{\boldsymbol{E}}$, and $\boldsymbol{e}$ for this:

E	-	absolute exergy of a substance (for closed systems)	(J)
$\dot{E}$	-	absolute exergy flow of a material stream (for open systems)	(kW)
$e = E/m$ $e = \dot{E}/\dot{m}$	-	specific exergy	(J/kg)
$e = E/N$ $e = \dot{E}/\dot{N}$	-	specific molar exergy	(J/mol)

The exergy can be derived from known enthalpy and entropy values as follows:

$$E = (H - H_{amb}) - T_{amb} \cdot (S - S_{amb}) \quad \text{(for conversion of heat } Q \text{ in mechanical work } W\text{)}$$
$$\dot{E} = (\dot{H} - \dot{H}_{amb}) - T_{amb} \cdot (\dot{S} - \dot{S}_{amb}) \quad \text{(for conversion of heat flow } \dot{Q} \text{ in mechanical power } P\text{)}$$
$$e = (h - h_{amb}) - T_{amb} \cdot (s - s_{amb}) \quad \text{(for specific value)} \tag{4.5}$$

where

H_{amb}, $\dot{H}_{amb}$, h_{amb}	-	enthalpy values at ambient conditions
S_{amb}, $\dot{S}_{amb}$, s_{amb}	-	entropy values at ambient conditions

4.5.5 Reference States

Since the ambient conditions depend on geography, climate, and season, for practical reasons it is sometimes easier to use a standardized reference state with $h^o(T^o, p^o)$ and $s^o(T^o, p^o)$ instead of actual ambient conditions.[7] as follows:

$$E = (H - H^o) - T_{amb} \cdot (S - S^o) \quad \text{(for conversion of heat } Q \text{ in mechanical work } W\text{)}$$
$$\dot{E} = (\dot{H} - \dot{H}^o) - T_{amb} \cdot (\dot{S} - \dot{S}^o) \quad \text{(for conversion of heat flow } \dot{Q} \text{ in mechanical power } P\text{)}$$
$$e = (h - h^o) - T_{amb} \cdot (s - s^o) \quad \text{(for specific value)} \tag{4.6}$$

whereby the so-called normal conditions or standard conditions may be used as reference states, defined as follows:

$T^o = 0\,°\mathrm{C}$, $p^o = 1.013\,\mathrm{bar}$	normal conditions
$T^o = 25\,°\mathrm{C}$, $p^o = 1.013\,\mathrm{bar}$	standard condition

The exergy of a material/material stream includes at least[8] the following:

- exergy of the mechanical energy contained in given material/material stream: if a material is under pressure, it can be expanded to ambient pressure in an expander in order to generate mechanical power,
- exergy of the thermal energy contained: if the considered material/material stream is cooled or heated to ambient temperature, a portion of heat (heat flow) $Q = \Delta H$ will be released; this heat/heat stream can be converted into mechanical energy/mechanical power by means of a reversible engine.

In this sense, the exergy value of a material/material stream indicates how much mechanical energy can be gained from a material/material stream if this material/material stream is brought/transferred to ambient conditions by means of a reversible process.

Since a reversible (loss-free) process is used for this conversion, it is also correct to say that

> **the exergy of a material/material stream at given conditions (pressure, temperature, phase state) represents a minimum amount of energy/power required to process this material/material stream from ambient conditions to the given state.**

For example, this means that the exergy of a liquid cryogenic fluid (liquid helium, liquid nitrogen, etc.) is equal to the minimum energy required to liquefy this fluid.

---

7 Do not forget the deviation between these standardized reference conditions and real ambient parameters during the analysis.

8 If chemical reactions of physical effects leading to change of composition of material streams are studied, then the chemical exergy (combustion heat, heat of reaction, etc.) must be considered. This sort of conversion process is rare in cryogenics, therefore usually the considerations are limited to thermomechanical effects.

4.5.6 Exergy for Multicomponent Materials/Material Flows

Obviously, mixing effects are not considered by Eq. (4.6), which reflects only pressure and temperature changes. Therefore, this equation is not applicable to processes and process units where the composition of working fluid changes, such as mixers, separation columns, phase separators of multi-component fluids, and similar processes.

The losses caused by irreversible mixing or de-mixing can be evaluated by an extended version of this equation:

$$e_{mix} = h_{mix}(T,p) - \Sigma_i (z_i \cdot h_i [T_{amb}, p_{amb}]) - T_{amb} \cdot \{s_{mix}(T,p) - \Sigma_i (z_i \cdot s_i [T_{amb}, p_{amb}])\} \tag{4.7}$$

Basically, Eq. (4.7) is very similar to Eq. (4.6). The normalized sum of enthalpies and the normalized sum of entropies of individual mixture components at ambient conditions are used here instead of the enthalpy and entropy values of the mixture:

$\Sigma_i (z_i \cdot h_i [T_{amb}, p_{amb}])$ instead of $h_{mix}(T_{amb}, p_{amb}))$

$\Sigma_i (z_i \cdot s_i (T_{amb}, p_{amb}))$ instead of $s_{mix}(T_{amb}, p_{amb}))$

This equation defines the specific exergy of a mixture at the given temperature $\boldsymbol{T}$ and pressure $\boldsymbol{p}$ as energy (or power) required

- to compose (mix) this mixture from pure components at ambient conditions and
- to bring the resulting mixture to the given pressure $\boldsymbol{p}$ and temperature $\boldsymbol{T}$

by means of a reversible process.

4.6 Application of Exergy Concept for Thermodynamic Analysis, Exergy Loss

Energy, energy flows, and energy-containing material flows are managed in every thermodynamic process: material flows are heated, cooled, compressed, expanded, mixed, and rectified; heat is transferred, or mechanical power is converted into other kinds of energy. Therefore, material and energy flows at the outlet of a process unit differ from material and energy flows at the inlet.

The exergy concept makes it possible

(a) to assess the value of all material and energy flows at the inlet and outlet of a system/unit in the same "currency" – exergy – and
(b) to perform a comparison of both these values. This procedure is referred to as the **exergy analysis**.

The difference between the exergy at the inlet and the exergy at the outlet of a process (unit) is referred to as **exergy loss** of the considered process (unit).

The most important feature of the "exergy loss" concept is that the "exergy loss" (caused by irreversibility of a process step and the corresponding entropy increase):

1. is translated directly into Joules and Watts,
2. corresponds to increased power consumption of the compressor that drives the entire system.

   If the loss can be reduced (by better process design) by X kilowatts, the overall process will need at least X kilowatts less driving power for the same task; therefore, the exergy loss becomes visible and tangible.
3. The exergy loss analysis for a complex system delivers information on
   - where (in which process unit) the exergy is lost,
   - how much exergy is lost,
   - how to improve the thermodynamic efficiency of the process.

The formal procedure for determination of exergy losses in a system consists of three steps:

1. Definition of system boundary.
2. Exergy assessment for all energy (energy flows) and materials (material flows), whereby the exergy for the given energy (energy flow) and material (material flow) will be determined via calculation, as described in Section 4.5.
3. Calculation of exergy loss by means of exergy balance sheet as a difference between the sum of all exergies (exergy flows) at the inlet and the sum of the exergies (exergy flows) at the outlet of the system being considered:

$$\dot{D} = \Sigma_i \dot{E}_i^{\ in} - \Sigma_i \dot{E}_i^{\ out} \text{(for open system)}$$
$$D = \Sigma_i E_i^{\ in} - \Sigma_i E_i^{\ out} \text{(for closed system)} \tag{4.8}$$

where

$\dot{D}$	–	loss (exergy loss) as exergy flow
$\Sigma_i \dot{E}_i^{\ in}$	–	the sum of the exergies of all energy flows and material flows entering the system
$\Sigma_i \dot{E}_i^{\ out}$	–	the sum of the exergies of all energy flows and material flows leaving the system
D	–	loss (exergy loss)
$\Sigma_i E_i^{\ in}$	–	the sum of the exergies of all energies and materials entering the system
$\Sigma_i E_i^{\ out}$	–	the sum of the exergies of all energies and materials leaving the system

5

Unit Operations for Cryogenic Processes

The most important unit operations used in cryogenic processes are compression, expansion, and heat transfer by means of a heat exchanger. Generally, the common literature about thermodynamics addresses and thoroughly covers all these unit operations. But the unit operations for cryogenic processes have some specifics. The current section serves two purposes: (i) it summarizes thermodynamic basics for typical process unit operations, and (ii) it explains some special aspects related to cryogenic process design. Another important aspect, that considerations given in this section are focused on thermodynamic losses.

5.1 Isenthalpic Expansion

Two physical effects causing a temperature drop by means of expansion of a fluid from a high-pressure to a low-pressure level are used in industrial cryogenics today: (i) isenthalpic expansion by means of relatively simple expansion devices like valve, orifice, capillary, and porous plug and (ii) expansion in an expander usually considered to be a near-isentropic expansion. The isenthalpic expansion is linked to the so-called Joule–Thomson effect.

5.1.1 Joule–Thomson Effect

The Joule–Thomson effect may be illustrated by following experiment as shown in Figure 5.1:

- Consider a conventional gas cylinder filled with nitrogen.
  The gas inside of a cylinder is compressed to a pressure p_1 (usually 200 bar if a standard freshly delivered cylinder is used), while the temperature of the gas T_1 corresponds to the room temperature.
- Carefully open the cylinder valve and release the gas for a relatively short[1] period into the environment.

1 If this experiment is executed for a relative long time (exceeding one minute), the pure Joule–Thomson effect will be overlaid and crossfaded by an additional cooling effect called Joule effect, which results in a cooling of the gas inside of the gas cylinder during the emptying of the cylinder.

Process Design for Cryogenics, First Edition. Alexander Alekseev.

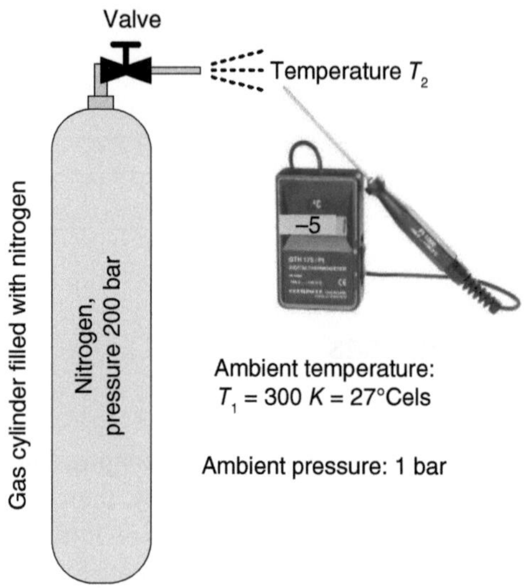

Figure 5.1 Isenthalpic expansion of a high-pressure gas. Source for measurement device: GHM Messtechnik GmbH Standort Greisinger.

- Measure the temperature of the released/expanded gas T_2. The measurement shows that the temperature of the released gas T_2 is lower than T_1. More precisely, the temperature drop will be about 30° if the pressure in the nitrogen cylinder is close to 200 bar.

This temperature drop ΔT is the so-called Joule–Thomson effect:

$$\Delta T = T_1 - T_2$$

This temperature drop is the result of expansion from high pressure p_1 (inside of the gas cylinder) to a relatively low pressure p_2 (outside of the cylinder), in current case corresponding to the ambient pressure. The pressure drop is converted into the temperature change: $\Delta p \rightarrow \Delta T$.

In this experiment, the gas simply flows and expands through the cylinder valve. There are no heats/heat flows or energies/energy flows applied to or removed from the gas during the expansion process. Therefore, the **enthalpy of the fluid remains constant** (Figure 5.2).

This cooling effect is independent of the expansion device design. Every device able to cause an appropriate pressure drop – orifice, capillary, porous plug, etc. – may be used for this purpose instead of the valve. The pressure drop is converted into the temperature change. The valve is the most convenient expansion device because it allows us to control the flowrate.

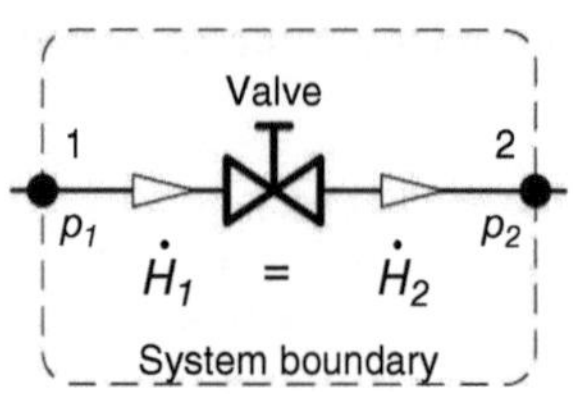

Figure 5.2 Energy conservation law for expansion in a valve.

Figure 5.3 Cryogenic valve, made by WEKA.

5.1.2 Expansion Valve

The mechanical design of a cryogenic valve does not differ much from that of conventional valves. The only difference is that the valve spindle is long, see Figure 5.3. It is designed as a thin-wall construction in order to reduce thermal conductivity and hence the heat flow from the warm valve head to the cold valve body.

In terms of materials, austenitic stainless steels, nickel alloys as well as copper and aluminum are used for cold parts, with PTFE (Polytetrafluorethylen)-based materials for sealing.

Cryogenic control valves are usually driven by a pneumatic actuator, since it is a more reliable drive than the alternative, an electric actuator. In some cases, a simple manual valve may be used.

5.1.3 Joule–Thomson Coefficient

Scientists introduced the so-called Joule–Thomson coefficient μ for characterization of the local temperature change for an isenthalpic change in pressure, which is a derivation:

$$\mu = \left(\frac{\partial T}{\partial p}\right)_{h=const}$$

The Joule–Thomson coefficient can be used for calculation of the whole Joule–Thomson effect ΔT by means of integration in required pressure range:

$$\Delta T = \int_{p_1}^{p_2} \mu \; dp \tag{5.1}$$

5.1.4 Isenthalpic Expansion in Temperature–Enthalpy Diagram

Figure 5.4 shows the process of isenthalpic expansion described in Section 5.1.1 – expansion of nitrogen from 200 to 1 bar; the initial state (300 K and 200 bar) is marked as point 1 at the top of the diagram. During the isenthalpic expansion, the enthalpy, by definition, remains constant. The expansion process is described by the vertical line starting from point 1. The state at the end of the expansion process – marked as point 2 – can be found as the crossing point of this vertical line with 1 bar – isobar. The temperature difference between 1 and 2 amounts to roughly 30 K.

5.1.5 Isenthalpic Expansion in Temperature–Entropy Diagram

Figure 5.5 shows the same expansion processes (1 → 2). This kind of diagram usually consists of two ensembles of lines, which are the well-known isobars and isenthalps (lines of constant enthalpy). Mostly, they run from top left to bottom right.

The shape of isenthalps relates to the Joule–Thomson coefficient: since the isobars (in the temperature–entropy diagram) are almost equidistant, the inclination of the isenthalps (more precisely the angle of inclination) can be interpreted as the Joule–Thomson coefficient $\mu = (\partial T/\partial p)_{h=const}$. Studying these lines in *T,s*-diagram reveals valuable information about the Joule–Thomson coefficient and corresponding cooling effect of the given fluid.

5.1.5.1 Learnings About the Joule–Thomson Coefficient from *T,s*-Diagram

1. The shape of isenthalps is complex. On the left-hand side, one may find area where the temperature increases, achieves a maximum, and decreases along this line; on the right-hand side the temperature remains almost always constant, Figure 5.5.
2. The Joule–Thomson coefficient is a material property like density, specific heat capacity, or other properties. It depends on thermodynamic conditions like pressure, temperature, and phase state.
3. For an ideal gas, the Joule–Thomson coefficient is always zero. By definition, an ideal gas has no interactions between molecules, independent of pressure and temperature. Consequently, a pressure change does not impact the internal energy and enthalpy of the ideal gas and does not cause any thermal effects.
   In a *T,s*-diagram, the above would be evident by means of a horizontal isenthalp because no temperature change during the expansion occurs. This behavior may be identified on the right-hand side in Figure 5.5 in the pressure range below 10 bar; here, the nitrogen may indeed be considered an ideal gas.

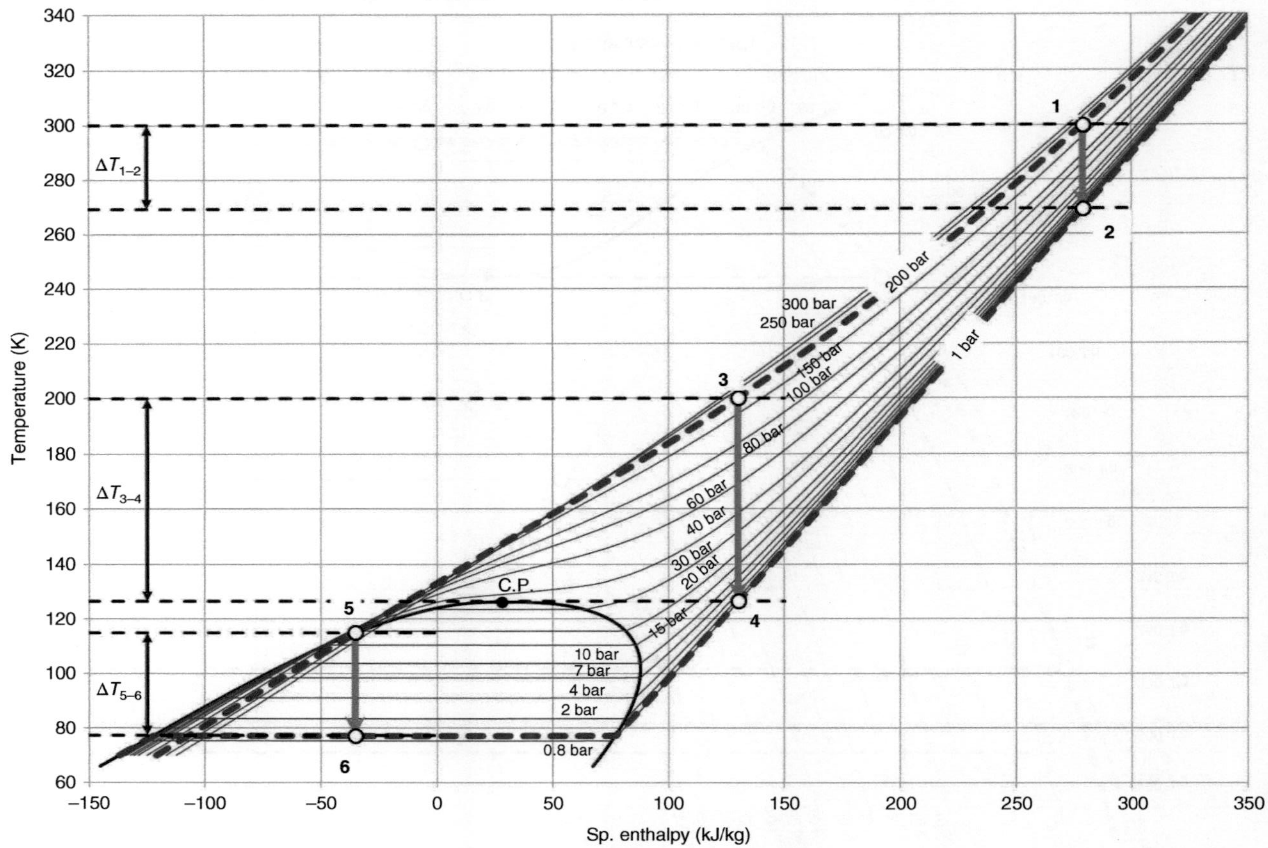

Figure 5.4 Isenthalpic expansion in the temperature–enthalpy diagram (nitrogen).

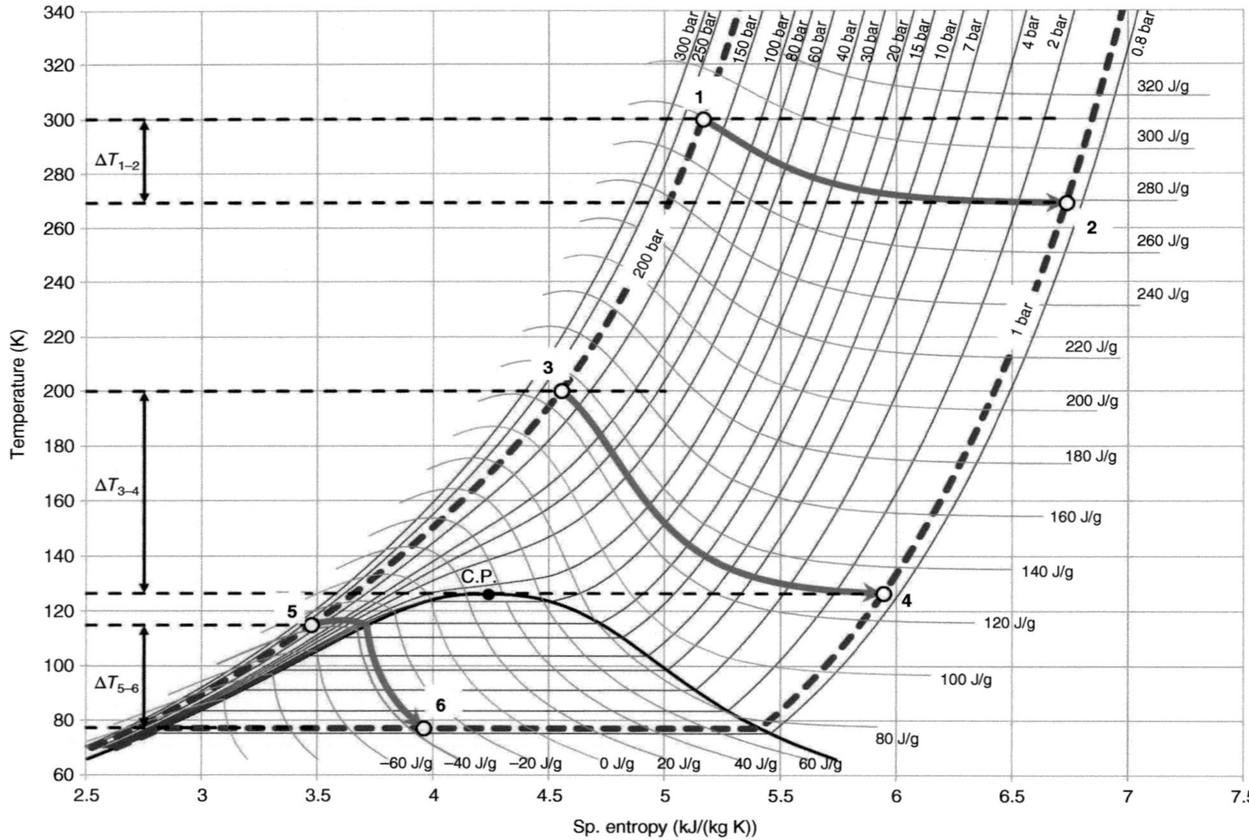

Figure 5.5 Isenthalpic expansion in the temperature–entropy diagram (nitrogen).

4. A non-zero value of Joule–Thomson coefficient means a deviation from the ideal gas state. In this sense, μ characterizes the non-ideality of the gas at given conditions. The deviation from the ideal gas condition is maximal in the vicinity of the two-phase area – in the two-phase area itself, as well as to the left, above, and right of it. Correspondingly, the Joule–Thomson coefficient achieves high values in all these areas. In a *T,s*-diagram this high non-ideality is expressed in the inclination of isenthalps. These lines can be seen rather as vertical lines than horizontal, which results in a high temperature drop ΔT at moderate Δp.
   Example: If the nitrogen in the gas cylinder is slightly precooled to a temperature $T_3 = 200$ K, marked as point 3 in Figure 5.5 prior to expansion – and the expansion process starts from this relatively low-temperature level – then the temperature at the outlet of the valve will be lower than 130 K (marked as point 4 in Figure 5.5). The temperature difference achieved by this kind of isenthalpic expansion amounts to more than $\Delta T > 70$ K. This is a significant improvement compared to the expansion from the room temperature ($1 \rightarrow 2$).
5. Another important fact is that the Joule–Thomson coefficient is a material-specific property. It behaves differently for different gases. If the experiment described before in Section 5.1.1 (release of a gas from a high-pressure gas cylinder into surrounding area) is executed with helium instead of nitrogen, the temperature of the helium after expansion will increase. Figure 5.6 shows the corresponding process in the *T,s*-diagram. In terms of refrigeration, helium has a negative Joule–Thomson effect at room temperature. Therefore, cooling of helium by isenthalpic expansion at this temperature level is not possible.
   It is, however, still possible to cool helium by isenthalpic expansion. As previously discussed, the Joule–Thomson effect becomes positive (in terms of refrigeration) at conditions close to the two-phase area. Thus, the gas being used must be cooled down to an appropriate temperature level (where the Joule–Thomson effect is positive) before the expansion. This is valid for helium expansion, too. Figure 5.7 (*T,s*-diagram for helium below 14 K) illustrates this case. The thick line shows the isenthalpic expansion process from pressure 20 to 1 bar starting with a temperature of approximately 7.4 K. At the end of this expansion process, the dew point is achieved, and the first droplet of liquid helium may be generated. In order to achieve a higher liquid output (liquefaction rate), it is necessary to reduce the temperature upstream of the valve.

5.1.6 Loss in an Isenthalpic Expansion Process

In this book, a loss is always referring to loss of exergy.[2] Exergy loss leads to an increase of power consumption of the compressor, which drives the cryogenic system. The procedure for calculating losses in a process unit is described in Section 4.6. It is based on the so-called exergy balance sheet around the corresponding process unit (Figure 5.8).

2 An exception is the term "pressure loss," which refers to pressure drop caused by friction.

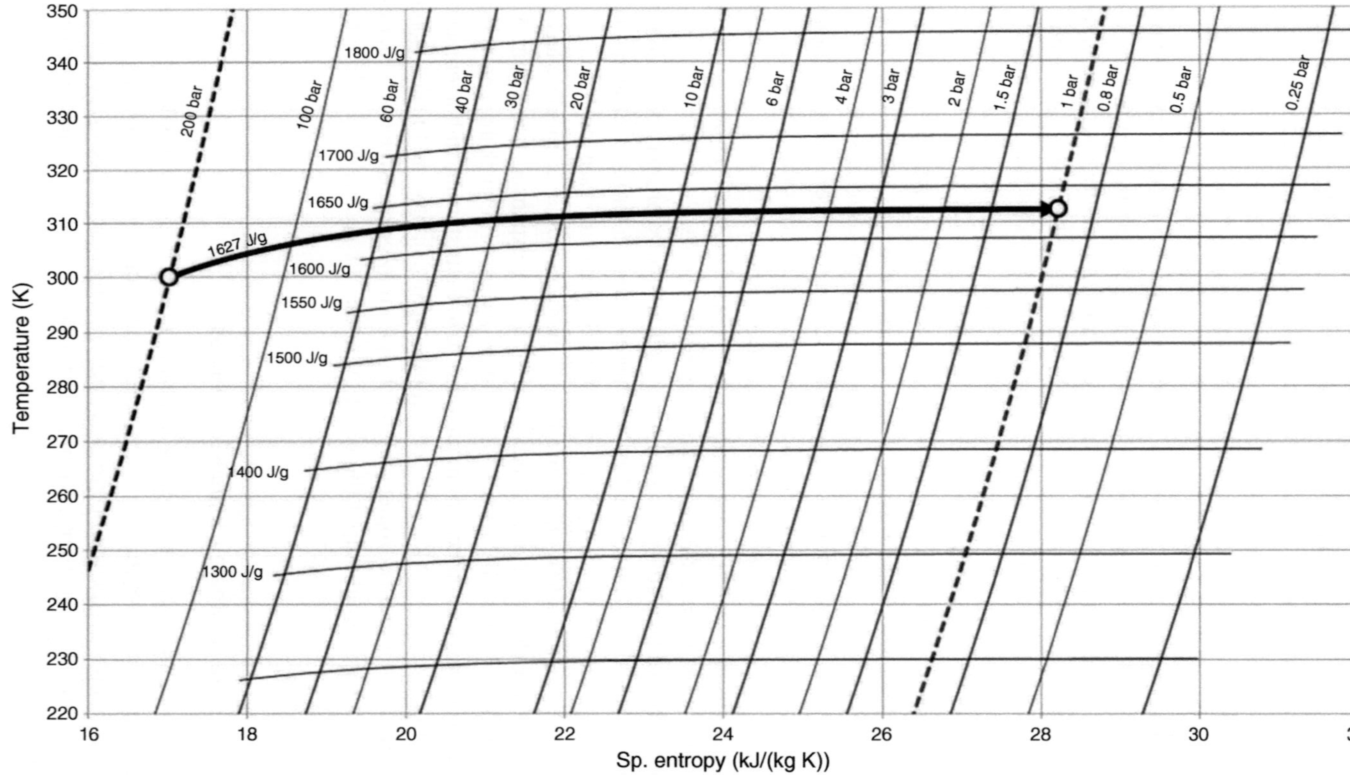

Figure 5.6 Temperature–entropy diagram for helium above 220 K, the 200 bar-isobar and the 1 bar-isobar are shown as dotted lines.

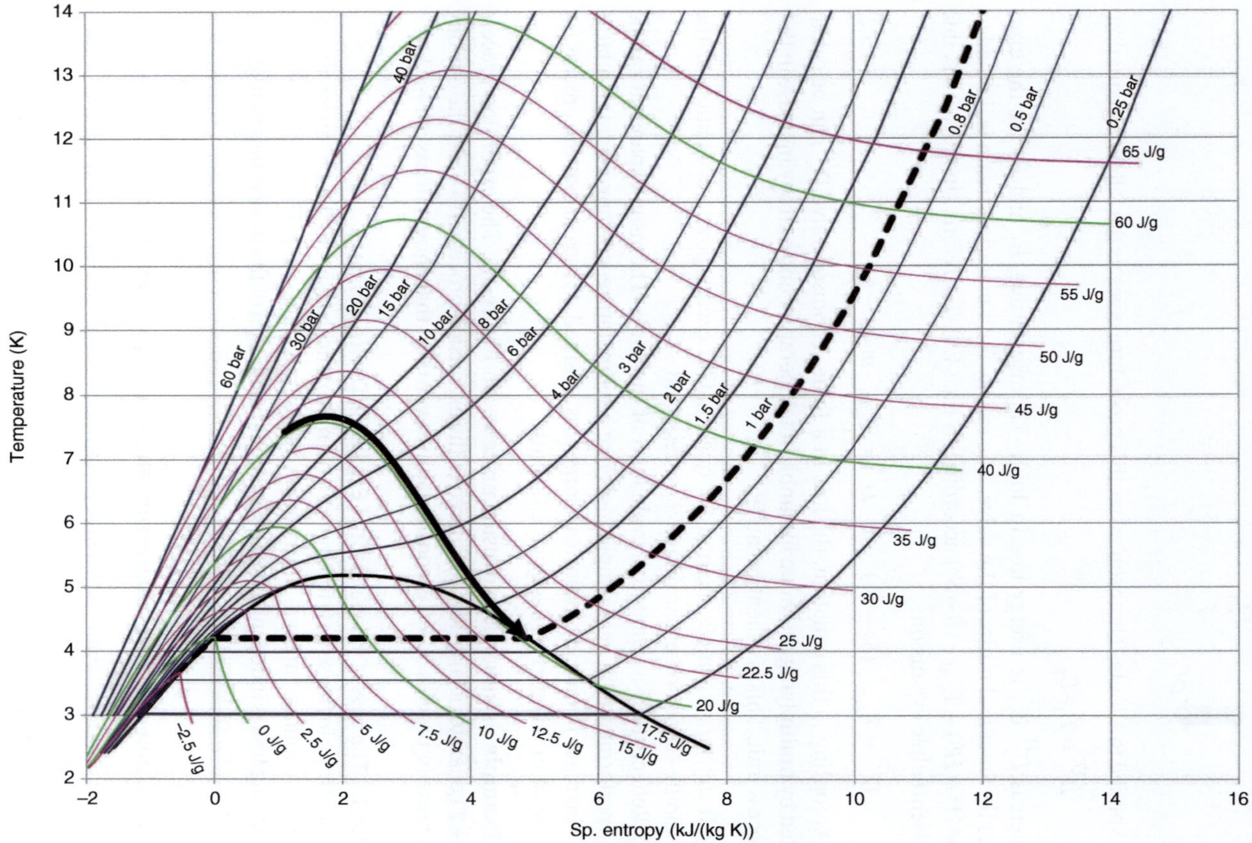

Figure 5.7 Temperature–entropy diagram for helium below 14 K (two-phase area).

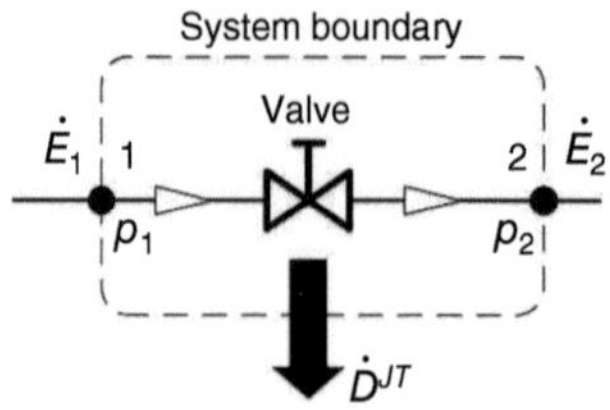

Figure 5.8 Exergy balance for a valve.

Applied to a Joule–Thomson valve, this procedure gives us the following equation:

$$\dot{D}^{JT} = \dot{E}_1 - \dot{E}_2$$

whereas $\dot{D}^{JT}$ is the exergy loss in Joule–Thomson valve, $\dot{E}_1$ and $\dot{E}_2$ are the exergies the inlet and outlet of the valve. The exergy for a material stream is defined as $\dot{E} = (\dot{H} - \dot{H}^o) - T_{amb} \cdot (\dot{S} - \dot{S}^o)$ according to Eq. (4.5). Therefore, since $\dot{H}_1 = \dot{H}_2$ due to isenthalpic expansion:

$$\dot{D}^{JT} = T_{amb} \cdot (\dot{S}_2 - \dot{S}_1) \quad \text{or} \quad \dot{D}^{JT} = T_{amb} \cdot \dot{m}\,(s_2 - s_1) \tag{5.2}$$

According to this equation, the loss in a Joule–Thomson valve $\dot{D}^{JT}$ can be calculated by multiplying the value for ambient temperature and entropy increase in this process unit. This is salient because:

(1) It delivers a simple method for quantitative exergy/power evaluation of the Joule–Thomson expansion in a process, and
(2) Reflects how a qualitative evaluation of the Joule–Thomson expansion can be performed with very simple tools like a temperature–entropy diagram: a large increase in entropy indicates a large loss. A small Δs means that the process (at least in this part) is designed efficiently.

Example: Isenthalpic expansion processes in Figure 5.5: The expansion process $1 \rightarrow 2$ generates a large loss compared to the expansion process $5 \rightarrow 6$, for the entropy increase in process $5 \rightarrow 6$ is considerably lower than that shown in cases $1 \rightarrow 2$.

5.1.6.1 The Loss is Independent of Expansion Device Design

The entropy increase Δs is independent of mechanical design of the expansion device (orifice, capillary, manual or driven valve, porous plug, etc.). Therefore, the corresponding loss is independent of the mechanical design of the device, depending only on fluid and process parameters.

5.1.7 Joule–Thomson Expansion and Isenthalpic Expansion

Thermodynamics scientists heavily emphasize that the definition of the Joule–Thomson coefficient ($\mu = (\partial T/\partial p)_{h=const}$) is valid only for gaseous fluids, and it is not correct to use it if a phase change occurs during the expansion process.

However, process design engineers avoid the isenthalpic expansion from gaseous to gaseous state, due to extremely large losses caused by this kind of expansion, see Section 5.1.6. In contrast, an isenthalpic expansion from the supercritical state into the two-phase area (whereby a phase change inevitably occurs) can be designed (in terms of process design) in a very efficient way, with only minor loss. Therefore, this type of isenthalpic expansion is used in many cryogenic processes.

A special term was introduced and is used for this kind of expansion – the so-called Joule–Thomson expansion.[3]

Example: Three isenthalpic expansion processes are shown in Figure 5.5:

1 → 2 (starting from 300 K),
3 → 4 (starting with 200 K), and
5 → 6 (starting with ca. 118 K).

The last one is considered a Joule–Thomson expansion because this expansion follows from the supercritical state into the two-phase area. The processes 1 → 2 and 3 → 4 are not Joule–Thomson expansion processes.

5.1.8 Flashgas

To reiterate, a portion of fluid evaporates during the Joule–Thomson expansion and is available at saturated vapor conditions. This vapor is referred to as the "flashgas." The fraction of the flashgas after the Joule–Thomson expansion depends on fluid and process parameters (such as pressure and temperature). Process design engineers always seek to target the minimization of the flashgas. The most beneficial and therefore the most common method to achieve this goal is to use all available means to lower the temperature before the Joule–Thomson expansion device. In case of a very low temperature at the inlet of expansion valve, the flashgas fraction can be reduced even to zero. Usually, for such a behavior, a temperature in front of expansion device must be equal to or even lower than the temperature after the expansion.

5.1.9 Summary of Isenthalpic Expansion

Table 5.1 summarizes the most important definition and equations around isenthalpic expansion.

Table 5.1 Summary of terms around isenthalpic expansion.

Isenthalpic expansion	Expansion (throttling) at *h* = *const*
Joule–Thomson coefficient	$\mu = \left(\frac{\partial T}{\partial p}\right)_{h=const}$
Joule–Thomson effect	Temperature change during isenthalpic expansion $\Delta T = \int_{p_1}^{p_2} \mu \; dp$
Joule–Thomson expansion	Expansion (throttling) at $h = const$, with following constraints: • Inlet condition: single phase: supercritical fluid or liquid • Outlet condition: two-phase area: a fraction of refrigerant is liquefied, but usually a partial evaporation occurs during this expansion process and a share/fraction of refrigerant is available as saturated vapor
Flashgas	Gas generated after the Joule–Thomson expansion
Loss in an isenthalpic expansion process	$\dot{D}^{JT} = T_{amb} \cdot \dot{m} \cdot (s_2 - s_1)$ It does not depend on the device's mechanical design

3 The name "Joule–Thomson expansion" originates from the so-called Joule–Thomson process, where the Joule–Thomson expansion plays an important role, see Section 7.2.

5.2 Expansion in Expander

Generally, a turboexpander works like a water mill[4] – the pressurized gas drives the turbine wheel (impeller), which is equivalent to producing mechanical power *P* in the form of a rotating shaft.

Figure 5.9 shows a typical turboexpander[5] for an air separation plant: the expander part is located on the left, the gas is supplied via inlet nozzle (numbered as 1) and is guided radially into the turbine wheel (marked as 2), and after expansion, the gas leaves this device axially to the left. The generated mechanical power is used to compress a fluid in a single-stage centrifugal compressor – the turbine wheel (marked as 8) of this machine can be seen on the right. This expander required some additional subsystems, like oil system for oil-bearings, sealing gas for seals, and instrumentation (not pictured). The key takeaway is that an expander is a very complex machine; it is an expensive device, at least compared to an expansion valve. We are then left to question **why such an expensive and complex equipment is used in cryogenics.**

There are two answers to the above question. First, expansion in an expander inevitably leads to a temperature drop regardless of the initial state and working

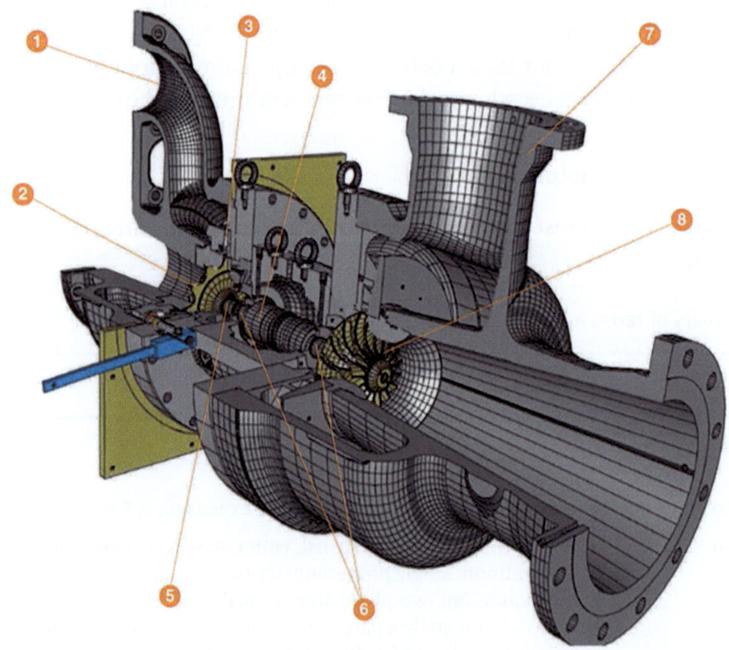

Figure 5.9 Turboexpander for an air separation plant: 1-expander casing, 2- expander wheel, 3- inlet guides vanes, 4- shaft/rotor, 5- seals, 6- bearings, 7- compressor casing, 8- compressor wheel. Source: Turbo-Expanders for Cold Production and Enery Recovery/Cryostar.

4 The process is more complicated, see Section 6.3 for details.

5 The terms expander, expansion machine, turboexpander, or turbine are used synonymously.

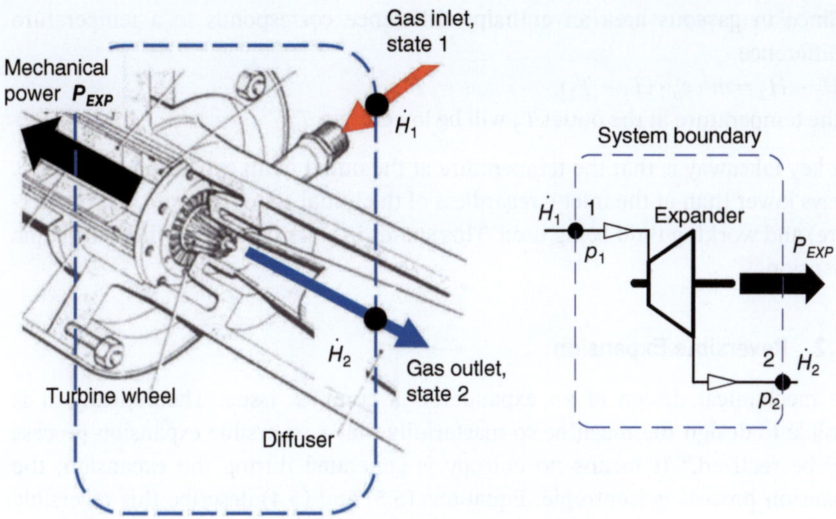

Figure 5.10 Expansion in an expander, energy conservation law.

fluid being used. The second reason is a significantly higher efficiency of the expansion process, at least compared to the isenthalpic expansion. These two aspects are described and further considered in this section. Often when a turboexpander is discussed, all considerations are also valid for reciprocating (piston) expansion machines.

5.2.1 Energy Conservation Law for Expander

Figure 5.10 shows an expansion process realized by means of a turboexpander in simplified form:

(a) The gas (pressure p_1, enthalpy $\dot{H}_1$) enters the device.
(b) The pressurized gas drives the turbine wheel (impeller), which is equivalent to producing mechanical power P in form of rotating shaft.
(c) The gas leaves the turbine wheel in axial direction with enthalpy $\dot{H}_2$ and lower pressure p_2.

According to energy conservation law, the generated power P_{EXP} is equal to the decrease of enthalpy $\dot{H}_1 - \dot{H}_2$:

$$P_{EXP} = \dot{H}_1 - \dot{H}_2 \quad \text{or} \quad P_{EXP} = \dot{m}\,(h_1 - h_2) \tag{5.3}$$

Therefore:

- In case no mechanical power is generated (the shaft does not rotate and $P_{EXP} = 0$), the expansion process may be considered as an isenthalpic process ($\dot{H}_1 = \dot{H}_2$).
- In the event that expander works properly, the turbine wheel rotates and a certain power $P_{EXP} > 0$ is generated, which leads to the enthalpy at the outlet $\dot{H}_2$ being automatically lower than the enthalpy on the inlet $\dot{H}_1$.

- Since in gaseous area an enthalpy difference corresponds to a temperature difference
  $\dot{H}_1 - \dot{H}_2 = \dot{m} \cdot c_p \cdot (T_1 - T_2)$,
  the temperature at the outlet T_2 will be lower than T_1.

A key takeaway is that the temperature at the outlet of an expansion machine is always lower than at the inlet – regardless of the initial state (pressure and temperature) and working fluid being used. This stands in stark contrast to the isenthalpic expansion.

5.2.2 Reversible Expansion

The mechanical design of an expander is a complex issue. Theoretically, it is possible to design the machine so masterfully that a reversible expansion process may be realized.[6] It means no entropy is generated during the expansion; the expansion process is isentropic. Equations (5.5) and (5.4) describe this reversible expansion process. Strictly speaking, they are valid for ideal gas, although they are still applicable to real gases if the deviation from ideal gas behavior is minimal. This is true for cryogenic gases, too.

$$\frac{T_{out}}{T_{in}} = \left(\frac{p_{in}}{p_{out}}\right)^{\frac{1-\kappa}{\kappa}} \tag{5.4}$$

$$P_{EXP\_rev} = \dot{N} R\, T_{in} \cdot \frac{\kappa}{\kappa - 1} \cdot \left[1 - \left(\frac{p_{in}}{p_{out}}\right)^{\frac{1-\kappa}{\kappa}}\right] \quad \text{in molar formulations} \tag{5.5}$$

$$P_{EXP\_rev} = \dot{m} \frac{R}{M_r}\, T_{in} \cdot \frac{\kappa}{\kappa - 1} \cdot \left[1 - \left(\frac{p_{in}}{p_{out}}\right)^{\frac{1-\kappa}{\kappa}}\right] \quad \text{in mass formulations} \tag{5.6}$$

with

R universal gas constant
$\dot{N}$ molar flowrate of fluid being expanded, in mol/s
$\dot{m}$ mass flowrate, in kg/s
T_{in} temperature at the inlet to the expansion machine, in K
κ isentropic exponent (kappa).

Since, in molar formulations, $P_{EXP} = \Delta\dot{H} = \dot{N} \cdot c_p \cdot \Delta T$, the temperature drop in a reversible expansion process can be found as

$$\Delta T_{rev} = \frac{R}{c_p} \cdot T_{in} \cdot \frac{\kappa}{\kappa - 1} \cdot \left[1 - \left(\frac{p_{in}}{p_{out}}\right)^{\frac{1-\kappa}{\kappa}}\right] \tag{5.7}$$

with molar specific heat capacity as c_p.

Figure 5.11 shows an expansion process of nitrogen in a reversible expander in a temperature–entropy diagram. If the starting point is close to the ambient temperature (300 K) marked by point 1, then the temperature at the outlet is

6 the target is to avoid all kinds of friction: mainly friction in form of turbulences in the gas flow, but other kinds of friction like mechanical friction, too.

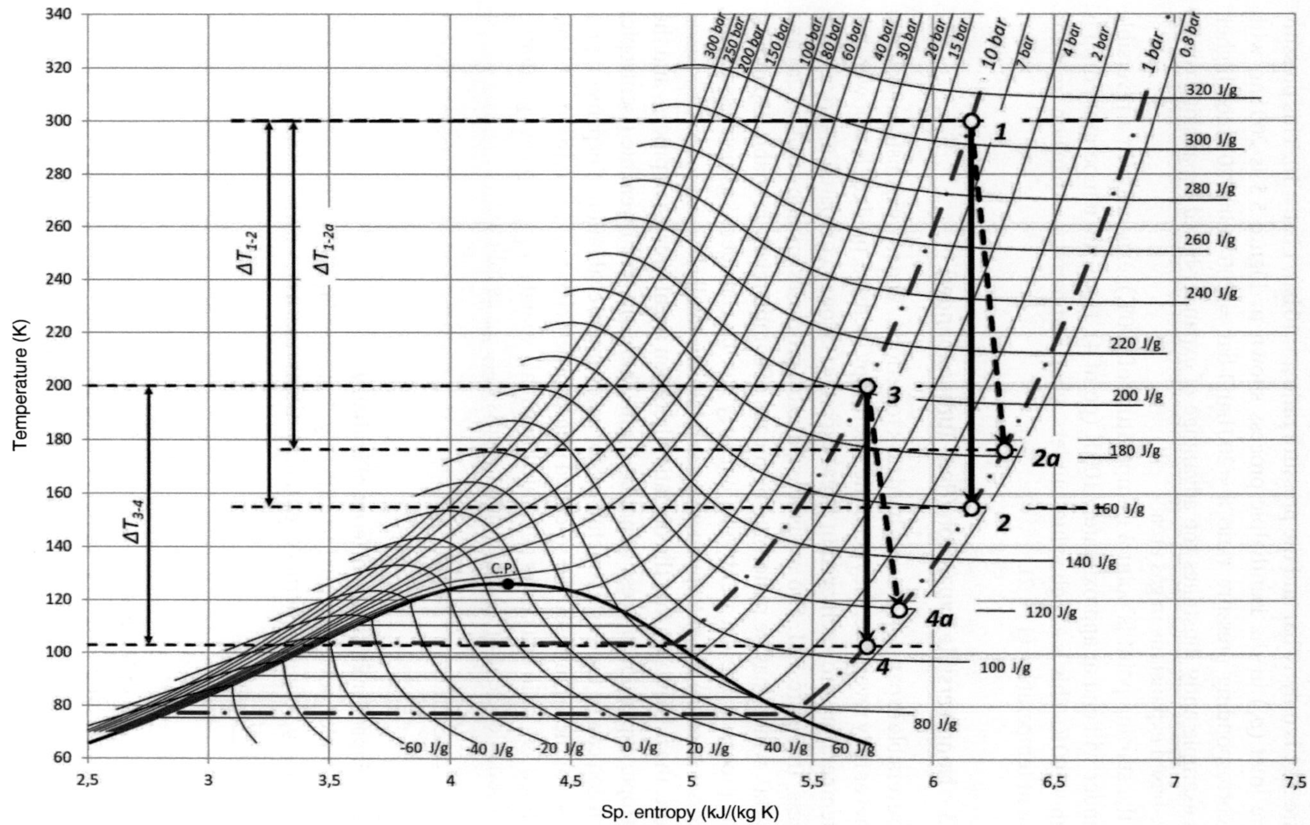

Figure 5.11 Expansion in an expander in T,s-diagram.

lower than 160 K (point 2). This corresponds to a temperature drop of more than $\Delta T_{1\to 2} > 120$ K. This is significantly more compared to an isenthalpic expansion (see Figure 5.5), where the pressure at the outlet in the expansion process shown in Figure 5.11 is only 10 bar and the pressure ratio $\pi = 10$ bar/1 bar = 10. This pressure at the inlet (p_1) in the isenthalpic process shown in Figure 5.5 is 200 bar, with the corresponding pressure ratio $\pi = 200$ bar/1 bar = 200 being 20 times higher. This example aptly illustrates the advantage of expander-based expansions versus valve-based expansions in gas area.

If the starting point is point 3 (temperature of 200 K), expansion in a reversible expander will end at approximately 100 K (point 4). Here, the temperature drop is roughly 100 K. This is a high value, but less when compared to expansion from the ambient temperature (1 → 2).

5.2.3 Nonreversible Expansion (Real Turboexpander)

The reversible expansion is a very hard task to perform. The real expansion machines are never fully reversible; they generate an entropy increase, produce less power, and the temperature drop achieved in a real expander is lower compared to a reversible process. In Figure 5.11, two realistic expansion processes are shown (1 → 2a) and (3 → 4a), which may be performed by means of commercially available expanders. These processes are not isentropic, but they can be considered as near-isentropic processes because the entropy increase is relatively small.

Here, the takeaway is that the deviation between a real expansion process and the isentropic expansion process is not so high. Therefore, the reversible (isentropic) expansion can be used as a benchmark for assessment/evaluation of expansion processes. For these purposes, the so-called isentropic efficiency η_s was introduced. This is defined as a ratio of a power generated in a real expander P_{EXP} to power generated in a reversible (isentropic) expansion process P_{EXP_rev}, as follows:

$$\eta_s = \frac{P_{EXP}}{P_{EXP\_rev}} \tag{5.8}$$

5.2.3.1 Expander Efficiency in Terms of Enthalpy

Since expander power is equal to enthalpy drop (see Eq. (5.3)), the isentropic efficiency can be defined in terms of enthalpy differences as

$$\eta_s = \frac{\Delta h_{real}}{\Delta h_s} \tag{5.9}$$

whereas

Δh_{real} - enthalpy difference in a real (nonreversible) expansion process
Δh_s - enthalpy difference in a reversible = isentropic expansion process

Using the isentropic efficiency η_s Eq. (5.5) as well as Eq. (5.7) can be extended and applied to real expanders as follows:

$$P_{EXP} = \eta_s \cdot \dot{m}\frac{R}{M_r} \cdot T_{in} \cdot \frac{\kappa}{\kappa - 1}\left[1 - \left(\frac{p_{in}}{p_{out}}\right)^{\frac{1-\kappa}{\kappa}}\right] \tag{5.10}$$

$$\Delta T_{EXP} = \eta_s \cdot \frac{R}{c_p} \cdot T_{in} \cdot \frac{\kappa}{\kappa - 1} \cdot \left[1 - \left(\frac{p_{in}}{p_{out}}\right)^{\frac{1-\kappa}{\kappa}}\right] \tag{5.11}$$

Typically, isentropic efficiency is higher than $\eta_s = 70\%$ but lower than 93%. It depends on the size and mechanical design of expander: small devices are less efficient; large machines achieve values exceeding 90%.

According to Eqs. (5.10) and (5.11), the temperature drop ΔT and the power P_{EXP} released during expansion in an expander depend on:

- pressure ratio $\pi = p_{in}/p_{out}$,
- isentropic exponents κ,
- isentropic efficiency η_s,
- temperature at the inlet T_{in}.

5.2.3.2 Impact of the Inlet Temperature

Equation (5.10) demonstrates the linear relationship between the released power and the inlet temperature. Essentially, an expander generates less power if the inlet temperature is reduced. For example, an expander with an inlet temperature of 150 K will produce only one-half of the power released in a similar process with starting temperature of 300 K (with the same fluid and the same pressure drop).

According to Eq. (5.11), the temperature drop ΔT depends linearly on the inlet temperature with similar consequences – the temperature drop generated in an expander is considerably reduced at low temperatures: ΔT for an expander with inlet temperature of 30 K will be 10 times lower than ΔT achieved in an expander with an inlet temperature of 300 K.

Both observations demonstrate a significant complication for engineers working on designing a process with an expander at low temperatures (below 30 K, for instance). This is challenging in terms of power: a relatively high flowrate of working fluid is required to generate power at very low temperatures, as well as in terms of temperature drop: it is difficult to achieve a significant temperature drop in an expander at very low temperatures.

5.2.3.3 Impact of Kappa

The isentropic exponent κ was introduced for characterization of temperature increase by compression/expansion of a gas in an isentropic process (see Eq. (5.4)). Generally, it is a material property and must be measured in an appropriate experimental setup or derived from available p-v-T-Data; it depends on the substance, temperature, and pressure. For diatomic (two atoms) molecules like nitrogen and oxygen, the kappa is approximately 1.4. For single-atom molecules like helium or neon, it is approximately 1.67. A relatively large value of 1.67 for helium means that even a relatively small increase or decrease in pressure causes a large temperature change.

For **an ideal gas,** the isentropic exponent is equivalent to the ratio of isobaric heat capacity to isochoric heat capacity, which is often abbreviated in the literature as a small Greek gamma: $\kappa_{\text{ideal gas}} = \gamma = c_p/c_v$

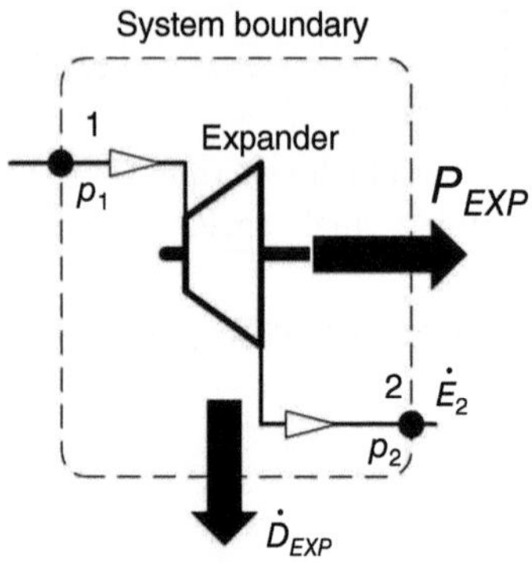

Figure 5.12 Exergy balance for an expander.

5.2.3.4 Impact of the Isentropic Efficiency

Equation (5.10) demonstrates the linear relationship between the released power and the isentropic efficiency of expander.

An important aspect is that the mechanical design of the expander is usually done and optimized for the specified set of process parameters (fluid, pressures, and temperatures at the inlet and outlet for normal operation case). If the real or current operating conditions differ from the specified conditions, the targeted isentropic efficiency η_s is unachievable and the loss will rise.

5.2.4 Exergy Loss in Expander

The procedure for calculation of losses in a process unit is described in Section 4.6. It is based on exergy balance around the corresponding process unit. Applied to an expander (see Figure 5.12), this method delivers the following equation:

$$\dot{D}^{EXP} = \dot{E}_1 - \dot{E}_2 - P_{EXP}$$

whereas $\dot{D}^{EXP}$ is the exergy loss in Joule–Thomson valve, $\dot{E}_1$ and $\dot{E}_2$ are the exergy flows at the inlet and outlet of expander.

The exergy for material streams is defined as $\dot{E} = (\dot{H} - \dot{H}^o) - T_{amb} \cdot (\dot{S} - \dot{S}^o)$ (see Eq. (4.5))

It is also known that $P_{EXP} = \dot{H}_1 - \dot{H}_2$ (energy conservation law).

Consequently $\dot{D}^{EXP} = T_{amb} \cdot (\dot{S}_2 - \dot{S}_1)$ or

$$\dot{D}^{EXP} = T_{amb} \cdot \dot{m}\,(s_2 - s_1) \tag{5.12}$$

The comments given in Section 5.1.5 for an expansion valve are also valid for the expander:

1. The loss in an expander $\dot{D}^{EXP}$ can be calculated by multiplication of value for ambient temperature and entropy increase in this process unit.
2. It delivers a simple method for quantitative loss evaluation in an expander, and can be performed with simple tools like a temperature–entropy diagram: a large increase in entropy indicates a large loss. A minor Δs means that the process (at least in this part) is designed well.

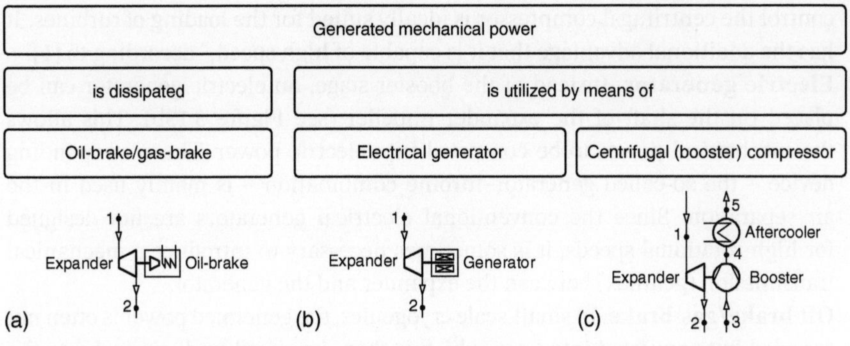

Figure 5.13 Loading devices for turboexpander (a) oil-brake, (b) electrical generator, (c) centrifugal (booster) compressor.

Example: Isenthalpic expansion processes in Figure 5.11: The expansion process $1 \rightarrow 2$ as well as the expansion process $3 \rightarrow 4$ generate only minor losses because the entropy increases in processes there are relatively low.

5.2.4.1 The Loss in Expander Depends on Mechanical Design of the Expander

The loss is defined by entropy increase Δs, and the entropy increase in given case depends on isentropic efficiency η_s, which ultimately reflects the quality of the entire machine's mechanical design. Therefore, the loss in an expander depends on mechanical design of the machine. This is different compared to the process of isenthalpic expansion.

5.2.5 Generated Power, Loading Devices

The mechanical power generated during expansion in an expander must be removed and disposed of. It would be ideal to utilize this power and implement it into the entire process. Three loading devices for turboexpanders are used in cryogenics:

1. oil-brake/or gas-brake
2. electric generator
3. centrifugal compressor (booster compressor).

Centrifugal compressor (Booster compressor): This is the most popular device, often used in large-scale systems like air separation plants, nitrogen liquefiers, and natural gas liquefiers. A special machine was developed for this purpose, called the expander–booster unit. In this machine, the expander turbine wheel and the impeller of a single-stage centrifugal compressor (booster) are placed on the same shaft (see Figures 5.9 and 5.13c). The mechanical power generated by the expander is used directly to drive the booster stage. Since the gas compressed in the booster stage becomes hot during the compression process, it is usually cooled to the ambient temperature in an aftercooler (heat exchanger where the gas is cooled by means of cooling water or cooling air) placed downstream of the booster stage. "Because of its simplicity and ease of control the centrifugal compressor is ideally suited for the loading of turbines. It has the additional advantage that it is capable of high speed," according to [1].

Electric generator: Instead of the booster stage, an electric generator can be placed on the shaft of the expander impeller (see Figure 5.13b). This allows the mechanical power to be converted into electric power. The corresponding device – the so-called generator–turbine combination – is mainly used in the air separation. Since the conventional electrical generators are not designed for high rotational speeds, it is sometimes necessary to introduce a mechanical transmission (gearbox) between the expander and the generator.

Oil-brake/gas-brake: In small-scale cryogenics, the generated power is often not recycled but converted into heat, which is then, irreversibly, dissipated into the environment. Usually, this is because of techno-economic reasons: for instance, if the power saving does not cover the additional cost of loading device. The corresponding device – the so-called oil-brake (see Figure 5.13a) – is based on a small impeller (similar to impeller of a small centrifugal pump) driven by expander. This impeller is placed in a closed reservoir filled with oil (or gas or another suitable fluid). The rotation of the impeller in a viscous fluid leads to heat generation (by friction), which is removed by means of cooling water. For this purpose, a heat exchanger (often a simple spirally wound tube) is placed in the fluid reservoir. The cooling water flows through this heat exchanger and cools the fluid down to the required temperature.

5.2.6 Comparison, Isenthalpic Expansion Versus Expansion in Expander

A comparison of isenthalpic expansion versus expansion in an expander is summarized in Table 5.2. It is given for didactic purposes only. In real processes/systems, it is not possible to replace an expander by a valve (and vice versa) without consequences. Such an action will require a substantial redesign of the whole process, including replacement of nearly all hardware components.

Example: Figure 5.14 exhibits three expansion processes in *T,s*-diagram. The starting point, marked as point 1, is characterized by pressure $p_1 = 6.5$ bar and temperature $T_1 = 133$ K; the gas is expanded to pressure $p_2 = 1.5$ bar. These values are typical for an air separation plant.

- In a reversible isentropic expansion process, a temperature of $T_2 = 86.8$ K will be achieved, therefore

$$\Delta T_{rev} = 46.3\text{ K},\quad \Delta h_{rev} = 43.1\text{ kJ/kg}$$

- In a real expansion machine, the temperature at the outlet will be $T_3 = 92.7$ K

$$\Delta T_{real} = 40.3\text{ K},\quad \Delta h_{real} = 36.6\text{ kJ/kg}$$

the isentropic efficiency is therefore

$$\eta_s = \frac{\Delta h_{real}}{\Delta h_{rev}} = \frac{36.3}{43.1} = 0.85 = 85\%$$

- For comparison: an isenthalpic expansion (1 → 4) would deliver a temperature drop of
$\Delta T_{h=const} = 5.2$ K.

Table 5.2 Comparing isenthalpic expansion to expansion in an expander.

	Isenthalpic expansion (valve)	**Expansion in expander**
Design	Simple	Sophisticated
Costs	Relatively low	Relatively high
Reliability	Very high	High, but less reliable in comparison to valve
Pressure ratio	No limitations for pressure ratio	Pressure ratio is limited by thermodynamics and flow mechanics; usually, it is less than 10 per stage
Phase change	No limitations	Preferably in single-phase area
Cooling effect	Cooling effect is not always given. It depends on refrigerant, temperature, pressure, and phase state	Cooling effect is always given
Temperature drop ΔT	Relatively low, depends on pressure difference Δp	Relatively large, depends on pressure ratio $\pi = (p_{in}/p_{out})$
Loss	Does not depend on mechanical design of the device Depends on the process (e.g. selected fluid and process parameters), loss is low in the vicinity of the two-phase area, loss is large in the gas area	Depends on mechanical design of the machine; there are some efficient and some less efficient designs available The mechanical design is optimized for the specified process parameters (normal operation case) The loss will rise if the real operating conditions differ from the optimal conditions

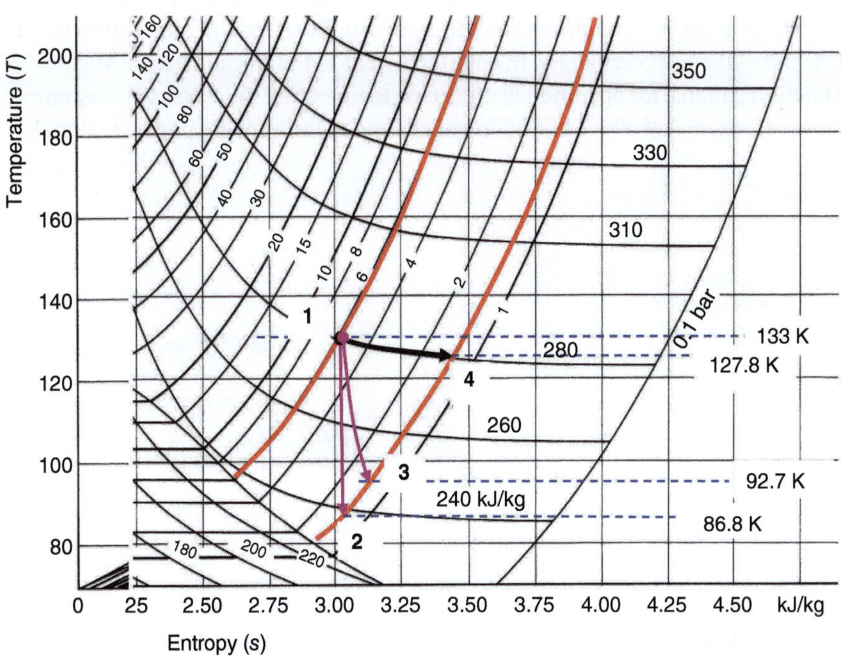

Figure 5.14 Expansion processes in temperature–entropy diagram, s in kJ/(kg K).

5.3 Heat Exchanger

5.3.1 Double-Tube (Multitube) Heat Exchanger

The heat exchanger is one of the most important components of a cryogenic plant. The classic book about this component is listed as [2], where almost all topics around heat transfer and heat exchangers are encyclopedically collected and addressed in detail.

In this current section, four aspects are described which are important for the process design. They are:

1. Definitions for temperature difference in heat exchanger (logarithmic temperature difference LTD, mean temperature difference MTD).
2. Definition of Q/MTD value.
3. Multi-stream heat exchanger.
4. Loss in heat exchanger.

In the simplest "laboratory" version, the heat exchanger can be designed as the so-called double-tube heat exchanger shown in Figure 5.15. The main elements are two metallic tubes with different diameters. The tube with smaller diameter (inner tube) is placed inside the tube with larger diameter (outer pipe). The resulting "double tube" is wound into a spiral in order to achieve a more compact construction. Special nozzles for feeding and removal or discharge of fluids are located at the ends of the "double tube."

The first (warm) stream is fed into one end of the heat exchanger; it is provided through the inner tube and removed at the other end of the heat exchanger. The second material stream (cold stream) is fed into the other end of the heat exchanger and flows through the channel between the outer and inner tubes to the warm end of the heat exchanger, in countercurrent to the warm material stream.

Materials suitable for operation at cryogenic temperature must be used: austenitic stainless steels, nickel alloys as well as copper and aluminum are proper materials.

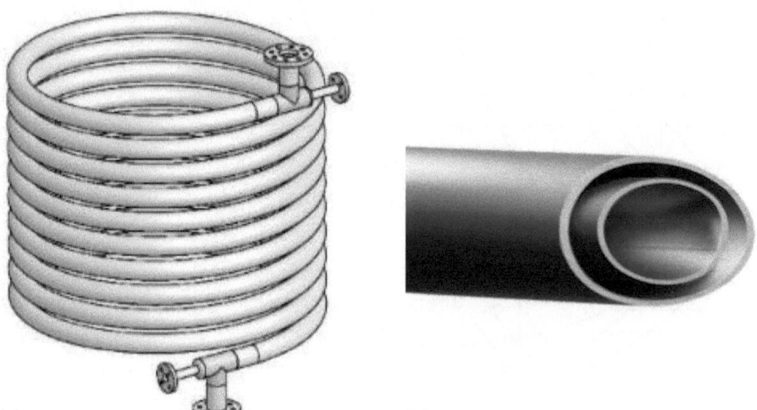

(a) (b)

Figure 5.15 (a) Double-tube heat exchanger, (b) cross-section of double tube.

Figure 5.16 Heat exchanger with two material streams.

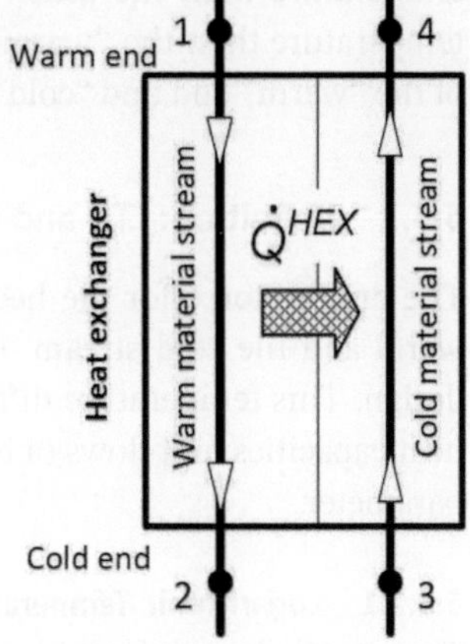

A similar design with multiple smaller tubes inside of a larger tube (referred to as a multitube heat exchanger) – may be considered as an alternative. It has a slightly larger specific surface and, consequently, is more compact. But both these constructions are primitive; the performance (specific surface, compactness, etc.) cannot meet the current requirements of hardware components for modern industrial systems (the state-of-the-art heat exchangers are described in section 6.2), but both these designs – double-tube heat exchanger and multitube heat exchanger – do visualize the heat exchanger principle in very distinct way.

Figure 5.16 shows schematically a heat exchanger consisting of channels for two material streams in close thermal contact. The first (warm stream) is fed on the one end of the heat exchanger (referred to as "warm" end), provided through the heat exchanger, and removed on the other end of the heat exchanger, referred to as "cold" end. It is cooled down from state 1 (temperature T_1) to state 2 (temperature T_2). The second material stream (cold stream) is fed on the "cold" end of the heat exchanger and flows through the heat exchanger to the "warm" end of the heat exchanger, in countercurrent to the warm material stream. It is warmed up from state 3 to state 4. A heat flow $\dot{Q}^{HEX}$ (often referred to as "heat exchanger duty") is transferred from the "warm" stream to the "cold" stream through the heat exchange surface A.

In cryogenics, the temperature difference between the "warm" end and the "cold" end of the heat exchanger is usually high (roughly 200 K for nitrogen-based processes), whereas the targeted temperature difference between the warm material stream and the cold material stream in every cross-section should be small.

Caution: The terms "warm" (stream) and "cold" (stream) are used often in habitual language. Depending on context (ambient air, sea water, drinks, etc.), a different magnitude is implied. In this conventional sense, all material streams used in a cryogenic process are cold, very cold, or even extremely cold. But some material streams have a higher temperature than others. The term "warm," used here and subsequently in the book, means that the corresponding stream has a bit higher temperature than the other. Correspondingly, the "cold" stream has a bit lower temperature than the "warm" stream. The same principle is applied for definition of the "warm" end and "cold" end of heat exchanger.

5.3.2 Definition: LTD and MTD

The driving force for the heat transfer is the temperature difference between the warm and the cold stream, making this parameter important for heat exchanger design. This temperature difference is usually not constant but rather depends on heat capacities and flows of both streams. It is a challenging task to determine this parameter.

5.3.2.1 Logarithmic Temperature Difference LTD

The so-called mean logarithmic temperature difference *LTD* is often used to characterize an average temperature difference between warm and cold stream in a heat exchanger. It is defined as follows:

$$LTD = \frac{\Delta T^{WE} - \Delta T^{CE}}{ln\,(\Delta T^{WE}/\Delta T^{CE})} \tag{5.13}$$

where

ΔT^{WE} - temperature difference at the warm end of the heat exchanger
ΔT^{CE} - temperature difference at the cold end of the heat exchanger

This equation provides a statement about the average temperature difference in a heat exchanger, based only on two input parameters: temperature difference at the cold end and temperature difference at the warm end. This equation does not consider the real temperature profile inside of the heat exchanger. Therefore, this is not a very accurate method for estimation of temperature difference; however, it is advantageous in its simplicity. Theoretically, it is valid only for streams with constant heat capacities. For such tasks, it delivers reliable results with minimal calculation efforts.

5.3.2.2 Mean Temperature Difference MTD

Unfortunately, there are only a limited number of cryogenic processes where the heat capacity of streams (being managed in a heat exchanger) remains constant. Most processes are based on fluids with a relatively complex heat capacity profile; Eq. (5.13) is, therefore, not directly applicable there. However, the method may be modified to achieve high accuracy in a procedure consisting of three steps:

(1) The first step aims to virtually divide the heat exchanger into number of single pieces called zones (usually numbered as 1, 2, … *i*, … *n*; see Figure 5.17). The number and the size of sections are chosen so that heat capacity of material streams in every zone remains more or less constant.
(2) Second, the logarithmic temperature difference LTD_i is calculated for every zone using Eq. (5.13), applicable in this case.
(3) The last step includes the calculation of the real mean (average) value referred to as MTD using an appropriate mathematical averaging method.

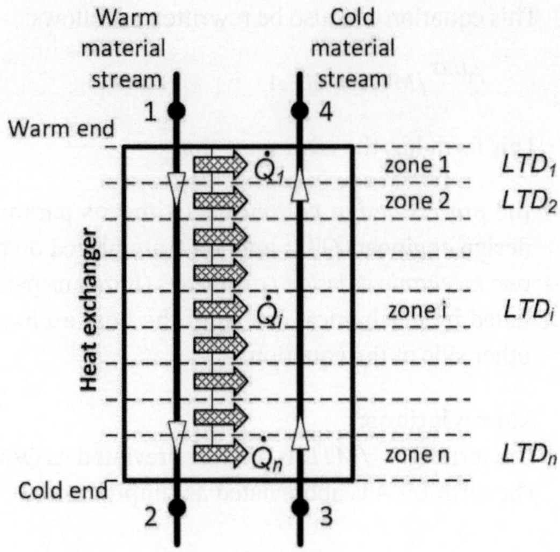

Figure 5.17 Heat exchanger divided in virtual zones.

If the zones are evenly distributed ($\dot{Q}_1 = \dot{Q}_2 = \dot{Q}_i = \dot{Q}_n$) the following equation may be used:

$$MTD = n/[\Sigma_i(1/LTD_i)] \tag{5.14}$$

In case the zones are not evenly distributed, a corresponding weighted averaging procedure should be applied.

Usually, no more than 50 zones are required to achieve a high accuracy. Often, 10 virtual zones deliver an acceptable level of precision. But even in the case of 10 zones, the calculation effort is relatively large. In commercial process simulators, this procedure is an integral part of the heat exchanger calculation. The user must specify the number of zones only, and the software delivers the result.

Example for deviation between *MTD* and *LTD*: In Section 7.2.3, an example of a nitrogen Joule–Thomson process is given. It includes a specification for the heat exchanger. The mean logarithmic temperature difference according to Eq. (5.13) calculated their amounts to $LTD = 17.3$ K. The MTD is calculated by using 20 zones. It delivers the value of $MTD = 12$ K. This number differs from the *LTD* by more than 30%. This deviation is explainable by the fact that the heat capacity of the high-pressure nitrogen varies in the given temperature range.

5.3.3 Definition: *Q/MTD* and *UA*

The heat flow $\dot{Q}^{HEX}$ being transferred in a heat exchanger is defined by

$$\dot{Q}^{HEX} = U \cdot A \cdot MTD \tag{5.15}$$

This equation states that the heat exchanger duty $\dot{Q}^{HEX}$ is defined by

(1) intensity of the heat transfer characterized by heat transfer coefficient U [W/(m^{2}K)],
(2) available heat transfer surface A [m^2], and
(3) MTD [K] between the warm and cold material stream

This equation can also be rewritten as follows:

$$\dot{Q}^{HEX}/MTD = U \cdot A \tag{5.16}$$

This form has the advantage that

- the *process design parameters* (it means parameters defined mainly by process design engineer) $\dot{Q}^{HEX}$ and MTD are placed on one side and
- two *mechanical design parameters* (it means parameters more or less predetermined by mechanical design of the heat exchanger) U and A are placed on the other side of the equation.

Abbreviations:
The term $\dot{Q}^{HEX}/MTD$ is often abbreviated as *Q/MTD*.
The term $U \cdot A$ is abbreviated as simply *UA*

5.3.3.1 What is the Q/MTD Good For?

The specific *UA*-values (in kW / (K m³)), as well as the specific surface (m² / m³) and the specific cost (€/m³) for all common heat exchanger types, are well known.[7] Therefore, the total heat exchanger volume V^{HEX} and the total heat exchanger cost C^{HEX} can be estimated based on $\dot{Q}^{HEX}$ and MTD – two parameters available as direct output of a process simulation. This allows the process design engineer to estimate size, volume, and cost of the heat exchanger during the process simulation and take corresponding appropriate actions if the process development falters.

Example: The typical specific volumetric surface A/V^{HEX} (sometimes referred to as "compactness") for aluminum plate heat exchangers is usually in the range of 350–500 m²/m³. The typical heat transfer coefficient is in the order of magnitude $U = 80\,\mathrm{W/(m^2K)}$ for air gases. Using Eq. (5.16), the following equation can be attained:

$$V^{HEX} = \frac{1}{C}(\dot{Q}^{HEX}/MTD) \tag{5.17}$$

where the constant $C = 30\,000\text{–}40\,000\,\mathrm{W/(K\,m^3)} = 30\text{–}40\,\mathrm{kW/(K\,m^3)}$.

Question: Section 9.2.3 includes a set of process parameters for a nitrogen liquefier based on the Claude process. How can we estimate the heat exchanger volume V^{HEX}?

Answer: The total *Q/MTD* for the heat exchanger is approximately 800 kW/K. According to Eq. (5.17), the heat exchanger volume:

$$V = \frac{\dot{Q}^{HEX}/MTD}{C} = \frac{800\,kW/\mathrm{K}}{30\ldots40\frac{kW}{\mathrm{K\cdot m^3}}} = 20\ldots27\,\mathrm{m^3}$$

7 The specific UA value is usually given for a defined process case characterized by a set of process parameters. It is called a reference case. The UA value for every other case (set of process parameters) may be derived from this reference case by rules based on dimensionless quantities like Reynolds number and Nusselt number and other.

5.3.4 Multi-Stream Heat Exchanger

Most textbooks on heat exchangers consider a heat exchanger for two material streams, one cold and one warm. Much information is available for this kind of apparatus. However, modern cryogenic processes and plants include heat exchangers with more than two streams in a single process unit. For example, a typical air separation plant usually requires a heat exchanger with more than five streams. In some cases, the process design includes up to 16 different material streams thermally integrated in a single unit.

In terms of mechanical design, these multi-stream heat exchangers are designed either as brazed aluminum plate-fin heat exchangers (see Section 6.2.1) or as spiral-wound heat exchangers (see Section 6.2.2).

5.3.4.1 Simplified Calculation Procedure

Figure 5.18 (left) shows a heat exchanger with four material streams (two hot and two cold streams). A rigorous process design of a multi-stream heat exchanger is a complex, time-consuming procedure. In many cases, however, a simplified calculation procedure is possible and is based on the following assumptions:

1. The temperatures of all cold material streams in every cross-section are equal.
2. The temperatures of all warm material streams in each cross-section are also equal.

In a real multiple-streams heat exchanger the temperatures of cold (or warm) streams in a cross-section are never equal. But the deviation is relatively small (less than 2–3 K), so this assumption is typically valid.

The consequence of these assumptions is that multiple warm material streams may be combined to a single virtual material stream – the so-called warm composite

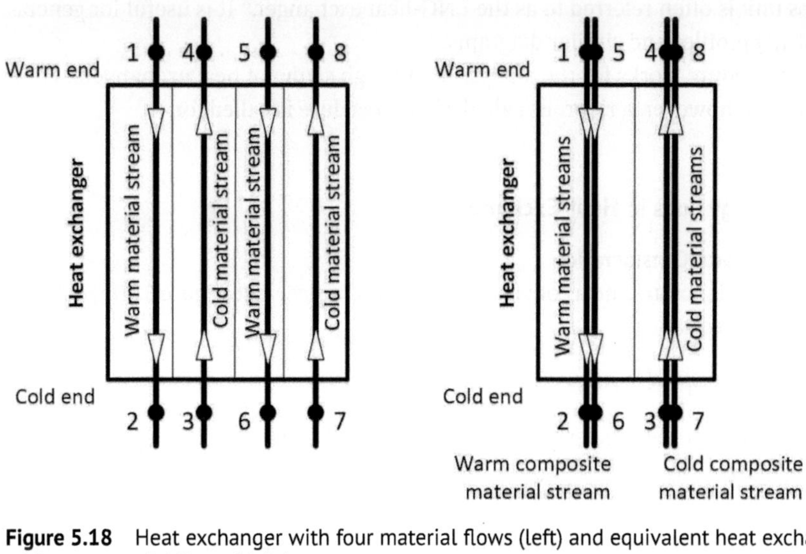

Figure 5.18 Heat exchanger with four material flows (left) and equivalent heat exchanger with two material flows (right).

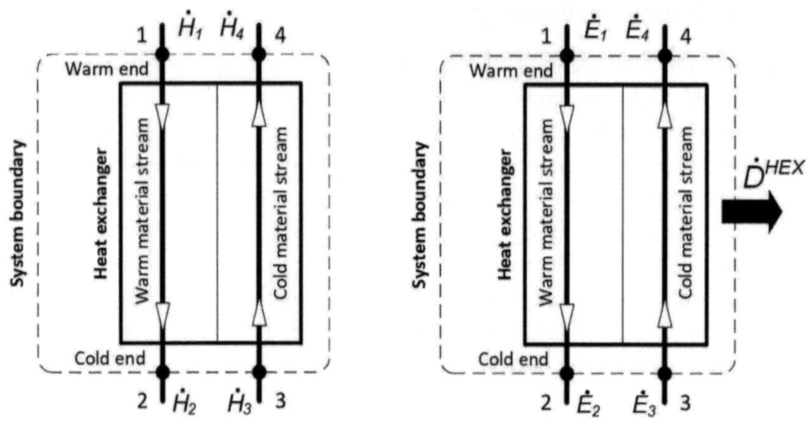

Figure 5.19 Balances for heat exchanger with two material flows (a) energy conservation law, (b) exergy balance.

stream, whereby the enthalpies/heat capacities of all warm material streams are simply added, so that at the end the enthalpy/heat capacity of the warm composite stream represents the sum of enthalpies/heat capacities of all warm material streams. Applying the same procedure to cold material streams the second virtual material stream – the cold composite stream – may be generated. At the end of this procedure, a multi-stream heat exchanger can be reduced to a conventional heat exchanger with two material streams – warm composite stream and cold composite stream (see Figure 5.18 right).

The calculation procedure described above is implemented in most of commercial process simulators (Aspen, Hysis, Unisim, gPROMS, Pro2, etc.). The corresponding process unit is often referred to as the LNG-heat exchanger.[8] It is useful for generation of *T,Q* profiles and similar diagrams.

This procedure works for the analysis and rough sizing of heat exchangers. For a final design, however, a rigorous calculation procedure is called for.

5.3.5 Exergy Loss in Heat Exchangers

5.3.5.1 General Consideration

The exergy balance around a conventional heat exchanger (see Figure 5.19b) delivers the following equation:

$$\dot{D}^{HEX} = \dot{E}_1 - \dot{E}_2 + \dot{E}_3 - \dot{E}_4$$

The exergy for material streams is defined as $\dot{E} = (\dot{H} - \dot{H}^o) - T_{amb} \cdot (\dot{S} - \dot{S}^o)$ (see Eq. (4.5)).

8 This is because LNG processes are based on multi-stream heat exchangers.

Moreover, it is also known that $\dot{H}_1 - \dot{H}_2 + \dot{H}_3 - \dot{H}_4 = 0$ according to energy conservation law for heat exchanger (see Figure 5.19a). Consequently,

$$\dot{D}^{HEX} = T_{amb}\,(\dot{S}_2 - \dot{S}_1 + \dot{S}_4 - \dot{S}_3) \text{ or} \tag{5.18}$$

If the mass flow $\dot{m}$ on the warm side and on the cold side are equal ($\dot{m} = \dot{m}_1 = \dot{m}_3$)

$$\dot{D}^{HEX} = T_{amb} \cdot \dot{m}\,(s_2 - s_1 + s_4 - s_3) \tag{5.19}$$

For a multi-stream heat exchanger, the following equation is valid:

$$\dot{D}^{HEX} = -T_{amb} \cdot \Sigma\,(\dot{S}_i) \tag{5.20}$$

The loss in a heat exchanger $\dot{D}^{HEX}$ is determined by ambient temperature T_{amb} and the total entropy increase by $\Sigma\,(\dot{S}_i)$. This is very similar to both unit operations discussed in Sections 5.1.5 and 5.2.4.

5.3.5.2 Exergy Loss in Heat Exchanger, Deep Dive

Since the temperature profile in a cryogenic heat exchanger usually varies considerably, it is better to consider a single zone of a heat exchanger (described in Section 5.2.1, as opposed to the whole heat exchanger). The performance for the overall heat exchanger is reachable by means of an integration procedure through all these zones.

In a single zone i, a portion $\dot{Q}_i$ of the whole $\dot{Q}$ is transferred in the heat exchanger from the temperature level T_{w_i} to the temperature level T_{c_i}. With this number of zones, we assume that both these temperature levels may be considered as constants. The local temperature difference ΔT_i is given as:

$$\Delta T_i = T_{w\_i} - T_{c\_i}$$

The exergy of the cold/cooling capacity $\dot{Q}$ – the exergy $\dot{E}(\dot{Q})$ – depends on the temperature level T. According to Eq. (4.4), the portion of a cold $\dot{Q}_i$, on the cold side of heat exchanger is characterized by exergy

$$\dot{E}_{c\_i}\,(\dot{Q}_i) = \dot{Q}_i \,\frac{(T_{amb} - T_{c\_i})}{T_{c\_i}}$$

The same portion of cold on the warm side of the heat exchanger is characterized by exergy

$$\dot{E}_{w\_i}\,(\dot{Q}_i) = \dot{Q}_i \,\frac{(T_{amb} - T_{w\_i})}{T_{w\_i}}$$

The difference between these two exergies $\dot{E}_{c\_i}\left(\dot{Q}_i\right)$ and $\dot{E}_{w\_i}\left(\dot{Q}_i\right)$ is considered as exergy loss $\dot{D}_i^{HEX}$ in a heat exchanger zone: $\dot{D}_i^{HEX} = \dot{E}_{c\_i}\left(\dot{Q}_i\right) - \dot{E}_{w\_i}\left(\dot{Q}_i\right)$

Example (based on process calculation for nitrogen Joule–Thomson process for $T_o = 100$ K; see Section 7.2):

- At the cold end of the given heat exchanger, a very cold nitrogen stream (temperature of roughly 100 K) is used to cool the warm nitrogen stream (temperature of roughly 160 K).

- A portion of $\Delta\dot{Q}_1 = 1\,\text{kW}$ at the temperature level of 100 K is valuable; the exergy of this cold is $\dot{E}_{c\_1} = 1\,\text{kW} \cdot (300\,\text{K} - 100\,\text{K}) / 100\,\text{K} = 2\,\text{kW}$.
- The exergy of the same amount of cold at the temperature level of 160 K is considerably lower: $\dot{E}_{w\_1} = 1\,\text{kW} \cdot (300\,\text{K} - 160) / 160 = 0.875\,\text{kW}$.
- The transfer of cold $\Delta\dot{Q}_1 = 1\,\text{kW}$ from temperature level of 100 K to the temperature level of 160 K means a loss of $\dot{D} = \dot{E}_{c\_1} - \dot{E}_{w\_1} = 1.125\,\text{kW}$.

Correspondingly, at least an additional 1.125 kW power is required to drive the cycle compressor, due to this exergy loss.

The following equation can be derived directly:

$$\dot{D}_i^{HEX} = \dot{Q}_i \cdot T_{amb} \frac{\Delta Ti}{T_{c_i}(T_{c_i} + \Delta T)} \tag{5.21}$$

for $\Delta T << T_c$ it can be simplified to:

$$\dot{D}_i^{HEX} = \dot{Q}_i \cdot T_{amb} \cdot \frac{\Delta T_i}{(T_{c\_i})^2} \tag{5.22}$$

According to this equation, the loss in the heat exchanger depends on:

1. Heat flow $\dot{Q}^{HEX}$ being transferred in the heat exchanger, linearly.
2. Temperature difference ΔT. This function is a linear one: if the temperature difference between the warm and the cold material streams is small, then the loss is relatively low. A large temperature difference in the heat exchanger is indicative of irreversibility and consequently increased loss.
3. Temperature level on the cold side T_{c_i}, whereby the function $\sim 1/T^2$ describes this relationship.

Observations:

1. The function $\sim 1/T^2$ may reach high values at low temperatures: even at temperatures below 200 K, this function may reach a high value but if $T \to 0$ this function has potential to reach exceedingly high values.
2. The heat transfer at the low temperature level T_{low} (such as at the cold end of the heat exchanger) driven by a temperature difference ΔT leads to considerably higher losses compared to the transfer of the same amount of heat at the same temperature difference ΔT but at the higher temperature level $T_{high} > T_{low}$ (for example in the middle of the heat exchanger).

The following points are important takeaways:

Learning 5.1 The total loss depends on temperature profile.
A large temperature difference at the cold end of a cryogenic heat exchanger indicates massive exergy loss; it is the worst-case scenario, ever. Therefore, a low temperature difference at the cold end of the heat exchanger must be always targeted during the process design.

Table 5.3 visualizes the findings following from Eq. (5.22). It is a simple tool for analysis of loss in cryogenic heat exchangers.

Table 5.3 Impact of temperature difference in heat exchanger on exergy loss depending on temperature level.

	Small temperature difference, less than 0.02 T	Medium temperature difference, (0.02–0.1) T	Large temperature difference, more than 0.1 T	Very large temperature difference
$T < 5$ K	<0.1 K Small loss	0.1–0.5 K Moderate loss	>0.5 K Large loss	>2 K Very large loss
$T = 20$ K	<0.4 K Small loss	0.4–2 K Moderate loss	>2 K Large loss	>5 K Very large loss
$T = 100$ K	<2 K Small loss	2–10 K Moderate loss	>10 K Large loss	>20 K Very large loss
$T = 200$ K	<4 K Small loss	4–20 K Moderate loss	>20 K Large loss	>50 K Very large loss

5.3.6 Efficiency of Heat Exchanger

Equation (5.23) quantifies the efficiency of single heat exchangers, whereby the so-called "exergy efficiency" is used. It is defined as ratio of benefit to effort, whereby the benefit here is the exergy increase on the warm side of the heat exchanger; the effort is the exergy decrease on the cold side:

$$\eta_e^{HEX} = (\dot{E}_2 - \dot{E}_1)/(\dot{E}_3 - \dot{E}_4) \tag{5.23}$$

where the subscripts in this equation refer to the Figure 5.16.

5.4 Single Adiabatic Compression Stage

The mechanical design of machinery is laid out in Section 6.5 (key hardware components). The current section describes basics required for understanding these machines, which are important for future process design considerations.

An adiabatic compressor stage is a device where the working gas is adiabatically compressed from an inlet (low) pressure p_1 to an outlet (high) pressure p_2; mechanical power P_{COMP} is required to drive this device (see Figure 5.20). The key word is **adiabatic**. In this case, the compression process is performed without heat exchange with environment; there are no heat flows entering or leaving this process unit. Figure 5.21 shows a corresponding symbol used in process flow diagrams (PFDs) or in piping and instrumentation diagrams (P&IDs) for an adiabatic centrifugal compressor stage.

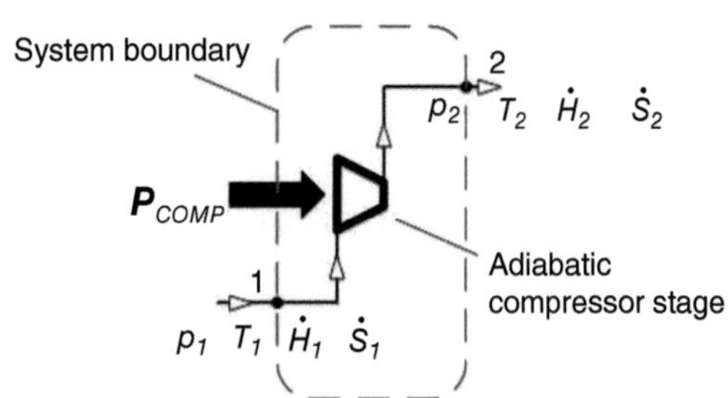

Figure 5.20 Adiabatic compression stage.

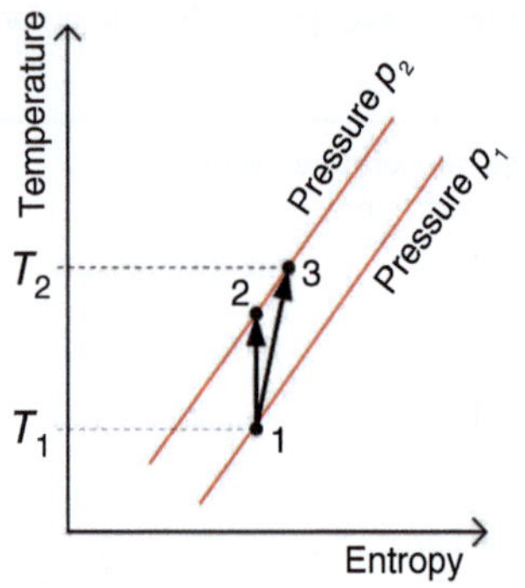

Figure 5.21 Adiabatic compression in *T,s*-diagram, 1 → 2 – reversible (isentropic) compression process, 1 → 3 – nonreversible (real) compression process.

The following compression devices may be considered as an adiabatic compressor stage:

- Single radial stage of a centrifugal compressor (turbocompressor), see Section 6.5.2.
- Axial turbocompressor (or axial stage of the turbocompressor), which usually consists of several impellers connected in series without intermediate cooling and intermediate fluid extractions or injections, see Section 6.5.3.
- Single cylinder of a reciprocating (piston) compressor, see Section 6.5.5.
- Dry (without oil lubrication) screw compressor.

A counterexample for adiabatic compression is an oil-lubricated (oil-flooded) screw compressor, see Section 6.5.4. During the compression process, a large amount of heat is transferred from the compression device (screw) to the compressor oil (and further to the environment), which is used as a sort of coolant fluid in this case. Therefore, it is a non-adiabatic machine.

Of course, all real compression devices are not fully adiabatic devices. Some amount of heat is always dissipated into the environment simply because the hot compressor parts are not thermally insulated. However, the amount of this dissipated heat is minor in comparison to the mechanical power converted during the compression process inside these devices.

5.4.1 Energy Conservation Law for Adiabatic Compression Stage

The energy conservation law[9] applied to an adiabatic compression stage (see Figure 5.20) delivers the following equation:

$$\dot{H}_1 + P_{COMP} = \dot{H}_2 \quad \text{or} \tag{5.24}$$

$$P_{COMP} = \dot{H}_2 - \dot{H}_1 = \Delta\dot{H}_{2-1}$$

It means, in an adiabatic stage the mechanical power is converted into the increase of the gas enthalpy.

9 It is assumed that the kinetic energy of the gas stream at the inlet and at the outlet can be neglected here.

5.4.2 Reversible Adiabatic Compression Stage

The mechanical design of a compressor stage is a challenging issue. Technically, it is possible to design the machine in such a way so that the compression process is performed in a reversible way – or, in other words, isentropically (zero-entropy generation). Figure 5.21 shows such a reversible adiabatic compression in *T,s*-diagram. It is shown schematically as process 1 → 2.

Equations (5.25) and (5.26) describe the reversible compression process. Strictly speaking, these equations are valid only for ideal gas. However, they are applicable to all cryogenic gases at ambient temperature because the deviation from the ideal gas behavior there is negligible.

$$T_2 = T_1 \cdot \left(\frac{p_2}{p_1}\right)^{(\kappa-1)/\kappa} \tag{5.25}$$

$$P_{comp\_rev} = \dot{N} \cdot R \cdot T_1 \cdot \frac{\kappa}{\kappa-1}\left[\left(\frac{p_2}{p_1}\right)^{\frac{\kappa-1}{\kappa}} - 1\right] \text{ in molar formulations} \tag{5.26}$$

$$P_{comp\_rev} = \dot{m}\frac{R}{M_r} \cdot T_1 \cdot \frac{\kappa}{\kappa-1}\left[\left(\frac{p_2}{p_1}\right)^{\frac{\kappa-1}{\kappa}} - 1\right] \text{ in mass formulations} \tag{5.27}$$

5.4.3 Nonreversible Adiabatic Compression, Isentropic Efficiency

The real compression processes realized by means of commercially available machines are never reversible. An entropy increase is generated during the compression process, the temperature at the outlet of a compressor is higher than it would be in the reversible process, and the machine needs more driving power. Figure 5.21 shows a nonreversible process schematically, marked as 1 → 3. The degree of irreversibility depends on mechanical design of used compression device – some devices are better, and some are worse in this sense. But the deviation from the reversible (isentropic) compression process is not high.[10]

The compression power for the reversible process is used as a benchmark for evaluation of real compressors. For these purposes, the term isentropic efficiency η_s was introduced. It is defined as ratio of power required for the reversible compression process to power required for the real compression:

$$\eta_s = \frac{P_{COMP\_rev}}{P_{COMP}} \tag{5.28}$$

$$P_{COMP} = \frac{1}{\eta_s} \cdot \dot{m}\frac{R}{M_r} \cdot T_1 \cdot \frac{\kappa}{\kappa-1}\left[\left(\frac{p_2}{p_1}\right)^{\frac{\kappa-1}{\kappa}} - 1\right] \tag{5.29}$$

According to these equations, the required driving power P_{COMP} depends mainly on:

- pressure ratio $\pi = p_2/p_1$,
- isentropic exponents κ,

10 If the deviation is too large, the use of the isentropic efficiency is no longer reasonable. In this case, another key figure for process characterization must be defined and used.

- isentropic efficiency η_s,
- temperature at the inlet T_1.

5.4.3.1 Impact of the Inlet Temperature

Equation (5.28) visualizes the linear relationship between the required driving power and the inlet temperature. It shows that the required power can be reduced by starting the compression process from a lower temperature. Since this temperature T_1 is normally quite close to 300 K, every reduction of the inlet temperature by 3 K (1% of 300 K) would lead to an energy saving of roughly 1%. That is an important information for process design engineers since it opens a way to reduce the power consumption.

5.4.3.2 Impact of the Isentropic Efficiency

Equation (5.28) visualizes the relationship between the required driving power and the isentropic efficiency. The mechanical design of compressor stage is, usually, done and optimized for the specified set of process parameters (fluid, pressures, and temperatures at the inlet and outlet for normal operation cases). If the real/current operating conditions differ from the specified conditions, the targeted isentropic efficiency η_s will not be achieved, and the compressor stage will require more power.

5.4.3.3 Impact of the Isentropic Exponent

Table 5.4 illustrates the impact of the isentropic exponent on the required power and outlet temperature. In this example, a reversible (isentropic) compression from ambient pressure ($p_1 = 1.013$ bar) to an outlet pressure of $p_2 = 2$ bar is calculated. The gas flowrate is chosen as 100 mol/s, and the inlet temperature is given as $T_1 = 300$ K. The calculation is carried out for two substances with different isentropic exponents. It shows a relatively high outlet temperature for helium

Table 5.4 Compression of air and helium.

			Air	Helium
Universal gas constant	R_m	J/(mol s)	8.31	8.31
Molar flowrate	$\dot{m}$	mol/s	100	100
Inlet temperature	T_{in}	K	300	300
Inlet pressure	p_{in}	bar	1.013	1.013
Outlet pressure	p_{out}	bar	2	2
Kappa	κ		1.4	1.67
Pressure ratio	$\pi = p_{out}/p_{in}$		2.0	2.0
Outlet temperature	T_{out}	K	364	394
Outlet temperature	T_{out}	°C	91	121
Compressor power	P_{COMP}	kW	187	195

(121°Cels), which is higher compared with the outlet temperature of the reversible air compressor ($T_2 = 91$°Cels). The power required for helium compression is approximately 195 kW, roughly 4% more than the corresponding number for air compression (approx. 187 kW). This example shows that

1. Temperature increase during the compression may be a critical issue.
2. Compression of gases with high kappa is especially challenging, and
3. Helium compression requires effective cooling.

The isentropic efficiency of real compressors is characterized by relatively high values. See the following examples:

- Isentropic efficiency of $\eta_s = 84$–86% is usually given for a radial stage of medium-size centrifugal compressor.
- Large-scale centrifugal compressors are more efficient devices – an isentropic efficiency of $\eta_S = 89$% and higher can be realized in these devices.
- Even relatively small- and medium-scale devices with $\eta_s = 80$–84% are still very efficient.

5.4.4 Exergy Loss, Adiabatic Compressor Stage

Generally, the procedure described in the current section and the results are very similar to the text section about the exergy loss in an expander (Section 5.2.4).

Figure 5.22 shows the exergy balance around the single adiabatic compression stage. The corresponding equation is as follows:

$$\dot{D}^{COMP} = \dot{E}_1 - \dot{E}_2 - P_{COMP}$$

The exergy for material streams is defined as $\dot{E} = (\dot{H} - \dot{H}^o) - T_{amb} \cdot (\dot{S} - \dot{S}^o)$ (see Eq. (4.5))

It is also known that $P_{COMP} = \dot{H}_1 - \dot{H}_2$ acc. to energy conservation law

Consequently $\dot{D}^{COMP} = T_{amb} \cdot (\dot{S}_2 - \dot{S}_1)$ or

$$\dot{D}^{COMP} = T_{amb} \cdot \dot{m}\,(s_2 - s_1) \tag{5.30}$$

All comments given in Section 5.2.4 for an expander are also valid for the single adiabatic compressor stage:

1. The loss in an adiabatic compressor stage $\dot{D}^{COMP}$ can be calculated by multiplication of value for ambient temperature and entropy increase in this process unit.

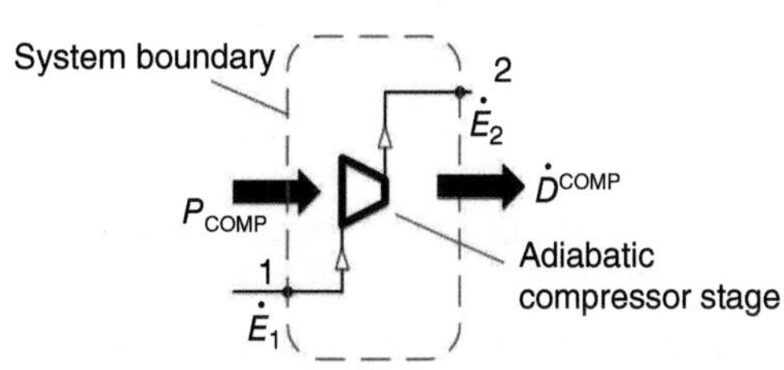

Figure 5.22 Exergy balance for single adiabatic compressor stage.

2. A qualitative evaluation of compression in an adiabatic compressor stage can be performed with very simple tools like a temperature–entropy diagram: a large increase in entropy indicates a large loss. A minor Δs means that the process (at least in this part) is designed well.
3. The loss is defined by entropy increase Δs and the entropy increase depends on isentropic efficiency η_s, which essentially reflects the quality of mechanical design of the entire machine.
   Therefore, the loss in an adiabatic compressor stage depends on mechanical design of the machine.

5.5 Multi-Stage Compression, Isothermal Compression

Multi-stage compressors are used in cryogenics often: e.g. centrifugal compressors for air separation, nitrogen liquefaction, as well as systems for natural gas liquefaction are built based on multi-stage machines.

Figure 5.23 shows the common symbol for a multi-stage compressor, used in PFDs. In this process unit the working gas is compressed from an inlet (low) pressure p_{in} to an outlet pressure p_{out}; mechanical power P_{COMP} is required to drive this device. The compression process is not adiabatic - a heat flow Q_{amb} is permanently removed from the gas into environment during the compression. This symbol visualizes this device's two functions: compression of the gas (visualized by compressor symbol) and integrated cooling (visualized by cooler/heat exchanger symbol).

The device for realization of the multi-stage compression process itself is considerably more complex than this symbol. It includes a chain of adiabatic compression stages with intermediate coolers (see Figure 5.24, where a four-stage device is shown as example). The gas is subsequently compressed in dedicated adiabatic compressor stages and cooled by means of intermediate coolers (intercoolers) several times. It would be more accurate to use the term "compressor system" instead of the term "compressor" when describing this device.

5.5.1.1 Special Case: Main Air Compressor

In the case of a main air compressor (MAC) – a machine for compression of ambient air containing natural humidity – the design of the compressor systems is even more complex. It includes additional devices, known as liquid separators, installed upstream of the intermediate coolers (see Figure 5.25). These separators are required

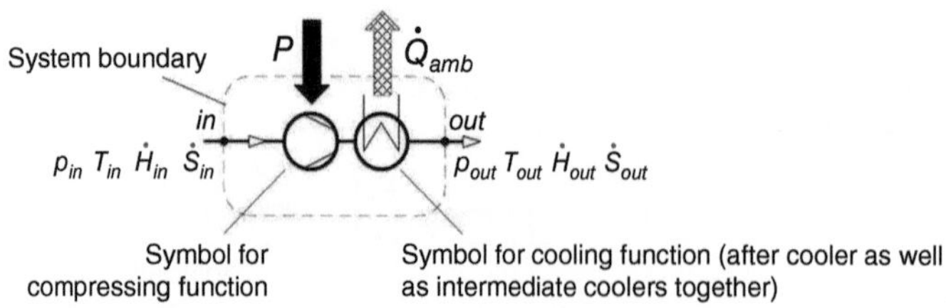

Figure 5.23 Symbol used in process flow diagrams for a multi-stage compressor

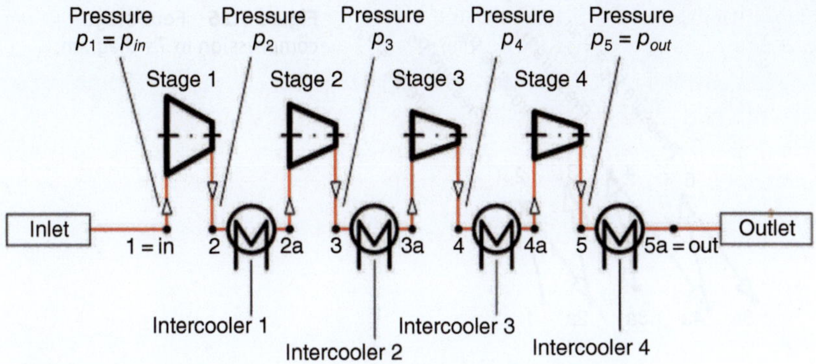

Figure 5.24 Four-stage centrifugal compressor, $p_1 < p_2 < p_3 < p_4 < p_5$.

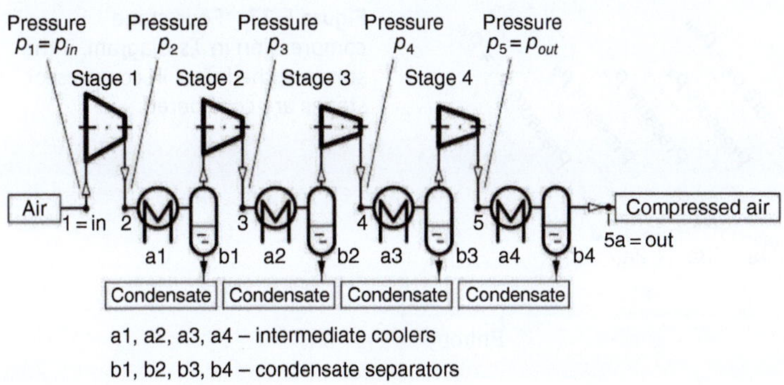

Figure 5.25 Example of a four-stage main air compressor with integrated condensate separation.

for removal of condensate – a liquid formed from natural humidity (after compression and cooling of the air to the ambient temperatures in the entire intercooler).[11] In terms of mechanical design, the separator is usually integrated into the corresponding intercooler.

The current section considers only a multi-stage compressor compressing a dry fluid (like that shown in Figure 5.24). However, these considerations are also applicable to the MAC.

Figure 5.26 shows a four-stage compression process in *T,s*-diagram, which corresponds to the compressor device shown in Figure 5.24. The gas is compressed in the first single adiabatic stage (1 → 2), cooled to temperature 2a (2 → 2a) in the intercooler 1, compressed in the second single adiabatic stage (2a → 3), cooled in the intercooler 2, and so on and so forth. At the outlet of the compression device, the gas is pressurized to outlet pressure $p_{out} = p_5$; the temperature at the outlet depends on the cooling media (cooling water or cooling air), but ultimately it remains close to ambient temperature.

11 The droplets of condensate as well as a liquid film in the inlet may damage the internals of compressor device.

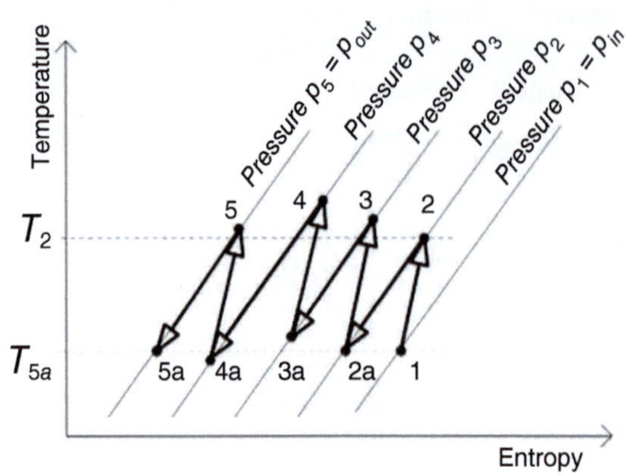

Figure 5.26 Four-stage compression in *T,s*-diagram.

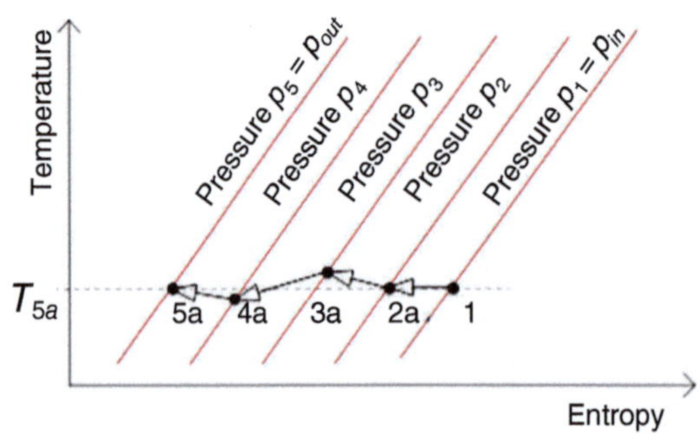

Figure 5.27 Four-stage compression in *T,s*-diagram, if the states at the inlets in compressor stages are considered.

The temperature at the outlet of every single adiabatic stage depends on the process parameters and isentropic efficiency of the corresponding stage. The temperature at the outlet of every intercooler depends on process parameters and the intercooler design.

The overall compression process is very complex. A certain simplification can be achieved if the process is considered as a chain: 1 → 2a → 3a → 4a → 5a (see Figure 5.27). This process looks more like an isothermal compression.

5.5.2 Reversible Isothermal Compression

Thermodynamic literature thoroughly covers the process of isothermal compression. The main idea is that the compression process is carried out extremely slowly in order to avoid any increase in fluid temperature caused by compression. This requires efficient heat exchange with the environment, as described in Section 4.1. Figure 5.28 shows this process in *T,s*-diagram. The gas is compressed from pressure p_1 to pressure p_5 isothermally. The process is considered as a reversible operation if the temperature between the gas being compressed and the cooling media (air or water) is indefinitely small.

Ideal gas: Eq. (5.31) is used for calculation of power for reversible isothermal compression of an ideal gas. It is applicable to cryogenic gases at ambient temperature,

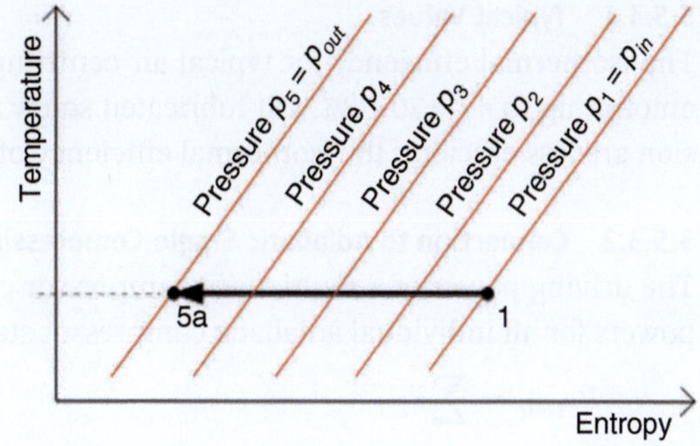

Figure 5.28 Isothermal compression in *T,s*-diagram.

as previously discussed.

$$P_{rev\_isotherm} = \dot{N} \cdot R \cdot T_{in} \cdot \ln\left(\frac{p_{out}}{p_{in}}\right) \text{ in molar formulations} \tag{5.31}$$

$$P_{rev\_isotherm} = \dot{m} \cdot (R/M_r) \cdot T_{in} \cdot \ln\left(\frac{p_{out}}{p_{in}}\right) \text{ in mass formulations} \tag{5.32}$$

According to this equation, the required driving power $P_{rev\_isotherm}$ depends only on

- pressure ratio $\pi = p_{out}/p_{in}$ and
- compression temperature T_{in}.

It does not depend on kappa and other thermodynamic properties of the substance.

Real gas: The power required for driving a reversible isothermal compressor can be calculated by means of exergy values – put simply, as exergy difference between inlet and outlet state. This method is applicable to real gases, too. It follows directly from Eq. (5.47):

$$P_{rev\_isotherm} = \dot{E}_{out} - \dot{E}_{in} = \dot{m} \cdot (e_{out} - e_{in}) \tag{5.33}$$

5.5.3 Nonreversible Isothermal Compression, Isothermal Efficiency

The power required for reversible isothermal compression is considered **as a benchmark** for the characterization of **multi-stage compressors.** The so-called **isothermal** compressor efficiency η_T indicates the deviation from the reversible process. It is defined as a quotient of the power required for a reversible compression process $P_{rev\_isotherm}$ and the power required for real compression P_{real}:

$$\eta_T = \frac{P_{rev\_isotherm}}{P_{real}} \tag{5.34}$$

For the ideal gas, the power required for compression in a multi-stage isothermal compressor is given by the following equation:

$$P_{real} = \frac{1}{\eta_T} \dot{N} \cdot R \cdot T_{in} \cdot \ln\left(\frac{p_{out}}{p_{in}}\right) \tag{5.35}$$

$$P_{real} = \frac{1}{\eta_T} \dot{m} \cdot (R/M_r) \cdot T_{in} \cdot \ln\left(\frac{p_{out}}{p_{in}}\right) \tag{5.36}$$

5.5.3.1 Typical Values

The isothermal efficiency for typical air centrifugal compressors is relatively high, amounting to $\eta_T = 70–78\%$. Oil-lubricated screw compressors for helium compression are less efficient; the isothermal efficiency of these devices is usually 55–65%.

5.5.3.2 Connection to Adiabatic Single Compression Stage

The driving power of a multi-stage compressor can be defined as a sum of driving powers for all individual adiabatic compressor stages P_i:

$$P_{COMP} = \sum_i P_i \tag{5.37}$$

Assuming an equal pressure ratio, equal inlet temperature, equal isentropic efficiency for every single adiabatic stage η_{S_i}, as well as zero pressure loss in intercoolers, the following equation for compressor consisting of n stages may be obtained:

$$P_{COMP} = \frac{n}{\eta_{S\_i}} \cdot \dot{N} \cdot R \cdot T_{in} \cdot \frac{\kappa}{\kappa - 1} \left[(\pi)^{\frac{\kappa-1}{n \cdot \kappa}} - 1 \right] \tag{5.38}$$

The comparison of Eqs. (5.35) and (5.38) delivers:

$$\frac{1}{\eta_T} \cdot ln(\pi) = \frac{n}{\eta_{S\_i}} \cdot \frac{\kappa}{\kappa - 1} \left[(\pi)^{\frac{\kappa-1}{n \cdot \kappa}} - 1 \right] \tag{5.39}$$

Therefore, the isothermal efficiency depends on the isentropic efficiency of single compressor stages:

$$\eta_T = \eta_{S\_i} \frac{a \cdot ln(\pi)}{n \cdot [(\pi)^{a/n} - 1]}, \text{where } a = \frac{\kappa - 1}{\kappa} \tag{5.40}$$

The mathematic analysis of this function shows that

- The isothermal efficiency is always smaller than the isentropic efficiency of a single stage.
- The higher the number of stages, the less the deviation between η_T and η_S.

The isothermal efficiency represents a sort of integral parameter for the entire ensemble of adiabatic stages. It only gives a rough estimation, but in many cases, it is enough to use when making some decisions.

5.5.3.3 Rigorous Calculation of Multi-Stage Compressor

Equation (5.38) describes a sort of a virtual process only. The real process looks more like that shown in Figure 5.26. Essentially, every stage has an individual pressure ratio, isentropic efficiency, and inlet temperature.

A rigorous calculation of a multi-stage compressor is no trivial task. A step-by-step approach must be pursued whereby every single compression stage is calculated separately using Eq. (5.29) and correct process parameters. These correct process parameters include individual inlet temperature, individual pressure ratio, individual isentropic efficiency for every stage, and pressure drop in every intermediate cooler.

If the MAC (for wet air) is considered, the local composition of the air must also be considered.

Observation:
The driving power of the isothermal compressor is a sum of powers required to drive every single adiabatic stage; see Eq. (5.38). This depends on the inlet temperature of each adiabatic stage, which corresponds to the gas temperature after the intermediate cooler is used directly. Consequently, the required power can be reduced if the temperature after the intermediate cooler is reduced (for example, by means of colder cooling fluids (air or water) or better heat exchange conditions in the intercooler). The achievable energy savings correspond to roughly 1% for every 3 K temperature reduction after the intercooler since the normal operation temperature is always close to the ambient temperature (300 K).

5.5.4 Rule of Thumb for Isothermal Compression

The thermodynamics of a multi-stage isothermal compressor as a process unit are described already in Sections 5.5.2 and 5.5.3. The power required for compression can be calculated by means of Eq. (5.35)

$$P_{COMP} = \dot{N}RT_{in} \cdot ln\,[p_{out}/p_{in}]/\eta_T$$

Commonly, the isothermal efficiency for a typical centrifugal compressor is $\eta_T = 70\text{–}76\%$, depending on size and design of the machine. This equation can be simplified, assuming following values:

- $T_{in} = 300\,\text{K}$
- $R = 8.31\,\text{J/mol/K}$
- $\eta_T = 71\%$
- 80.7 for conversion from [nm^3/h] to [mol/s] for air or nitrogen
- 2.3 for conversion from (ln) to (lg).

It results in an equation which is easy to remember

$$P_{COMP}\,[\text{kW}] = 0.1 \cdot \dot{N}[nm^3/\text{h}] \cdot \text{lg}\,[p_{out/}p_{in}] \tag{5.41}$$

Example Calculation: The required power of a nitrogen compressor must be defined, for a flowrate of 60 000 nm^3/h, inlet pressure 5.5 bar, and outlet pressure 55 bar;

$$P_{COMP} = 0.1 \cdot 60\,000 \cdot \text{lg}[55/5.5] = 6000\,\text{kW} = 6\,\text{MW}$$

Clearly, Eq. (5.41) delivers a very rough estimation only. However, it is very useful if a quick assessment is required.

For a more reliable prediction, equation (5.29) with correct values for T_{in}, p_{out}, p_{in}, and η_T can be applied. The next level of accuracy can be achieved if the total power is calculated as a sum of power requirements for every single compression stage. In this case, Eq. (5.29) and corresponding individual parameters T_{in}, p_{out}, p_{in}, and η_S for every single stage are required.

5.5.5 Energy Conservation Law For Multi-Stage Compressor

The energy conservation law for a multi-stage compressor (see Figure 5.23) can be formulated as:

$$\dot{H}_{in} + P = \dot{H}_{out} + \dot{Q}_{amb} \tag{5.42}$$

The enthalpy of an ideal gas does not depend on pressure, as described in Section 5.1.3. Therefore, for ideal gases the following statement is valid:

$$\dot{H}_{in} = \dot{H}_{out} \tag{5.43}$$

In this case, the energy conservation law looks even simpler:

$$P = \dot{Q}_{amb} \tag{5.44}$$

The cryogenic gases at the ambient temperature may be considered ideal gases if the pressure is relatively low (<30 bar). At higher pressures, where the Joule–Thomson coefficient deviates from zero, this is probably not always correct. However, even in the case of high pressure the enthalpy difference $\dot{H}_{out} - \dot{H}_{in}$ is still considerably smaller than the mechanical power processed in the compressor ($P >> \dot{H}_{out} - \dot{H}_{in}$); therefore

$$P \approx \dot{Q}_{amb} \tag{5.45}$$

5.5.6 Second Law for Multi-Stage Compressor

Equation (5.45) emphasizes that almost all mechanical power invested in compression is converted into the waste heat. That is a bit confusing message, since the compression requires a lot of mechanical power. According to this equation, this very valuable mechanical power is converted into the waste heat and at the end it is dissipated into the environment. This fact somewhat contradicts the common logic. Therefore, this is a legitimate question: **what is the isothermal compression good for?**

An exergy-based analysis helps answer this question. Figure 5.29 shows an isothermal compressor from the exergy perspective. A material stream with corresponding exergy flow $\dot{E}_{in}$ enters the machine where it is compressed, and the exergy of the material stream leaving the compressor is marked as exergy flow $\dot{E}_{out}$. The mechanical power P is used for compression of this fluid, and the corresponding exergy of

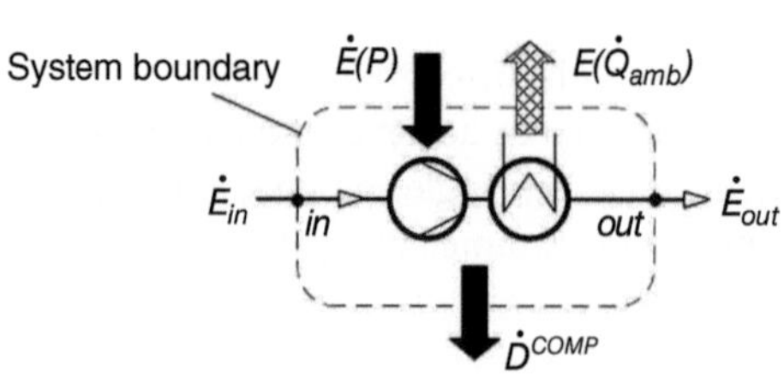

Figure 5.29 Exergy balance for multi-stage compressor.

this mechanical power is marked as $\dot{E}(P)$. The exergy of the waste heat $\dot{E}(Q_{amb})$ must also be considered, in addition to exergy loss $\dot{D}^{COMP}$.

The exergy balance equation can be written in following form:

$$\dot{E}_{in} + \dot{E}(P) = \dot{E}_{out} + \dot{E}(\dot{Q}_{amb}) + \dot{D}^{COMP} \tag{5.46}$$

This equation may be simplified by using the definitions for exergy of mechanical power and waste heat:

1. The exergy of mechanical power is equivalent to the mechanical power itself according to Eq. (4.2): $\dot{E}(P) = P$.
2. The exergy of waste heat is zero. This is due to the heat flow being available at the ambient temperature ($T_w = T_{amb}$), therefore $\dot{E}\left(\dot{Q}_{amb}\right) = 0$.

Consequently, we can derive following equation:

$$P = \dot{E}_{out} - \dot{E}_{in} + \dot{D}^{COMP} \tag{5.47}$$

This equation shows that the mechanical power is converted partially into exergy increase of the gas $(\dot{E}_2 - \dot{E}_1)$ and partially dissipated in form of exergy loss $\dot{D}^{COMP}$.

By using Eq. (4.5) (definition of exergy for material streams $\dot{E} = (\dot{H} - \dot{H}^o) - T_{amb} \cdot (\dot{S} - \dot{S}^o)$ as well as energy conservation law (see Section 5.5.5), a following equation may be obtained:

$$P = -T_{amb} \cdot (\dot{S}_{out} - \dot{S}_{in}) + \dot{D}^{COMP} \quad \text{or}$$

$$P = -T_{amb}\, \Delta\dot{S} + \dot{D}^{COMP} \tag{5.48}$$

For a reversible multi-stage compressor, where $\dot{D}^{COMP} = 0$, the following equation is valid:

$$P_{rev} = -T_{amb}\, \Delta\dot{S} \quad P > 0, \quad T_{amb} > 0, \quad \text{therefore } \Delta\dot{S} < 0 \tag{5.49}$$

This equation contains new, salient information. It addresses and answers the question, "what is isothermal compression good for?". The answer is:

isothermal compressor converts mechanical power in entropy reduction

This is an extremely important feature of this device. All other unit operations previously discussed (heat transfer, expansion, etc.) are characterized by an entropy increase. For realization of a closed cycle, a process unit is required, which reduces the entropy of the working fluid to the original start value (at the beginning of the process chain). The isothermal compression is the device that realizes this function, making closed cycles possible.

5.5.7 Exergy Loss, Isothermal Compression

In a reversible isothermal compression process, the loss amounts to zero: $\dot{D} = 0$. The corresponding required power $P_{rev\_isotherm}$ is minimal and may be calculated using Eq. (5.31) or (5.33).

Table 5.5 Comparison, adiabatic versus isothermal compression.

	Single-stage adiabatic compression	**Multi-stage isothermal compression**
Thermodynamic process is	Adiabatic, in reversible case: isentropic	Quasi-isothermal, in reversible case: isothermal
Efficiency is characterized by	Isentropic efficiency $\eta_s = P_{rev\_isentrop}/P_{real}$	Isothermal efficiency $\eta_T = P_{rev\_isothermal}/P_{real}$
Mechanical power is used for	Enthalpy boost: $P = \Delta\dot{H}$	Entropy reduction $P_{rev} = -T_{amb}\,\Delta\dot{S}$
Application examples	• Single Booster stage of booster–expander unit, • Axial turbocompressor or axial stage of a turbocompressor	• Main air compressor, • Multi-stage booster air compressor or recycle compressor, • Product compressor for external compression

A real compression process is not reversible; therefore, $\dot{D}^{COMP} > 0$. The power required for the compression is higher than power required for reversible process: $P_{real\_isotherm} > P_{rev\_isotherm}$. Equation (5.35) takes this fact into account by means of isothermal efficiency η_T as $P_{real} = P_{rev\_isotherm}/\eta_T$. The corresponding exergy loss can be identified as the difference between these two values:

$$\dot{D}^{COMP} = P_{real} - P_{rev\_isotherm} \tag{5.50}$$

$$\dot{D}^{COMP} = P_{rev\_isotherm}\left(1 - \frac{1}{\eta_T}\right) \tag{5.51}$$

Equation (5.47) may also be used for calculation of loss if the parameters at the inlet and outlet of the entire compressor (temperature, pressure, and therefore exergy) are known. The loss is determined as follows:

$$\dot{D}^{COMP} = P_{real\_isotherm} - (\dot{E}_{out} - \dot{E}_{in}) \tag{5.52}$$

5.5.8 Comparison, Adiabatic Versus Isothermal Compression

Table 5.5 presents a comparison of both compression processes, adiabatic and isothermal.

References

1 Haselden, G.G. (1971). *Cryogenic Fundamentals*, 444. London and New York: Academic Press. ISBN: 012 3305500.

2 Stephan, P., Kabelac, S., Kind, M. et al. (2019). *VDI Wärmeatlas, Verein Deutscher Ingenieure, VDI-Gesellschaft Verfahrenstechnik und Chemieingenieurwesen*, 11e. Berlin: VDI e.V. Springer. ISBN: 3642199801 EAN: 9783642199806.

6

Key Hardware Components

Process design is a scientific art on the border between thermodynamics and engineering. Its primary aim is to design the most energy-efficient and cost-efficient system possible based on available hardware components. The performance (efficiency) and cost of available hardware components are important points of consideration during the process design phase. The unavailability of desired components or low efficiency of available components hinders the creativity of process design engineers. The process design engineer normally communicates the needs and requirements for desired hardware components to the corresponding engineering departments, groups, and engineers. This is an exciting moment, for this interaction (with ensuing discussion and debate) is a great trigger for further improvements and innovative solutions. In this manner, the targets for the development of new hardware components or for the improvement of existing hardware (required for the realization of favorable process concepts) are indicated, and the development process is initiated.

Since the design of cryogenic processes depends heavily on key hardware components, this section deals with such key components as compressors, heat exchangers, thermal insulation, and devices for the expansion of fluids (e.g., expanders and valves).

6.1 Thermal Insulation/Coldbox

In terms of mechanical design, the cold part of the system (heat exchanger, expansion valve, and evaporator) is placed in a rectangular box (more typical for large cryogenic systems), as shown in Figure 8.19, or in a cylindrical vessel (for small or medium-size systems), as shown in Figures 8.27–8.29. Both of these assemblies (box and cylinder) are referred to as coldboxes, even if the cylinder is not really box-shaped. The coldbox is filled with an appropriate insulation material.

There are two different means for heat inleak into the coldbox. The first relates to the mechanical design of the equipment inside the coldbox, especially heat bridges caused by the inappropriate mechanical design of the hardware components, support, suspensions and piping. The second group relates to heat flow through thermal insulation (see Figure 6.1).

Process Design for Cryogenics, First Edition. Alexander Alekseev.

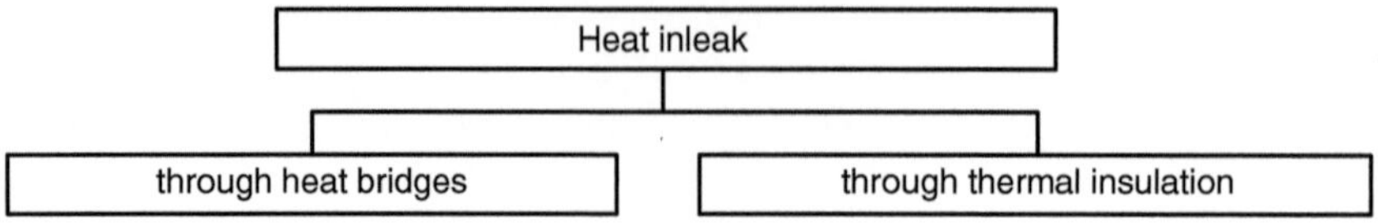

Figure 6.1 Heat inleak into the coldbox.

Since cooling capacity at cryogenic temperatures is very valuable in terms of power (it requires a lot of mechanical/electrical power to generate cold at cryogenic temperatures), all additional generation of cold for covering heat inleak has a considerable impact on the power consumption of the refrigerator. Therefore, it always makes sense to develop a strategy for minimization of heat inleak. For example, thorough minimization of heat bridges is an important criterion for the proper design of the coldbox. The second objective is the proper selection of optimal thermal insulation.

Generally, there are four obvious physical mechanisms for heat transfer:

1. The heat conductivity of solid materials,
2. The heat conductivity of the gas filling the space inside the coldbox (if any),
3. Thermal convection of that same gas,
4. Thermal radiation: heat can be transferred from an object (mechanical part) with a higher temperature T_{warm} to another object (mechanical part) with a lower temperature $\boldsymbol{T}_{cold}$ even in a vacuum, solely via radiation. According to the Stefan–Boltzmann law, heat flow depends on both of these temperatures: $\dot{Q}_{radiation} \sim (T_{warm})^4 - (T_{cold})^4$. More information about heat transfer by radiation can be found in Ref. [20] and [16].

Heat flow through heat bridges: This heat flow is driven mainly by the thermal conductivity of the construction materials (metallic and non-metallic), such as piping that connects elements with different temperatures or the metallic hardware components themselves. Typical examples include:

- The own heat conductivity of the aluminum plate-fin heat exchanger (PFHE), which conducts heat from warm end to cold end,
- Supporting structures for the heat exchanger
- Copper leads that connect a sensor (placed on an object being cooled) to a measurement device located in the surroundings at ambient temperature.

But sometimes, other mechanisms must be considered. Some examples are:

- The heat conductivity of the gas filling the space inside of a tubing/piping connecting a colder element with a warmer element,
- The thermal radiation through the straight piping connecting a colder element with a warmer element.

For illustrative purposes, in Section 6.1.3, a typical element of cryogenic equipment (neck tube of a Dewar vessel) is considered as well as the corresponding heat inleak mechanisms are expounded.

Heat flow through thermal insulation: Various types of thermal insulation are available; see Ref. [1] (p. 359) and Ref. [2] for further information. In terms of performance, two categories of insulation can be distinguished:

- High-performance insulation, which is usually vacuum-based technology,
- Low-performance insulation, which usually works at normal ambient pressure conditions.

6.1.1 Low-performance Insulation

In the simplest case, this can be a polyurethane foam or other foam made from a suitable low-heat-conducting material. Another material often used in large-scale systems is perlite, a lightweight powder-like substance. Both materials (foams and perlite) help to reduce the thermal convection and, partially, radiation, but the heat conductivity of the gas within the foam and the solid conductivity of the material itself remains; therefore, the overall heat conductivity corresponds roughly to the heat conductivity of the gas within the insulation, which is usually 10–40 mW/(m K).

Large cryogenic systems for liquid nitrogen temperatures (e.g., air separation units, nitrogen liquefiers) and cryogenic systems for higher temperatures (e.g., natural gas liquefiers and similar) utilize perlite (without vacuum) for insulation purposes. The main advantages of this insulation type are its low cost and easy maintenance.

6.1.2 High-performance Insulation

The next level is vacuum insulation. In its simplest version, the space inside the coldbox is evacuated by means of a vacuum pump to a level lower than 10^{-3} mbar.[1] This space does not contain any liquid or solid substances, the single gas molecules do not interact with each other. Obviously, under these conditions; the first three physical mechanisms – thermal conductivity of solid materials, gas heat conductivity, and thermal convection – are completely eliminated, and only heat transfer via radiation remains.

The most well-known device based on this technology is the "thermos" flask, originally invented by James Dewar. The thermos flask consists of two shells in the form of bottles, one (smaller one) placed within the other (bigger one) and joined at the neck. The gap between two of these shells is evacuated of air, creating a vacuum. For more information, see Ref. [19].

The highest level of thermal insulation is achieved by means of multilayer insulation (MLI), see Ref. [3]. The MLI is a further improvement of vacuum insulation: (i) the space inside the coldbox is still evacuated, but (ii) additionally, supplementary measures are taken to minimize the heat transfer by radiation in the

1 At this condition, the gas inside of the volume cannot be considered as continuum anymore since molecular interaction virtually no longer occurs, molecules collide with the walls more often than with each other. Therefore, this kind of gas movement is defined as molecular flow.

form of radiation shields, i.e. additional surfaces positioned between the cold and the warm objects, see Ref. [20].

In terms of mechanical design, the radiation shields are based on very thick (5–100 μm) aluminized polyester foil (mylar) or a dual-material sheet – a combination of aluminum foil and glass-fiber sheet, as shown in Figure 6.2. Instead of the glass-fiber sheet, various materials can be applied, starting from conventional paper sheets to mesh from suitable plastic materials. All these materials can be used as flat or as corrugated sheets, targeting optimal geometry and density of the whole MLI package. Often, the aluminum sheets are perforated, which helps to shorten the evacuation procedure. The internals inside the coldbox (hardware components, piping, etc.) are wrapped in this material, with multiple (40–200) parallel layers.

The purpose of the thermal shield can be well understood considering a single radiation shield, which is held at an intermediate temperature T_{interm}. This

Figure 6.2 An example of multilayer insulation based on the combination of aluminum foil and glass-fiber paper. Source: Herose UK.

temperature is warmer than the cold surface T_{cold}, but colder than the warm surface T_{warm}: $T_{cold} < T_{interm} < T_{warm}$. This single radiation shield (one layer of MLI) will allow to reduce the radiation heat flow, described well by the Stefan- Boltzmann equation, as follows

from $\dot{Q}_1 \sim (T_{warm})^4 - (T_{cold})^4$ (without thermal shield)

to $\dot{Q}_2 \sim (T_{warm})^4 - (T_{interm})^4 = (T_{interm})^4 - (T_{cold})^4$ (with thermal shield)

For example, for $T_{warm} = 300$ K (ambient temperature level) and $T_{cold} = 80$ K (liquid nitrogen temperature level), the resulting shield temperature corresponds to $T_{interm} = 252.6$ K, the radiation heat flow is reduced by factor 2: $\dot{Q}_2 = \dot{Q}_1/2$, which is a remarkable improvement. This effect can be multiplied by applying multiple radiation shields (at corresponding temperature levels), so that in the end, the heat transfer by radiation can be considerably reduced, by one order of magnitude or even further.

The aluminum layer provides the functionality of the radiation shield: (i) it reduces the radiation heat flow from layer to layer and (ii) it secures the required high reflection rate. Therefore, the high reflectivity of the material used for the shield is an important requirement for the material. The glass-fiber material works as a thermal and mechanical spacer between the individual radiation shields (metal layers) to ensure the correct temperature profile.

Properly working MLI insulation is extraordinarily efficient; the measured "effective" thermal conductivity of MLI in good vacuum conditions amounts to less than 10^{-4} W/(m K). In comparison to low-performance insulation like polyurethane foam, the MLI insulates better by 3–4 orders of magnitude.

But from a long-term perspective, it is not easy to maintain the high level of vacuum required for high performance. In terms of cost, the vacuum-based thermal insulation (including MLI) is relatively expensive, since it requires either expensive vacuum-tight assembly or additional equipment for evacuation (vacuum pumps, valves, and piping), and relatively high effort to maintain the vacuum inside the coldbox over decades of operation.

Commonly, the MLI is used for:

- systems for liquid helium, liquid hydrogen, and liquid neon temperature levels,
- medium-size vessels for long-term storage of liquid nitrogen, liquid oxygen, and liquid argon, and
- small to medium-size systems for liquid nitrogen temperatures.

6.1.3 Dewar Vessel

A good example of a device with high-performance insulation is a vessel for storing cryogenic fluids, referred to as "Dewar" [as a tribute to this famous British engineer and innovator]. Figure 6.3 shows a small vessel for storing liquid nitrogen. It is a very efficient device; the liquid cryogen can be stored in such a container for weeks and even longer since both parasitic heat flows (through thermal insulation and through heat bridges) are minimized.

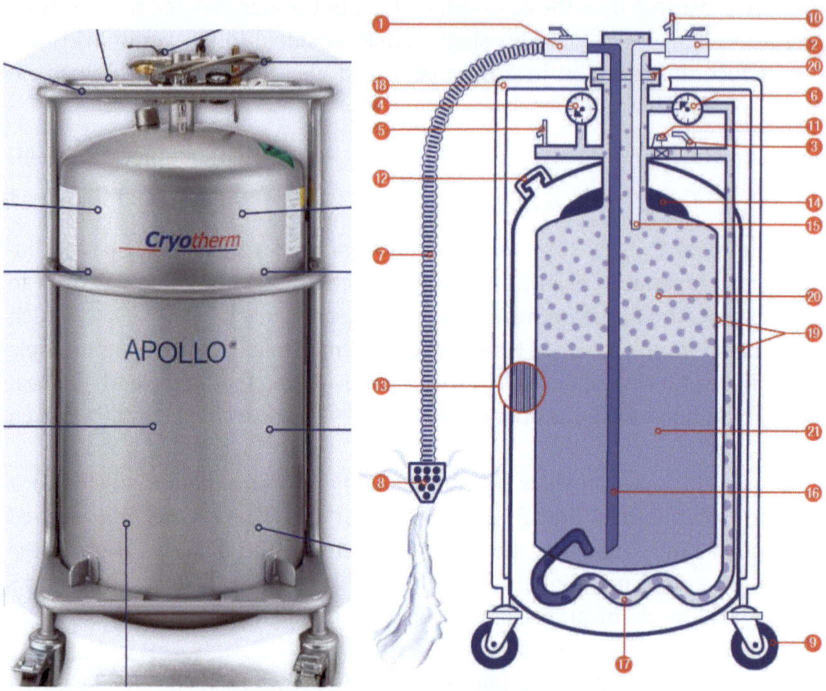

Figure 6.3 Design of vessel for storage of cryogenic liquid. Source: © Cryotherm.

In terms of mechanical design, the Dewar vessel is a further development of the thermos flask concept. It consists of two shells in the form of cylindrical vessels, one (inner vessel) placed within the other (outer vessel) and connected by means of the neck tube. The inner vessel and neck tube are wrapped in MLI (referred to as "Super insulation" in Figure 6.3). The whole assembly is vacuum-tight. The maintenance of the vacuum inside the vessel over decades of operation is the key to the best performance. For this purpose, high-performance adsorptive materials as well as getter materials are integrated; commonly, the adsorber material is placed at the outer surface of inner vessel. Further specific details are usually not publicly available, as this is considered as a trade secret by manufacturers.

The design of the neck tube is shown schematically in Figure 6.4. The neck tube is made usually from a thin (0.5–1 mm) stainless steel tube, the typical diameter is 2″, and the length is longer than 150 mm. It connects the outer vessel (at the top) at the

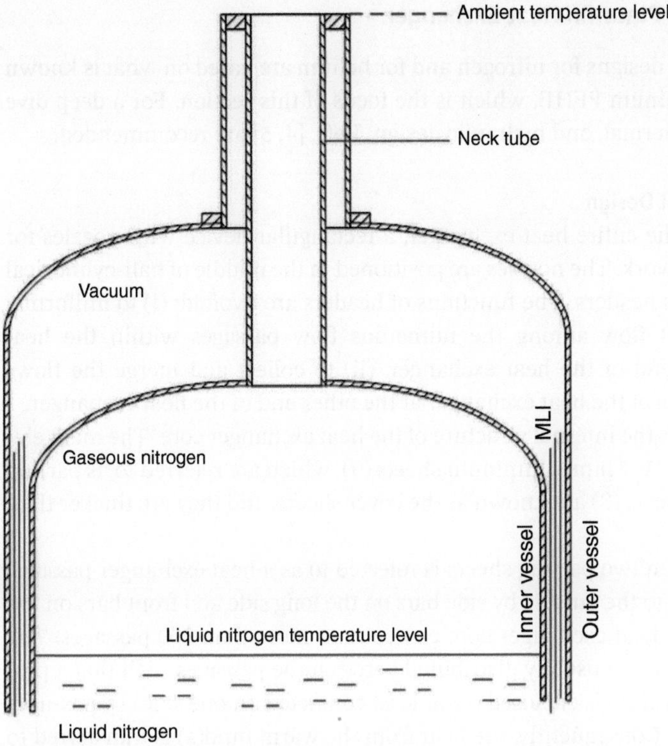

Figure 6.4 Neck tube connection, schematically.

ambient temperature and the inner vessel (at the lower end) with the temperature of liquid cryogen directly. Due to this temperature difference, the neck tube is a source of three parasitic heat flows:

1. Thermally, the neck tube itself is a heat bridge. A little heat flow $\dot{Q}_{\lambda\_MATERIAL}$ leaks continuously from ambient to the inner vessel and further to the liquid through this element simply by thermal conductivity of the material.
2. Additionally, this tube is filled with nitrogen gas, which will conduct some heat flow $\dot{Q}_{\lambda\_GAS}$ by his own heat conductivity.
3. The next parasitic heat flow is the heat flow by radiation $\dot{Q}_{RADIATION}$ through the free cross section of the neck tube. The surface of the cold liquid cryogen is directly visible from outside through this hole. Consequently, it is accessible for direct thermal radiation.

6.2 Heat Exchanger

Primarily, two kinds of heat exchangers are used in industrial cryogenics: aluminum PFHEs and spiral-wound heat exchangers (SWHEs).

6.2.1 Aluminum Plate-fin Heat Exchanger

Industrial liquefier designs for nitrogen and for helium are based on what is known as the brazed aluminum PFHE, which is the focus of this section. For a deep dive into mechanical, thermal, and hydraulic design, Refs. [4, 5] are recommended.

6.2.1.1 Mechanical Design

Figure 6.5 shows the entire heat exchanger, a rectangular device with nozzles for connection to pipework. The nozzles are positioned in the middle of half-cylindrical shells referred to as headers. The functions of headers are twofold: (i) to uniformly distribute the inlet flow among the numerous flow passages within the heat exchanger at one end of the heat exchanger, (ii) to collect and merge the flows leaving the passages of the heat exchanger at the other end of the heat exchanger.

Figure 6.6a shows the internal structure of the heat exchanger core. The main elements are flat, thin (1–4 mm) aluminum sheets (9), which are referred to as parting sheets; the outer sheets (8) are known as the cover sheets, and they are thicker than the parting sheets.

The space between two parting sheets is referred to as a heat exchanger passage. Passages are sealed to the outside by side bars on the long side and front bars on the front side. A single heat exchanger core can contain more than 200 passages. The cold and warm fluids are usually distributed between the passages such that a passage with a warmer fluid is enclosed (or at least contacted on one side) by passages with a colder fluid. Consequently, the heat from the warm fluid(s) is transferred to the cold fluid(s) through the parting sheets.

The space inside a passage is filled with the fins (see Figure 6.7, or position 10 in Figure 6.6), which are thin (0.1–0.2 mm) aluminum sheets corrugated in a special way such that the fluid flows through the passage mainly in longitudinal direction, following the corrugation. The parting sheets and fins are brazed together and form a rigid structure with good thermal contact between passages. The fins have three functions:

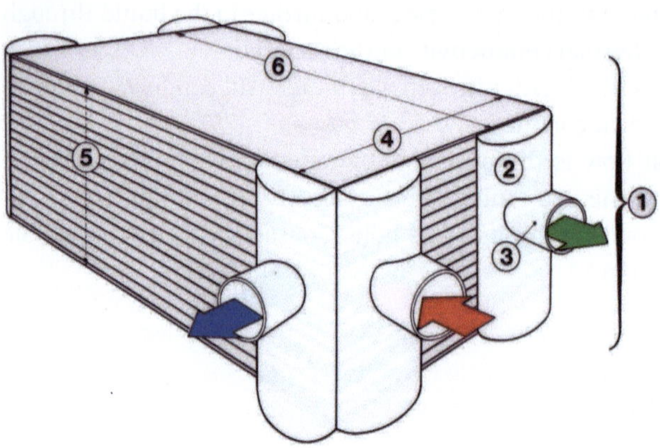

Figure 6.5 Aluminum plate-fin heat exchanger, 1 – Core, 2 – Header, 3 – Nozzle, 4 – Width, 5 – Height, 6 – Length.

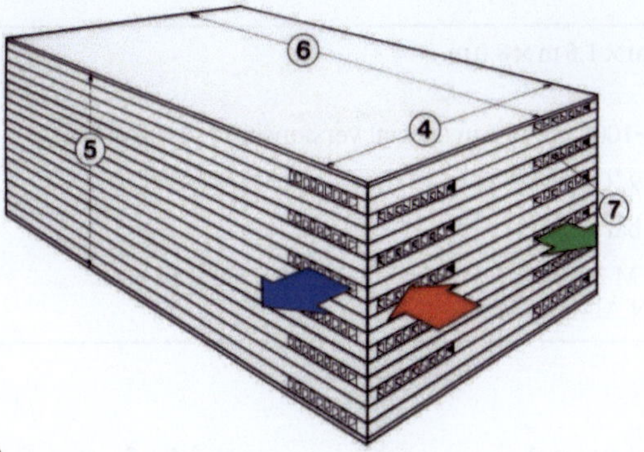

(a)

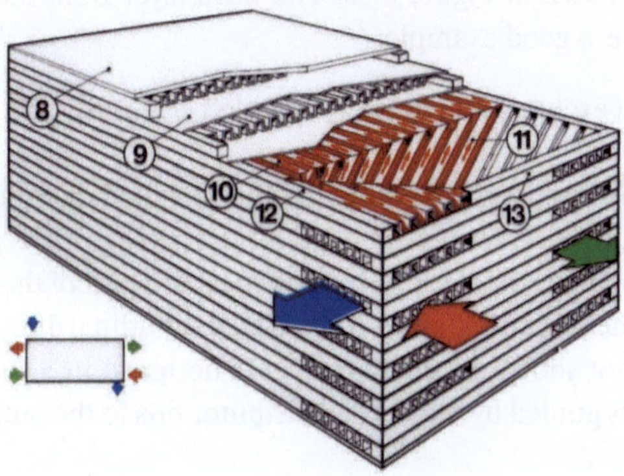

(b)

Figure 6.6 Aluminum plate-fin heat exchanger, 4 – Width, 5 – Height, 6 – Length, 7 – Passage inlets/outlets, 8 – Cover sheet, 9 – Parting sheet, 10 – Heat transfer fins, 11 – Distribution fins, 12 – Side bar, 13 – Front bar.

Figure 6.7 Typical fin designs.

(a) They improve the mechanical stability of the heat exchanger core, so that the aluminum PFHE is able to manage high-pressure fluids.
(b) They increase the heat transfer surface and improve the thermal performance of the heat exchanger.
(c) They guide the fluid flow from the passage inlet to the passage outlet.

The last function is difficult to execute since the fluid enters a passage through a relatively small channel (7) on the front or on the side of the passage. Therefore, the fluid must first be directed from the inlet to the entire passage and distributed

Table 6.1 Typical characteristics of brazed aluminum plate-fin heat exchanger.

Maximum size (width × height × length)	1.8 m × 1.5 m × 8.0 m
Specific surface	300–1000 m^2/m^3, in special versions up to 1500 m^2/m^3
Temperature range	−269 °C to +65 °C
Maximum pressure	115 bar
Typical materials	ASTM 3003 (DIN AlMnCu), ASTM 5083, and 6061 (DIN AlMg4.5Mn)

homogeneously over the whole width of the passage. Consequently, the fin structure is complex. Several details are shown in Figure 6.6b. The third layer from the top, with material flow marked red, is a good example:

- The "red" fluid enters the heat exchanger core at the front left side through a tight inlet.
- It is guided first longitudinally and later diagonally by the so-called distribution fins (11); this way it is distributed over the entire width of the passage.
- At the outlet of the distribution fins (11), the fluid enters the main part of the heat exchanger with conventional heat transfer fins and flows in longitudinal direction.
- The outlet of the passage is not shown in this figure. It is designed in a similar manner as the inlet: the flow is guided by means of distributor fins to the outlet at the rear right side.

The main part of the heat exchanger with longitudinal heat transfer fins is usually considerably longer (by one order of magnitude) than the part with distribution fins. The heat exchanger example presented in Figure 6.6b is a bit confusing in this sense, since in terms of length, the main part of this heat exchanger appears to be comparable to the distribution section. This is usually not the case in actual practice. Typical characteristics of brazed plate-fin heat exchangers are given in Table 6.1.

6.2.1.2 Thermal Design, Hydraulic Design

The fin geometry bears a great impact on the heat transfer coefficient as well as the pressure drop in the heat exchanger. From the perspective of flow mechanics, the fin geometry defines the hydraulic diameter. This is typically 1–5 mm. The corresponding Reynolds numbers will be between 500 and 10 000. According to this, the flow of gaseous substances is turbulent rather than laminar, or it falls within the transition zone between laminar and turbulent flow.

Numerous designs for fin geometry have been developed by original equipment manufacturers (OEMs). They differ in height, thickness, perforation, cut pattern, and other parameters. Two popular designs (plain perforated and serrated) are shown in Figure 6.7. The fin height is typically 4–10 mm. In principle, the fin is a sort of spacer between two parting sheets; therefore, the height of the heat exchanger passage is defined by the fin height.

The heat transfer can be intensified by the perforation and dislocation of fins. However, these measures lead to higher pressure losses. Therefore, the optimum design is always a compromise between heat transfer performance and pressure loss.

The heat exchanger design procedure is complex. Usually, the engineer receives a process-based specification, where the temperatures, vapor fractions, flowrates, and compositions for all inlet and outlet streams are given, as well as the targeted pressure drops. During the design procedure, the engineer optimizes the length and width of the heat exchanger, the number of passages and their arrangement (i.e., which stream flows through which passages), and the fin geometry. The objectives are to fulfill the specified requirements and, simultaneously, to minimize the cost.

6.2.1.3 Attributes

Versatility: A unique feature of the aluminum PFHE is its unlimited capacity in terms of the number of material streams and the direction of the flows inside (countercurrent, co-current, or cross-countercurrent). Additionally, the material streams can be introduced into the heat exchanger or withdrawn from the heat exchanger at virtually any location.

Figure 6.5 shows an example of a heat exchanger for an air separation plant where three material streams are managed, whereby two of them are low-pressure streams and one is a high-pressure stream:

- the high-pressure air stream is marked with a red arrow,
- the low-pressure oxygen stream is marked with a blue arrow, and
- the low-pressure nitrogen stream is marked with a green arrow.

Such a configuration is either not realizable at all by means of other kinds of heat exchangers (e.g., SWHEs (described in Section 6.2.2) or shell-and-tube heat exchangers), or the corresponding re-design leads to a gigantic increase in cost. The implementation of additional passages for a further material stream in a PFHE causes only an incremental cost increase. In this sense, the aluminum PFHE is the champion in terms of versatility.

Compactness/large specific surface: The next important feature is the large specific exchange surface. Usually, this lies within the range of 300–1000 m^2/m^3, but in some special cases, it can be increased to 1500 m^2/m^3 or even higher. Consequently, processes with very small temperature differences between hot and cold streams can be realized by relatively compact units.

Pressure drop: One particularly favorable feature of the brazed aluminum PFHE is a relatively low pressure drop. For cryogenic processes, for cooling from 300 K to 100 K, the targeted pressure drop is 80–400 mbar, whereas the lower number is used for energy-optimized systems, the higher number is valid for cost-optimized design. The PFHE is able to fulfill this requirement.

Cleanliness: Since the heat exchanger is brazed using a vacuum furnace, it is absolutely clean after manufacturing. It is "ready-to-use" for processing of high-purity fluids.

Specific costs: The specific cost of a PFHE is defined by material cost, manpower cost, power/utility costs required for manufacturing, and equipment depreciation costs, e.g. for the vacuum furnace.

The material costs are relatively low since the average heat transfer surface is quite thin. For economical design, parting sheets of 1.5–2 mm thickness are used in combination with 0.1–0.2 mm fins. This results in an average thickness of less than 0.5 mm. By comparison, the heat transfer surfaces in other kinds of heat exchangers, such as SWHEs or shell-and-tube heat exchangers, are formed by tubes with a wall thickness of more than 1.5 mm.

The production costs are moderate. The manufacturing sequence is shown in Figure 6.8. The majority of production steps at the beginning of manufacturing – stamping, cutting, washing, and surface checking – are executed mainly by machines, but there are still many operations to be performed by qualified personnel. This manpower is required for

- stacking, where the sheets, fins, and bars are assembled to form the heat exchanger core,
- manufacturing of headers, which later will be welded to the heat exchanger core,
- welding of heat exchanger modules to large blocks if required,
- testing (X-raying, helium leakage test, pressure, and pressure drop test).

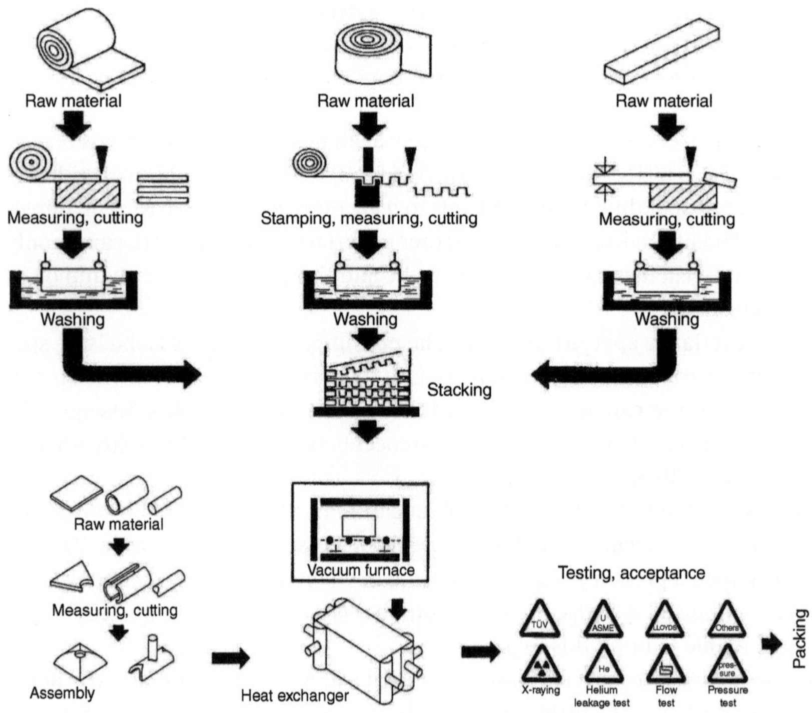

Figure 6.8 Manufacturing of aluminum plate-fin heat exchangers.

Brazing in a vacuum furnace requires a large quantity of power, since this process step takes 6–10 days or even more depending on the size of the PFHE. During this procedure, the assembled heat exchanger core is (i) heated to the temperature of approximately 650 °C required for the melting of solder coating, then (ii) brazed, and finally (iii) cooled down. To avoid thermal stress, all three of these operations require highly homogeneous temperature distribution inside the PFHE, with a targeted temperature deviation of only a few degrees. Such high homogeneity is achieved by performing these processes very slowly. The rate of heating as well as the cooling rate are limited to several degrees per hour. Therefore, the overall brazing procedure takes days to weeks to perform and requires meticulousness, time, and patience.

6.2.1.4 Restrictions and Limitations

The main components used in building PFHEs are **plain** parting sheets. These sheets do not include curved or wave-shaped elements, which could and would absorb any mechanical stress that may occur. Due to this rigid mechanical design, the PFHE is sensitive to inhomogeneous mechanical and thermomechanical loads. In comparison to other heat exchanger types (SWHEs or shell-and-tube heat exchangers), PFHEs must be handled with the utmost care. Certain rules must be observed for safe operation of PFHEs over longer periods of time:

- Thermomechanical stress caused by rapid and/or inhomogeneous temperature changes must be minimized.
  - The temperature profile inside the heat exchanger changes, for example, during the cool-down or warm-up of the process plant during planned maintenance procedures or unplanned outages; therefore, these processes must be performed very slowly.
  - Even during normal operation, temperature changes may occur during load changes (overload-underload) of the process plant. For example, large temperature differences may occur if multiple heat exchangers are used in parallel and a material flow is distributed incorrectly among them during a load change.
  - Processes with a high temperature difference between warm and cold streams (e.g., the classical Joule–Thomson process) cause thermomechanical stress by nature, which PFHEs are unable to manage. The upper limit for maximum temperature difference is roughly 30 K.
- Mechanical vibrations caused by pumps or compressors can damage PFHEs. Therefore, proper piping design is required to prevent the propagation of vibration from machines to PFHEs.
- In terms of fluids used, only media compatible with aluminum can be processed. In this sense, fluids with a pH higher than 8 or lower than 5 are not acceptable. Also, sea water or fluids containing mercury or copper (e.g., flue gas) are not tolerable since they will attack aluminum and cause corrosion.
- Due to the tight fin structure inside, the PFHE cannot be properly cleaned if it becomes contaminated. Therefore, any material flow containing mechanical particles, such as dust, must be avoided. The dust will deposit inside the heat exchanger, reduce the free cross section, and lead to excessive pressure drop.

6.2.2 Spiral-wound Heat Exchanger

SWHEs are often used for large-scale liquefaction of natural gas. Sometimes middle-scale apparatuses of this kind are applied in air separation, especially if the requirements on pressure level or temperature difference are higher than the PFHE is able to manage. The next application area relates to small-scale cryogenics, where such heat exchanger are used for miniature Joule–Thomson cooler, see Ref. [7]. For further reading, Refs. [8, 9] as well as [10] are recommended.

6.2.2.1 Mechanical Design

Tube Bundle: As shown in Figure 6.9, a SWHE consists of helices of thin tubes that are wrapped around a core cylinder, referred to as the mandrel, in a number of layers. Right-wound helices alternate layer by layer with left-wound helices. Each layer normally consists of a number of tubes (helices) in parallel. The required distance between tubes in a single layer, as well as the distance between layers, is provided by means of distance spacers and distance strips. The whole tube bundle is wrapped by a metal sheet called a shroud (for prevention of bypass flow) and inserted into a cylindrical jacket, referred to as a shell.

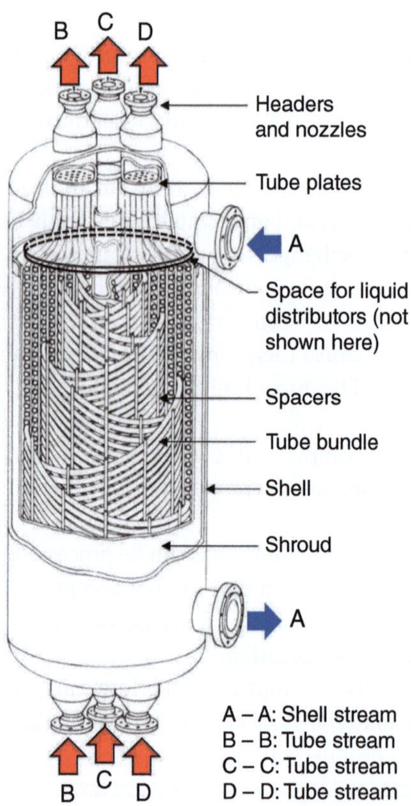

Figure 6.9 Spiral-wound heat exchanger, mechanical design. Source: Ref. [8]/ Springer Nature.

Figure 6.10 Spiral-wound heat exchanger, tube plate end. Source: Ref. [8].

Tube plates, headers, and nozzles: The pipe ends are welded into devices referred to as tube plates or collectors, which are connected to the nozzles, as shown in Figure 6.10. Each tube plate (with corresponding nozzle) can be fed by different material streams (fluids with different compositions, pressures, and temperatures). Therefore, on the tube side, several material streams (B, C, and D in Figure 6.9) can be integrated into the same heat exchanger.

The nozzles for tube-bundle streams can be located on the top and bottom of the heat exchanger, as shown in Figure 6.9, as well as on the side, as shown in Figure 6.14. The same principle applies to a shell-side stream, which can be introduced through a nozzle on the top or bottom, as shown in Figure 6.9, or on the side, as shown in Figure 6.14.

Shell side: On the shell side, only one material stream can be processed; in Figure 6.9, this is marked as A. This restriction bears a great impact on process design for use of an SWHE. An appropriate process is described in Section 8.4.6.1.

Materials: Both basic cryogenic metallic materials – aluminum and stainless steel – can be used for manufacturing SWHEs.

Low-pressure fluid on shell side: Commonly, the high-pressure working fluid being cooled in the SWHE is provided through the tubes. The colder, low-pressure fluid is provided through the shell side; it flows across the wound tubes in the space between the mandrel and shell. A reverse arrangement (high-pressure fluid on the shell side, low-pressure fluid on the high-pressure side) is theoretically conceivable. However, such a configuration is rather a curiosity.

Degrees of freedom: A number of degrees for optimization of mechanical design are available, with the aim of obtaining the best possible thermal performance. These are primarily related to

- the geometry of the tube itself: diameter, fin geometry, material,
- the geometry of the bundle: mandrel diameter, number of tubes per layer, number of layers, distance between the tubes in a layer, distance between the layers, etc.

All these parameters impact the thermo-hydraulic performance on both sides of heat exchangers, i.e. on the shell side as well as on the tube side.

Constraints: Unfortunately, SWHEs do not offer many possibilities for provision of fluid with an intermediate temperature to an intermediate position or location inside the tube bundle, or for removal of a fluid stream from the middle of the heat exchanger (at an intermediate position inside the tube bundle) since all tubes and layers start and end more or less simultaneously at either end of the heat exchanger. This is the second restriction that bears a major impact on SWHE-based process design. An appropriate process is described in Section 8.4.6.1.

Table 6.2 summarizes some typical data and properties for SWHEs. Table 6.3 shows the data for the spiral-wound heat exchangers used for the Snohvit (Norway) plant as an example.

The manufacturing procedure is illustrated in Figures 6.11–6.16, which are more or less self-explanatory.

6.2.2.2 Thermal and Hydraulic Design

Flow arrangement: From an external perspective, the flow arrangement appears to be a countercurrent flow – the warmer fluids flow from bottom to top, and the cold fluids flow from top to bottom. However, in reality, it is definitely a cross-counter

Table 6.2 Spiral-wound heat exchanger for LNG applications.

Typical diameter of tubes	<20 mm
Typical wall thickness (for tubes)	<2 mm
Bundle diameter	<6 m
Bundle height	<60 m
Surface area	<40 000 m^2
Mass	<250 t
Pressure on tube side	<200 bar
Compactness (specific heat transfer surface)	$100 \pm 50\ m^2/m^3$

Table 6.3 Examples of spiral-wound heat exchangers for LNG applications.

	Snohvit middle heat exchanger (liquefier)	**Snohvit lower heat exchanger (subcooler)**
Heat transfer surface	11 330 m^2	16 470 m^2
Shell diameter	3.4 m	3.7 m
Bundle height	10.4 m	12.7 m
Compactness	120 m^2/m^3	120 m^2/m^3

Source: Data from Ref. [9].

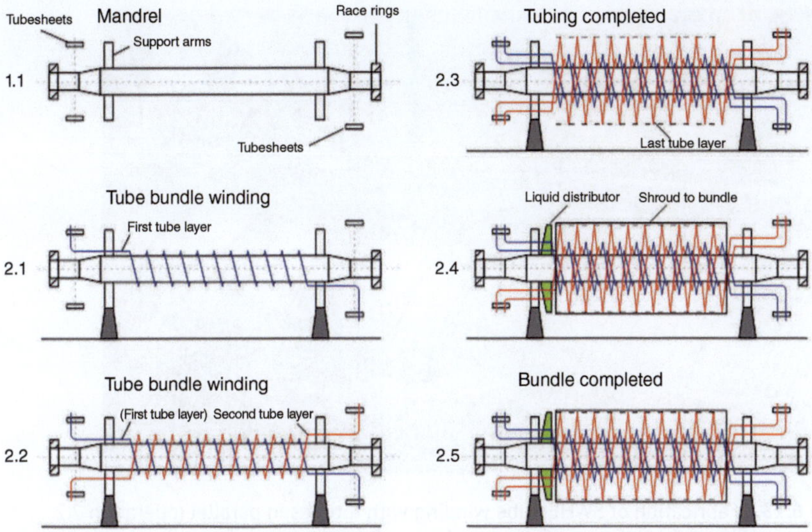

Figure 6.11 Spiral-wound heat exchanger, manufacturing of tube bundle acc to Ref. [6]. Source: Courtesy of Linde Engineering.

Figure 6.12 Fabrication of SWHE, mandrel (operation 1.1 according to Figure 6.11). Source: Courtesy of Linde Engineering.

flow arrangement since the fluid in the tubes crosses the flow of fluid on the shell side at every single point.

Maldistribution may occur on the tube side as well as on the shell side. On the tube side, in order to avoid flow maldistribution and resulting performance deterioration, all tubes and helices carrying a particular stream must be of the same hydraulic resistance, which depends highly on the tube length and slightly on the diameter of the helices (the smaller the diameter, the higher the pressure drop). Therefore, for each individual material stream, quite similar tube lengths are chosen, but tubes

Figure 6.13 Fabrication of SWHE, tube winding with 4 tubes in parallel (operation 2.2 according to Figure 6.11). Source: Courtesy of Linde Engineering.

Figure 6.14 Fabrication of SWHE, tube winding complete (operation 2.5 according to Figure 6.11). Source: Courtesy of Linde Engineering.

for outer layers may be a bit longer than those for inner layers. Generally, this principle is well-suited to the mechanical design of SWHEs since the outer layers have a larger diameter and require longer tubes for the same bundle height. In practice, exact compensation of longer length by larger diameter of winding is difficult to implement. The manufacturers of SWHEs have developed their own approaches for managing this problem. Usually, these solutions are an important part of their know-how portfolio and are not available for public review.

On the shell side, the maldistribution-compatible design of SWHE depends on the process and fluid used.

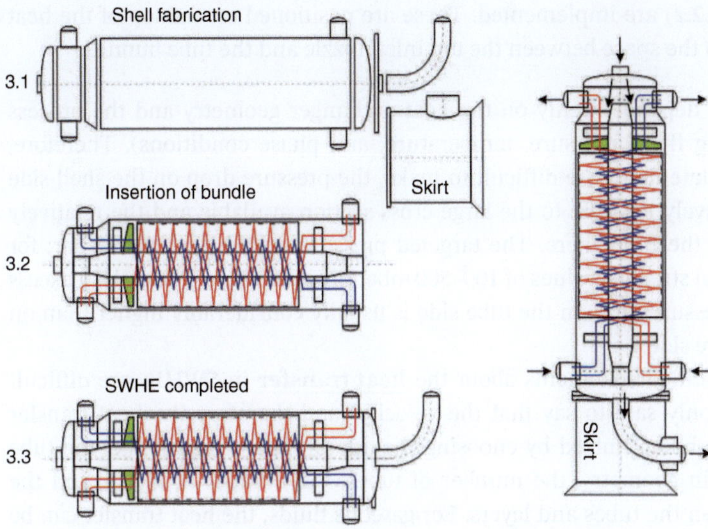

Figure 6.15 Spiral-wound heat exchanger, manufacturing procedure according to Ref. [6]. Source: Courtesy of Linde Engineering.

Figure 6.16 Fabrication of SWHE, heat exchanger complete (operation 3.3 according to Figure 6.15). Source: Courtesy of Linde Engineering.

- If a gaseous fluid is used, either no special measures at all are taken or relatively simple measures are applied, such as baffles behind the inlet nozzles. Another approach is based on the use of multiple (two, four, etc.) nozzles symmetrically distributed around the shell perimeter to guarantee the distribution of the material stream over the entire cross section of the heat exchanger in the required quality.
- If a liquid or a two-phase fluid is processed (usually injected at the top of the SWHE to flow from top to bottom), special devices for homogeneous distribution (see

Section 8.4.5.2.2) are implemented. These are positioned on the top of the heat exchanger, in the space between the top inlet nozzle and the tube bundle.

Pressure drop depends highly on the heat exchanger geometry and the process design (working fluid, pressure, temperature, and phase conditions). Therefore, while general statements are difficult to make, the pressure drop on the shell side is usually relatively low due to the large cross section available and the relatively low velocity of the fluid there. The targeted pressure drop is less than 1 bar; for gaseous material streams, values of 100–500 mbar are achievable. At levels in excess of 1 bar, the pressure drop on the tube side is usually considerably higher than on the low-pressure side.

Similarly, qualified statements about the **heat transfer** in SWHEs are difficult to make. It is only safe to say that the velocity (and therefore the heat transfer coefficient) can be optimized by choosing the appropriate geometry, i.e. the tube diameter and fin geometry, the number of tubes, the number of layers, and the distance between the tubes and layers. For gaseous fluids, the heat transfer can be well estimated by using conventional methods based on Re and Nu numbers. For a mixed-fluid-based process, this becomes a complex task, involving (i) **multi-phase** flow of (ii) a **multi-component mixture over** (iii) a wide temperature range, and (iv) a wide range of phase states, from almost 100% liquid at the cold end of the heat exchanger up to 100% gas at the warm end. Therefore, for multicomponent fluids in two-phase conditions, the conventional Re/Nu methods are not applicable.

6.2.2.3 Restrictions and Limitations Relevant for Use and Operation

The main elements used in SWHEs are spirally wound tubes. Due to the high elasticity of the tubes, the tube bundle can absorb mechanical forces caused by mechanical or thermal stress. Therefore, the SWHE is a highly robust device. There are practically no limitations for the steady-state temperature profile (temperature difference) within the heat exchanger itself, which may result from the process design, and there are no hard limits concerning the ramping rate (change of temperature profile due to load change) or even higher stress during cooling and warming procedures.

Mechanical vibrations caused by pumps or compressors can, however, cause damage to SWHEs. Therefore, proper piping design is required to prevent propagation of vibrations from machines to the heat exchanger.

In terms of fluid used, only media that are compatible with the materials used can be processed. For aluminum SWHEs, the restrictions described in Section 6.2.1.4 (aluminum plate-fin heat exchangers) apply.

Likewise, due to its numerous thin tubes, the SWHE cannot be properly cleaned if it becomes contaminated. Therefore, any material flows containing problematic contaminations (e.g., mechanical particles such as dust) must be avoided.

The SWHE tube helix cannot be repaired if a single tube is damaged since it is not accessible from outside the bundle. The typical repair action includes removing the damaged tube from service by means of plug-in tube plates mounted on both ends of the tube.

6.2.2.4 Summary: Spiral-wound Heat Exchanger

Versatility: In comparison with aluminum PFHEs, the spiral-wound design is less versatile.

- The number of material streams on the shell side is limited to one, and the number of material streams on the tube side is limited to one digit.
- The flow arrangement is fixed (cross-countercurrent).
- Material streams cannot be introduced or withdrawn from the heat exchanger at intermediate points in a simple way.

Compactness/large specific surface: in terms of specific heat transfer surface, the SWHE with $100 \pm 50\,m^2/m^3$ is not a compact device. The PFHE with 300–1000 m^2/m^3 surpasses it by one order of magnitude.

Specific costs: The specific costs of SWHEs are defined by material cost, manpower cost, power/utility costs required for manufacturing, as well as depreciation of equipment, e.g. welding tools and other machines.

- The material costs are relatively high, since the heat transfer surface on average is relatively thick, because it is formed by tubes with a wall thickness of 0.7–1.5 mm, which is a bit higher than the average thickness of materials used in PFHE.
- The production costs are rather high, since the main manufacturing operations (winding, welding in tube plates, testing, etc.) are not well-suited to automation and are performed by qualified personnel.
- The manufacturing does not require a lot of power.
- The winding operation takes a lot of time, since in principle every single tube is handled individually, at least where all the procedures and manipulations around tube plates are concerned. Therefore, the winding procedure alone may take several months. The overall manufacturing time is even longer.

In direct comparison with PFHEs, the SWHE always performs poorly. Therefore, the question is: Why is it still applied? Apparently, the SWHE does ultimately offer several features that the PFHE cannot match. The first advantageous feature is its exceptional robustness – SWHEs can withstand an extremely high level of mechanical and thermal stress. The second beneficial feature relates to virtually unlimited scalability: gigantic heat exchangers can be manufactured, transported, and installed as a single piece of equipment, which affords a significant savings in construction effort and cost. The next important criterion is that a lot of experiences and a lot of references are available for the SWHE. Therefore, the technical risks related to this type of heat exchanger are considered to be low.

6.3 Expanders

The thermodynamics of the expander as a process unit are described in Section 5.2. The current section deals primarily with mechanical design of these machines. But, since the expanders bear an impact on process design, two sections related to process design are inserted:

1. Section 6.3.2: Requirements for design of a process with expander
2. Section 6.3.5: Joule–Thomson Expander

6.3.1 General Issues

This section examines several general issues regarding all types of expanders.

6.3.1.1 Expansion Process in Detail

The core part of the expander section is shown in Figure 6.17. It consists mainly of

(1) rotating part referred to as turbine wheel or impeller,
(2) static (not-rotating) device referred to as inlet nozzle assembly or simply inlet nozzles (see Figure 6.25).

The inlet nozzle assembly consists of a series of teardrop-shaped vanes, which are mounted around the circumference of the expander wheel. They are swept in the direction of expander wheel rotation. There are different versions of this device available: in small-scale turboexpanders, inlet nozzles with fixed geometry are applied, whereas in large-scale versions, the inlet nozzles are usually adjustable, as shown in Figure 6.17.

The expansion process in a turboexpander consists of four steps:

Step 1: The gas enters the inlet (volute) casing or scroll, where it is distributed to the nozzles.

Step 2: In the nozzles, by flowing through narrowing channels between the vanes, referred to as throat, the gas flow is accelerated and directed at the correct angle into the turbine wheel. During this second step the potential energy of the gas (pressure p_1) is converted into kinetic energy $\Delta p \rightarrow \dot{E}_{kin} \approx \rho {v_2}^2/2$ according to the Bernoulli equation (see Eq. (6.1)), whereby the pressure is reduced to $p_2 < p_1$

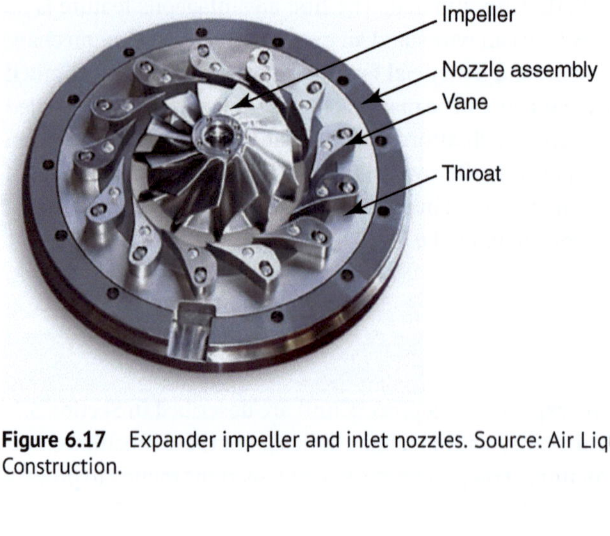

Figure 6.17 Expander impeller and inlet nozzles. Source: Air Liquide Engineering & Construction.

Step 3: The accelerated gas is directed into the turbine wheel. It hits the blades of the impeller, and consequently, the gas velocity is converted into rotation (mechanical power) of the shaft. This way the kinetic energy of the gas is partially converted into mechanical power ***P*** in form of rotation: $\Delta\dot{E}_{kin} \rightarrow P$. The flow paths of the fluid may be different. The majority of cryogenic turbines have wheels with blades that are arranged entirely in the radial plane and are curved to form the exducer section. The flow enters the blading radially (at the outer edge of the turbine wheel) and leaves it axially [close to the eye of the wheel and therefore reduced velocity].

Step 4: Finally, during deceleration, the gas passes from the wheel to a diffuser (a channel with a widening cross-section), whereby the residual kinetic energy is converted into pressure increase, so that the pressure at the outlet of the expander section is even a bit higher than at the expander wheel outlet. The gas leaves the diffuser at a reduced speed. The kinetic energy of this flow is relatively low.

$$\left(\frac{\rho v^2}{2}\right)_1 + p_1 = \left(\frac{\rho v^2}{2}\right)_2 + p_2 \qquad \text{(for horizontal flow)} \tag{6.1}$$

Generally, for the conversion of gas velocity (kinetic energy) into rotation (mechanical energy), two mechanisms/methods may be applied:

(1) The first and most obvious way is via impulse transfer. The direction of the high-velocity gas is changed inside the impeller. The resulting impulse spins the turbine and diminishes the kinetic energy of the fluid flow. There is no pressure change of the fluid or gas in the turbine blades; the entire pressure drop takes place in the inlet nozzles.

(2) The second method uses reaction forces. The most famous engine of this kind is Hero's engine (see https://en.wikipedia.org/wiki/Aeolipile and Figure 6.18). In original version, it consists of a spherical vessel with oppositely bent or curved nozzles projecting outward. The vessel is designed to rotate on its axis. When the vessel is pressurized with steam, the gas is expelled out of the nozzles, which generates thrust (or, in other words, reaction) due to the "rocket principle," as a consequence of Newton's third law of motion. Therefore, this kind of expansion device is referred to as a "reaction turbine." The modern design of expansion devices is vastly different from that of Hero's engine, but the principle is the same – the blades are designed such that the gas leaving the blades "torques" the wheel in the opposite direction. Normally, the pressure of the fluid changes as it passes through the turbine rotor blades.

The modern cryogenic turbine design uses both reaction and impulse concepts to varying degrees; the pressure drop takes place partially in the nozzles and partially in the blades.

Figure 6.18 Hero's engine. Source: Heronin aeolipiili.

6.3.1.2 Isentropic Efficiency

For characterization of expander performance, a plot of isentropic efficiency versus $\boldsymbol{U/C_o}$ is often used. $\boldsymbol{U}$ is the wheel tip speed, calculated as $\boldsymbol{U = \pi\ D\ N}$, where $\boldsymbol{D}$ is impeller diameter and $\boldsymbol{N}$ is rotation speed. The definition of velocity $\boldsymbol{C_o}$ is more complex. It relates to the reversible (i.e., isentropic) expansion process. This parameter is the virtual velocity achieved if expander power is completely converted into kinetic energy. Since the specific expander power is equal to the enthalpy difference $\boldsymbol{\Delta h_s}$ (see Section 5.2.3), the formal definition for $\boldsymbol{C_o}$ is given by:

$$\dot{m}\frac{C_o^2}{2} = \dot{m}\Delta h_s \quad \text{or} \quad C_o = \sqrt{2 \cdot \Delta h_s} \tag{6.2}$$

A typical plot is shown in Figure 6.19. The expander is normally operated within the range $\boldsymbol{U/C_o} = 0.6–0.8$. Therefore, the performance in this area is of great interest. The curve reaches its maximum at $\boldsymbol{U/C_o} = 0.7 \pm 0.05$. Logically, the efficiency is 0 at $\boldsymbol{U/C_o} = 0$.

Since the gas velocity is a derivate of the volumetric flowrate and the geometry of the flow channel, the expander efficiency can be defined for every given set of process parameters (mass flowrate, temperature, and pressure). Using this kind of plot and assuming that the gas velocity is close to $\boldsymbol{U}$.

6.3.1.3 Bearing System

Given that the mechanical design of an impeller is a complex task, the biggest challenge is the design of the bearing system for the rotor. According to Wikipedia, "a bearing is a machine element that constrains relative motion to only the desired motion, and reduces friction between moving parts."

According to Ref. [11], the main "functions of the bearing system are: (i) to support the rotor in the correct position and permit the rotor to run stably at the

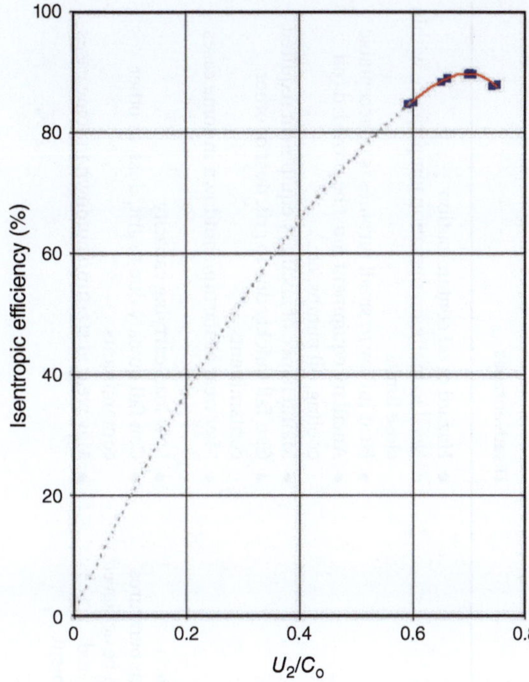

Figure 6.19 Typical efficiency curve for a centrifugal expander.

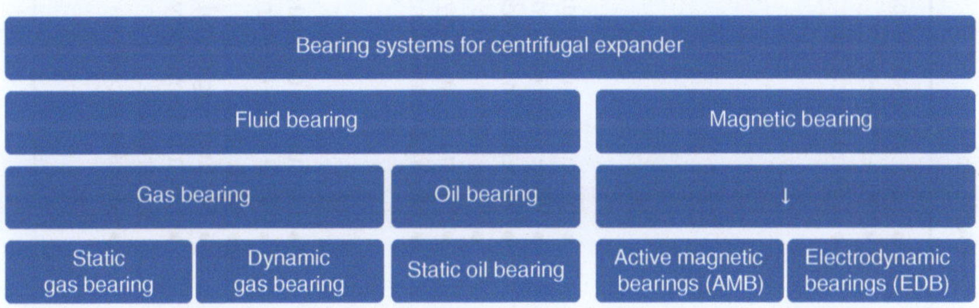

Figure 6.20 Classification of bearings for centrifugal expander.

design speed within the specified operation range, and (ii) to withstand the axial and radial forces imposed on the rotor by the working fluid. In order to perform these functions, the bearing system normally consists of two radial bearings and two axial bearings (sometimes referred to as thrust bearings). The radial bearings are high-speed, lightly loaded components, whereas the axial bearings are subject to considerable load variation since axial forces cannot be balanced completely at all operation conditions."

A classification of bearing systems applicable for cryogenic expanders is given in Figure 6.20. A comparison is given in Table 6.4.

Fluid bearings (see Ref. [17]) are bearings in which the load is supported by a thin layer of rapidly moving pressurized liquid or gas between the bearing

Table 6.4 Comparison of bearing systems.

Bearing system	Advantages	Disadvantages
Oil-lubricated bearings (oil is forced between two faces and held in by edge seal). Application: nitrogen liquefaction, air separation, natural gas liquefaction, and large-size hydrogen liquefier.	• High axial load capacity • High stiffness • Low maintenance	• Hazard of oil contamination • Need to maintain operation temperature within close limits • Heat inflow in small turbines is unacceptable • Ancillary equipment must be provided (oil cooling, oil pumps, etc.) • Maintenance of ancillary equipment required • Can fail quickly due to grit, dust or other contaminants • May wear at startup/shutdown in some cases
Gas-lubricated bearings (gas is forced between two faces). Application: small turbines, for example in helium liquefaction (see Section 6.3.4 on helium expanders) and medium-scale hydrogen liquefaction.	• Immune to oil contamination • Wide range of operating temperatures • Extremely high speeds can be achieved • Heat inflow can be minimized • Not much ancillary equipment • Negligible maintenance	• Low load-carrying capacity • Can fail quickly due to grit, dust, or other contaminants • May wear at startup/shutdown in some cases
Magnetic bearings (bearing faces separated by magnets). Application: Magnetic bearings perform well in medium-scale systems. However, their competitiveness is still questionable due to their relatively high cost. In the future, if the cost issue is resolved, magnetic bearings could replace oil-lubricated bearings.	• Immune to oil contamination • Wide range of operating temperatures • High speed • Medium-high axial load capacity • Not much ancillary equipment (mainly electronics) • Maintenance free	• Complex control system • Relatively high initial cost • AMBs required constant [considerable] power for levitation even if the expander is not in operation • Low stiffness

surfaces.[2] Since there is no contact between the moving parts, there is no sliding friction, allowing fluid bearings to have lower friction, wear, and vibration than many other types of bearings. Thus, it is possible for some fluid bearings to have near-zero wear if operated correctly. Fluid bearings can be broadly classified into two types: fluid dynamic bearings and fluid static bearings. The static bearings are externally pressurized fluid bearings, where the fluid is usually oil or gas and the pressurization is performed by a pump or compressor. Dynamic bearings rely on the extremely high speed of the journal (the part of the shaft resting on the fluid) to pressurize the fluid in a wedge between the faces. Fluid bearings are frequently used in high-load, high-speed, or high-precision applications.

A magnetic bearing (see Ref. [18]) is a type of bearing that supports a load using magnetic levitation. Magnetic bearings support moving parts without physical contact. They are able to levitate a rotating shaft and permit relative motion with very low friction and no mechanical wear.

For cryogenic expanders, two further aspects must be considered during the design phase:

1. The first aspect relates to the large temperature difference between the cold gas in the expander wheel and the warm (above ambient) temperature in the bearings. For mechanical stability purposes, a short distance between the bearing and the wheel is usually targeted, but the large temperature difference may lead to undesirable effects like increased heat leakage into the cold gas or cooling of the bearing to a temperature level lower than expected.
2. Another aspect relates to the relatively high rotation speed (10 000–100 000 rpm) required to achieve high tip speed since the diameter of the impeller in cryogenic applications is relatively small (see Table 6.6).

6.3.2 Requirements for Design of a Process with Expander

The current section addresses two aspects that bear an impact on process design; they are:

1. Limited pressure ratio (expansion ratio) in a single-stage expander,
2. No phase state change in the expander.

6.3.2.1 Limited Pressure Ratio

This is a highly simplified description of the process inside a turboexpander. It gives an idea of the connection between the velocity (reached at the outer edge of the turbine wheel) and the pressure ratio in an expander: the power [and therefore the pressure ratio of the expander according to Section 6.3.1.1] simply cannot be higher than the kinetic energy of the gas at maximum velocity. Therefore, the maximum speed effectively limits the pressure ratio.

The maximum speed (the result of the rotating frequency and diameter of the wheel) is limited by the mechanical strength of the material, which has to withstand

2 Ice skates are a well-known example of a fluid bearing system: the blade forms a hydrodynamic fluid bearing where the blade and the ice are separated by a layer of water.

Table 6.5 Constraints for pressure ratio, helium.

Wheel material	Aluminum	Titanium
Maximum speed, m/s	400–500	600
Maximum pressure ratio at $T = 150$ K	1.47–1.89	2.52
Maximum pressure ratio at $T = 50$ K	3.88–17.3	>20

the centrifugal force. Aluminum alloys can withstand such forces, which correspond to the maximum speed of roughly 400 m/s. Titanium alloys can be used for higher speeds, but these also fail at approximately 600 m/s.

It is important to note that the sound speed of the gas at given conditions can limit the maximum speed of the expander wheel. The sound speed of air gases (air, nitrogen, oxygen, and argon) is relatively low, amounting to 200–350 m/s.

Both of these limitations on wheel speed constrain the pressure ratio $\pi = (p_{in}/p_{out})$ manageable in a single-stage turboexpander. These constraints depend on the working fluid and thermodynamic conditions. The maximum pressure ratios for air are often given as follows:

- $\pi = (p_{in}/p_{out}) = 8\text{–}10$ for gas inlet temperatures of approx. 150 K,
- $\pi = (p_{in}/p_{out}) = 4\text{–}5$ for gas inlet temperature near room temperature (300 K).

The rough values for helium are given in Table 6.5.

Multi-staging

If the process design requires a pressure ratio higher than that which a single-stage expander can offer, it is always possible to realize the expansion process by means of two or more stages in series. Depending on the hardware available, this may be two single-stage machines connected serially or one expansion device consisting of two expander wheels on a common shaft.

6.3.2.2 No Phase State Change

Turboexpanders are **sensitive to phase changes of the fluid** during the expansion process (liquid droplets flying at a velocity of more than 200 m/sec may damage or even destroy the turbine blades). Therefore, the conservative process design approach (i.e., the choice of temperature and pressure at the inlet and outlet) presumes that the expansion occurs in one phase area only, i.e. either

- in the gaseous phase: from the gas into the gas area or
- in the liquid phase: from the liquid into the liquid area.

Expander manufacturers currently offer gas expanders that are able to manage some liquid fraction at the outlet (usually it is specified to be lower than 10 mol%). Therefore, process design engineers are currently attempting to implement this new degree of freedom into their process designs. This new generation of expanders requires a very stringent and sophisticated process control system that is able to guarantee that the liquid fraction at the outlet is always lower than the maximum specified value, even during the transient operation modes (such as fast ramping (load change) of the system, which presents a very challenging task).

This book pursues a more conservative, "old-fashioned" approach, which presumes no phase change inside the expander.

6.3.3 Large Turboexpanders with Oil Bearings

Medium-size expanders are used for nitrogen liquefaction, with power usually in the range 0.5–5 MW. This same type of expander is used for air separation or for small-scale natural gas liquefaction (for large-scale natural gas liquefaction, the power requirements are higher, typically 20 MW or more).

The loading devices for receiving the mechanical power generated during expansion are described in Section 5.2.6. In nitrogen liquefaction, a single-stage centrifugal compressor (often referred to as a booster) is used as the loading device, according to Ref. [11] "because of its simplicity and ease of control, the centrifugal compressor."

From a process design perspective, such an expander-booster unit consists of three process elements: expander, booster, and aftercooler – a heat exchanger required for cooling of the hot gas from the compressor outlet. In this section, only the machine – the expander-booster unit – is considered. An example of such a machine is given in Figure 5.8. Another example is given in Figure 6.21, whereby only the housing, the rotor with both impellers, and the oil bearing in the middle are shown. Here, the expander section is located on the left side, with the compressor section on the right side of the device.

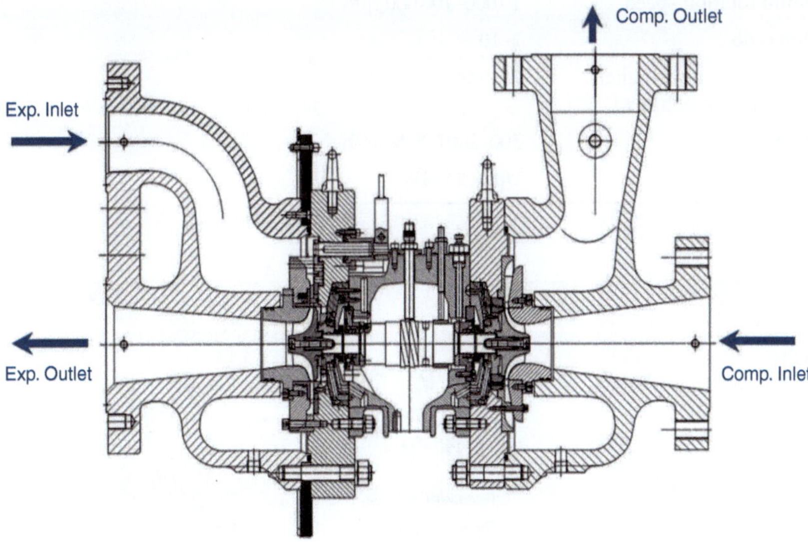

Figure 6.21 Expander-booster unit.

6.3.3.1 Mechanical Design

Generally, a cryogenic expander has more similarities with a water mill than with a steam turbine or windmill. The main elements are:

- Housing with inlets and outlets for both devices (expander and booster),
- Rotor with expander impeller at one end and compressor impeller at the other end,
- Bearing system for rotor, including corresponding oil management system (cooling, filtering, pumping, injecting) and sealing gas system,
- Integrated instruments for control of temperatures, rotational speed, vibration, and other parameters.

Expander section

Similarly to the compressor wheel design, the primary target of the expander wheel design is to realize reversible isentropic expansion or to approximate the isentropic process as closely as possible. All comments in Section 6.5.2.2 for impeller design of a centrifugal compressor apply to the design of the expander impeller as well.

Table 6.6 summarizes some typical values.

Several typical designs of expander impellers are shown in Figure 6.22. The closed wheel design is often better than the open wheel, which in turn is better than the scalloped wheel in terms of efficiency. The scalloped wheel seems to be more reliable if a phase change occurs inside the expander.

Table 6.6 Typical values/parameters for gas expanders.

Impeller diameter	80–600 mm
Tip speed	200–500 m/s,
Maximum rotation speed	10 000–100 000 rpm
Pressure ratio	2–10
Isentropic efficiency (higher values for large machines)	75–92%
Flowrate	500–200 000 Nm^3/h
Power	Up to 25 MW

Closed wheel Open wheel Scalloped wheel

Figure 6.22 Impeller designs for expander. Source: © Cryostar.

Compressor section:

The compressor stage is described in Section 6.5.2.1

Section between expander wheel and compressor wheel

Unfortunately, further details are not evident in Figure 6.21. Logically, the oil bearing is located in the middle part of this device, with nozzles for oil injection and for the seal gas, which prevents the leakage of oil into the working gas in the expander and compressor parts.

6.3.3.2 Bearings for Nitrogen Liquefiers, Labyrinth Sealing, and Seal Gas

In nitrogen liquefaction, cryogenic expanders with oil-lubricated bearings are widely used. Since lubrication oil and cryogenic temperatures are not compatible items, it is important to prevent the oil from contaminating the working fluid. The standard proven technique is the labyrinth seal in combination with seal gas. The principle is shown in Figure 6.23.

1. The idea of the first measure – the labyrinth seal (also called a gap seal) – is to separate the space containing the working fluid (in the impeller) and the space containing oil (in the bearing) by means of a channel with correspondingly high flow resistance (e.g., a long, narrow channel or one designed to cause high turbulence). Thus, it is ensured that only a small amount of fluid (leakage mass flow) can escape through the labyrinth seal. The mechanical design depends on the manufacturer, but the principle is always the same as that shown in Figure 6.23. Absolute tightness is not possible with such seals. The decisive advantage here is that this type of seal is contactless and therefore free of wear.
2. The idea of the second measure – seal gas – is to introduce a non-contaminated fluid between the impeller side and the oil bearing at a pressure that ensures flow primarily in the direction of the oil bearing to prevent the oil from moving in the opposite direction, see Figure 6.24.

Figure 6.23 Labyrinth seal, principle.

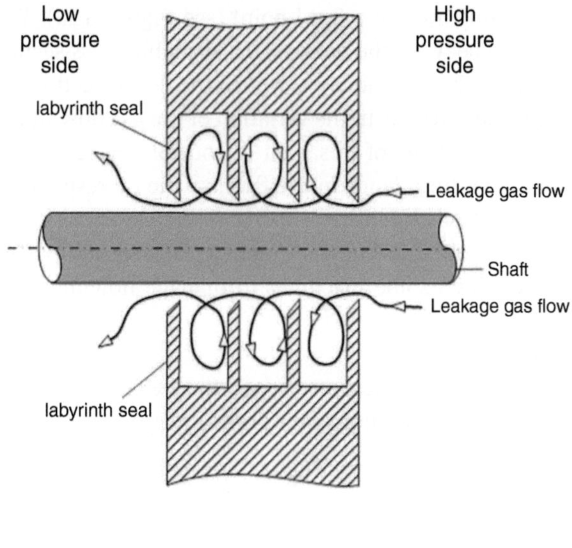

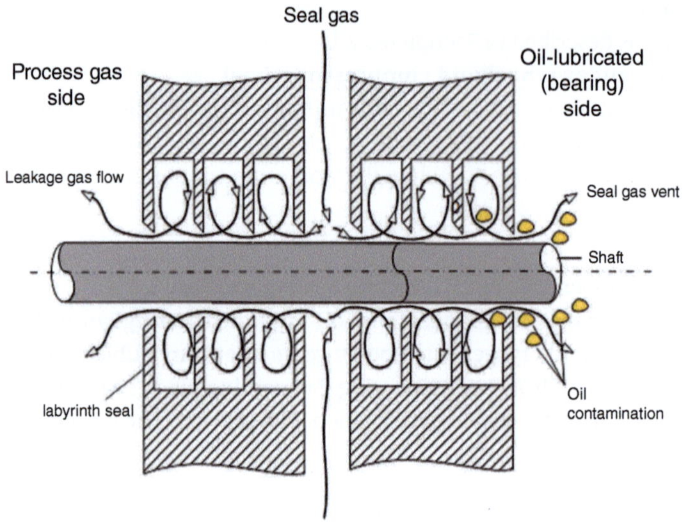

Figure 6.24 Combination of labyrinth seal and seal gas.

6.3.3.3 Control and Instrumentation

Since the efficiency of the Claude-like process depends on flowrates of both expander flows (warm expander and cold expander), fine tuning of the expander flow is required to achieve best performance. This can be done by a number of means – adjustable nozzles, outlet guide vanes, throttling at the inlet or outlet of the expander by corresponding valves/devices, or a combination of these methods.

In modern expanders for nitrogen liquefiers, adjustable inlet nozzles are used mainly for control. The adjustable nozzles are often referred to as inlet guide vanes (IGVs), which is the same term as that used for the control device of a centrifugal compressor; however, each is a distinctly different mechanical device. This is an integrated mechanism that allows smooth and fine adjustment of the inlet nozzles by rotating around a fixed point (see Figure 6.25). That way, the geometry of the channel between the nozzles, the available cross-section at the nozzle outlets – and therefore the gas velocity as well as the flow direction – can be slightly changed. This adjustment results in the variation of the gas flowrate through the machine.

The advantage of this kind of control device is that it does not generate much additional turbulence in the flow, so the corresponding loss is minimized or even avoided. Consequently, the expander efficiency is not impacted much, and the range of maximum efficiency is widened. This second positive effect of wider operation range allows some machines to manage flowrates between 30% and 150% of the design case.

Instrumentation

Centrifugal expanders are very costly devices, and they are critical hardware components for the operation of the entire system. It is therefore of paramount importance

Figure 6.25 Adjustable inlet guide vanes without expander wheel.

that damage be avoided and that critical situations with damage potential be prevented. Therefore, every expander is equipped with extensive instrumentation and auxiliary devices, the most important of which are listed below:

- Speed monitoring and overspeed protection,
- Vibration monitoring
- Axial thrust measurement
- Emergency quick-shutoff valves
- Pressure and temperature control
- Temperature in bearings and in the auxiliary oil system
- Sealing gas supply and pressure

6.3.4 Helium Expander

For helium expanders, the conventional engineering approach of applying the hardware components available on the market does not work, since no appropriate expanders are available. Therefore, OEMs have always been forced to develop a corresponding expansion machine of their own design. This means that access to helium expander technology is a de facto prerequisite to entering the helium liquefier market.

The first commercial reciprocating devices were developed in the 1940s and 1950s. The first centrifugal expanders were introduced in the 1960s and 1970s and were very successful. Recently, centrifugal machines have dominated the market. This is

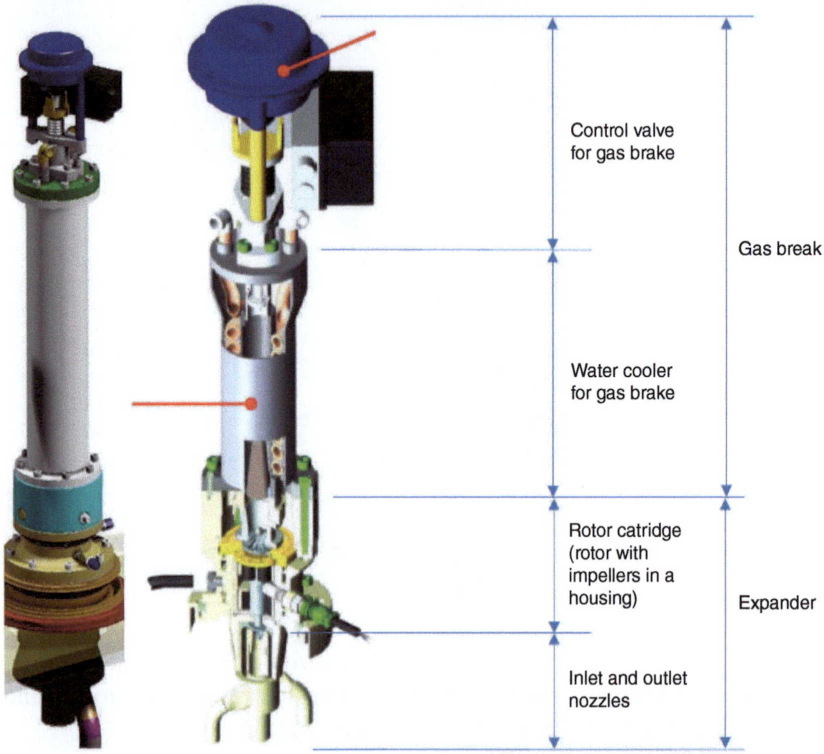

Figure 6.26 Helium expander, overall design. Source: Copyright Linde.

a unique technology, and a considerable portion of helium liquefaction know-how relates to the expansion turbine, since other hardware components (compressors and heat exchangers) are more or less available on the free market. The good book about the design of helium expander is Ref. [11].

6.3.4.1 Mechanical Design

Figure 6.26 shows an example of a helium expander. The device consists of two main sub-units –an actual expander and a gas brake. In terms of size, the gas brake is the larger unit; it consists mainly of a heat exchanger (cooler) for cooling of compressed brake gas by means of cooling water and a control valve on the top of this device. The actual expander is located in the lower part and consists primarily of a cartridge with a rotor and bearings inside a connection piping (inlet and outlet fittings) for gas.

The working principle of centrifugal expanders for helium is similar to the nitrogen expander described in Section 6.3.3: the gas enters the expander part through a lateral connection fitting, then is accelerated in the inlet nozzles, expanded in the expander wheel, and finally leaves the machine axially through the central connection fitting.

Another similarity is that the power produced by the expander is used to drive a small single-stage centrifugal compressor. The corresponding compressor impeller is located at the other end of the rotor.

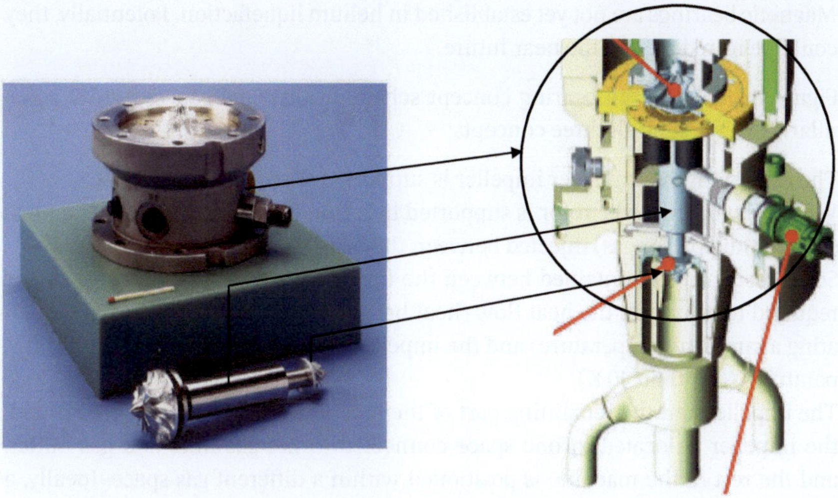

Figure 6.27 Rotor cartridge of a helium expander, expander wheel on the right side of the rotor, up to 9000 revolutions per second. Source: Copyright Linde.

But the mechanical design of the overall device is slightly different. These differences relate to:

- Size: expanders for helium applications are very compact devices; they are by one order of magnitude smaller than the wheels for nitrogen expansion. Figure 6.27 shows an example of an expander cartridge for a laboratory helium liquefier: the impeller (the smaller wheel) diameter is less than 20 mm.
- Rotational speed: as mentioned above, the diameter of the helium expander is relatively small. To achieve a gas velocity of 200–300 m/s, which is typically required for the conversion of a pressure drop into kinetic energy, an extremely high rotation speed is required and may achieve 5000–9000 revolutions per second.
- Arrangement of the rotor: In the case of a helium expander, the rotor shaft is arranged vertically, whereas the rotor of a nitrogen expander is arranged horizontally.
- To maintain this high rotational speed, a special design for bearings is required, which is discussed in Section 6.3.4.2.

6.3.4.2 Bearing System

Virtually, all three bearing systems described in Section 6.3.1.3 (i.e., gas-lubricated bearings, oil-lubricated bearings, and magnetic bearings) are applicable. In terms of their historical development,

- oil bearings were introduced first,
- but the state-of-the art is now that gas-lubricated bearings, especially static gas-lubricated bearings have become an established technology, but the newer variant – dynamic gas-lubricated bearings – is a relatively recent development, introduced during the last decade.

– Magnetic bearings are not yet established in helium liquefaction. Potentially, they could be introduced in the near future.

Figure 6.28 shows the bearing concept schematically. One can recognize some similarities shared by all three concepts:

- The rotor with the expander impeller is supported at two or more points.
- Within the bearing, the rotor is supported by a thin layer of rapidly moving pressurized fluid (oil or gas) injected between the bearing surfaces.
- Some distance is maintained between the impeller and the first bearing. This is required to minimize the heat flow (heat bridge) from the bearing (usually operating at ambient temperature) and the impeller, which operates at cryogenic temperatures (less than 40 K).
- The impeller and the remaining part of the machine are also spatially separated: the impeller is located in one space connected to the gas inlet and gas outlet, and the rest of the machine is positioned within a different gas space. Ideally, a gas connection between these two spaces must be avoided. However, in real-world designs, this connection is always present due to the thin gap around the impeller shaft. This gas connection is a challenging issue, since the gas from bearing space can become contaminated (e.g., by the oil in oil-lubricated bearings). To avoid propagation of such contaminants inside the device (and liquefier), the pressure inside the bearing space is held lower than the pressure in the expander wheel space. Therefore, a small fraction of helium gas flows continuously from one space into the other. The drawback is obvious: this is a kind of built-in permanent leakage, which exacts a corresponding penalty in terms of power consumption. On the other hand, this arrangement prevents contamination of the working fluid, which is an invaluable feature.

Several differences between the concepts are also evident in Figure 6.28. For example, the oil-lubricated rotor is mounted with the expander wheel located at the top, whereas both gas-lubricated versions use the opposite set-up.

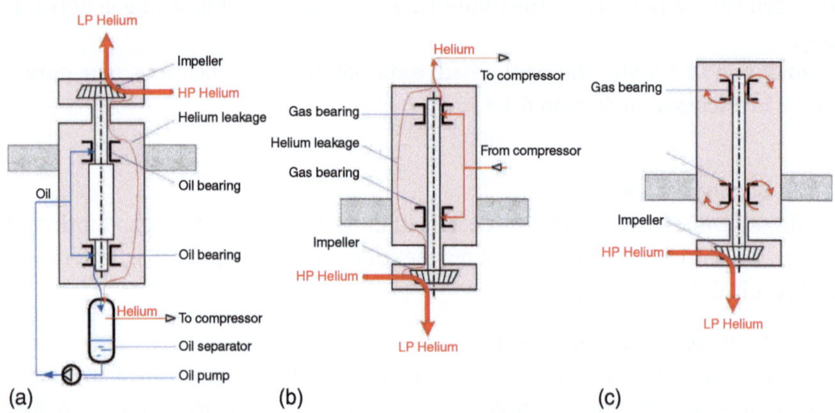

Figure 6.28 Bearing concepts for helium expander (a) oil bearings, (b) static gas bearings, (c) dynamic gas bearings.

In an oil-lubricated bearing, the oil is pressurized by means of a separate oil pump. It is injected at the bearing points, providing the required supporting function. When it collects at the bottom of the device, this oil is removed and returned to the oil pump through an oil separator, where the helium gas (the intended leakage shown in Figure 6.28a) is separated and returned to the main compressor inlet. The main drawbacks of oil-lubricated bearings are:

(a) additional expenditure for the oil-management system – oil pump, oil separator, oil inventory system, and potentially a redundant oil pump, and
(b) additional operational expenditure due to
   - maintenance efforts for the oil system and
   - power required for re-compression of the gas leakage flow (in the main compressor).

In a static gas-lubricated bearing, the compressed helium gas from the main compressor is used as the lubrication fluid. It is injected at the bearing points and returned to the main compressor. An oil system and its related disadvantages are eliminated completely. The main drawback is the power required for compression of the helium gas used for lubrication purposes.

A dynamic gas-lubricated bearing does not require any external supply of pressurized gas at all. The lubrication gas layer is generated automatically at the bearing points by means of a special integrated device when the rotor rotates. The details of this device are not publicly available, as this is a trade secret. Further know-how relates to maintaining bearing function when the rotor is not rotating slowly or not at all. Under such circumstances, the lubrication gas layer might not be sufficiently stable to guarantee proper lubrication; therefore, several additional measures are required for this situation.

For all fluid-lubricated bearings, a particularly challenging issue is how to ensure a "soft landing" of the expander if the bearing fluid supply is ever interrupted due to a failure in the system. A comparison of bearing systems is given in Table 6.7.

Table 6.7 Comparison of bearing systems for cryogenic expanders (extension of Table 6.4).

	Advantages	Disadvantages
Oil-lubricated bearings	• Proven concept, many references	• Oil is not compatible with cryogenic temperatures • Power required for operation of oil system • Power required for re-compression of seal gas
Gas-lubricated bearings	• Compact design • Low loss / high efficiency	• For static design: power required for re-compression of lubrication gas

6.3.4.3 Utilization of Expander Power

In helium liquefaction, small-size expanders are applied, and the required power is usually less than 0.1 MW. This amount is by one order of magnitude lower than the power required by typical expanders for nitrogen liquefaction.

In helium liquefaction, the gas brake is used for the utilization of mechanical power generated during expansion (see Section 5.2.6). Figure 6.26 (upper part) shows an example of a gas brake device. This is principally a closed volume filled with helium, as shown in Figure 6.29. Within this space, an integrated closed-cycle process is implemented:

(a) In the first step, the gas is compressed by means of a single-stage compressor; the temperature of the gas is increased.
(b) In the second step, the gas is throttled in a brake valve.
(c) In the final step, this hot gas is returned to the compressor inlet, where it is routed through a heat exchanger, where it is cooled by means of cooling water (the heat is transferred to the cooling water).

This means the generated mechanical power is converted into heat, which is dissipated into the environment or, in other words, simply destroyed (or lost). From an

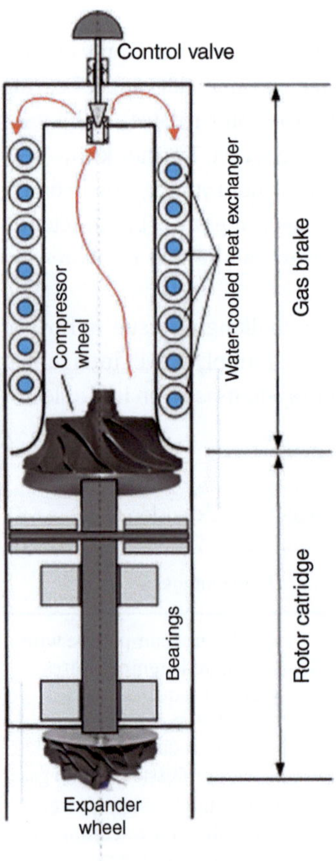

Figure 6.29 Principle of gas brake.

energy efficiency perspective, this method does not seem to be optimal. The reasons for this decision are described in Section 5.2.5.

6.3.4.4 Control and Instrumentation

The main control strategy and methods are described in Section 6.3.3.3 for nitrogen expanders. Basically, the same logic applies to control of the helium expander. But several methods, such as the adjustable inlet nozzles commonly used in nitrogen liquefaction, are not employed in helium expanders. The reliability requirements for helium liquefaction are higher than for nitrogen liquefaction. This device is considered critical since it may affect the reliability of the entire device. Therefore, this control method is avoided. The common method is throttling of the gas flow at the inlet (or outlet) by a corresponding valve. This method is not efficient, but it works well and is reliable.

The level of instrumentation required is not as high as for nitrogen expanders. Rotation speed monitoring and overspeed protection are mandatory, but the number of other instrumentation devices (vibration control, pressure, and temperature in bearings) is reduced to a minimum.

6.3.4.5 Assessment of Helium Expander

Due to its simple design, the gas-lubricated expander is a very reliable, robust, and durable device. It normally requires no maintenance as long as the working fluid is not contaminated.

The efficiency of this device depends on the bearing system used. Expanders with oil-lubricated bearings have a moderate isentropic efficiency of less than 75%. The isentropic efficiency of expanders with static gas-lubricated bearings is somewhat better but still lower than 80%. Only machines with dynamic gas-lubricated bearings can achieve efficiency slightly greater than 80%.

It is difficult to make a statement about the cost since this device is practically unavailable on the open market. Since this machine is pre-engineered and does not require much material, the cost is primarily driven by manufacturing expenses. This cost component depends on access to appropriate economical machining technology and tools. A reliable prediction is therefore difficult to make.

6.3.5 Joule–Thomson Expander

In modern large-scale cryogenic plants (mainly air separation plants with integrated nitrogen liquefiers but with stand-alone nitrogen liquefiers as well), a special expansion machine – the Joule–Thomson expander – is used. The device name reflects the main application of this device, as it is used for expansion of the Joule–Thomson stream.

The classical isenthalpic Joule–Thomson expansion is not directly replaced by an expander but by a combination of non-isenthalpic expansion in an expander and isenthalpic expansion in a valve, as shown in Figure 6.30b. Correspondingly, the expansion occurs in two steps: the cold supercritical fluid is expanded from

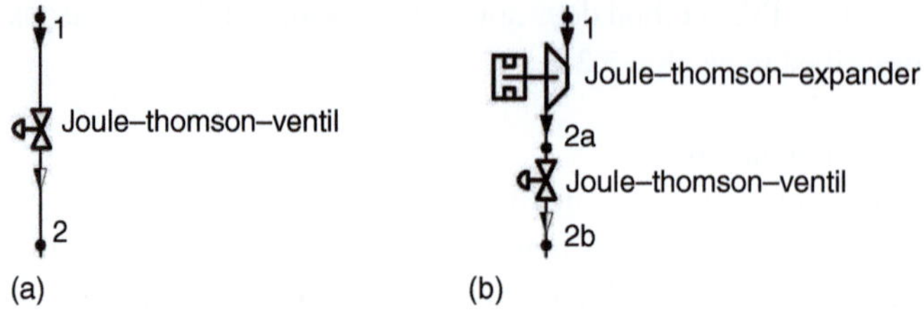

Figure 6.30 Expansion of supercritical liquid by means of (a) expansion valve, (b) combination of centrifugal expander and expansion valve.

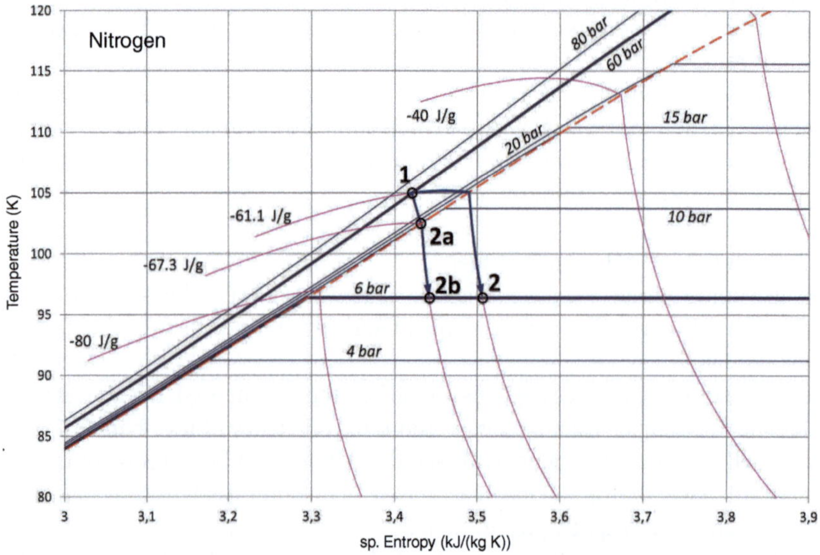

Figure 6.31 T,s diagram of expansion in a Joule–Thomson expander: the dashed line denotes the saturation line.

high pressure to intermediate pressure in the Joule–Thomson expander before it is expanded further to low pressure in an expansion valve. The main criteria for the selection of the intermediate pressure level is that phase change must not occur during the expansion in the expander, so that the fluid at the outlet of the Joule–Thomson expander remains liquid.[3] Flash gas is generated only during the expansion of the valve.

Figure 6.31 shows an example of an expansion process in a T's diagram. Here the state of the nitrogen gas at the expander inlet is marked as point 1 (60 bar, 105 K, h = −61.1 J/g), the state at the expander outlet is marked as 2a (10 bar, 102.5 K, h = −67.3 J/g), and the state at the outlet of the expansion valve is marked as 2b (6 bar, h = −67.3 J/g). For comparison, the conventional isenthalpic expansion

3 This topic is discussed in Section 6.3.2.2.

process starting from the same initial state 1 is also shown in Figure 6.31. This process ends at the point marked as state 2 (6 bar, $h = -61.1$ J/g).

In this diagram, the advantage of the expander-based process can easily be discerned in the form of a minor entropy increase of 0.02 J/(g K). For comparison, classical isenthalpic expansion yields an entropy increase of 0.09 J/(g K). In the end, this miniscule entropy increase results in a higher liquid fraction (i.e., a lower vapor fraction and correspondingly less flash gas) following expansion. In terms of process design:

1. The most obvious effect is that for the same liquid production, the flowrate of the Joule–Thomson stream can be reduced, and some energy savings can be achieved in compression of the Joule–Thomson stream.
2. In the case of the Claude process, the positive effect is amplified further: Since the flowrate of the Joule–Thomson stream is reduced, less cooling power is needed to cool this material stream. Therefore, the requirements for cooling capacity produced by both Brayton stages are reduced as well. Correspondingly, the flowrates of both the other compressed streams – the warm expander stream and the cold expander stream – can be reduced, and further energy savings for compression of both streams can be realized. Since the flowrates of both expander streams are considerably higher (by a factor of 2 to 5) than the flowrate of the Joule–Thomson stream, these additional power savings are significant.

In the end, implementation of a Joule–Thomson expander improves the efficiency of the overall process by 4–5% (depending on process design and parameters), which represents a considerable improvement.

6.3.5.1 Capacity

In terms of compressibility, the liquid nitrogen behaves like water, i.e. it is practically incompressible. Therefore, the pressurization of liquid nitrogen (by means of a pump) from an intermediate pressure level to high pressure does not require much power.

In principle, the expansion of liquid nitrogen from a typical high pressure $p_{HP} = 60 \pm 20$ bar to a typical intermediate pressure of $p_{IP} = 15 \pm 5$ bar without phase change can be considered to be a reversed pumping process. The power generated during this kind of expansion is relatively small; it is usually only 50–300 kW. This is considerably less (by one order of magnitude) than power generated in conventional gas expanders, as described in Section 6.3.3.

6.3.5.2 Techno-economics

The advantage in efficiency and consequently in operational expenditures, as mentioned above, is achieved at the cost of higher capital expenditures, since the Joule–Thomson expander is a complex and therefore costly device. This is the typical trade-off. Generally, for a decision "pro/contra Joule-Thomson expander," the additional expenditures for the expander must be compared with the resulting savings in the power bill for a defined payback duration.

Generally, from an economic perspective, if the local power price is high, the implementation of a Joule–Thomson expander for medium- and large-scale plants always makes sense. If power is cheap and the cryogenic plant is small, the classical Joule–Thomson expansion through a valve is probably the better solution.

6.3.5.3 Mechanical Design

The mechanical design of the Joule–Thomson expander is similar to the gas expanders described in the previous Section 6.3.3. The difference is that an electric generator is used as a loading device. Since conventional electric generators are not designed for the high rotational speed of typical expanders, a mechanical transmission (gear box) is introduced between the expander shaft and the generator.

6.4 Expansion Valve

The mechanical design of a cryogenic valve is described already in Section 5.1.

6.5 Compressors

The compressor is the most cost-intensive component of a cryogenic plant. The cost of the compressor (including its motor) amounts to 1/5–1/3 of the total capital expenditures. In terms of operational expenditures, the compressor is even more prominent, since the power consumption of a liquefaction or refrigeration plant is determined mainly by the power consumption of the compressor, and the largest share of maintenance and service expenditures relates to the maintenance and servicing of the compressor. All these facts reflect the extraordinary importance of the compressor for the overall performance and techno-economics of the plant.

The operation and application ranges for the three most important compressor types – reciprocating, screw, and centrifugal (sometimes referred to as radial or turbo compressor) – are shown in Figure 6.32. In the past, reciprocating machines were commonly used due to their availability. Later, screw and centrifugal compressors were developed and introduced.

Applications

Modern large industrial nitrogen or natural gas liquefaction plants are based on centrifugal machines because of their capability to compress large gas flows in a single unit and thanks to their better techno-economics. Unfortunately, centrifugal compression is not applicable to helium due to the low molecular weight involved, as described in Section 6.5.2.4.2. The screw compressor has become established in recent years as the standard main compressor device for helium liquefaction. Reciprocating machines are used mainly for hydrogen compression as well as for auxiliary functions, for example, for compression of helium to pressure levels in excess of 20 bar for storage purposes.

6.5.1 General Issues

This section examines several general issues regarding all types of compressors.

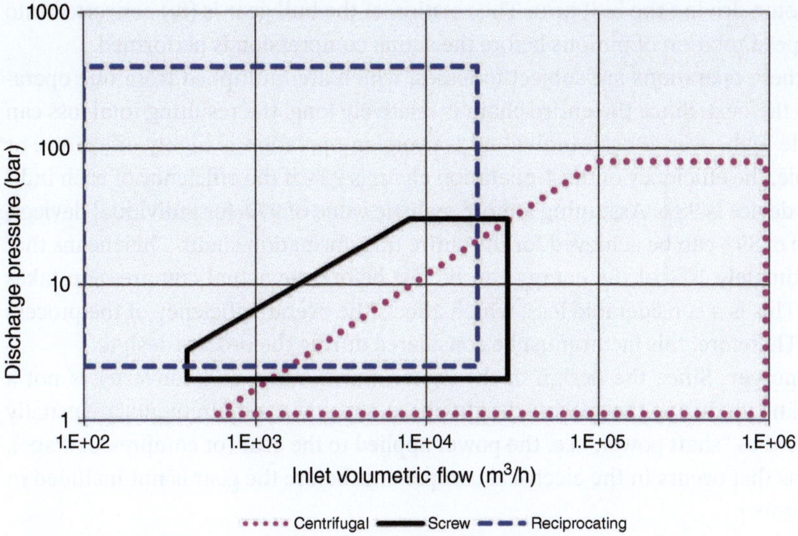

Figure 6.32 Operation ranges for three compressor types.

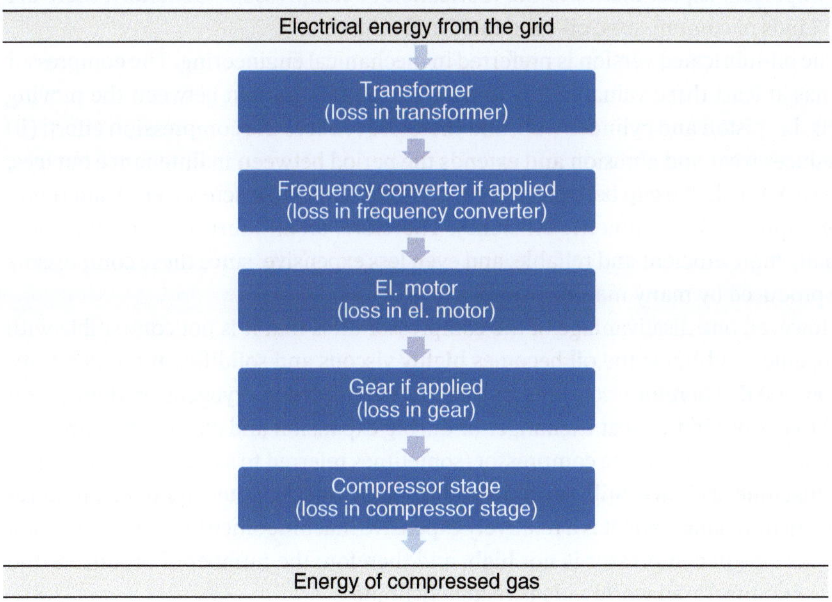

Figure 6.33 Energy conversion and transformation steps in a centrifugal compressor.

6.5.1.1 Energy Transformation Chain for Compression

The entire energy transformation chain, from the electrical grid to the energy of the compressed gas, is shown in Figure 6.33. The electrical power from the grid is converted (i) in a transformer to the proper voltage and (ii) in a converter to the right frequency. After that, the electrical energy is (iii) converted into mechanical energy in a motor, driving the bull gear. The rotation of the bull gear is (iv) converted into high-speed rotation of pinions before the actual compression is performed.

All these operations are subject to losses, which are multiplied from one operation to the next. Since the entire chain is relatively long, the resulting total loss can be quite high, even if each individual conversion operation is highly efficient. For example, the efficiency of the 4-operation chain is 92% if the efficiency of each individual device is 98%. Assuming a more realistic value of 97% for individual devices, a value of 89% can be achieved for the entire transformation chain. This means that approximately 10% of the energy can be lost before the actual compression takes place. This is a considerable loss, which affects the overall efficiency of the process plant. Therefore, this factor must be considered during the process design.

Shaft power: Since the design of the motor, transformer, and converter is not a priori known during the process design phase, the power requirements are usually expressed as "shaft power," i.e. the power applied to the gear (or compressor stage). The loss that occurs in the electrical components before the gear is not included in this number.

6.5.1.2 Compressor Lubrication

An important aspect relates to the lubrication of compressors. Generally, there are two kinds of compressors: oil-lubricated and oil-free.

The oil-lubricated version is preferred in mechanical engineering. The compressor oil has at least three valuable functions: (i) it reduces friction between the moving parts, i.e. piston and cylinder wall, and therefore reduces the compression effort; (ii) it reduces wear and abrasion and extends the period between maintenance outages; and (iii) it seals the gap between the piston and cylinder (or screws, etc.) and therefore improves the volumetric efficiency. Therefore, an oil-lubricated compressor is usually more efficient and reliable, and even less expensive, since these compressors are produced by many manufacturers.

However, one disadvantage of the compressor oil is that it is not compatible with cryogenic conditions: the oil becomes highly viscous and solidifies at temperatures below 200 K. Therefore, compressor oil must be avoided in cryogenic systems, for it will freeze out in the heat exchanger or during expansion and clog the system.

In this sense, an oil-free compressor (sometimes referred to as a dry compressor) is the machine of choice. Still, several disadvantages must be noted: for one, it requires more maintenance, and it is a relatively expensive machine since the market demand for this type of compressor is not high, and therefore the number of producers and units manufactured worldwide is relatively limited.

An alternative solution could be based on an oil-lubricated compressor in combination with an additional oil removal system (ORS)). Since the requirements for oil purity in cryogenics are high, the ORS consists of at least three filtration stages: (i) pre-filter device for removal of large oil droplets, (ii) fine oil filter (usually a filter-coalescer) for removal of fine, aerosol-like oil droplets, and (iii) adsorber for removal of vaporous oil. Therefore, this is a relatively complex unit. It bears an impact on the cost of the system and may need some maintenance, which means that additional operation expenditures are incurred.

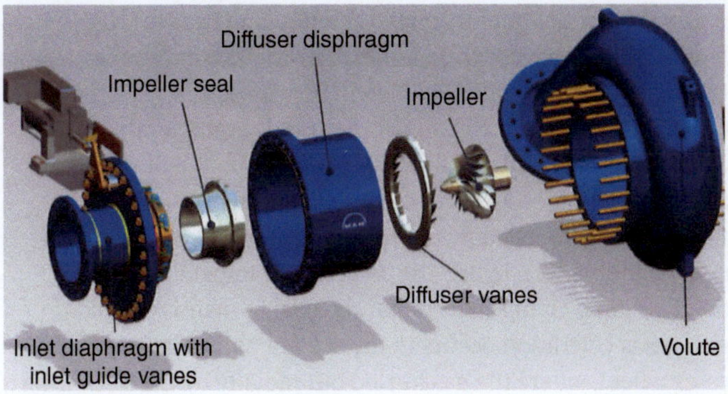

Figure 6.34 Stage of a radial compressor. Source: Copyright MAN Energy Solutions.

6.5.2 Centrifugal Compressor

Centrifugal compressors (i.e., turbocompressors) are the preferred compressor type for large-scale plants. This is mainly due to their better economic value, since these compressors are efficient, the maintenance effort is low, and a relatively large number of manufacturers and suppliers (and hence strong competition) results in a reasonable cost. Centrifugal radial (water mill-like) and axial (wind mill-like) compressors are the two main types of turbocompressors. The section is focused on radial machines. For a deep dive into this matter, Ref. [12] is recommended.

6.5.2.1 Mechanical Design and Principle of Compression in a Centrifugal Compressor Stage

Centrifugal radial compressors typically consist of one or more stages. In this section, a single compression stage is considered. Figure 6.34 shows main elements of a centrifugal compressor stage:

(1) inlet assembly with integrated IGVs,
(2) diffuser, which include diffuser diaphragm (tapering inlet channel) and diffuser vanes assembly at the end of the tapering channel,
(3) impeller (compressor wheel) and
(4) volute.

Figure 6.35 shows a separate view of two key elements of a compressor stage: an impeller (rotating element) and surrounding diffuser vanes assembly (stationary element). The diffuser vane is a device that acts in reverse to the inlet nozzles in the centrifugal expander described in the section regarding centrifugal expander 6.3.1.1. The main difference is that the expander nozzles of a large expander can be designed as an adjustable device, whereas the compressor diffuser vanes have a fixed geometry.

Generally, compression of a gas in a compressor is the reverse of the expansion of a gas in an expander. Therefore, principally, the core part of a compressor is very similar to the core part of the expander, but it works in reverse directions. The compression process in a single adiabatic compressor stage consists of four steps:

Step 1: The gas passes from the inlet through the inlet guide vanes (IGV is the device for flowrate control) and further to the compressor wheel through a channel with a narrowing cross section (unfortunately not recognizable in this figure). It is accelerated slightly there and provided axially and centrally into the impeller.

Step 2: The main acceleration occurs during the second step, where the gas is mechanically impelled by the rotating compressor wheel (i.e., the impeller blades). Inside the compressor impeller, the direction of gas flow changes from mainly axial to mainly radial; the largest gas acceleration occurs in the second half of the compressor wheel, where the gas hurled out radially by the rotating impeller blades. The maximum gas velocity is reached at the outer edge (the tip) of the compressor wheel. During this process, the mechanical energy of the wheel is converted into the kinetic energy of the gas ($\boldsymbol{P} \rightarrow \boldsymbol{E}_{kin} \sim \rho v^2/2$).

Step 3: From the impeller outlet, the gas passes through the diffuser (diffuser vanes), which is a circular chamber with a widening cross-section, following a spiral path. The main purpose of this device is to decelerate of the gas. The velocity head of the gas (the kinetic energy of the gas) gets converted to pressure head (potential energy) $\boldsymbol{E}_{kin} \sim \rho v^2/2 \rightarrow \Delta p$ as per Bernoulli's principle, see Eq. (6.1).

Step 4: In the final step, the gas is collected in the volute and passed to the intermediate cooler or the next compression stage. Further deceleration occurs during this processing step. Therefore, the pressure increases slightly further.

Figure 6.35 Impeller and diffuser of a radial compressor Source: EnggCyclopedia.

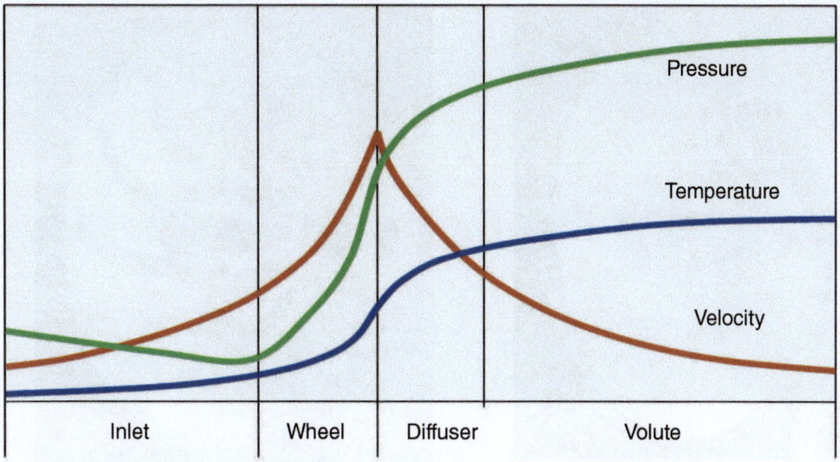

Figure 6.36 Pressure, temperature, and velocity profiles in an adiabatic compression stage of a centrifugal compressor.

The extent of pressure rise in a centrifugal compressor mainly depends on the rotational speed of the impeller and the impeller size. However, the maximum allowable speed is limited by the strength of the structural material of impeller blades and sonic velocity of the gas. As a result, limitations on maximum achievable pressure rise exist.

Figure 6.36 shows the principal distribution of pressure, temperature, and gas velocity within a compressor stage.

6.5.2.2 Impeller Design

The primary target is to realize reversible isentropic compression or get as close to isentropic process as possible. This target is hard to achieve; consequently, the real compression process deviates from the isentropic process. Usually, it is because of friction phenomena, whereby the mechanical energy is converted into heat and dissipated further into the environment.

There are two sorts of friction phenomena. The most obvious is the classical mechanical frictions, caused by relative lateral motion of two solid surfaces in contact (directly or with a layer of lubricant in-between). The second kind of friction is the fluid friction between layers of fluid that are moving relative to each other. In a gas flow, this friction becomes partially visible in the form of fluid viscosity. But the largest source of internal friction is associated with turbulence, where an eddy rotates faster than the surrounding fluid; it rubs on a layer of the surrounding fluid, and consequently, the kinetic energy of the eddy is converted/dissipated into heat.

Since turbulence occurs in all elements of the compressor but mainly in the impeller, the impeller geometry is extensively optimized. Modern manufacturing methods are so well developed that practically every imaginable mechanical design may be made with the required precision at a moderate cost. Therefore, it is theoretically possible to achieve an extremely isentropic efficiency for a single fixed

Closed impeller   Open impeller

Figure 6.37 Options for compressor wheel design. Source: S&D Precision Toolmakers.

set of given parameters. In reality, compressor wheels are designed not for a single operation point but for a range of pressures, temperatures, and flowrates. Therefore, the target is to achieve high efficiency over a relatively wide operating range.

The mechanical design of the impeller is a science in itself. It depends on many parameters, like pressure, flowrate, power requirements, fluid type, and special operating conditions. Some typical designs are shown in Figure 6.37, whereby the first line shows the wheels with closed wheels and the second line – with open wheels. In terms of efficiency, commonly, the closed wheel design is better than the open wheel.

6.5.2.3 Multi-staging

The recycle compressor is a multistage machine. Usually, it consists of 4–6 compressor stages and intercoolers in between. All these stages are driven by a single driver, which is usually an electric motor.

Integrally geared compressor

The most popular version is the so-called integrally geared compressor. Figure 6.38 shows the machine without intercoolers. An integrally geared compressor consists of 4, 6, or more compressor stages attached to both ends of high-speed gear pinions; for 4 compressor stages, two pinions are required; for 6 stages – three gear pinions. The pinions are mounted in a housing that contains a large low-speed bull gear, which drives the individual pinions mentioned above. The large central bull gear is driven by a motor.

The integrally geared design has some advantages. The most important one relates to the ability to convert a relatively low rotation speed of the motor/bull gear into a high rotation speed of the gear pinions. Even more, the rotational speed of pinions can be adapted individually by choosing the best gear ratio (number of teeth on the pinion). Usually, this degree of freedom gives a chance to optimize the machine and, consequently, achieve a better efficiency in the compression process.

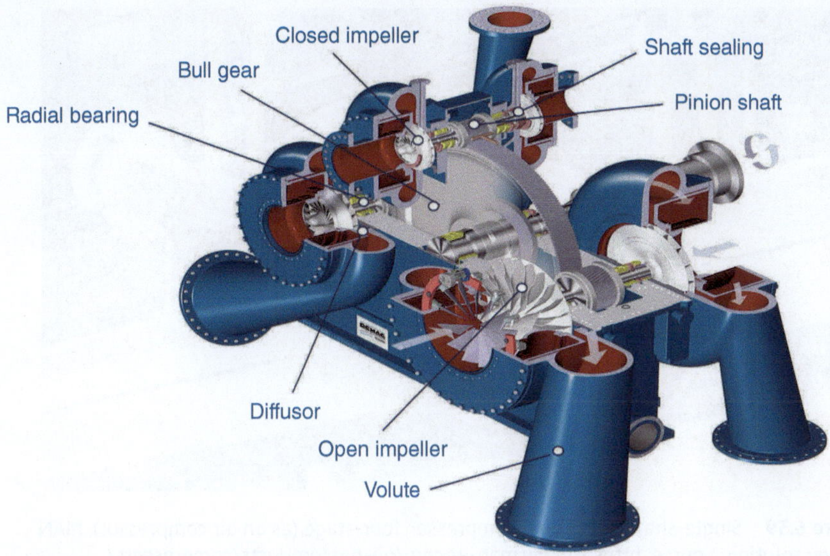

Figure 6.38 Integrally geared compressor, MAN Energy Solution.

The main disadvantage is that the transmission of mechanical energy from the motor to the pinions through the gear itself is an additional operation with corresponding internal loss caused by mechanical friction in the gear. Commonly, the gear efficiency amounts to 94–98%, depending on the mechanical design. That is a factor, which must be considered during the process design, since it affects the overall efficiency of the machine.

Another aspect is that the gear train itself is a relatively complex assembly with a number of auxiliary sub-systems (e.g. oil bearings) and instrumentation. It uses high-performance materials. This complexity becomes visible in the form of high cost.

<u>**Single shaft design**</u>

Another popular mechanical design is the so-called single-shaft compressor, whereas all compressor stages are mounted on a single shaft, as shown in Figure 6.39. They can be radial stages, as shown in this picture, or a combination of a single axial stage (as the first stage) and several radial stages. This design is used for large-scale systems with high volumetric flowrates.

The single-shaft design has the advantage of direct transmission of mechanical power from the motor to the wheels, the gear as a source of losses is eliminated completely. But, since the rotational speed of single wheels cannot be adapted individually, the optimization of the machine (targeting a better efficiency of the compression process) is a challenging task.

6.5.2.4 Thermodynamics of Centrifugal Compressors

The thermodynamics of the compressor as a process unit are described in Section 5.5. This section describes the specifics of centrifugal machines as well as their

Figure 6.39 Single-shaft centrifugal compressor, four-stage (as an air compressor), MAN Energy solution. Source: https://www.man-es.com/oil-gas/products/compressors/centrifugal.

typical restrictions and limitations. Especially, this section outlines the thermodynamic challenges related to the compression of light gases in centrifugal compressors.

6.5.2.4.1 Pressure Ratio in a Single Stage

From the process design perspective, it is salient to note that the pressure ratio in a single compression stage is limited. The main drawback is the limited mechanical strength of materials used for the compressor wheel. The stress caused by centrifugal forces at high speed may damage the wheel. Therefore, the velocity and the total kinetic energy managed within a single stage, and consequently the pressure ratio produced there, are capped.

The thermodynamics behind this: The correlation between the maximum possible pressure increase and the impeller speed can be established, assuming that the mechanical power (one stage, isentropic compression)

$$P = \frac{1}{\eta_s} \dot{N} \cdot R_m \cdot T_{in} \cdot \frac{\kappa}{\kappa - 1} \left[\left(\frac{p_{out}}{p_{in}} \right)^{\frac{\kappa-1}{\kappa}} - 1 \right]$$

is converted into kinetic energy/power of the gas (in a first step) with an efficiency EFF[4]:

$$P_{kin} = \frac{\dot{m} \cdot v_{tip}^2}{2} \qquad \dot{m} = \dot{N} \cdot \frac{M_r}{1000}$$

where v_{tip} – is the gas velocity at the impeller outlet.

4 The factor $M_r/1000$ is required for conversion into kg/s of mol/s, where M_r – molecular weight.

The maximum achievable pressure ratio [in one stage] can then be calculated as

$$\pi = \left[1 + EFF \cdot \eta_s \cdot \frac{\kappa - 1}{\kappa} \cdot \frac{M_r \cdot v_{tip}^2}{2000 \cdot RT_{in}}\right]^{\frac{\kappa}{\kappa-1}} \tag{6.3}$$

6.5.2.4.2 Pressure Ratio for Light Gases

According to Eq. (6.3), the pressure ratio in one stage depends on the

1. isentropic exponent,
2. impeller speed and
3. molecular weight of the used gas.

Figure 6.40 illustrates this correlation for typical cryogenic gases. For light materials (gases with low molecular weight) such as helium and hydrogen, the achievable

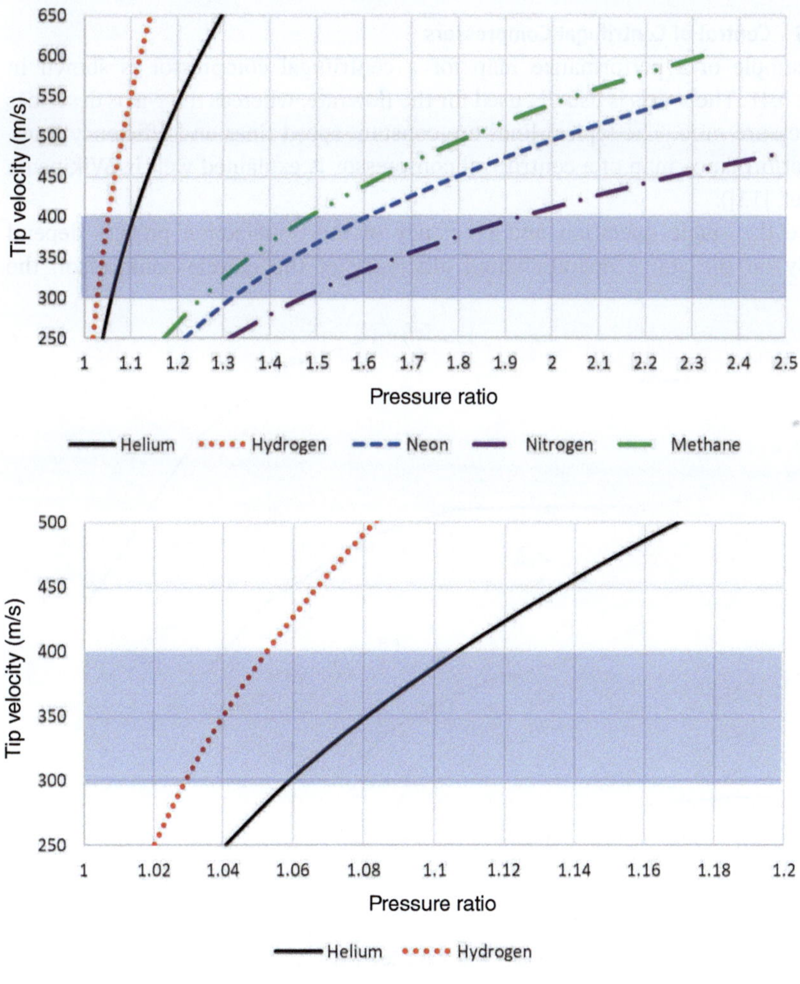

Figure 6.40 Pressure ratio in a single compression stage depending on impeller speed, EFF = 95%, $\eta_s = 85\%$, $T_{in} = 300$ K.

pressure ratio per stage is very small: 1.06–1.08 at today's usual impeller speeds of 300–350 m/s. Therefore, many isentropic stages are required to achieve a total pressure ratio of over 6 (which is required, for example, for helium liquefaction).

To our consternation, the multistage compressor includes a corresponding number of intercoolers placed between the single compressor stages. These intercoolers, then, cause a pressure loss. Therefore, the effective pressure ratio for a combination "compression stage + intercooler" is lower than that proposed by Eq. (6.3). Therefore, the real number of required stages exceeds what we originally anticipated. Ultimately, an appropriate turbo helium compressor (for a total pressure ratio of over 6) would be a very complicated and capital-intensive machine because of the great number of compressor stages.

For comparison, a pressure ratio of 1.6–1.9 for nitrogen can be easily achieved in a single centrifugal stage.

6.5.2.5 Control of Centrifugal Compressors

An example of a performance map for a centrifugal compressor is shown in Figure 6.41. The x-axis is usually used for the flowrate, whereas the y-axis describes the pressure ratio. The typical lines are constant-speed lines and efficiency lines. The performance map of a centrifugal compressor is explained well in Wikipedia (see Ref. [13]).

Since the stable operation and efficiency of the liquefaction process depend directly on the stable operation and efficiency of the recycle compressor, the

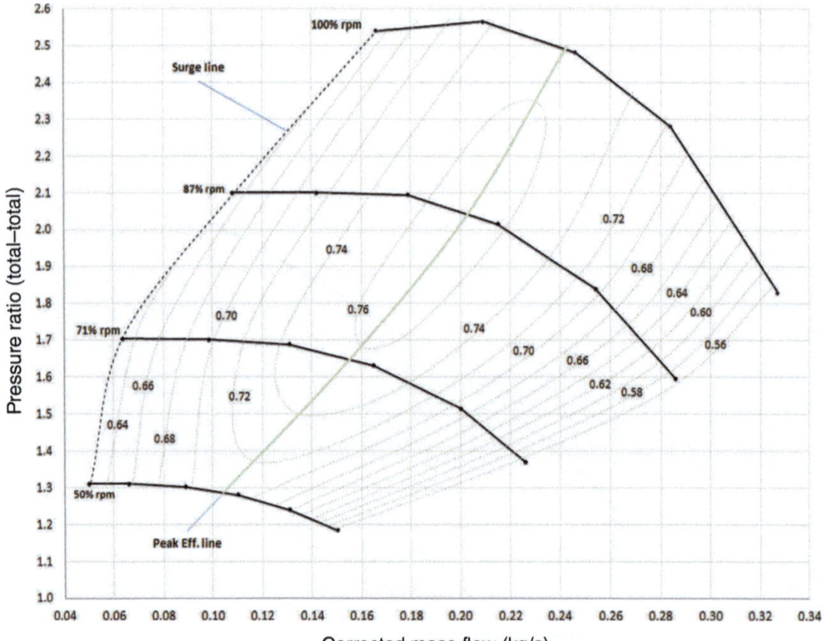

Figure 6.41 Performance map of a centrifugal compressor. Source: Mkoronowski/Wikipedia/Public Domain.

compressor flow and discharge pressure must be controlled and tuned in order to achieve the optimum performance of the process. There are a number of means available for this:

- inlet or outlet guide vanes,
- throttling at the inlet or at the outlet of the compressor via corresponding throttling valves,
- bypass valve,
- variable-frequency motor as driver

as well as various combinations of these.

Bypassing, Throttling

In the past, bypassing or throttling of the compressed fluid (at the inlet or outlet of the compressor) was often used for control purposes, mainly because of its simplicity. In terms of efficiency, such methods are disadvantageous since the mechanical energy is simply dissipated in the corresponding valves. This method is still used for cost-optimized small-scale plants. However, in energy-optimized large- and medium-sized plants this loss is inacceptable.

Inlet guide vanes (IGV)

With modern nitrogen liquefiers, adjustable IGVs are commonly used for control purposes. The IGV is a mechanism integrated in the inlet channel of the compressor. An example is shown in Figure 6.34. It consists of blades that guide the gas into the eye of the impeller. Figure 6.42 illustrates how this device works: by changing the blade inclination angle, the rotation of the inlet gas flow can be regulated, which contributes (according to the degree of inclination) to the impeller rotation. This adjustment results in the variation of the gas flowrate through the machine.

The advantage of this control device is that it does not generate much additional turbulence in the flow, and the corresponding friction loss is minimized or even avoided altogether. Consequently, the compressor efficiency is not impacted by much. The range of maximum efficiency has been widened. The second positive

Figure 6.42 Inlet guide vanes, left: closed, right: open. Source: Matthew Oswald, Inlet Guide Vane Animation in www.youtube.com/watch?v=hDdd1hbbArg.

effect is that the operation range itself is wider. The common requirement for the load range of a conventional centrifugal compressor lies between 70% and 100%. With a well-sized and well-controlled IGV, some machines are able to manage flowrates between 60% and 110% of the design case.

Variable-frequency driver

Another control method is based on the application of a variable-frequency electric motor as a driver. Obviously, the rotation speed of this device is controlled by frequency. Therefore, a corresponding frequency converter is an important element of this mechanism.

Generally, it is a very efficient method that allows precise control and stable operation of the compressor. An additional advantage relates to its soft-start capability – the driver can gradually ramp the motor up to operating speed to lessen mechanical and electrical stress, reduce maintenance and repair costs, and extend the life of the motor and the driven equipment. In contrast, a conventional electric motor subjects the motor to a high starting torque and to current surges that are up to eight times the full-load current, with all corresponding drawbacks.

The main disadvantages of the variable frequency driver are the following:

(1) Additional cost for the frequency converter, whereby
   - in the low power range, these devices are cost-efficient because they are produced in bulk due to high demand, resulting in lower prices;
   - in the high power range (megawatts), the demand is quite limited, and therefore the corresponding large-scale devices are designed and manufactured practically individually for each individual motor (customer). This results in higher specific cost.

(2) Efficiency loss: the transmission of electrical energy from the transformer or grid to the motor through the frequency converter is an additional process operation with corresponding energy transformation loss. The efficiency depends on technology, motor type, and size; commonly, it is higher than 95%.

Surge phenomenon

Operation engineers have respect for the so-called surge phenomenon, which relates to operation of centrifugal compressors. It occurs at low flowrates, if the impeller cannot add enough energy to overcome the system flow resistance: the high pressure downstream of the impeller hinders the gas from flowing in the right direction, and so, for a very short time, it pushes the gas in reverse direction, i.e. back over the tips of the rotor blades towards the impeller eye (inlet); as a result, the backpressure is slightly reduced and the impeller can push the gas in the proper direction again.

These cyclic events cause high vibration, temperature increases, and rapid changes in axial thrust. These phenomena can damage the rotor seals, the rotor bearings, and the compressor driver, which can degrade the cycle operation. Turbomachines are designed to withstand occasional surging, but if the machine is forced to surge repeatedly over an extended period of time, such repeated surges can result in catastrophic failure.

In the performance map, the operation area where the surging phenomenon may occur is marked by a line referred to as the surge line (see Figure 6.41). Compressor operation near this line must be avoided, and compressor operation to the left of this line is not recommended at all.

6.5.2.6 Driver for Compressor

Electric motor: The centrifugal compressors are the preferred option for medium-scale and large-scale applications. If the required power is lower than 25 MW, the conventional electric motor is commonly used as a driver. The background is that electric motors for this power range are usually pre-engineered; they are produced in high numbers in framework of an optimized standardized manufacturing process. Consequently, it is a cost-efficient option. Unfortunately, for higher power (above 25 MW), the standardized electric motors are not available. They are engineered and manufactured individually for specific customer requirements, which leads to higher specific costs.

Steam turbine: An alternative driver technology in this power class is a steam turbine. This device has the same advantages as a medium-scale electric motor mentioned above: the unit is usually pre-engineered, and sub-suppliers are pre-qualified, manufacturing process is well established. Therefore, in terms of cost and efficiency, this option is very competitive, even if steam turbine units are not produced in large quantities. In terms of operation, steam turbines can easily be speed matched to the compressor, which is able to operate over a relatively wide speed range. The decision pro electric motor or pro steam turbine depends often on infrastructure available on site (electricity or steam supply, cooling water access) or other local conditions.

In natural gas liquefaction, a **gas turbine** (aeroderivative gas turbine or industrial gas turbine depending on power required) may be applied as driver instead of a steam turbine, using the advantage of availability of natural gas [as a fuel] on-site. Such a gas turbine-driven LNG plant does not require much electric power for operation. This fact means a high degree of freedom in terms of autonomy and choice of geographical location – virtually, such a plant can be built anywhere, even in places without existing electrical grid. The power-to-weight ratio is superior. The package concept minimizes the need for additional/support equipment. Therefore, the gas turbine is the preferred option for use in remote locations and offshore applications. Disadvantageously, the gas turbine is an expensive option, and the maintenance and service costs are relatively high.

Other options are combination of devices described above, like an electric motor driven by power generated by gas turbine, a steam turbine, or combination of both.

In terms of mechanical design, an electric motor or a steam turbine have the advantage that the shaft is available on both sides of the device. Therefore, these drivers can propel two compressors simultaneously – one on the left side and one on the right side of the driver. For gas turbine, this option is principally not available since the hot flue gas is released on one side there.

6.5.3 Screw Compressor

Screw compressors are the preferred compressor type for small- and medium-scale cryogenic plants, in air separation, nitrogen liquefaction, and even in small-scale

Figure 6.43 Helical screws of a screw compressor. Source: Ingersoll Rand.

Figure 6.44 Screw compressor, principle of compression. Source: unknown

natural gas liquefaction. But the most salient application area is helium-based cryogenics.

6.5.3.1 Screw Compressor, Design, and Working Principle

The rotary-screw compressor is well described on Wikipedia.[5] The following text section stems mainly from there but has been slightly adapted: "Rotary-screw compressors use two very closely meshing helical screws, known as rotors, to compress the gas. Modern screw compressors are made with different profiles on the two rotors: the male rotor has convex lobes that mesh with the concave cavities of the female rotor, as shown in Figure 6.43.

Usually, the male rotor has fewer lobes than the female rotor, so it rotates faster. Originally, screw compressors were made with symmetrical rotor cavity profiles, but modern versions use asymmetrical rotors, with the exact rotor designs being the subject of patents.

Working principle

The compression principle of a screw compressor is difficult to explain in words; it is always better to find and watch a video on the internet. Figure 6.44 attempts to explain this through illustrations. Gas enters on the suction side and moves through the threads as the screws rotate. The meshing rotors force the gas through the compressor, and the gas exits at the end of the screws, whereby the lobes of the female rotors functions like a piston pressing the gas into concave cavities of

---

5 https://en.wikipedia.org/wiki/Rotary-screw_compressor.

the female rotor. The working area is the inter-lobe volume between the male and female rotors, which is larger at the intake end and decreases along the length of the rotors to the exhaust port. This reduction in volume causes the pressure of the system to rise.

Generally, the compression does not take place absolutely continuously but rather in pulses, corresponding to the time frames when the inter-lobe volume between the male and female rotors opens to the discharge side. However, due to the high rotation speed of the screws and the multiple pulses per turn (depending on the number of cavities in the female rotor), the gas is compressed virtually continuously.

Oil lubrication

Generally, there are two types of screw compressors:

1. "dry" compressors (whereby the word "dry" means "without oil lubrication"),
2. oil-lubricated (or oil-flooded) compressors.

"In a dry-running rotary-screw compressor, timing gears ensure that the male and female rotors maintain precise alignment without contact, which would produce rapid wear. In an oil-flooded rotary-screw compressor, lubricating oil bridges the space between the rotors, both providing a hydraulic seal and transferring mechanical energy between the rotors, allowing one rotor to be entirely driven by the other."

With helium liquefaction, only oil-lubricated compressors are used. The oil serves at least three useful purposes, see Figure 6.45:

- Obviously, the first is the lubrication function itself. It reduces the friction and corresponding abrasion and wear of the screws and of the bearings. This is the main reason for the high reliability and long life of oil-lubricated machines; these

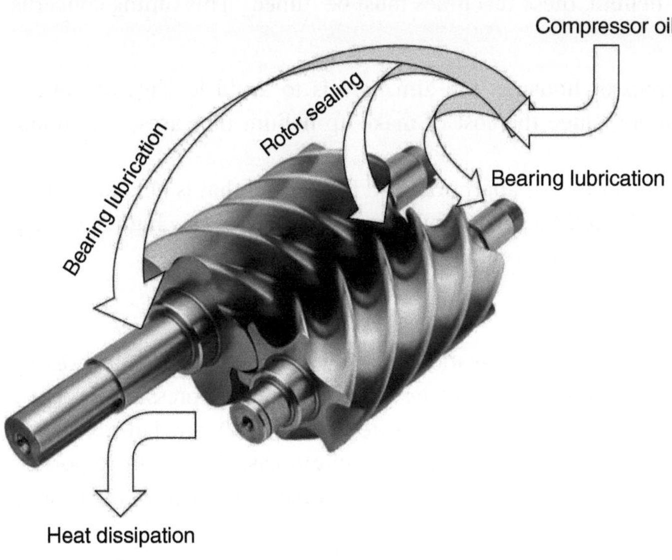

Figure 6.45 Screw compressor, functions of compressor oil.

machines are able to operate for years or even decades without any apparent performance decrease.

- Second is the sealing function, which ultimately affects efficiency: the effectiveness of the screw mechanism is dependent on precise fitting clearances between the helical rotors and between the rotors and the chamber for sealing of the compression cavities. The compressor oil provides hydraulic sealing and reduces internal leakage; hence, efficiency is improved.
- The third function relates to the transfer of heat from the compressor (cooling function). Some heat does develop within the compressor due to friction and inefficiencies in mechanisms, but it arises mainly due to the process of gas compression per se. Isentropic compression was examined in Section 5.3.1.1. Due to its high isentropic exponent (kappa), the compression of helium leads to extremely high temperatures (see Table 5.5) even at moderate pressure ratios, which presents a challenging issue. The oil-flooded machine is able to manage this problem: since the screws are flooded with oil, the compressor oil works as a kind of cooling fluid, which absorbs and removes compression heat. Consequently, the compression process in an oil-flooded compressor becomes non-adiabatic. It is still not isothermal, but the extremely high temperatures inside the device are avoided. The hot compressor oil is cooled by means of an additional oil cooling sub-system, which includes an oil-circulation pump, heat exchanger, and filter and is an integrated part of a screw compressor (see Figure 6.45).

6.5.3.2 Tuning of Available Screw Compressors

Conventional oil-flooded screw compressors were originally developed for applications in air compression, refrigeration, and air conditioning. They do not meet the exact requirements on machines for cryogenics, particularly for helium liquefaction. For compression of helium, these machines must be "tuned." This tuning concerns mainly

- tightness of compressor housing: the aim here is to avoid leakage of helium into the environment, since the cost of make-up helium increases operational expenditures,
- selection of compressor oil: an appropriate oil is required that is able to guarantee tribology, viscosity, and thermal stability even at the relatively high operating temperatures typical for helium compression.

6.5.3.3 Process Parameters

An important criterion for the application of screw compressors is that existing air screw compressors are developed and designed for a suction pressure level of 1 bar (ambient pressure) and a discharge pressure level of 8–10 bar. This discharge pressure is higher than the critical pressure of helium gas. That is an important enabler, since it allows a liquefaction process to be designed based on the Claude process.

The screw compressors developed originally for conventional refrigeration may work with higher inlet pressure and higher outlet pressure level. That is important for the design of Brayton-type processes, since it gives more degrees of freedom for optimization of the cycle.

6.5.3.4 Oil Removal System (ORS)

Since the compressed fluid at the outlet of the compressor is contaminated with compressor oil, efficient oil separation is required. This includes at least the following components:

(a) oil separator,
(b) rough filter-coalescer for filtering of large oil droplets,
(c) fine filter-coalescer for filtering of fine oil droplets,
(d) adsorber for removal of vaporous oil components

In some cases, an additional super-fine coalescer is integrated to filter out the finest aerosol-like oil droplets. The ORS is a reliable process unit if sized and maintained properly. In normal operation, it does not require external power.

6.5.3.5 Compressor Control

Reliable control of the compressor is a prerequisite for stable operation and efficiency of the liquefaction process. The most important parameters to be controlled are the flowrate and discharge pressure at a given inlet pressure. There are a number of various means available for this. Some of those, described in Section 6.5.2.5 (control of centrifugal compressor), are applicable to screw compressors. These require no further explanation:

- throttling at the compressor inlet or outlet by corresponding throttling valves,
- bypass valve between inlet and outlet (often an integral part of screw compressors),
- variable-frequency motor as a driver,

Slide valve: However, a typical screw compressor has an additional, unique feature not available with other compressors. It is the so-called slide valve, an integrated mechanism that allows the inlet gas to be directed not only toward the front end of the screws but to any position along the screws. During maximum-load operation, the gas is provided to the front end of the screw, and the entire interlobular space is used for compression. In minimum-load operation, the gas is provided to the middle of the screws, with the front-end space bypassed so that only the second half of the screws (from gas inlet to discharge chamber) is used for the compression process.

6.5.3.6 Assessment of Screw Compressors

Thanks to oil lubrication, the screw compressor is a very reliable device. In closed systems, it does not require much maintenance; the usual recommendation is to replace compressor oil after 20000 to 25000 hours of operation. In open systems like helium liquefiers, the feed helium may contain some impurities. In such cases, the maintenance intervals may be shorter.

Screw compressors are commonly very robust and durable devices.

In terms of cost, screw compressors are very attractive machines. They are produced in large quantities by multiple manufacturers all over the world for application in conventional refrigeration, air conditioning, etc. Due to this bulk production and high competition, the specific cost level is low.

The main drawback relates to low efficiency: the isothermal efficiency is usually less than 60%.

6.5.4 Reciprocating Compressor

In the compressor family, the reciprocating compressor is the oldest member. Since this technology was established more than 100 years ago, there are many experiences and extensive literature about this device available. Reference [14] is just one example of worthy literature; it gives an extensive overview as well as a detailed description of reciprocating machines, starting from thermodynamic basics and mechanical design to operational aspects and maintenance issues.

A reciprocating compressor is a versatile machine; it is used for various applications ranging from consumer-scale (examples are small-size hermetic compressors used in home fridges and air conditioners) to industrial-scale applications in refineries, petrochemistry, pipeline transport, etc. In gas industry, reciprocating compressors are used for compression of oxygen, hydrogen, and helium in process plants, as well as auxiliary machines for all imaginable services like loading of gas cylinders or tube trailers, pressurization for medium- and large-size gas storages, caverns, etc.

The popularity of reciprocating compressors declined in early 1960s due to incredible fast ascension of centrifugal compressors, characterized by higher capacity and lower maintenance cost when compared to reciprocating machines. But currently, reciprocating compressor technology enjoys a great deal of attention, since it represents a viable and feasible option for compression of hydrogen – a fluid considered as the key substance for future green power infrastructure.

6.5.4.1 Design and Working Principle

The reciprocating compressor is a positive displacement machine: it compresses and moves a gas by repeatedly enclosing a fixed volume, with the aid of seals or valves; the cyclic moving action is driven by a piston. A typical stage design for a classic single-acting[6] piston stage is shown in Figure 6.46. The main elements are:

1. Piston and cylinder, sealed by means of piston rings,
2. Spring-loaded inlet and exit valves, placed in an element referred to as cylinder head with corresponding channels for gas entering and compressed gas leaving the compressor stage,
3. Moving elements: crankshaft, crank, connecting rod, and piston pin,
4. Elements for cooling of compressor stage: cooling jackets for cylinder wall and cylinder head.

6 The compressor is double-acting when compression takes place on each side of the piston.

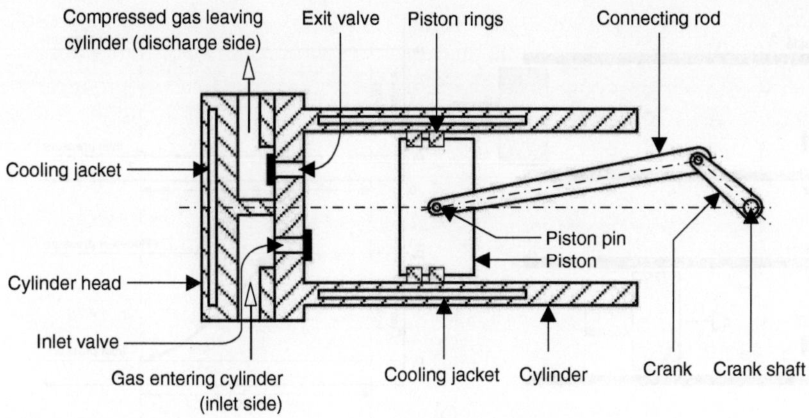

Figure 6.46 Reciprocating compressor, main elements.

The valves are spring-loaded devices, which means they are controlled by pressure difference across the valve:

- inlet valve opens when the pressure in the cylinder is slightly below the intake pressure,
- the discharge valve opens when the pressure in the cylinder is slightly above the discharge pressure.

Figure 6.47 illustrates the working cycle for a compressor stage with perfect valves – devices requiring infinitely small pressure difference for acting. On the left side, the main process steps are shown. On the right size, this process step is visualized by means of plot referred to as "indicator diagram" or "p–V diagram," whereby the y-axis shows the pressure change, depending on piston position [stroke] as x-axis, which is directly linked to the volume of the gas inside of the cylinder. Figure 6.47a and f show the basic configuration at the beginning of compression (state 1): cylinder is filled with working gas at the intake pressure p_1, and both valves are closed.The main process steps are as follows:

1. Compression stroke (process step 1 → 2, see Figure 6.47b and g):
   - valves remain closed,
   - the piston is moved to the left, reducing the volume of working gas with a corresponding pressurization from pressure p_1 to discharge pressure p_2.
2. Discharging stroke (process step 2 → 3, see Figure 6.47c and h):
   - the discharge valve opens just beyond pressure p_2 and remains open,
   - the piston is moved further to the left, forcing the working gas through the discharge valve to the outlet until reaching position 3, ideally at constant pressure p_2,
   - at this position, the discharge valve will close, and the clearing space (cylinder) will be filled with working gas at pressure p_3.
3. Expansion stroke (process step 3 → 4, see Figure 6.47d and i):
   - both valves remain closed,

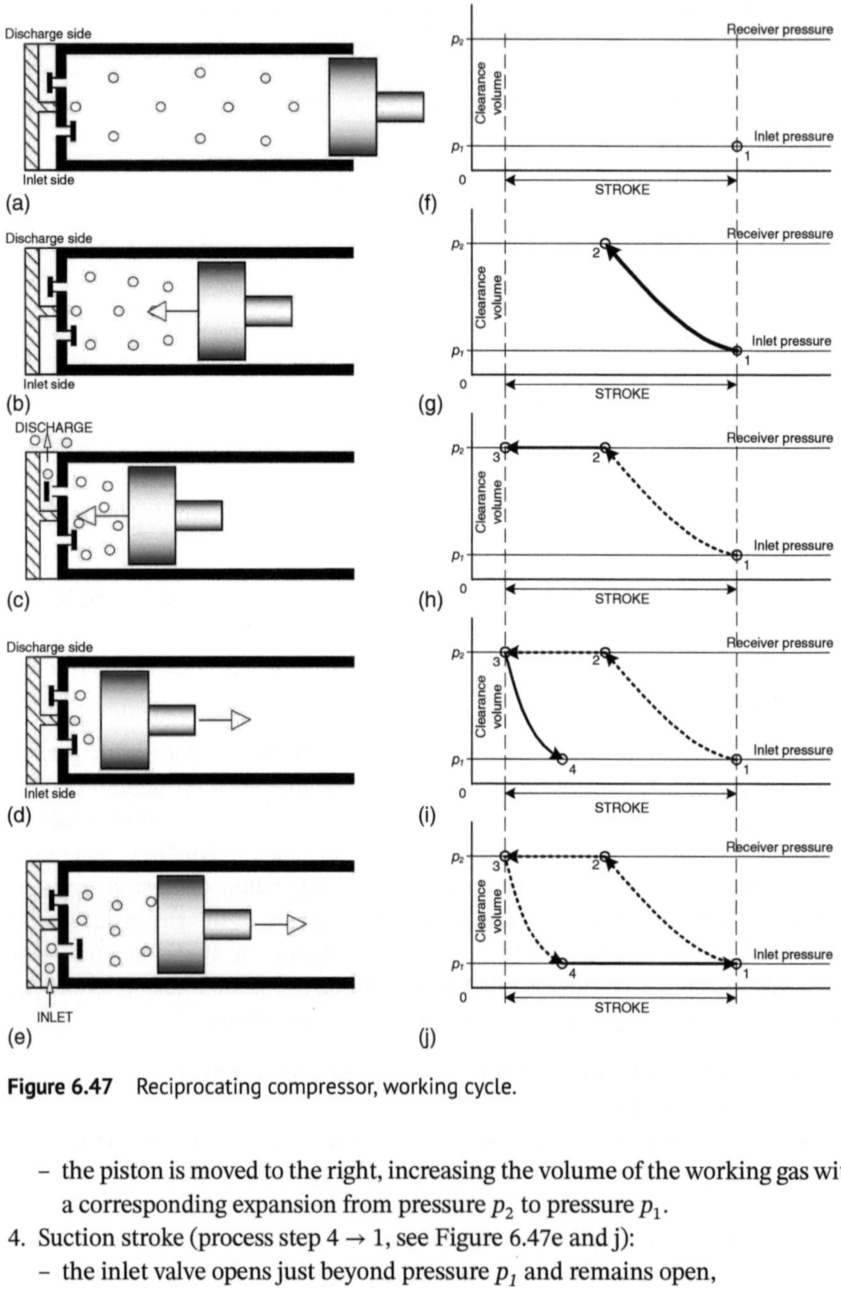

Figure 6.47 Reciprocating compressor, working cycle.

- the piston is moved to the right, increasing the volume of the working gas with a corresponding expansion from pressure p_2 to pressure p_1.

4. Suction stroke (process step 4 → 1, see Figure 6.47e and j):
   - the inlet valve opens just beyond pressure p_1 and remains open,
   - the piston is moved further to the right,
   - working gas flows through the inlet valve into the cylinder until reaching position 1, ideally at constant pressure p_1,
   - at this position, the inlet valve will close, and the clearing space (cylinder) will be filled with working gas at pressure p_1, the cycle is closed. On the next revolution of the crank, the cycle will be repeated.

Observation: Considering the compression process, one can realize that every reciprocating compressor is simultaneously an expansion machine: It compressed the fluid during the compression stroke, but during the expansion stroke, the machine is acting as an expansion device – the gas is expanded from high-pressure level to low-pressure level recovering some mechanical power. This duality (compressor and expander) is an outstanding attribute of every piston compressor. Other kinds of compressors, such as screw or centrifugal machine, do not provide this feature.

Depending on mechanical design of the compression device, real compression process deviates from the compression cycle shown in Figure 6.47j. Usually, these deviations lead either to reduced capacity (flowrate) or increased power for compression. The most important effects caused by valves, cylinder seals, and friction are discussed in next sections.

Impact of valves and gas passages

The valves are devices for controlling compression process. The main tasks are as follows: (i) to open widely at the right moment, (ii) to close tightly at the right moment. It is hard job to perform both these tasks in an ideal way.

Real valves require a tangible pressure difference for acting. Therefore, the real compression cycle looks as shown in Figure 6.48: discharge valve opens at pressure p_{2a}, which is higher than ideal discharging pressure p_2, as the intake valve opens at pressure p_{4a}, which is lower than ideal intake pressure p_1. Therefore, the real compressor stage operates between pressures p_{4a} and p_{2a} (instead of p_4 and p_2), the effective pressure ratio is higher than the specified p_2/p_1. Consequently, the compressor stage requires more power for compression task. Pressure drops in gas passages before suction valve and after discharging valve have similar effects on the power and capacity of compression stage.

The secondary effects caused by non-ideal valves are that (i) the mean gas density during suction phase (4a → 1) is lower than in ideal process (4 → 1) due to lower actual pressure, (ii) the mean gas density during discharging phase (2a → 3) is higher than expected in process with perfect valves (2 → 3) due to higher actual pressure. Both these effects result in reduced capacity of compression stage.

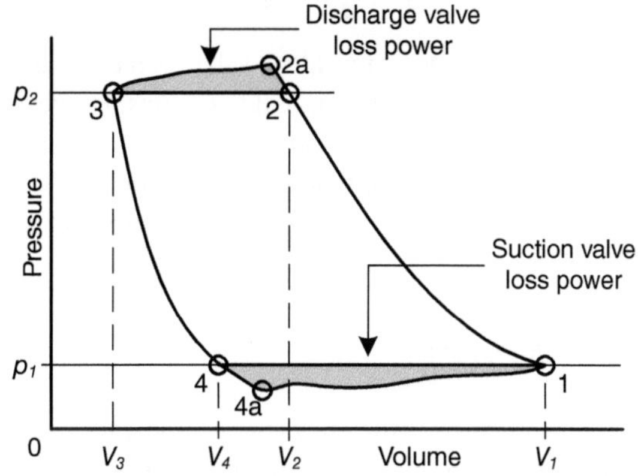

Figure 6.48 Reciprocating compressor, impact of valves on indicator diagram.

The tight closing of the valves is the next important requirement:

- If the discharge valve does not close properly, the fluid flows back into the cylinder from the discharge gas passage. This returned gas is still hot; therefore, this leakage badly impacts the capacity of the compressor stage.
- If the suction valve does not close properly, the gas flows from the cylinder bore to the suction passage. The effects on compressor capacity are quite similar to those of discharge valve.

The valve leakages increase with increasing pressure ratio, and decrease with rotation speed, which is self-explanatory. For fluids with low viscosity, the leakage effects are more critical than for more viscous gases.

Impact of internal leakage through piston seal

The leakage through the piston ring(s), which are used as seal between piston and cylinder wall, has a negative impact on compressor capacity. Obviously, the capacity is reduced by loss of mass through the seal during the compression and discharging stroke. The leakage becomes worse at higher pressure ratios, since the pressure difference increases as the driving force increases, too. For low-molecular-weight fluids like hydrogen and helium, due to low viscosity of these fluids, ring leakage is more critical than for more viscous fluids.

Theoretically, the ring leakage can be reduced by shortening of the time slot for compression and discharging phase, for example, by increasing rotation speed. This measure can be considered during the design phase in combination with other measures like improved sealing design (more rings, better materials, etc.).

Generally, oil-lubricated compressors can manage internal leakages better than dry compressors since the compressor oil works as a sealing fluid.

Volumetric efficiency

For the perfect single-cylinder compression cycle shown in Figure 6.47, the cylinder displacement $\dot{V}_{id}$ follows directly from the geometry. It is product of three parameters: the piston face area $\frac{\pi D^2}{4}$ (with cylinder diameter D), the stroke length S_L, and the number of strokes in a minute or second n_{strokes}:

$$\dot{V}_{\text{GEOM}} = S_L \cdot \frac{\pi D^2}{4} \cdot n_{\text{strokes}} \tag{6.4}$$

Due to leakages, non-ideal valve behavior, and other effects, this ideal value $\dot{V}_{\text{GEOM}}$ is never reached in real machines. However, this theoretical value is a good benchmark for comparison of real compressor stages in terms of flowrate. The so-called volumetric efficiency η_{V} is a good indicator for this purpose. It is defined as ratio of actual capacity $\dot{V}_{\text{real}}$ of a compressor stage to the capacity derived from cylinder-piston geometry as follows:

$$\eta_{\text{V}} = \frac{\dot{V}_{\text{real}}}{\dot{V}_{\text{GEOM}}} \tag{6.5}$$

The volumetric efficiency is not a constant number. It depends on inlet pressure, pressure ratio π, properties of gas being compressed, speed of the compressor, and other parameters. Usually, η_V approaches values of 0.95–0.97 at pressure ratio $\pi \to 1$, at higher pressure ratio, it becomes lower and lower.

The term "volumetric efficiency" is a bit misleading in this sense that it imparts a connection to power efficiency. But this number does not characterize power efficiency of the compressor stage; it reflects the quality of mechanical design only. The volumetric efficiency is useful for prediction of what will be the flowrate (compressor capacity) at different operating conditions.

6.5.4.2 Process-relevant Aspects

6.5.4.2.1 Impact of the Isentropic Exponent on the Compression Process

Figure 6.49 shows the impact of the isentropic exponent on the indicator diagram. For fluids with higher κ, both – compression and expansion – phases become shorter (the corresponding line in indicator diagram rises steeply), and the whole diagram becomes broader.

6.5.4.2.2 Isentropic Efficiency

The compression process in a reciprocating stage can be considered as an adiabatic process. For calculation of the power, Eq. (5.29) can be used. The deviation from reversible isentropic process is considered in this equation by a factor referred to as isentropic efficiency η_s, see Eq. (5.28). For large reciprocating machines, for pressure ratio 3–6, the isentropic efficiency achieves a high value of 85% (for lubricated design) or even higher.

The majority of reciprocating compressors are developed for air, gases, and substances. But in cryogenics, these machines are considered mainly for compression of light gases such as hydrogen and helium. Figure 6.50 shows a plot of η_s versus pressure ratio for a given cylinder when compressing two different gases, hydrogen and natural gas, as given in Ref. [15]. The text to this graph is as follows: "note how efficiency increases as compression ratio increases. The efficiency curve will have this shape for any reciprocating compressor cylinder. Also note how much higher the efficiency is for hydrogen (with a very low mole weight of two) as compared to natural gas (with a medium mole weight of 18). Everything else being equal, compressing hydrogen will have one ninth (11% less) of the valve loss power, hence much better efficiency."

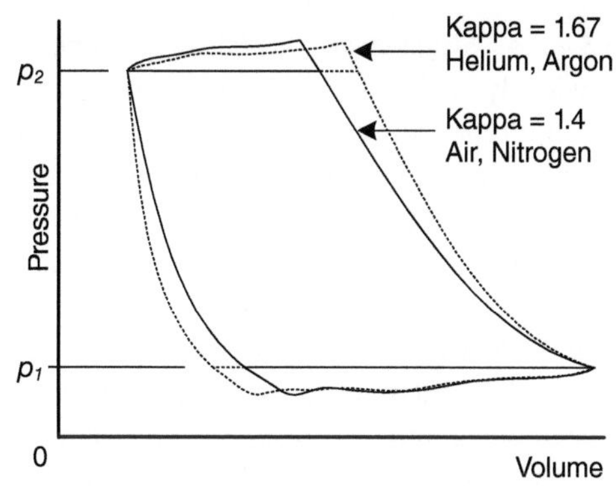

Figure 6.49 Reciprocating compressor, impact of isentropic exponent on indicator diagram.

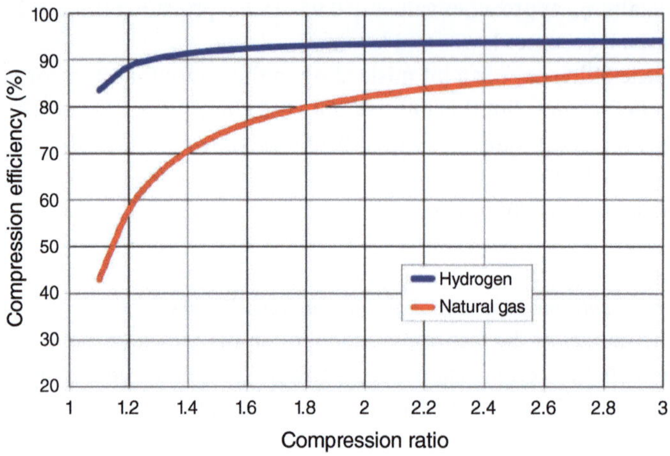

Figure 6.50 Compressor efficiency versus compression ratio for hydrogen and natural gas.

6.5.4.2.3 Mechanical Efficiency

Note, that isentropic efficiency describes the efficiency of the compression process in a stage/cylinder only. But some loss occurs between electric motor and compression cylinder. In Section 6.5.1.1, where the entire energy transformation chain for compressors is considered, this loss is referred to as mechanical loss. This loss is caused mainly by friction, whereby mechanical energy is converted into heat. Depending on mechanical design (see Section 6.5.4.3), the sources of friction may be different. The most obvious examples are friction caused by piston rings rubbing against cylinder bore and crankshaft turning in the bearings.

This mechanical loss is accounted by a factor referred to as mechanical efficiency η_M. The value of $\eta_M = 95\%$ is used often for preliminary (very rough) estimations. But the exact estimation of η_M is a complex item. The real value may vary from 93% to 97%. The mechanical efficiency η_M is not provided as a separate number by original equipment manufacturer (OEM). Usually, it is included in the isentropic efficiency as follows: $\eta_s^{OEM} = \eta_s \cdot \eta_M$. This way, both kinds of losses – caused by irreversibilities in compression process as well as cause by mechanical friction – are reflected by a single parameter.

6.5.4.2.4 Single-stage and Multi-stage Compressors

Obviously, a single-cylinder machine is considered as a single-stage compressor. For multiple-cylinder machines, the definition of a single-stage compressor must be extended: if multiple cylinders on a common frame are connected in parallel, compressing a fluid from the same inlet pressure to common outlet pressure, the compressor is still referred to as single-stage compressor. However, when multiple cylinders are connected in series, usually through an interstage cooler, the arrangement is referred to as a multistage compressor.

The thermodynamics for multi-stage compressor are considered in Section 5.5. Just a quick reminder: for characterization of multistage-compression process, a special factor – isothermic efficiency η_T – is applied.

Almost all compressors required for driving cryogenic processes are multistage devices. In this sense, reciprocating machines are well suited for serving cryogenic processes.

6.5.4.3 Mechanical Design

Reciprocating compressor is a highly integrated device: all the required functions – compression in cylinders, transfer of mechanical power from shaft to cylinders, lubrication and cooling of cylinders and mechanisms are integrated in a common relative complex structure.

A large number of compressor designs were developed over the past 100 years by inspired engineers. It is not possible to consider all these developments in this section. Therefore, only some very common notes are given here. To get a better understanding or more detailed view, we would recommend reading a good textbook about compressors, for example, Ref. [28].

Note 1. Driver: reciprocating compressors can be driven by piston engine, gas turbine, or steam turbine. The most popular driver is an electric motor. In the past, flywheels were sometimes used to dampen torque pulsations, minimize transient torque absorbed by driver, and to tune torsional frequencies. Today, flywheels are practically not used. It means higher requirements on driver, since driver inertia must absorb torque pulses.

Note 2. Transfer of mechanical power from shaft to cylinders is realized either by a relatively simple method without crosshead, referred to as trunk or automotive piston type, or by a more complex method with an element referred to as crosshead:

- The trunk method presumes that driver is connected directly to the crankshaft, which is further connected to the piston through connecting road. The whole transfer chain works as follows:

  Driver → Crankshaft → Connecting rod → Piston

  This method is used rather for small-size compressors with single-acting cylinders.

- the crosshead type of compressors are more common for medium-size and large-size machines. The difference from the trunk method is that the connecting rod is linked to the crankshaft through a crosshead. The energy transfer chain is as follows:

  Driver → Crankshaft → Crosshead → Connecting rod → Piston

  The cylinder is often isolated from the crankcase by an additional section referred to as distance piece. This design has obvious advantages for hazardous fluids (hydrogen and natural gas). The cylinder can be either single- or double-acting.

Note 3. Cylinder arrangement: In reciprocating machine, the weights in form of pistons and gas are moved during each crankshaft revolution. Since the piston actually comes to a complete stop at the top of the stroke and at the bottom of the stroke, its speed and acceleration at any given moment are always changing. The piston is always either accelerating from or decelerating to zero speed. Therefore, the balancing of the compressor is an important task during the compressor design; it is crucial for maximizing the operating life. A classification of compressors according to cylinder arrangement is not often applied. The most common configurations are as follows:

1. Horizontal balanced-opposed design, as shown in Figure 6.51, is common for large-size machines: the cranks are arranged so that the motion of each piston is balanced by the motion of an opposing piston.
2. Figure 6.52 shows a compressor with vertically oriented arrangement of cylinders. They are the optimal solution for low-space locations and requirements.
3. Other popular arrangements are of "V"-, "L"- or "star"-type.

Note 4. Lubrication: Very generally, lubrication aspects are described already in Section 6.5.1.2.

Cylinder lubrication: The lubrication of cylinders can be tailored to the application; it can be designed as

(1) dry, it means without lubrication,
(2) oil-lubricated (oil-injected or oil-flooded), whereby the gas being compressed is mixed with lubricant (e.g., in compressors for refrigerating purposes, this method is very common),
(3) a compromise lubrication method is referred to as mini-lube system, which is based on dry design but requires a small amount of lubrication oil for some special purposes.

For frame lubrication two basic principles are applied:

1. for small-size compressors, an integrated so-called splash lubrication system is applied, whereby the oil is distributed/sprayed by the splashing of oil by existing moving parts (e.g., crankshaft with or without attached dippers).
2. for large-size machines, a dedicated pressurized closed oil lubrication system is applied, which contains an oil pump (usually crankshaft driven), oil cooler, oil filter, and corresponding instrumentation for safety and control (oil level, oil pressure, safety valves, etc.). Lubricant is provided to specified lubrication points on cylinders, packings, and other critical parts.

Note 5. Cooling: The small-size compressors can be designed as air-cooled machines. For this purpose, distinctive metallic fins are placed on the outer side of the cylinders. But medium-size and large-size compressors are water-cooled devices, whereby the pressurized cooling water is provided through cooling jackets implemented in cylinder wall and cylinder head (see Figure 6.46). If compressor oil is used, it functions as cooling fluid, too. Consequently, it becomes warm and needs to be cooled. In closed cooling systems, the cooling fluid is either water or a mixture of water and ethylene glycol.

6.5.4.4 Load Range

The operation and application ranges for reciprocating compressors are shown in Figure 6.32. Principally, this figure visualizes that the discharge pressure is practically unlimited, whereas the volumetric flow is restricted.

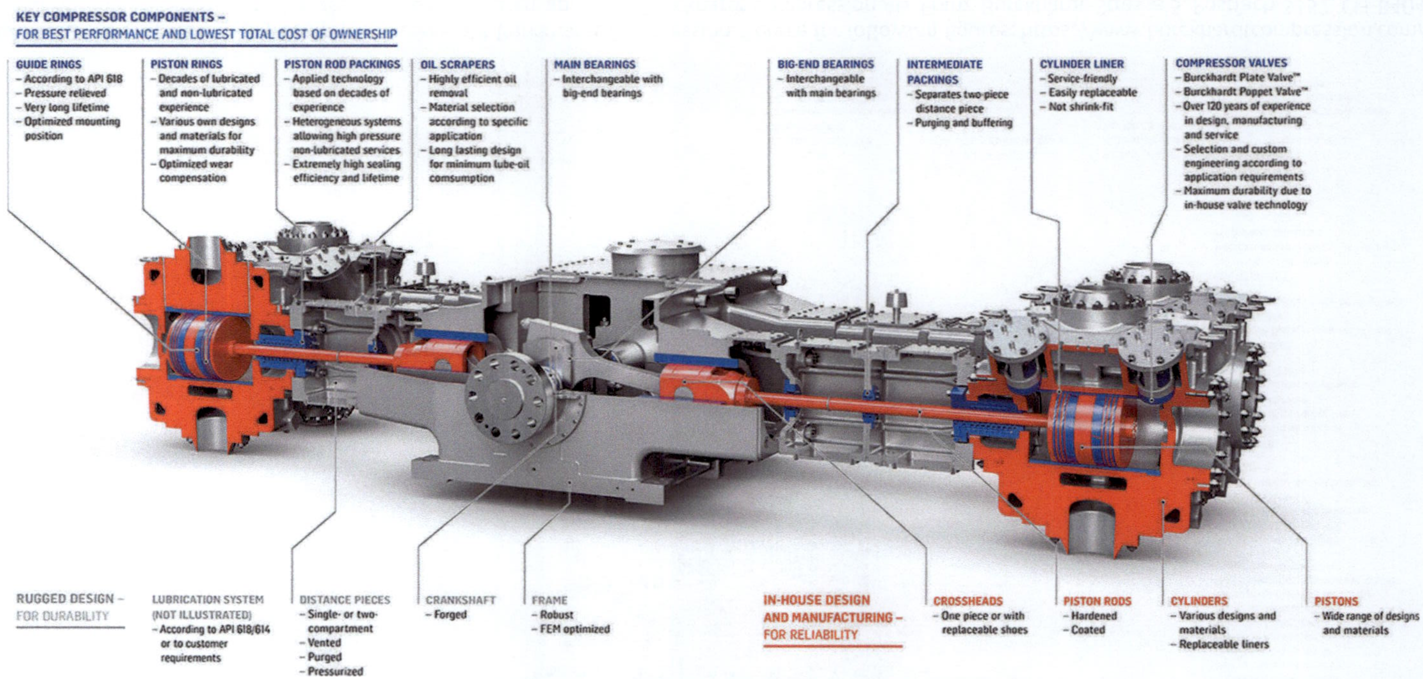

Figure 6.51 Process compressor, copyright Burckhardt Compression. Source for following figures: https://www.burckhardtcompression.com/de/solution/kompressorsysteme/prozessgas-kompressoren-api-618, Burckhardt Compression AG, Franz-Burckhardt-Strasse 5, Postfach 3352, CH-8404 Winterthur, Schweiz, Tel. +41 52 261 55 00, info@burckhardtcompression.com.

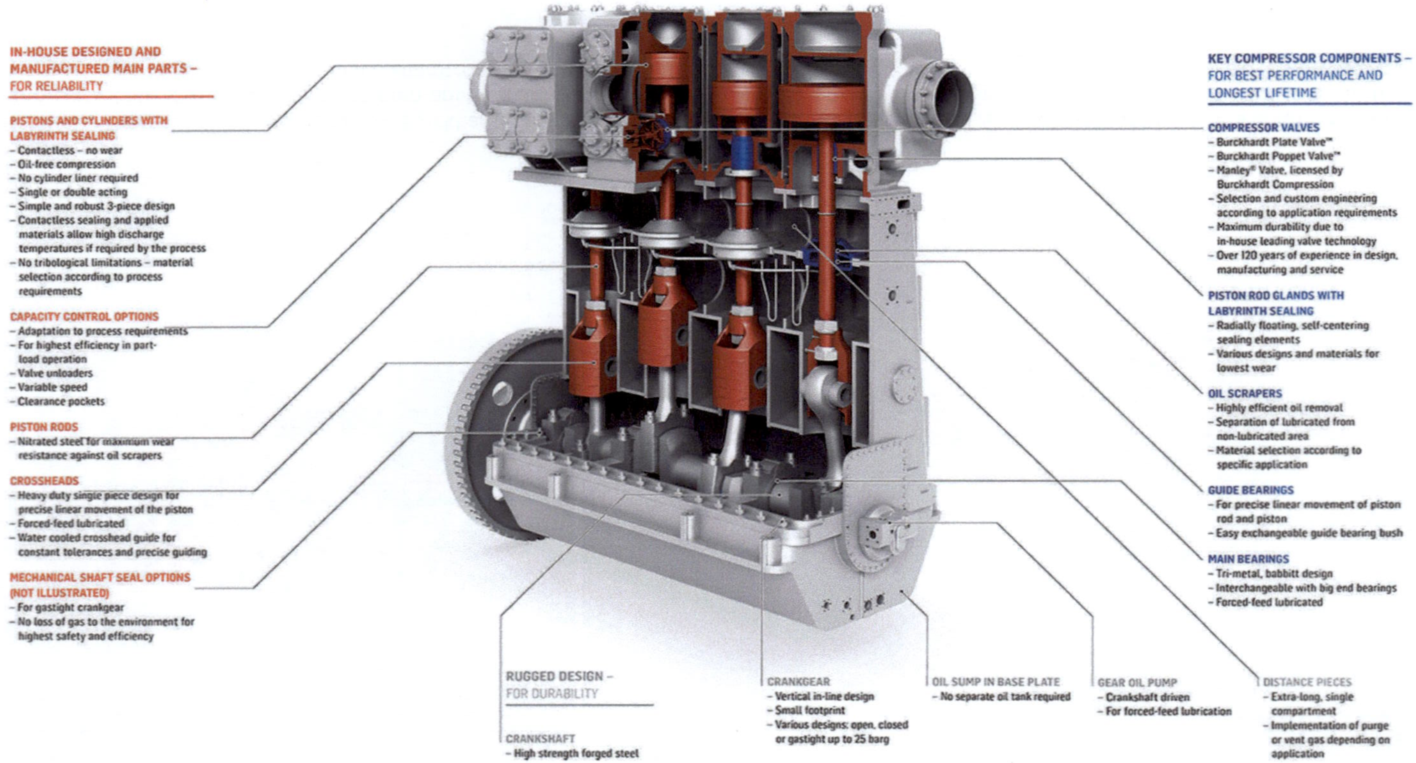

Figure 6.52 High-pressure dry compressor, copyright Burckhardt Compression. Source for following figures: https://www.burckhardtcompression.com/de/solution/kompressorsysteme/prozessgas-kompressoren-api-618/, Burckhardt Compression AG, Franz-Burckhardt-Strasse 5, Postfach 3352, CH-8404 Winterthur, Schweiz, Tel. +41 52 261 55 00, info@burckhardtcompression.com.

The last limitation is mainly due to the fact that the volumetric flow of piston compressor is defined by number and geometry of the cylinders as well as speed of the piston. For processing of high volumetric gas flowrates, larger cylinders and heavier pistons are required. But fast moving of heavy piston is a challenging issue. It means not only high requirements on mechanical design of the whole compressor and correspondingly high compressor cost, but it also leads to increased wear and tear. The accelerated abrasion means (a) short mean time between maintenances (MTBM) as well as higher expenses for materials and replacement parts during maintenance. The higher the piston speed the higher the maintenance and the lower the reliability. Therefore, high maintenance expense is an obvious reason to limit the piston speed. The upper limit for volumetric flow corresponds to a compressor design with cost level (CAPEX and maintenance expenditures), which is currently accepted by customers. Clearly, it depends on current technology level and will increase with progress in compressor technology.

6.5.4.5 Control

Reliable control of the compressor is a requirement for stable operation and efficiency of the refrigeration / liquefaction processes. The most important parameters to be controlled are the flowrate and discharge pressure at a given inlet pressure. There are a number of various means available for this. Some of those, described in Section 6.5.2.5 (control of centrifugal compressor), are applicable to screw compressors. These require no further explanation:

- throttling at the compressor inlet or outlet by corresponding throttling valves,
- bypass valve between inlet and outlet (often an integral part of screw compressors),

The reciprocating compressor operates with fixed volume. The most obvious method of volume variation is by speed modulation. Therefore, a variable-frequency motor as a driver is a very common option for control.

Additionally, a reciprocating compressor has a unique control feature. It is an integrated mechanism for controlling volumetric capacity/ flowrate referred to as adjustable pocket(s). The principle is shown in Figure 6.53, it based on control of clearance volume. In Ref. [28] the clearance pocket is described as follows: "it is essentially an empty volume, typically in the outer head of the cylinder, with a valved passage to the cylinder bore. During normal operation, the valve is closed, and the cylinder operates at full capacity. For reduced capacity operation, the valve is opened, and the cylinder capacity is reduced."

6.5.4.6 Pulsations and Pulsation Control

Reciprocating compressors are characterized by their fluctuating flow patterns, which repeat with each rotation of the crankshaft: the discharge flow is temporarily interrupted at the end of compression stroke until the next discharging stroke; a similar interaction occurs on the suction side of the system – the suction process is interrupted at the end of suction stroke until the next suction stroke. These interruptions in flowrate introduce pressure variations within the connecting pipe

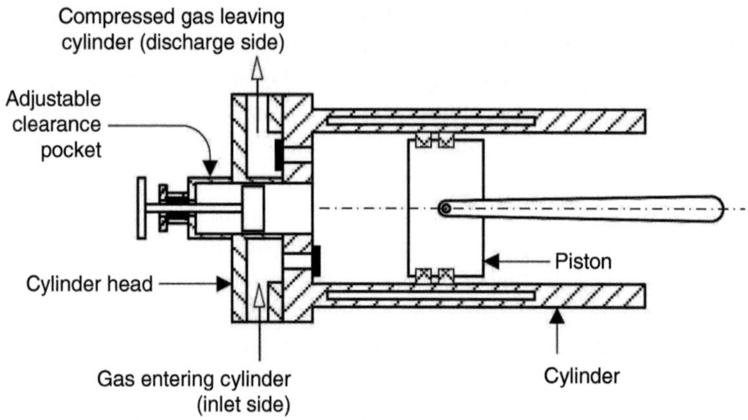

Figure 6.53 Cylinder with adjustable clearance pocket, principle.

work known as pulsations. It's important to note that these pulsations do not follow a smooth sinusoidal form; instead, they exhibit rather a binary pattern (either it is full flowrate or no flowrate) with short transient zones.

When these pulsations coincide with the vibration spectrum of the external system, resonance can occur, potentially leading to dangerous pressure peaks. Excessive pulsations directly impact compressor performance and, in turn, have indirect adverse effects on system performance. Over time, uncontrolled pulsations result in increased costs due to the repair or replacement of worn or failed components, extended downtime, and decreased operational efficiency.

Pulsation control is a regular engineering task. The objective is to reduce vibration level to an acceptable range by implementing pulsation control elements. These elements come in various forms, such as volume bottles, baffles, orifices, and pulsation snubbers / dampeners. From a process design perspective, it's important to consider that implementing pulsation control measures introduces a pressure drop into the system. Note that these devices require space for installation and necessitate regular service and maintenance. Therefore, they incur both capital (CAPEX) and operational (OPEX) expenses. Ultimately, ensuring proper pulsation control is essential for the safe and efficient operation of reciprocating machines and the systems they serve.

6.5.4.7 Short Assessment of Reciprocating Compressors

In the gas industry, reciprocating compressors stand as one of the most well-known compression devices, boasting an extensive reference list and a long history of use. They find application across a wide range of processes. The core technology has already reached a high level of maturity, ongoing developments include the integration of new materials, the adoption of modern control and instrumentation methods, implementation of predictive maintenance methods, and the incorporation of digitalization-based features to enhance their performance.

In terms of power efficiency, reciprocating compressors excel, offering impressive levels of efficiency. However, their mechanical design is inherently complex, comprising numerous moving parts integrated within the system. As a result, these compressors necessitate significant maintenance efforts, leading to elevated maintenance costs—this stands as their primary drawback. Additionally, they tend to have a relatively high specific capital expenditure (CAPEX), primarily due to their lower specific volumetric flowrates. Furthermore, the considerable noise generation associated with reciprocating compressors is often considered a disadvantage, although it's worth noting that compressors, in general, are known for their noise.

In recent times, the combination of properties represented by "high efficiency, high CAPEX, and high maintenance" has made reciprocating compressors less attractive when compared to alternatives such as centrifugal compressors or screw compressors. Consequently, reciprocating compressors are typically employed in a limited number of specific processes, particularly for the compression of light gases such as hydrogen and helium, or in cases where smaller-sized process plants are involved.

References

1 Flynn, T. (1996). *Cryogenic Engineering*. Taylor & Francis Inc., ISBN10 0824797248, ISBN13 9780824797249.

2 Kaganer, M.G. (1969). *Thermal Insulation in Cryogenic Engineering*, Israel program for Scientific Translation.

3 Nast, T. (1998). Multilayer insulation systems, Chapter 3-4-1. In: *Handbook of Cryogenic Engineering* (ed. J.G. Weisend II,). Taylor and Francis, ISBN: 1-56032-332-9.

4 The standards of the brazed aluminum plate-fin heat exchanger manufacturers' association, first edition, 1994, published by The Brazed Aluminum Plate-Fin Heat Exchanger Manufacturer's Association (ALPEMA), www.alpema.org.

5 Taylor, M.A. *Plate-Fin Heat exchangers – Guide to their Specification and Use, HTFS*. Harwell Laboratory, Oxon, OX11 ORA, *UK*.

6 Foerg, W., Bach, W., Stockmann, R., et al. (1998). A new LNG baseload process and production of the main heat exchanger, *Proceedings of the LNG12: 12th international congress on liquefied natural gas*.

7 Maytal, B.-Z. and Pfotenhauer, J.M. (2013). *Miniature Joule-Thomson Cryocooling, Principles and Practice*. Springer, ISBN: 978-1-4419-8284-1.

8 Linde, H. (1985). *Hausen H*. Tieftemperturtechnik: Springer Verlag. ISBN: 978-3-662-10553-5.

9 Presentation "LNG heat exchangers" by Jostein Pettersen from Statoil Hydro, TEP-10 (2008). https://folk.ntnu.no/cadorao/courses/TEP10/files/TEP10-LNG%20heat%20exchangers-2008.pdf.

10 Brochure "Coil-wound heat exchangers" Linde Engineering (2019). https://www.linde-engineering.com/en/images/Coil-wound-heat-exchanger-2019_tcm19-5793.pdf.

11 Blackford, J., Halford, P., and Tantam, D.H. (1971). Expanders and pumps. In: *Cryogenic Fundamentals* (ed. G.G. Hausen). Academic Press, ISBN: 0 12 330550 0.

12 Boyce, M.P. (2002). *Centrifugal Compressors: A Basic Guide*. Pennwell, ISBN: 0-87814-801-9.

13 https://en.wikipedia.org/wiki/Centrifugal_compressor#cite_note-wood-12.

14 Bloch, H.P. (2006). *A practical guide to compressor technology*, Wiley, ISBN: 978-0-471-72793-4.

15 Philippi, G. (2017). Basic Thermodynamics of reciprocating compression, 45th Turbomachinery and 32th Pump Symposia, Houston TX, *September* 12–15, 2016.

16 Howell, J.R., Mengüç, M.P., Daun, K., and Siegel, R. (2020). Thermal Radiation Heat Transfer. CRC Press. https://doi.org/10.1201/9780429327308.

17 Hamrock, B.J., Schmid, S.R., and Jacobson, B.O. (2004). Fundamentals of Fluid Film Lubrication. *CRC Press*. https://doi.org/10.1201/9780203021187.

18 Maslen, E.H., and Schweitzer, G. (Éds.). (2009). Magnetic Bearings: Theory, Design, and Application to Rotating Machinery. Springer Berlin Heidelberg. https://doi.org/10.1007/978-3-642-00497-1.

19 Britannica, The Editors of Encyclopaedia. "vacuum flask."*Encyclopedia Britannica*, April 9, 2019. https://www.britannica.com/technology/vacuum-flask.

20 Barron, R.F., *Cryogenic Heat transfer, Section 5.4: Radiant Heat Transfer between two gray surfaces, pp. 230–231*, Taylor and Francis, 1999, ISBN: 1-56032-551-8.

7

Cryogenic Refrigeration

This section consists of two parts:

(1) a section about principles of refrigeration (Section 7.1),
(2) descriptions and assessments of four standard cryogenic processes, in the versions, adapted for refrigeration purposes (Section 7.2, Section 7.3, Section 7.4, and Section 7.5). This description follows the principles defined in Section 2 (basics of process design).

The same basic processes, in slightly modified forms, are used for liquefaction purposes (see Section 8.1) as well as for gas separation purposes. Therefore, understanding of these basic processes is an essential requirement for understanding cryogenics. The standard cryogenic processes (see Figure 7.1) are:

- Joule–Thomson process,
- Brayton process,
- Claude process,
- mixed-fluid Joule–Thomson process.

In Section 7.2, the most popular version of the Joule–Thomson process – the nitrogen Joule–Thomson process – is described.

The Brayton process (Section 7.3) is realized often with air/nitrogen as a working fluid, or sometimes with helium. There are no differences (neither in morphology nor in thermodynamics) between the nitrogen-based and the helium-based processes. Therefore, the helium version only is discussed in detail.

The Claude process (Section 7.4) is considered in two versions: as a nitrogen Claude process and separately as a helium Claude process, because the morphology of the helium-based process differs from the morphology of the nitrogen-based version.

The Section 7.5 is about mixed-fluid Joule–Thomson process. Although similar in name and morphology, the mixed-fluid Joule–Thomson process represents a distinctly different kind of cryogenic process (in comparison to the nitrogen Joule–Thomson process).

Process Design for Cryogenics, First Edition. Alexander Alekseev.

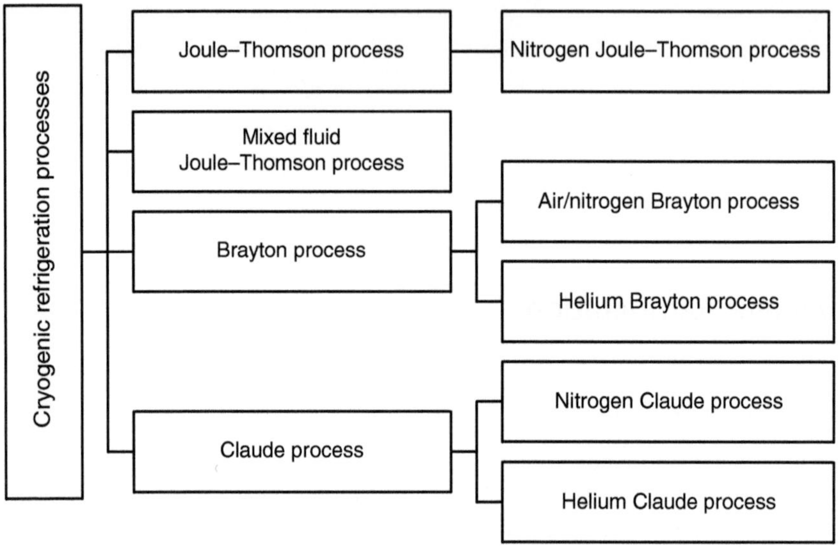

Figure 7.1 Overview, cryogenic refrigeration processes.

The first process – the nitrogen Joule–Thomson process – is considered in detail. This extended analysis includes:

- process description,
- hardware components description,
- example process calculations and process analysis,
- considerations about how this process can be improved,
- examples of process optimization,
- considerations concerning control concepts.

The description and analysis of the three other standard processes are given in shortened versions. An extended analysis of these processes can be performed by everybody using the Section 7.2 as a template, all relevant data is available in the Sections 7.3, 7.4, and 7.5.

Classification of Cryogenic Processes According to Pressure

The creativity of engineers and scientists is quite unlimited, as evidenced by new patent applications in the field of cryogenic processes being published monthly. Usually, these inventions relate to advantageous modifications of existing processes, but sometimes completely novel ideas are published. Consequently, the number of cryogenic processes and modifications of these processes is very high and remains on the rise. Maintaining an overview of the cryogenic process is a complex task that may be facilitated by introducing an appropriate structure.

The most accepted approach is based on ordering according to the maximum pressure level used in process, whereby the critical pressure of the working fluid $\boldsymbol{p_c}$ serves as a benchmark. Usually, two categories are defined:

1. supercritical processes, if the maximal process pressure is higher than the critical pressure ($\boldsymbol{p_{HP} > p_c}$);
2. subcritical processes, if the maximal process pressure is lower than the critical pressure ($\boldsymbol{p_{HP} < p_c}$).

Sometimes, three categories are defined: (i) the high-pressure process, (ii) the middle-pressure process, and (iii) the low-pressure process. The goal is to improve the resolution in subcritical areas. The methodology is quite the same – the critical pressure of the working fluid $\boldsymbol{p_c}$ serves as a benchmark. The subcritical processes are divided into two groups:

1. if the maximum process pressure is lower than 1/3 of the critical pressure ($\boldsymbol{p < 1/3\ p_c}$), the process is considered a "low-pressure process";
2. the "middle-pressure process" is the process within the pressure range of $\boldsymbol{1/3\ p_c < p < p_c}$.

The supercritical process ($\boldsymbol{p > p_c}$) is referred to as a "high-pressure process."

Table 7.1 summarizes these definitions and includes some examples. According to this definition, all modern helium-based processes are high-pressure processes.

Table 7.1 Ordering of cryogenic refrigeration processes according to pressure range.

	High-pressure process	**Middle-pressure process**	**Low-pressure process**
Pressure range	$p > p_c$	$1/3\ p_c < p < p_c$	$p < 1/3\ p_c$
Pressure range for nitrogen-based processes ($p_c = 34$ bar)	$p > 34$ bar	$p = 15$–25 bar	$p = 6$–9 bar
Pressure range for helium-based processes ($p_c = 2.3$ bar)	Usually $p > 8$ bar	—	—
Process examples	Joule–Thomson process, Claude process	Mixed-fluid Joule–Thomson process	Brayton process
Example of application	Nitrogen liquefier, helium liquefier	—	Cryogenic refrigeration

7.1 Principle of Cryogenic Refrigeration

If your espresso is too hot, you wait a minute so that the ambient air cools the coffee and you can enjoy the drink. If your Coca-Cola is too warm, you put some ice into the cup and your drink becomes suddenly colder. In both scenarios, a warm object contacts (directly or indirectly) the colder objects. This thermal contact is an essential requirement for cooling.

What happens on the interface between warm and cold objects in terms of thermodynamics? The heat from the warm object flows to the cold object: from espresso cup to the colder ambient air, or from the coke to the ice. We experience that an object can be cooled if some other material/object – that is already a bit colder – is available and a thermal contact between these objects is established. Two examples of natural cooling media are ice and snow. If a really low temperature is required (−100 °C or lower), then a special device – the so-called refrigerator – is used for cooling purposes.

7.1.1 Refrigerator and Definitions Around It

The most important feature of every refrigerator is an area (or surface) where the temperature is a bit lower than the required cooling temperature. To cool an object, a thermal contact between this object and the cold area of the refrigerator needs to be established; subsequently, the heat $\dot{Q}_o$ in Figure 7.2 flows from the object to the cold surface and the object becomes colder.

Cooling temperature: in this book, only one version of cooling task is considered: the main purpose of a refrigerator here is to keep the cooling object at the required cooling temperature $\boldsymbol{T_o}$.

Cold surface: the term "cold surface" used here requires a separate explanation: often it is a real solid (usually metallic) surface like a cold finger of a cryocooler. However, some refrigerators produce cold liquid cryogen (liquid helium or liquid nitrogen), which, while flowing to/through the cooling object (for example a superconducting magnet), evaporates. Then, the vapor is returned to the refrigerator. In this case, the liquid surface may be considered the cold surface. The cooling

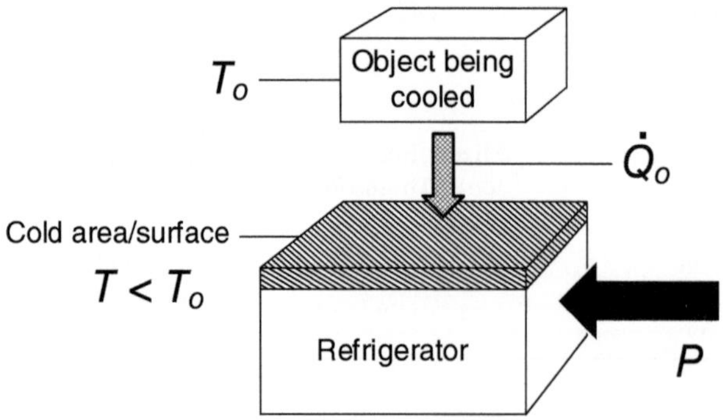

Figure 7.2 Refrigerator idea: cold surface (of a refrigerator) absorbs heat from an object.

surface of refrigerator itself has a temperature a bit lower than $\boldsymbol{T_0}$ (see Figure 7.2). This temperature difference is required as the driving force for the heat transfer between the object being cooled and the cold surface.

Cooling capacity: Theoretically, some cooling objects exist that do not generate their own heat; they simply need to be cooled to the required temperature. But in reality, almost all objects (magnets, sensors, current leads) need a permanent cooling because (i) they generate heat continuously during operation; (ii) some heat from the warm environment leaks through the imperfect thermal insulation; and (iii) they are connected by numerous electrical wires to the warm part of the system, and the heat is conducted through these metallic wires to the cooling object. Therefore, the "heat" in this case is a permanent heat flow from cooling object to refrigerator.

If a refrigerator is able to keep an object at temperature $\boldsymbol{T_0}$ and the heat flow being removed from the object amounts to $\dot{Q}_0$, then one says: "the refrigerator has cooling capacity $\dot{Q}_0$ at the temperature $\boldsymbol{T_0}$."

Driving power: Another property is that the refrigerator (conventional home fridge as well as a helium liquefier) consumes power $\boldsymbol{P}$ (mechanical or electrical power).

Refrigerator/Liquefier: Another cooling device – the so-called liquefier – is often used in cryogenics. A liquefier produces cryogenic liquid that is then drawn off. The thermodynamics for refrigerator and liquefier are quite similar. In this book, the refrigerator will be considered in the current section. Liquefiers will be discussed separately in Section 8.1–8.5.

7.1.2 Energy Conservation Law for Refrigerator

In very simplified version, the energy conservation law says that what goes into a system must either come out or accumulate inside: for steady-state conditions (without accumulation inside of system), the sum of energy streams entering the system equals the sum of energy streams leaving the system. An energy stream is for example, a heat flow, enthalpy flow (for material streams), or mechanical power for a driving compressor or pump.

7.1.2.1 Energy Conservation Law for Refrigerator

Let us apply this law to the refrigerator working in steady-state conditions as shown in Figure 7.3. The first energy stream – the heat flow $\dot{Q}_0$ from the object to refrigerator – is entering the system. The driving power $\boldsymbol{P}$ is the second energy stream entering the system. Consequently, according to the energy conservation law, at least one further energy stream must exist; this energy stream must leave the refrigerator to balance the system.

This energy stream exists indeed; it is the waste heat flow, abbreviated as $\dot{Q}_{amb}$. Usually, a refrigerator consists of a compressor, where the gaseous refrigerant is compressed from one pressure level to a higher-pressure level. During the compression process, the gas becomes hot and is cooled by means of cooling air or cooling water. This allows the heat $\dot{Q}_{amb}$ to be rejected to the environment.

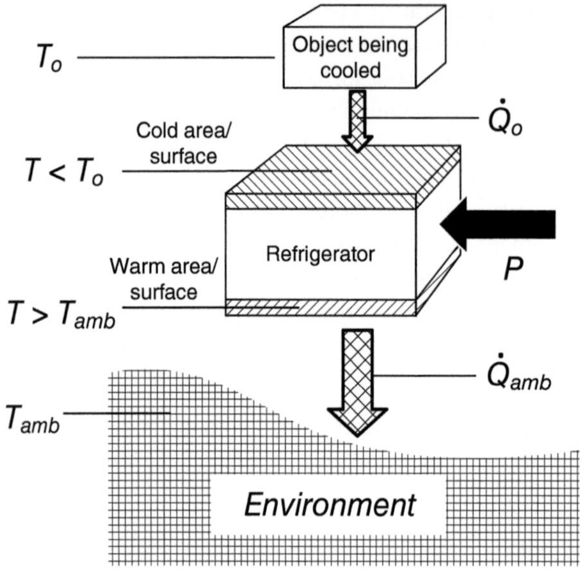

Figure 7.3 Energy conservation law for refrigerator.

A temperature difference is required to transfer $\dot{Q}_{amb}$ to the environment. Therefore, every refrigerator has another feature – the so-called warm surface – a surface with a temperature slightly warmer than the ambient temperature T_{amb}. Through this surface, the waste heat will be removed and transferred to the environment, as shown in Figure 7.3.

The energy conservation law can be expressed as:

$$\dot{Q}_o + P = \dot{Q}_{amb} \tag{7.1}$$

7.1.2.2 Three Interfaces

Every refrigerator has at least three interfaces:

1. "cold surface" to receive and absorb the heat flow $\dot{Q}_o$ from the object being cooled,
2. "warm surface" to reject the waste heat flow $\dot{Q}_{amb}$ to environment,
3. driving power (mechanical/electrical) P

All refrigerators have the above three interfaces. Figure 7.4 shows a conventional home refrigerator. The cold surface is located inside of the cabinet. Outside, on the rear, two devices are placed. The first one is a small compressor driven by electrical power and a black grid made from tubes. The tubes are warm if the compressor runs. Through this surface, the waste heat is removed from the fridge.

7.1.2.3 The Pump Analogy

Prof. Hans Quack found an analogy between a refrigerator and a water pump (see Figure 7.5), pumping the water from a deep source (depth $\boldsymbol{HH_o}$ in Figure 7.5) to the earth's surface $\boldsymbol{HH_{amb}}$ continuously:

(a) the pump is driven by an electric motor and consumes power $\boldsymbol{P}$,

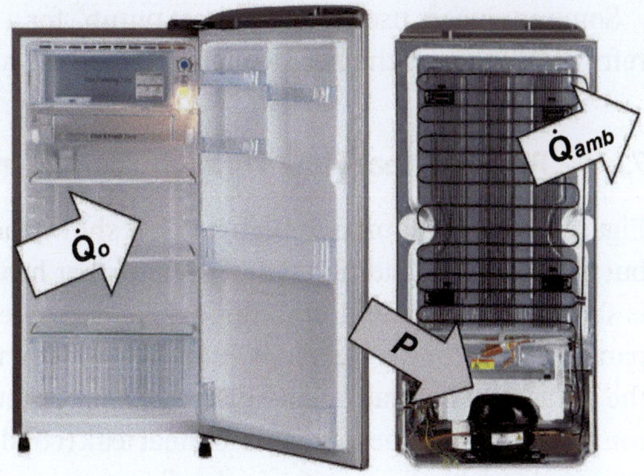

Figure 7.4 Home refrigerator (GL-B201SLLB), Source: LG.

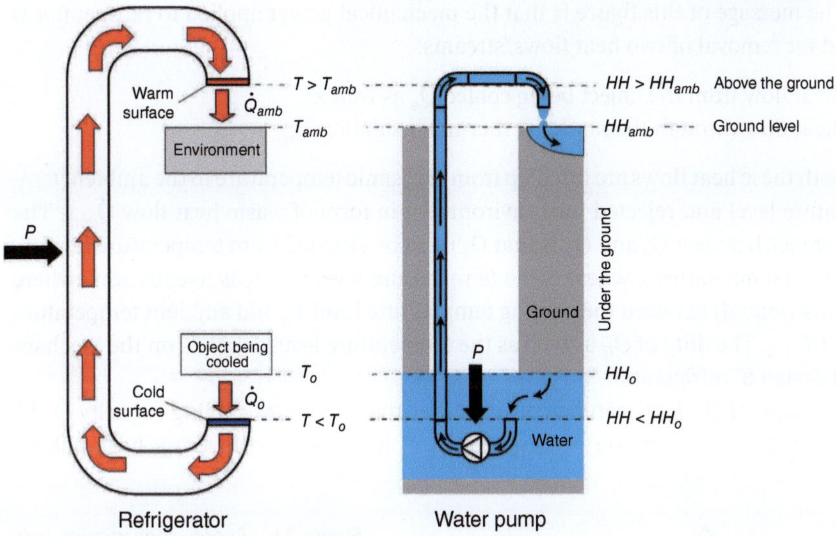

Figure 7.5 Refrigerator-pump analogy.

(b) it pumps the water from the deep source to the environment (surface),
(c) suction (inlet) nozzle of the water pump is located a bit lower than liquid level in the source, and is necessary to guarantee continuous flow to the pump,
(d) pressure at the discharge nozzle is a bit higher than the ambient pressure, otherwise, water could not flow out.

The refrigerator works in a very similar way:

(a) the heat flow $\dot{Q}_o$ is transferred from the object being cooled to the cold surface of the refrigerator,
(b) the refrigerator lifts (elevates) this heat from the cooling temperature to a temperature a bit higher than the ambient temperature; this process requires electrical/mechanical power ***P***,
(c) the resulting heat $\dot{Q}_{amb}$ is rejected to the environment.

Some engineers use wording "heat pump" for a refrigerator, since, functionally, a refrigerator pumps the heat from low temperature to the ambient temperature.

7.1.3 Parasitic Heat Flow Through Thermal Insulation

Figure 7.6 is very similar to Figure 7.3. It shows the same main idea of a refrigerator, but it includes an additional detail – a further heat flow referred to as heat leak $\dot{Q}_{is}$[1] is shown there in addition to three obligatory energy streams $\dot{Q}_o$, P, and $\dot{Q}_{amb}$. This parasitic heat flow $\dot{Q}_{is}$ develops automatically due to a temperature gradient between the ambient temperature and elements of refrigerator, which are colder than environment. Physical mechanisms for heat leak (conduction, convection, and radiation) are considered in Section 6.1 in detail.

The message of this figure is that the mechanical power applied to refrigerator is used for removal of two heat flows/streams:

1. heat flow from the object being cooled $\dot{Q}_o$ as well as
2. heat flow through the nonideal thermal insulation $\dot{Q}_{is}$.

Both these heat flows are lifted up from cryogenic temperature to the ambient temperature level and rejected into environment in form of waste heat flow $\dot{Q}_{amb}$. The difference between $\dot{Q}_o$ and $\dot{Q}_{is}$ is that $\dot{Q}_o$ must be elevated from temperature level T_o (cooling temperature), whereas the temperature level for $\dot{Q}_{is}$ is usually somewhere (or distributed) between the cooling temperature level T_o and ambient temperature level T_{amb}. The duty of $\dot{Q}_{is}$ as well as the temperature level depends on the mechanical design of refrigerator.

The sum of $\dot{Q}_o + \dot{Q}_{is}$ can be considered as the maximum cooling capacity $\dot{Q}_o^{max}$ generated by refrigeration process; a part of it is used for covering heat leakage

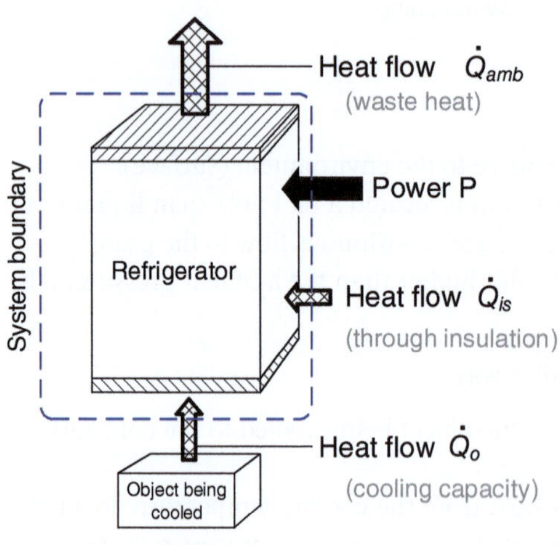

Figure 7.6 Energy conservation law for refrigerator with parasitic heat flow through insulation.

1 The subscript "is" derived from the term "insulation" since this heat flow reaches refrigerator mainly through thermal insulation.

through insulation $\dot{Q}_{is}$, the remaining fraction $\dot{Q}_o$ corresponds to the cooling capacity effectively available for the cooling purposes:

$$\dot{Q}_o^{max} = \dot{Q}_o + \dot{Q}_{is} \tag{7.2}$$

According to this approach, a corrected energy balance equation for a refrigerator can be formulated as follows:

$$\dot{Q}_o^{max} + P = \dot{Q}_{amb} \tag{7.3}$$

This equation is similar to Eq. (7.1), but it considers the heat leak.

7.1.4 Exercise: Energy Conservation for Refrigerator

Given: A refrigerator (see Figure 7.6) provides cooling capacity $\dot{Q}_{amb} = 450\,\text{W}$ at the temperature level $\boldsymbol{T_o} = 4.2\,\text{K}$. The thermal insulation of the cold part is nonideal, a small parasitic heat flow $\dot{Q}_{is} = 50\,\text{W}$ enters cold part of refrigerator from the environment. The plant is driven by a compressor with a consuming power $P = 154.5\,\text{kW}$.
Question: Quantify the heat flow rejected to environment (waste heat) $\dot{Q}_{amb}$ required for design of cooling water system.
Solution: Let's apply the energy conservation law to this system; driving power P, heat from the cooling object $\dot{Q}_o$, and heat leak through insulation $\dot{Q}_{is}$ are entering the system, and waste heat flow is leaving the system, meaning that:

$$\dot{Q}_{amb} = P + \dot{Q}_o + \dot{Q}_{is}$$

$$\dot{Q}_{amb} = 154.5\,\text{kW} + 0.45\,\text{kW} + 0.05\,\text{kW} = 155\,\text{kW}$$

An engaged observer will immediately notice that the heat flow $\dot{Q}_o$ as well as the $\dot{Q}_{is}$ are considerably smaller in comparison to the driving power P. Therefore, the waste heat flow roughly corresponds to the compressor power. That is very typical for cryogenic refrigeration.

7.1.5 Reversible Refrigerator and Carnot Equation

The energy conservation law delivers some information about the refrigerator's performance. However, this information is insufficient for analysis of the system. For example, it cannot answer the question: what would be the power consumption of a refrigerator for defined cooling capacity $\dot{Q}_o$ (for example, $\dot{Q}_o = 100\,\text{W}$) and defined cooling temperature $\boldsymbol{T_o}$ (for example $\boldsymbol{T_o} = 100\,\text{K}$)? The energy conservation law expresses only that generated waste heat flow is a sum of cooling capacity and driving power: $\dot{Q}_{amb} = \dot{Q}_o + P$. It will be

$$\dot{Q}_{amb} = 1100\,\text{W if the driving power is given as } \boldsymbol{P} = 1000\,\text{W}, \text{ or}$$

$$\dot{Q}_{amb} = 101\,\text{W if the driving power is given as } \boldsymbol{P} = 1\,\text{W},$$

despite how unlikely the last number may look.

7.1.5.1 Reversible Refrigerator

Concept of reversible refrigerator (see Section 4.2) is useful for answering the question about performance of a refrigerator: In a reversible refrigeration process, the driving mechanical power $\boldsymbol{P}$ is fully transformed in the generation of cooling capacity $\dot{Q}_o$. Literary, the wording "reversible" means that this process can be operated in reversed way, whereas the cooling capacity $\dot{Q}_o$ is transformed back into the mechanical power $\boldsymbol{P}$. The classical reversible process/engine is the well-known Carnot process/engine described in every thermodynamic textbook.

7.1.5.2 Carnot Equation

For the reversible refrigerator, the so-called Carnot equation, originally introduced by Rudolf Clausius, defines the required power as follows:

$$P_{min} = P_{rev} = \dot{Q}_o \cdot \frac{T_{amb} - T_o}{T_o} \tag{7.4}$$

whereas the term P_{min} means the minimal driving power required for generation of cooling capacity $\dot{Q}_o$ at the required temperature level T_o for given T_{amb}. The terms P_{rev} (power required for reversible refrigerator) and P_{min} are used synonymously.

7.1.5.3 Extension of Carnot Equation to Irreversible Refrigerator

For irreversible processes, the Carnot equation is usually defined in form of an inequation (with sign "≥") as follows:

$$P \geq \dot{Q}_o \cdot \frac{T_{amb} - T_o}{T_o} = \dot{Q}_o \cdot \left(\frac{T_{amb}}{T_o} - 1 \right) \tag{7.5}$$

7.1.5.4 What Carnot Equation Expresses?

This equation describes the relationship between four parameters: driving power, cooling capacity, cooling temperature, and ambient temperature. It contains four statements.

1. The power required for driving a refrigerator P depends on the cooling capacity $\dot{Q}_o$ linearly: the larger the required cooling capacity → the higher the required driving power,
2. The power required for driving a refrigerator P depends on the ambient temperature T_{amb} linearly: the higher the ambient temperature → the higher the required driving power,
3. The power required for driving a refrigerator P depends on the cooling temperature T_o: the lower the cooling temperature → the more power is required.
   The first three statements do not require further explanation. However, the fourth statement is a bit different:
4. The last function (correlation between driving power and cooling temperature T_o) is nonlinear, it is

   $$P \sim \frac{1}{T_o}$$

Exercise for Application of Carnot Equation:

What would be the power consumption of a refrigerator for cooling capacity of $\dot{Q}_o = 100\,\text{W}$ at $\boldsymbol{T_o} = 100\,\text{K}$?

Answer: assuming the ambient temperature of 300 K (≈27 °C) and according to Carnot equation:

$$P_{min} = 100\,\text{W} \cdot \frac{(300\,\text{K} - 100\,\text{K})}{100\,K} = 200\,\text{W}$$

A reversible refrigerator will require at least $P = 200$ W.

The real refrigerator will require more than 200 W.

A refrigerator with less than $\boldsymbol{P_{min}} = 200$ W for given specification cannot be realized at all.

7.1.6 The Second Law of Thermodynamics for Refrigerator

The formal definition of second law of thermodynamics for refrigerator follows from Carnot equation directly, it can be expressed as follows:

1. By nature, the heat can flow only in one direction: from objects with higher temperatures to objects with colder temperatures;
2. The heat transport in the opposite direction (from cold to hot) is possible by means of a refrigerator, which requires power/energy for this service;
3. In the case of a reversible machine, the power can be defined as

$$P_{min} = P_{rev} = \dot{Q}_0 \frac{T_{amb} - T_0}{T_0}$$

A real refrigerator will need more power due to irreversibility $P_{real} > P_{min}$.

7.1.7 Minimum Power Requirement for Typical Cooling Temperatures

A limited number of substances are used in cryogenics, with the most popular being nitrogen, hydrogen, and helium, with normal boiling temperatures of roughly 77 K, 20 K, and 4.2 K. Therefore, these temperatures are considered to be natural "standard" temperature levels. The impact of cooling temperature on min power required for refrigeration is investigated in this section for these three temperature levels.

Table 7.2 collects numbers for power consumption of three **reversible** refrigerators providing the same cooling capacity $\dot{Q}_0 = 100$ W for these three "standard" cooling temperatures. The refrigerator for liquid nitrogen temperature level needs $\boldsymbol{P} \approx 290$ W for this purpose. The refrigerator for liquid hydrogen temperature level requires roughly four times more $\boldsymbol{P} \approx 1400$ W for the same cooling capacity. The refrigerator for liquid helium temperature level requires considerably more power $\boldsymbol{P} \approx 7050$ W; this is 25 times more than the refrigerator for the nitrogen temperature level.

Table 7.2 Minimum power requirement for typical temperature levels, $T_{amb} = 300$ K.

	Liquid nitrogen (LIN)	Liquid hydrogen (LH_2)	Liquid helium (LHe)
Cooling temperature T_0, K	77.4	20.1	4.2
Required cooling capacity $\dot{Q}_0$, W	100	100	100
Min. required driving power P, W	288	1393	7043
$P/\dot{Q}_0$	2.9	13.9	70.4

***Exercise: Helium Refrigerator for* 1.8 K**

A scientific laboratory has an older helium refrigeration system with cooling capacity $\dot{Q}_0 = 450$ W at $\boldsymbol{T_0} = 4.2$ K. The scientists wish to upgrade this system and produce the same cooling capacity, but at $\boldsymbol{T_0} = 1.8$ K. The difference between 4.2 and 1.8 K is only 2.4 K; it seems to be very small. The expectation is that the upgrade would not be very costly, but it would yield a hugely positive impact on the lab's capabilities because a superfluid helium (at 1.8 K) would then be available.

First, the scientist wishes to check what this change means for a refrigerator in terms of power consumption. For example, they can find out the minimum power consumption of the existing system using Carnot equation. After that, they can determine the minimum power consumption for the future 1.8 K system and compare both numbers.

Assuming ambient temperature of $T_{amb} = 293$ K ($\approx$20 °C) the min power for 4.2 K system is:

$$P_{min} = 450\,\text{W} \cdot \frac{293\,\text{K} - 4.2\,\text{K}}{4.2\,\text{K}} = 30\,943\,\text{W} = 30.9\,\text{kW}$$

The minimum driving power for 1.8 K-system would be:

$$P_{min} = 450\,\text{W} \cdot \frac{293\,\text{K} - 1.8\,\text{K}}{1.8\,\text{K}} = 72\,800\,\text{W} = 72.8\,\text{kW}$$

Surprisingly, the new number of ca. 73 kW is considerably higher than the number for the 4.2 K-system by a factor of 2.4. Thus, the future 1.8 K-system will be 2.4 times larger than the existing plant. It will consume more power and probably require a significant investment. Therefore, it makes sense to analyze and check the upgrade idea once more.

7.1.8 Key Performance Indicators for Refrigerators

Three indicators used for characterization of refrigerator performance are described in this section: Coefficient of performance (COP), Carnot efficiency, and specific power.

7.1.8.1 Coefficient of Performance

A measure of the performance of a refrigerator is the COP, defined according to the well-known principle "benefit divided by effort." The benefit of the refrigerator is the cooling capacity $\dot{Q}_0$; the effort is power $\boldsymbol{P}$ required for production of cooling capacity. Therefore:

$$\boldsymbol{COP} = \frac{\dot{\mathbf{Q}}_0}{\mathbf{P}} \tag{7.6}$$

Let's try to calculate COP for refrigerator described in Section 7.1.4:

$$COP = \frac{\dot{Q}_0}{P} = \frac{0.45\,kW}{154.5\,kW} = 0.0029 = 0.29\%$$

That is a strange feeling saying that this system is a high-efficiency system, since the COP = 0.3 % seems to be very small. Let us try to estimate the COP for a reversible refrigerator for the same cooling capacity and temperature level using the Carnot equation:

$$P_{rev} = 450\,\text{W} \cdot \frac{293\,K - 4.2\,K}{4.2\,K} = 30943\,\text{W} = 30.9\,kW$$

$$COP_{rev} = \frac{\dot{Q}_o}{P} = \frac{0.45\,kW}{30.9\,kW} = 0.0145 = 1.45\%$$

Surprisingly, the reversible refrigerator also has a very low COP, at less than 1.5 %. This fact tells us that the COP of helium refrigerators is always very small.

The more general remark is the following: the generation of cold/cooling capacity at very low temperature requires a considerable power, even in a reversible process. The low COP value for liquid helium temperatures reflects this fact.

7.1.8.2 Carnot Efficiency

To compare the performance of practical systems, we may use another key performance indicator – the so-called Carnot efficiency η_e (sometimes referred to as Carnot fraction ***CF*** or Figure of Merit ***FOM***) defined as COP of considered refrigerator in relation to COP of a reversible refrigerator (calculated by means of Carnot equation). This parameter makes more sense than COP because it indicates potential improvements:

$$\eta_e = \frac{\mathbf{COP}}{\mathbf{COP_{rev}}} \tag{7.7}$$

For the refrigerator described in Section 7.1.4, the Carnot efficiency amounts to 20%: $\eta_e = 0.29\%/1.45\% = 0.2 = 20\%$. It is high efficiency for the 4.2 K temperature level.

7.1.8.3 Specific Power

The specific power $\boldsymbol{p_{sp}}$ is defined as follows:

$$\boldsymbol{p_{sp}} = \frac{\mathbf{P}}{\dot{\mathbf{Q}}_\mathbf{o}} \tag{7.8}$$

It is a value reversed to COP. This parameter is sometimes more convenient for using since $\boldsymbol{P}$ is usually higher than $\dot{Q}_o$ and – consequently – $\boldsymbol{p_{sp}}$ is a double-digit number.

7.2 Joule–Thomson Process (Nitrogen Joule–Thomson Process)

Historically, the Joule–Thomson process was the first process introduced for industrial cryogenic applications.[2] It is still used for cooling purposes at the liquid nitrogen temperature level. This process may be realized with gases like air, argon, oxygen, and nitrogen as working fluids. But most of these fluids have some drawbacks

2 In a version for liquefaction purpose, which differs a bit from the version for refrigeration purpose described in this section.

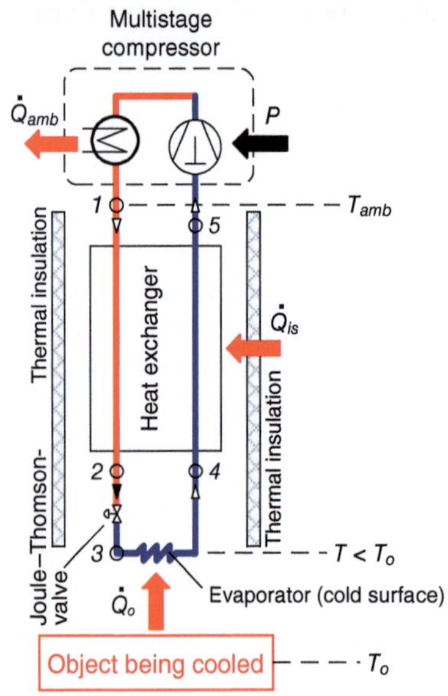

Figure 7.7 Joule–Thomson process, basic process flow diagram; left- thermal insulation is shown, right- thermal insulation is removed.

(argon is expensive, oxygen is dangerous, and air boiling temperature is not constant); therefore, the nitrogen-based Joule–Thomson process is the most popular version.

7.2.1 Main Elements

The basic process flow diagram (BPFD) is shown in Figure 7.7. The main elements of a Joule–Thomson refrigerator are:

1. Compressor unit, which is usually a multi-stage isothermal compressor.
2. Countercurrent heat exchanger.
3. Expansion valve, often referred to as a Joule–Thomson valve.
4. Evaporator.
5. Thermal insulation.

7.2.1.1 Thermal Insulation/Coldbox

In terms of mechanical design, the cold part of the system (heat exchanger, expansion valve, and evaporator) is placed in a rectangular box (more typical for large-scale systems) or in a cylinder-shaped vessel, which is filled with an appropriate insulation material, see section 6.1.

7.2.1.2 Heat Exchanger

Heat Exchanger as unit operation is described in Section 5.3.

Basically, there are no heat exchangers designed and developed specifically for Joule–Thomson refrigerators. The conventional approach is to look at the heat exchangers available on the market, for they may meet the requirements. The process design is usually adapted to the design of the heat exchanger.

The design of large industrial cryogenic plants is based on either aluminum-brazed plate-fin heat exchangers (see Section 6.2.1) or spiral-would heat exchangers (see Section 6.2.2). Unfortunately, the brazed aluminum plate-fine exchanger is not ideal for Joule–Thomson process, since (i) this kind of heat exchanger is sensitive to large temperature differences, which are typical for Joule–Thomson process; (ii) aluminum plate-fin heat exchangers for high pressures of above 100 bar, which are also typical for Joule–Thomson process, are relatively expensive. The so-called spiral-wound heat exchanger is considered as a suitable option for high-pressure processes with high-temperature differences in the heat exchanger, because it can manage these requirements well.

In the simplest "laboratory" version the heat exchanger can be designed as the so-called double-tube heat exchanger, see Section 5.3.1.

7.2.1.3 Compressor

The compressor is the most expensive and sophisticated part of the system. Since high pressure is an important feature of the Joule–Thomson process and the optimal pressure range is rather higher than 100 bar, an appropriate multi-stage compressor is required for realization of the process.

The multi-stage compressor is a complex system consisting of a chain of compressing stages and intermediate coolers after every compressing stage. The corresponding thermodynamics of multi-stage compression is given in Section 5.5.

Similar to the situation with heat exchangers, there are no compressors designed and developed specifically for Joule–Thomson refrigerators. The conventional approach is to look at the compressors available on the market, for they may meet the requirements. The process design is usually adapted to performance of compressor.

In terms of mechanical design: reciprocating compressor is a suitable option (see Section 6.5.4). Another suitable option for small-scale devices is the so-called diaphragm compressors, see Ref. [1].

7.2.1.4 Expansion Valve

Since the Joule–Thomson cooling effect is independent of the expansion device design (as mentioned in Section 5.1.1), every device able to cause an appropriate pressure drop – orifice, capillary, porous plug, etc. – may be used for expansion purposes; the pressure drop will be converted into the temperature change. These expansion devices mentioned above are used mainly if the low costs for the refrigerator are targeted. The issue is that all these devices have no control function. In this sense, a valve is a convenient expansion device because it allows control of the flowrate. Therefore, the valve is often used for isenthalpic expansion, see Section 5.1 for more information.

7.2.1.5 Evaporator

Evaporator provides the cold surface, mentioned in Section 7.1.1, which absorbs the heat from the object being cooled. The object needs to be thermally connected to

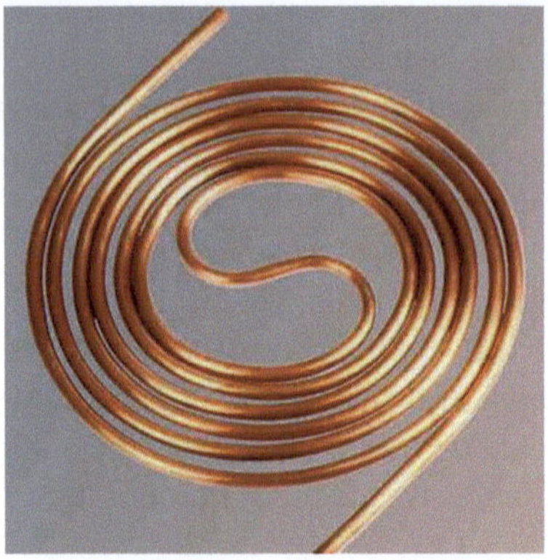

Figure 7.8 Copper tube as evaporator.

the evaporator. In the simplest version, it may be a metallic tube (like copper tube shown in Figure 7.8). The heat power from the object is transferred to a cold fluid flowing inside of the tube in this device. Another design example is described in Section 7.2.3.2.

7.2.1.6 Joule–Thomson Refrigerator Assembled

In this section, a version of a virtual Joule–Thomson refrigerator is shown, which can be easily built by volunteers in every university laboratory if a suitable compressor is available. If all required components are assembled, the final system may look like somewhat shown in Figure 7.9. I hope, this illustration helps establish a visual connection between theory (in form of BPFD) and reality (in form of a laboratory system).

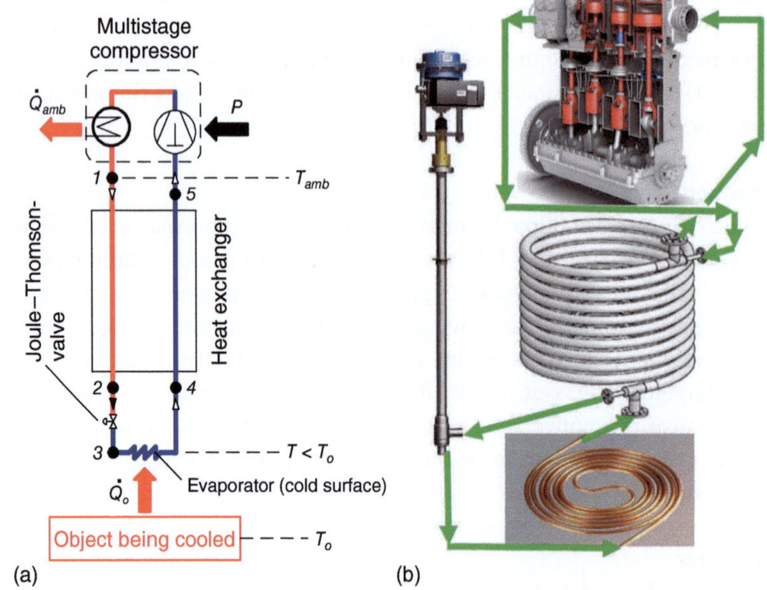

Figure 7.9 Joule–Thomson refrigerator, (a) basic process flow diagram, (b) "laboratory" version.

7.2.2 Process and Process Features

7.2.2.1 Cool-Down Procedure

If all relevant components are assembled as shown in Figure 7.9, tested, and the system is filled with nitrogen (e.g. it is connected to a corresponding nitrogen buffer tank), then it is ready for use. Let's look at what happens inside this system if the compressor is switched on, in a sort of slow motion.

Initial state: At the beginning of the experiment, all elements have the same temperature, which corresponds to the ambient temperature. It is assumed to be

$T_{amb} = 300\,\text{K}.$

The experiment starts if the compressor is switched on. The nitrogen is compressed there to high-pressure level. In this virtual experiment a very high pressure of

$p_{HP} = 200\,\text{bar}$

is assumed. The pressure level on the compressor intake (surge) is assumed to be

$p_{LP} = 1\,\text{bar}.$[3]

At the outlet compressor stage working fluid (nitrogen) is cooled (in corresponding intermediate coolers) to the ambient temperature level. Therefore, the state at the outlet of the compressor (marked as "1" in Figures 7.7 and 7.10) is characterized by:

$p_1 = p_{HP} = 200\,\text{bar}$

$T_1 = T_{amb} = 300\,\text{K}.$

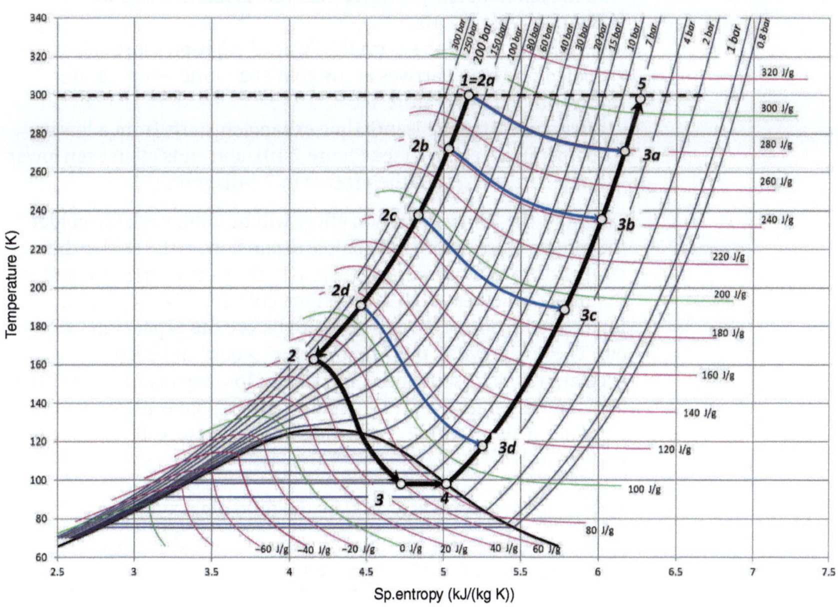

Figure 7.10 Cool-down procedure of a Joule–Thomson refrigerator.

3 LP means "low pressure," HP means "high pressure."

	The compressed refrigerant flows through the heat exchanger (inner tube of the double-tube heat exchanger) and arrives at the cold end without any temperature change (because the heat exchanger has the same temperature). Therefore, the condition at the cold end of the heat exchanger (marked as point 2 in Figure 7.7) is similar to the condition in point 1, only the pressure is slightly lower due to pressure losses in the heat exchanger: $p_2 < p_1$.
Cooling effect:	The cool-down process starts with isenthalpic expansion of the refrigerant by means of expansion valve from high-pressure level p_{HP} to low-pressure level p_{LP}. The expansion process itself is described in detail in Section 5.1.1. It results in a temperature drop due to the Joule–Thomson effect. The expansion from ambient temperature level is visualized as process step 1 → 3a in *T,s*-diagram (Figure 7.10). The resulting cold gas is passed through the evaporator, which is thermally connected to the cooling object. This way, the cool-down procedure of both of these parts – the object and the evaporator – is initiated. The cooling rate depends on nitrogen flowrate, material heat capacity, and heat transfer condition.
Enhancing of cooling effect by means of heat exchanger:	The key point is that after some time, this low-pressure gas arrives and enters the heat exchanger (at the cold end) with a temperature slightly colder than ambient temperature. This colder gas flows through the heat exchanger from the "cold end" to the "warm end." Due to the temperature difference between the "warm" and the "cold" material streams, the heat transfer is initiated: – the gas on the low-pressure "cold" side is warmed up quite close to the ambient temperature; this state is marked as 5 in Figures 7.7 and 7.10. – simultaneously, the gas on the "warm" high-pressure side is cooled down and arrives at the cold end at the temperature lower than temperature T_1. – after some time the isenthalpic expansion starts from a lower temperature, e.g. T_{2b} (see Figure 7.10), and ends at an even lower temperature T_{3b}. See the process in *T,s*-diagram. This way, the Joule–Thomson effect will be enhanced; the colder the temperature before isenthalpic expansion → the colder the temperature will be after expansion, both in the evaporator and at the inlet of the heat exchanger on the cold side. This causes a further reduction of the temperature before the expansion valve, etc. The cooldown of the whole system occurs and even accelerates. After some time, the isenthalpic expansion starts from a lower temperature T_{2c} (see Figure 7.10) and ends at temperature T_{3c}, later it starts with T_{2d} and ends at temperature T_{3d}, and so on and so forth.
Final state:	The final state is characterized by the following conditions: – The temperature at the inlet of the valve is in the range of 150–165 K[4] (point 2 in Figure 7.10).

4 The exact temperature depends on the high-pressure level used in process.

- At these conditions, the isenthalpic expansion (process step 2 → 3) ends in the two-phase area (point 3 in Figure 7.10). This means that a fraction of the nitrogen after expansion is liquefied. Actually, according to the process 2 → 3 shown in Figure 7.10, only a small part of the fluid is liquefied after expansion: roughly 10%, while the flash-gas amounts to roughly 90% in terms of mass fraction. In terms of volumetric flow, the liquid fraction is even lower because the density of liquid nitrogen is considerably higher than the density of the flash-gas. Visually, this two-phase flow looks like a gaseous flow with few tiny liquid droplets inside. Nonetheless, the liquid is available.
- This liquid–vapor mixture flows through the evaporator (process step 3 → 4) and cools the cooling object, whereby the liquid droplets evaporate, and the resulting cold vapor (point 4 in Figure 7.10) is provided to the cold end of the heat exchanger.
- This cold material stream flows through the heat exchanger (process step 4 → 5), where it is warmed up quite close to the ambient temperature (point 5 in Figures 7.7 and 7.10),
- It flows further to the compressor, where it is compressed to high-pressure level and cooled to ambient temperature level (process step 5 → 1 in Figures 7.7 and 7.10).
- This high-pressure stream is cooled in the heat exchanger (process step 1 → 2) to the temperature level of 150–165 K mentioned above by cold low-pressure nitrogen stream.

The typical temperature profile in the heat exchanger is given in Figures 7.15 and 7.16.

The most important lesson learned from this experiment is that

Learning 7.1 A nitrogen Joule–Thomson refrigerator can cooldown itself and the cooling object from the ambient temperature to the required cooling temperature. This is an important property for practical applications.

7.2.2.2 Joule–Thomson Process Description

The process sequence is given in Section 7.2.2.1 as the final state at the end of the cool-down procedure.

7.2.2.3 Cooling Capacity of the Joule–Thomson Process

In this section, the cooling capacity of a classic Joule–Thomson process is considered by means of energy balance for the cold part of the process (heat exchanger, valve, and evaporator).

As shown in Figure 7.11, four energy streams are crossing the system boundary:

$\dot{Q}_0$ – heat flow from the object being cooled
$\dot{Q}_{is}$ – heat flow from the environment into the coldbox through thermal insulation
$\dot{H}_1$ – enthalpy flow (in the form of material flow) on the warm end of the heat exchanger, high-pressure stream, inlet
$\dot{H}_5$ – enthalpy flow (in form of material flow) on the warm end of the heat exchanger, low-pressure stream, outlet

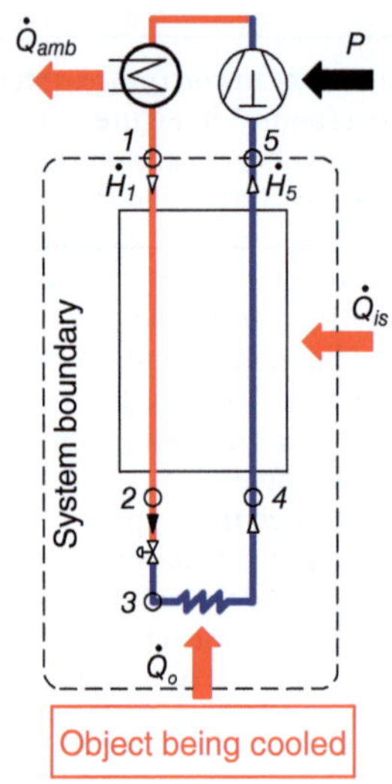

Figure 7.11 Joule–Thomson process, system boundary.

The energy balance equation is as follows:

$$\dot{Q}_o + \dot{Q}_{is} = \dot{H}_5 - \dot{H}_1 \tag{7.9}$$

According to approach described in Section 7.1.3, the sum of $\dot{Q}_o + \dot{Q}_{is}$ can be considered as the maximal cooling capacity $\dot{Q}_o^{max} = \dot{Q}_o + \dot{Q}_{is}$ generated in the process (a part of it is used for covering heat leakage through insulation $\dot{Q}_{is}$, the remaining fraction $\dot{Q}_o$ corresponds to the cooling capacity available for the cooling purposes).

The enthalpy difference $\dot{H}_5 - \dot{H}_1$ corresponds to the enthalpy difference at the warm end of the heat exchanger: $\Delta\dot{H}^{WE} = \dot{H}_5 - \dot{H}_1 = \dot{m}\,(h_5 - h_1)$, where $\dot{m}$ is the flowrate of the refrigerant. Therefore,

$$\dot{Q}_o^{max} = \Delta\dot{H}^{WE} \tag{7.10}$$

The Eq. (7.10) expresses

Learning 7.2 The maximal cooling capacity of the Joule–Thomson process depends only on the enthalpy difference at the warm end of the heat exchanger $\Delta\dot{H}^{WE}$.

For simplification purposes, it is often assumed[5] that the cold part of the process is thermally well-insulated, therefore the heat flow through insulation is negligibly small ($\dot{Q}_{is} \to 0$). The energy balance equation in this case is as follows:

$$\dot{Q}_o = \Delta\dot{H}^{WE} = \dot{H}_5 - \dot{H}_1 = \dot{m}\,(h_5 - h_1) \tag{7.11}$$

5 This assumption ($\dot{Q}_{is} \to 0$) is usually acceptable for cryogenic systems for liquid nitrogen temperature level and higher. For lower temperatures, especially for the helium temperature level, the heat inleakage through the thermal insulation is always considered since it may significantly reduce the available cooling capacity of the process.

7.2.2.3.1 Joule–Thomson Process with Infinitely Large Heat Exchanger

The term $\Delta\dot{H}^{WE}$ is remarkable, as it is often used when analyzing cryogenic processes. Therefore, it makes sense to consider it in detail.

Valuable knowledge can be generated by considering a process based on a heat exchanger with an infinitely large heat transfer surface.[6] In such a heat exchanger, the heat transfer may be realized even at an infinitesimal temperature difference. The main consequence of this assumption for the Joule–Thomson process is that the temperature of the cold stream at the outlet of the heat exchanger moves very close to the temperature T_1. This "new" state is marked by point "5id"[7] in Figure 7.12. The temperature T_{5id} is still lower than T_1, but the temperature difference $T_1 - T_{5id}$ is infinitesimal. The Joule–Thomson process with an infinitely large heat exchanger corresponds to the process chain 1 → 2 → 3 → 4 → 5id. Other distinguishing process points shown in Figure 7.12 in the temperature–entropy diagram are states 1, 3a, and 5 used/defined already in the previous Section 7.2.2.1. Table 7.3 summarizes the definitions of these states.

For considered process (1 → 2 → 3 → 4 → 5id), the Eq. (7.10) can be modified as follows:

$$\Delta\dot{H}^{WE} = \dot{H}_{5id} - \dot{H}_1 \tag{7.12}$$

Important definition	This enthalpy difference is often referred to as the isothermal enthalpy difference at the warm end $\Delta\dot{H}_T$, respectively Δh_T, since both these enthalpy values are given at the same temperature T_1, the corresponding equation is as follows:

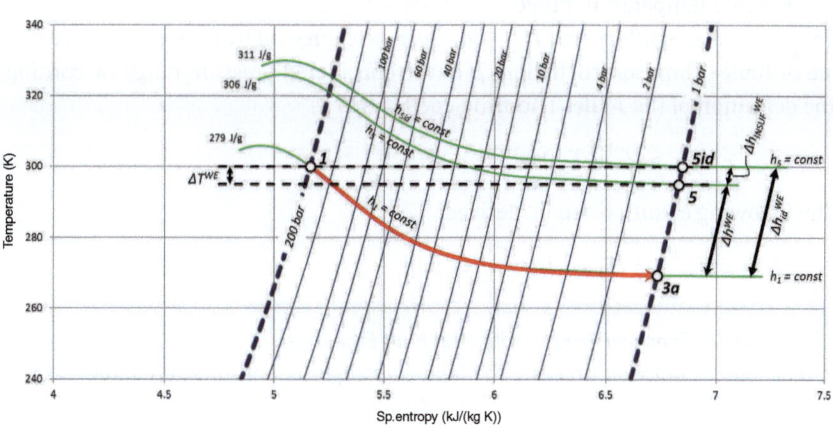

Figure 7.12 Warm end of heat exchanger in *T,s*-diagram.

6 On the one hand, a process with an infinitely large heat exchanger is only a virtual process. On the other hand, this process is not completely unrealistic because it is always possible to use an oversized heat exchanger and achieve a similar effect: the larger the heat exchanger, the closer the process is to the virtual process considered in this section.

7 Whereby the subscript "id" means "idealized." For more information concerning idealization see: https://en.wikipedia.org/wiki/Idealization_(science_philosophy).

Table 7.3 Definitions of some distinguishing process states.

State	Description	Corresponding process parameters
1 (point 1)	The high-pressure stream (p_{HP}) at the inlet of the heat exchanger	$T_1, p_1, \dot{H}_1, s_1$
5 (point 5)	The state of the low-pressure stream at the outlet of the heat exchanger	$T_5, p_5, \dot{H}_5, s_5$
5id (point 5id)	Characterized by: – temperature equal to the temperature in state 1: $T_{5id} = T_1$ – the pressure corresponds to the p_{LP}	$T_{5id}, p_{5id}, \dot{H}_{5id}, s_{5id}$
3a (point 3a)	characterized by: – pressure is equal to the pressure in state 5: $p_{3a} = p_5$ – the enthalpy is equal to the enthalpy in state 1: $\dot{H}_{3a} = \dot{H}_1$ (or $h_{3a} = h_1$)	$T_{3a}, p_{3a}, \dot{H}_{3a}, s_{3a}$

$$\dot{H}_{5id} - \dot{H}_1 = \Delta\dot{H}_T{}^{WE} = \dot{m}\,\Delta h_T{}^{WE} \tag{7.13}$$

The Eq. (7.13) can be rewritten as:

$$\Delta\dot{H}_T{}^{WE} = \dot{H}_{5id} - \dot{H}_1 = \dot{H}_{5id} - \dot{H}_{3a} = \dot{m}\,c_p\,(T_{5id} - T_{3a}) \tag{7.14}$$

where c_p is the specific heat capacity of the refrigerant on the low-pressure side in the considered temperature range.

The temperature difference $(T_{5id} - T_{3a})$ can be expressed by means of the average value of Joule–Thomson coefficient $\overline{\mu}$ in the considered pressure range (according to the definition of the Joule–Thomson coefficient) as

$$(T_{5id} - T_{3a}) = (T_1 - T_{3a}) = \overline{\mu}\,(p_1 - p_5)$$

The following equation can be derived:

$$\dot{Q}_o = \Delta\dot{H}_T{}^{WE} = \dot{H}_{5id} - \dot{H}_1 = \dot{m}\,c_p\,\overline{\mu}\,(p_1 - p_5) \tag{7.15}$$

7.2.2.3.2 Joule–Thomson Process with Real Heat Exchanger

In a real Joule–Thomson process, the heat exchanger is not infinitely large; rather, a finite temperature difference $\Delta T^{WE} > 0$ (usually it is 2–10 K) occurs at the warm end. Therefore, the available enthalpy difference $\Delta\dot{H}^{WE} = \dot{H}_5 - \dot{H}_1$ is smaller than the enthalpy difference $\Delta\dot{H}_T{}^{WE} = \dot{H}_{5id} - \dot{H}_1$. The link between these two terms is shown in the following equation:

$$\Delta\dot{H}^{WE} = \Delta\dot{H}_T{}^{WE} - (\dot{H}_5 - \dot{H}_{5id}) \tag{7.16}$$

The enthalpy difference $(\dot{H}_5 - \dot{H}_{5id})$ itself can be expressed as

$$(\dot{H}_5 - \dot{H}_{5id}) = \dot{m}\,c_p\,(T_{5id} - T_5) = \dot{m}\,c_p\,\Delta T^{WE} \tag{7.17}$$

Therefore, the real enthalpy difference at the warm end is as follows:

$$\Delta H^{WE} = \Delta \dot{H}_T{}^{WE} - \dot{m}\, c_p\, \Delta T^{WE} \quad (7.18)$$

Correspondingly, the cooling capacity of a Joule–Thomson process is defined as

$$\dot{Q}_o = \Delta \dot{H}_T{}^{WE} - \dot{m}\, c_p\, \Delta T^{WE} \text{ or} \quad (7.19)$$

$$\dot{Q}_o = \dot{m}\, c_p\, \overline{\mu}\, (p_1 - p_5) - \dot{m}\, c_p\, \Delta T^{WE} \text{ or} \quad (7.20)$$

$$\dot{Q}_o = \dot{m}\, c_p\, \left[\overline{\mu}\; (p_1 - p_5) - \Delta T^{WE}\right]$$

The Eq. (7.20) is important since it is a basis for

1. Process Optimization; see the Section 7.2.4;
2. Development of the control mechanism and
3. development of ideas and strategy, how the Joule–Thomson process may be further improved.

The following are learnings from Eq. (7.20):

Learning 7.3 This equation expresses that cooling capacity of Joule–Thomson process depends on:

1. difference between both process pressures $p_1 - p_5$,
2. average Joule–Thomson coefficient[8] $\overline{\mu}$,
3. heat capacity of used refrigerant c_p.

Learning 7.4 Temperature difference at the warm end of the heat exchanger ΔT^{WE} has a negative impact on the cooling capacity of the Joule–Thomson process. Therefore, minimization of ΔT^{WE} must be targeted during the process design phase.

Further Observations:

1. The cooling capacity of a Joule–Thomson process is independent of the cooling temperature achieved in the process. This is atypical for cryogenic refrigeration, given that the cooling capacity of a process normally decreases if required cooling temperature is reduced. This is not the case in the Joule–Thomson process.
2. According to Eq. (7.20), a reduction of the process pressure $\boldsymbol{p_5}$ (low-pressure level) can impact the cooling capacity of the process. However, this statement is not fully correct. Since the Joule–Thomson coefficient at low pressures is usually close to zero (cryogenic gases can be considered as an ideal gas at this condition), the decreasing of p_5 has no impact or only minimal impact on the cooling capacity of the cycle.

8 Probably, the term "Joule–Thomson process" originates from the fact that the cooling capacity of this process is defined by the Joule–Thomson coefficient.

7.2.2.3.3 How to Increase the Cooling Capacity of the Joule–Thomson Process

According to the Eq. (7.20), the cooling capacity of a Joule–Thomson process can be increased by the following measures:

1. Increase of refrigerant flowrate $\dot{m}$ (trivial solution),
2. Increase of process pressure on the high-pressure side p_1,
3. Use of other refrigerants (argon, oxygen, etc.) with higher Joule–Thomson coefficients and heat capacities,
4. Designing the process in such a way that the Joule–Thomson coefficient is maximized. The background here is that the cooling capacity depends on the **average** JT coefficient. Therefore it makes sense, to choose the process parameters (high-pressure level or temperature at the warm end of heat exchanger) in such a way that the value "on average" is as high as possible. Two corresponding process design approaches are discussed in Section 9.1.2.

7.2.2.4 Typical Process Pressure Levels

Nitrogen Joule–Thomson process uses two pressure levels only:

1. High-pressure level, defined by pressure at the outlet of compressor.
2. Low-pressure level, defined by pressure at the inlet of compressor.

7.2.2.4.1 High-Pressure Level

As already mentioned in Section 7.2.1, the typical high-pressure level p_{HP} is above 50 bar (therefore, a multi-stage compressor is required for the process). There are two reasons for this:

1. The cooling capacity of the Joule–Thomson process depends mainly on high-pressure level and Joule–Thomson coefficient of used refrigerant, according to Eq. (7.20). The available single-component refrigerants, such as nitrogen, oxygen, argon, and air, have a relatively low Joule–Thomson coefficient (e.g. nitrogen with less than 0.25 K/bar). Therefore, high pressure is required simply in order to generate a reasonably specific cooling capacity.
2. Another important fact is that the Joule–Thomson coefficient itself depends on the pressure. The reasonable cooling effect (ΔT) occurs only at pressures higher than 10–20 bar.

7.2.2.4.2 Low-Pressure Level

The low-pressure level p_{LP} is predefined by used refrigerant and required cooling temperature T_o, since the cooling temperature T_o is quite close to the boiling temperature of the liquid refrigerant T_b, which evaporates in the evaporator at given pressure p_{LP}:

$$T_o = T_b + \Delta T^{EVAP} \tag{7.21}$$

whereas the boiling temperature T_b is a function of p_{LP}: $T_b = T_b\,(p_{LP})$.

The temperature difference ΔT^{EVAP} (needed as driving force for heat transfer from object being cooled to refrigerant in evaporator) is minor only, usually, there are few degrees of Kelvin: $\Delta T^{EVAP} < 3$ K.

Table 7.4 Typical cooling temperature ranges as well as pressures ranges.

Cooling temperature (K)	Low-pressure level (bar)	Comment
≈80	≈1	Since many single cryo applications use liquid nitrogen as refrigerant, which boils at 77.4 K at atmospheric pressure, the first typical cooling temperature is T_0 ≈ 80 K, and the corresponding pressure level is close to 1 bar. In closed-cycle systems the pressure is usually slightly above the atmospheric pressure in order to avoid subatmospheric conditions (see comment below)
≈100	≈ 6	Another typical low-pressure level is ≈6 bar, and the corresponding cooling temperature is roughly 100 K. The background is that all liquid nitrogen is produced by means of either air separation plants or nitrogen liquefiers linked to air separation plants. Both these systems use processes with pressure levels close to 6 bar. Therefore, this pressure level is a sort of standard pressure level for nitrogen-based processes
Less than 70	Less than 1 (subatmospheric conditions)	For some applications, such as those based on high-temperature superconductors, a cooling temperature lower than 70 K would be advantageous. The corresponding boiling pressure would be less than the atmospheric pressure. That is a critical issue. The process design engineers are always trying to avoid subatmospheric conditions in process plants because of the significant disadvantages, they pose, which are described in next paragraph

Table 7.4 summarizes the typical cooling temperatures and corresponding low-pressure levels used in cryogenics.

7.2.2.4.3 Disadvantages of Subatmospheric ($p < 1$ bar) Conditions

The disadvantages of vacuum conditions can be ordered in two groups:

1. operation issues, and
2. issues affecting process design.

The operational challenge relates at first to the leakage issue: If a subatmospheric cryogenic system is not tight, the ambient air (contaminated with H_2O, CO_2, and other substances; see Section 3.1) will permeate into the system due to pressure difference between the outside (ambient pressure) and inside (subatmospheric condition). These contaminants will accumulate in the form of solid particles at cryogenic temperatures and cause clogging and blockage. Subsequently, the system will require a time-consuming maintenance procedure (typical sequence includes steps like switch-off → warm-up → de-contaminate → switch-on → cooldown).

The main issues related to process design are as follows:

1. The density of the gas at subatmospheric conditions is very low. Therefore, all hardware components (heat exchangers, piping, fittings, valves, etc.) need

large cross-sections/diameters to handle the low-density material streams. The hardware components and the whole system become bulky, more expensive, and require more space;

2. The next aspect relates to the impact of pressure loss in the heat exchanger (low-pressure side) on the compressor's performance: the subatmospheric pressure itself is low, and the pressure drop in the heat exchanger reduces this pressure further. Consequently, the compressor must compress the fluid from a very low pressure-level. The requirements on the pressure ratio (defined as $\pi = p_{outlet}/p_{inlet}$) to be delivered by the compressor are rising dramatically. Usually, it leads to (i) increased power demand for the compression, as well as to (ii) higher expenditures for the required machine, since it becomes large and requires more compressor stages.

7.2.3 Example Process Calculation, Nitrogen-Joule–Thomson Process

An example of a typical process calculation procedure is presented in this section.

7.2.3.1 Task Definition

A nitrogen Joule–Thomson process must be designed. The required cooling capacity is $\dot{Q}_o = 10\,\text{kW}$ at cooling temperature $T_o = 100\,\text{K}$; ambient parameters are given in Table 7.5. The target is to provide:

1. process parameters (flowrate, temperature, pressure, phase state, composition, enthalpy, entropy, exergy) at all relevant process points;
2. total power consumption, COP, efficiency, as well as exergy loss in every process component;
3. figures required for design of heat exchangers and compressor.

7.2.3.2 Process and Hardware Concept

The process design is always based on the hardware concept, which is usually developed simultaneously with the process idea (in the form of a process flow diagram). The main idea in this particular case is to develop a refrigerator based on the Joule–Thomson process with a so-called bath evaporator, spiral-wound heat exchanger, dry multi-stage reciprocating compressor, and efficient thermal insulation.

(a) A bath evaporator is a simple vessel, as shown in Figure 7.13. The two-phase nitrogen is injected into this vessel after isenthalpic expansion in the Joule–Thomson valve. The liquid portion of nitrogen collects in the lower part of

Table 7.5 Basis of design.

Cooling temperature	$T_o = 100\,\text{K}$
Cooling capacity	$\dot{Q}_o = 10\,\text{kW}$
Ambient temperature	$T_{amb} = 300\,\text{K}$
Ambient pressure	$p_{amb} = 1.01\,\text{bar}$

Figure 7.13 Joule–Thomson refrigerator with bath evaporator.

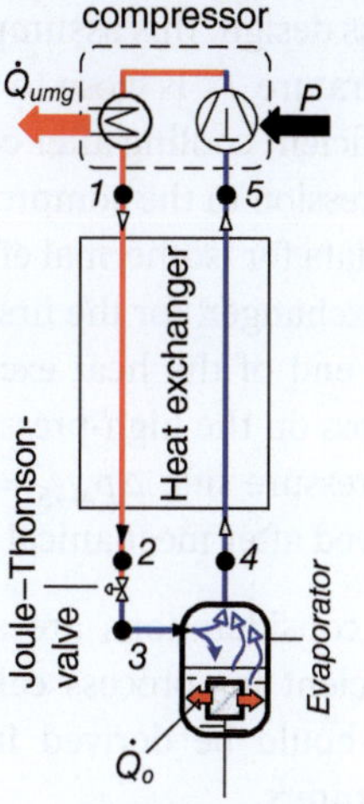

the vessel. The cooling object is placed on the bottom of the vessel in such a way that it is submerged in the normal operation case. Because of heat power ($\dot{Q}_o$ = 10 kW) generated continuously by the cooling object, the sump liquid continuously evaporates. This vapor as well as flash-gas generated after the Joule–Thomson expansion, are continuously removed from the vessel through a nozzle located in the upper part of the vessel. This vaporous materials stream (marked as point 4) is provided further to the heat exchanger.

(b) An efficient spiral-wound heat exchanger (see Section 6.2.2) is used. This kind of heat exchanger is characterized by:
- minimal temperature difference of 2 K may be achieved;
- pressure loss on the high-pressure side (tube side) is relative high (≈1 bar), whereas the pressure loss on the low-pressure side (shell side) may be small (≈0.1 bar);

(c) multi-stage reciprocating compressor is used. It includes water coolers after every compressor stage and is able to cool the compressed nitrogen to ambient temperature T_{amb};

(d) cold part is thermally insulated by an efficient multilayer insulation.

Some data required for process calculation can be directly derived from this process description. For example, it can be assumed that

1. pressure loss in the evaporator is negligibly small ($\Delta p_{3\rightarrow 4}$ = 0) because the nitrogen does not flow through this device;
2. it may be assumed that the cooling object is cooled in an efficient way because of direct contact between the object and liquid nitrogen, therefore a small temperature difference between the liquid and cooling object can be expected; for example, $(T_o - T_3) = 2\,\text{K}$ as a first estimation;
3. heat leakage through thermal insulation is negligibly small due to the efficient insulation; this assumption can be reviewed later, after mechanical design of the

cold part is finished and the real heat leakage will be calculated; but for the first process design, this assumption is acceptable;

4. temperature T_1 is close to the ambient temperature ($T_1 \approx T_{amb} = 300$ K) due to the efficient cooling after compression;
5. compression in the compressor can be calculated as a quasi-isothermal compression, data for isothermal efficiency of an appropriate machine are necessary;
6. heat exchanger: for the first process design a small temperature difference at the warm end of the heat exchanger ($T_1 - T_5 = 2$ K) may be assumed, and pressure loss on the high-pressure side $\Delta p_{1\to2} = 1$ bar as well as pressure loss on the low-pressure side $\Delta p_{4\to5} = 0.1$ bar may be assumed. These assumptions can be reviewed after mechanical design of the cold part.

These considerations are summarized in Table 7.6. However, these data are not sufficient for process calculation. Table 7.7 provides additional data, which usually should be derived from previous experiences or talks with equipment manufacturers.

7.2.3.3 Process Simulation

In this section, one of many possible calculation sequences is shown. The fluid property software "miniREFPROP[9]" is used for calculating thermodynamic properties.

Usually, the process calculation starts with calculation of pressures at all relevant process points (the so-called pressure table).

Table 7.6 Input process data derived from the hardware concept.

Pressure loss in evaporator	$\Delta p_{3\to4} = 0$
Temperature difference between cooling object and liquid in evaporator	$T_o - T_3 = 2$ K
Heat leakage through thermal insulation	$\dot{Q}_{is} = 0$
Temperature difference at the warm end of the heat exchanger	$T_1 - T_5 = 2$ K
Temperature at the outlet of compressor (=inlet heat exchanger)	$T_1 = 300$ K
Pressure loss in heat exchanger, high-pressure side	$\Delta p_{1\to2} = 1$ bar
Pressure loss in heat exchanger, low-pressure side	$\Delta p_{4\to5} = 0.1$ bar
Compression	Quasi-isothermal

Table 7.7 Input data known from previous calculations.

Isothermal efficiency of compressor	$\eta_T = 60\%$
Pressure at the outlet of compressor	$p_1 = 100$ bar

9 Default equation of state was used, reference state for specific enthalpy and specific entropy: 298.15 K, 0.01 bar.

Pressure Table

p_1	$p_1 = 100$ bar as given in Table 7.7
p_2	This pressure is lower than p_1 by pressure loss in the heat exchanger, high-pressure side: $p_2 = p_1 - \Delta p_{1\to 2} = 100\text{ bar} - 1\text{ bar} = 99\text{ bar}$
p_3	Pressure p_3 can be found as boiling pressure, which corresponds to given boiling temperature T_3 by means of "miniREFPROP" using temperature as an input parameter. The temperature T_3 amounts to $T_3 = 98$ K ($T_o - T_3 = 2$ K; see Table 7.6, $T_o = 100$ K). Consequently, $p_3 = 6.7565\text{ bar} \approx 6.76\text{ bar}$
p_4	This pressure is equal to the pressure p_3, ($\Delta p_{3\to 4} = 0$, according to Table 7.6). Therefore, $p_4 = 6.7565\text{ bar} \approx 6.76\text{ bar}$
p_5	This pressure is lower than p_4 by pressure loss in the heat exchanger, low-pressure side: $p_5 = p_4 - \Delta p_{4\to 5} = 6.7565\text{ bar} - 0.1\text{ bar} = 6.6565\text{ bar} \approx 6.66\text{ bar}$

Other Parameters, Also are Summarized in Table 7.8

State in point 1	The state 1 is explicitly defined by given temperature and pressure because it is in a one-phase area[10]: $T_1 = \mathbf{300}$ K, $p_1 = \mathbf{100}$ bar. All other values (enthalpy and entropy) can be found by means of "miniREFPROP" using pressure and temperature as input parameters. The output numbers are: specific enthalpy $h_1 = \mathbf{291.93}$ kJ/kg specific entropy $s_1 = \mathbf{5.4165}$ kJ/(kg K).
State in point 5	Temperature T_5 is 2 K lower than T_1 (see Table 7.6), therefore: $T_5 = T_1 - 2\text{K} = 300\text{ K} - 2\text{ K} = \mathbf{298\ K.}$ The pressure p_5 is already known. The state in point 2 is explicitly defined by temperature and pressure (because it is in a one-phase area). All other values (enthalpy and entropy) can be found by means of "miniREFPROP" using pressure and temperature as input parameters. The output numbers are: specific enthalpy $h_5 = \mathbf{307.85}$ kJ/kg, specific entropy $s_5 = \mathbf{6.2722}$ kJ/(kg K).
State in point 4	Thermodynamically, this state corresponds to the dew point at pressure p_4, temperature T_4 is equal to temperature T_3, (boiling curve, the same pressure) $T_4 = T_3 = 98$ K, specific enthalpy and entropy can be found by means of "miniREFPROP" using T_4 as boiling temperature as input parameters. The output numbers are: specific enthalpy: $h_4 = \mathbf{87.514}$ kJ/kg, specific entropy $s_4 = \mathbf{5.0181}$ kJ/(kg K).

10 In two-phase area, a specification by pressure and temperature is not sufficient. Usually, the state can be explicitly defined by (i) vapor fraction and temperature/pressure, (ii) enthalpy and pressure/temperature, and (iii) entropy and pressure/temperature.

State in point 2	The enthalpy h_2 can be found by means of enthalpy balance for heat exchanger: $h_1 - h_2 = h_5 - h_4$. Three values h_1, h_4, and h_5 are known already. Therefore $h_2 = h_1 - h_5 + h_4 = 291.93 - 307.85 + 87.514 = \mathbf{71.594}$ kJ/kg. The pressure p_2 is already known. The state in point 2 is explicitly defined by h_2 and p_2 (because it is in a one-phase area). The temperature T_2 and specific entropy can be found by means of "miniREFPROP" using the pressure and specific enthalpy as input parameters. The output numbers are: temperature: $T_2 = \mathbf{157.99}$ K, specific entropy $s_2 = \mathbf{4.3794}$ kJ/(kg K).
State in point 3	The specific enthalpy in point 3 is equal to the enthalpy h_2 (isenthalpic expansion) $h_3 = h_2 = \mathbf{71.594}$ kJ/kg. The temperature T_3 is already known $T_3 = 98$ K. The pressure p_3 is known, too: $p_3 = 6.7565$ bar ≈ 6.76 bar. Theoretically, the state in point 3 is explicitly defined by given temperature and specific enthalpy. But the calculation procedure is a bit more complex than may be expected because the software "miniREFPROP" is not able to perform this kind of calculation in a two-phase area directly. Therefore, a two-step approach is usually necessary: in the first step, the vapor fraction in state 3 will be defined, and in the second step, the specific entropy will be calculated. The Eqs. (4.3) and (4.4) may be used for this task (see Sections 3.9 and 3.10): $h_{2phase} = (1 - VF) \cdot h_d + VF \cdot h_b$ $s_{2phase} = (1 - VF) \cdot s_d + VF \cdot s_b$ the software "miniREFPROP" delivers following values if $T = 98$ K is used as input parameter: $h_b = -77.813$ kJ/kg $h_d = 87.514$ kJ/kg $s_b = 3.3311$ kJ/(kg K) $s_d = 5.0181$ kJ/(kg K) Therefore, $VF_3 = (h_3 - h_b)/(h_d - h_b) = (71.594-(-77.813))/(87.514-(-77.813)) = \mathbf{0.903}$ $s_3 = (1 - VF_3) \cdot s_d + VF_3 \cdot s_b = (1 - 0.903) \cdot 3.3311 + 0.903 \cdot 5.0181 = \mathbf{4.8545}$ kJ/(kg K)
Refrigerant flowrate	The heat transferred in evaporator $\dot{Q}_o = \dot{m}\,(h_4 - h_3)$, therefore the mass flowrate $\dot{m} = \dot{Q}_o/(h_4 - h_3) = 10/(87.514 - 71.594) = 0.628$ kg/s $= \mathbf{628}$ g/s. $\dot{N} = \dot{m}/M_r = 628$ (g/s)/28 (g/mol) $= 22.4$ mol/s

The results are shown in Table 7.8. The input parameters are underlined. The specific exergy as well as absolute exergy are calculated using Eq. (4.10); the reference state is referred to as $p = 1.013$ bar and $T = 300$ K. Figure 7.14 visualizes the process in *T,s*-diagram.

Table 7.8 Output process data, nitrogen Joule–Thomson process, $T_o = 100$ K.

State	Temperature (K)	Pressure (bar)	Vapor fraction (kg/kg)	Specific enthalpy (kJ/kg)	Specific entropy (kJ/(kg K))	Flowrate (kg/s)	Specific exergy (J/g)	Exergy (kW)
1	300	100	1.0	291.93	5.4165	0.628	406.6	255.41
2	158	99.0	1.0	71.59	4.3794	0.628	497.4	312.45
3	98	6.76	0.904	71.59	4.8545	0.628	354.9	222.92
4	98	6.76	1.0	87.51	5.0181	0.628	321.7	202.09
5	298	6.66	1.0	307.85	6.2722	0.628	165.8	104.16
Boiling point	98	6.76	0.0	−77.813	3.3311	—	—	—
Dew point	98	6.76	1.0	87.514	5.0181	—	—	—
Reference state	300	1.01	1.0	311.19	6.8361	—	0.0	0.0

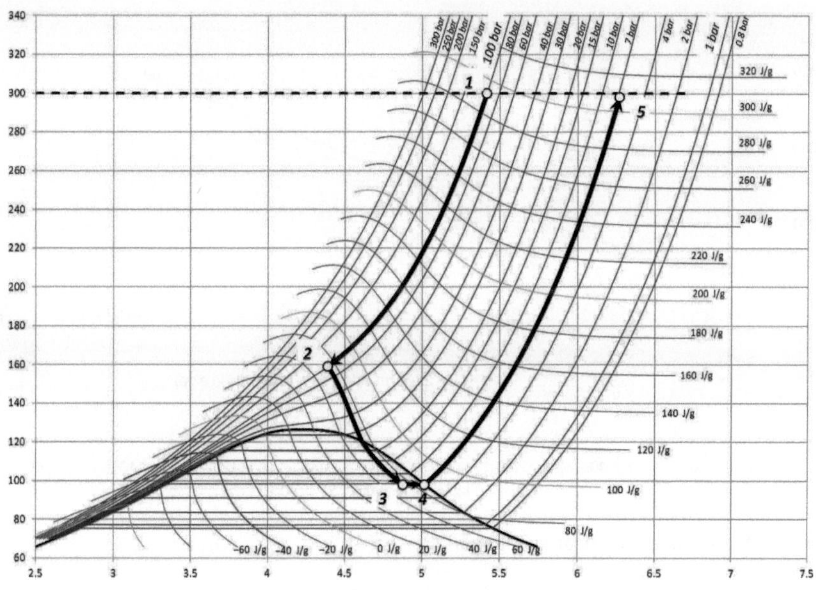

Figure 7.14 Nitrogen-Joule–Thomson-process in *T,s*-diagram.

7.2.3.4 Process Units

7.2.3.4.1 Heat Exchanger

Table 7.9 summarizes the process data for the heat exchanger.

The deviation between $LTD = 17.05$ K and $MTD = 11.9$ K is explained by the fact that the heat capacity of the high-pressure stream (warm side of the heat exchanger) is not constant; rather, it varies in a given temperature range. Figure 7.15 shows the temperature profile in the heat exchanger. The red dashed curve visualizes the

Table 7.9 Heat exchanger data, nitrogen-Joule–Thomson-process, $T_o = 100$ K.

The heat transferred in the heat exchanger (heat exchanger duty)	$q^{HEX} = h_5 - h_4$ $\dot{Q}^{HEX} = \dot{m} \cdot q^{HEX}$	$q^{HEX} = 307.85 - 87.51 = 220.34$ kJ/kg 0.628 kg/s · 220.34 kJ/kg = 138.4 kW
Temperature difference at the warm end	$\Delta T^{WE} = T_1 - T_5$	300 − 298 = 2 K
Temperature difference at the cold end	$\Delta T^{CE} = T_2 - T_4$	158 − 98 = 60 K
Max temperature difference in heat exchanger	ΔT_{MAX}	60 K
Mean logarithmic temperature difference *LTD*[a]	*LTD*	17.05 K
Real temperature difference MTD	*MTD*	11.9 K[b]
Q/MTD	$\dot{Q}_{HEX}/MTD$	138.4 kW/11.9 K = 11.6 kW/K

a) The definitions of MTD, LTD, and Q/MTD-value are given in Section 5.3.2.
b) The calculation of MTD = 11.9 K (as well as temperature profiles) was done by means of process simulator Unisim.

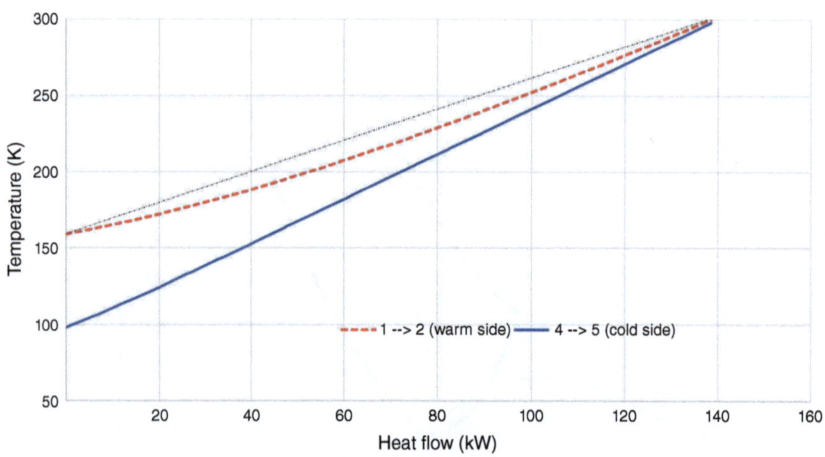

Figure 7.15 *T,Q*-diagram, heat exchanger.

temperature on the high-pressure side. It always runs below the black dot-dashed line, which directly connects the start and end points – a behavior assumed by the equation for LTD. Figure 7.16 shows the temperature difference profile in the heat exchanger, and the deviation mentioned above is seen even better there.

7.2.3.4.2 Evaporator

The *T,Q*-diagram for evaporator is shown in Figure 7.17. The liquid nitrogen evaporates here at the constant temperature $T_3 = T_4 = 98$ K. It is assumed, then, that the temperature of the cooled object equals 100 K. Cooling capacity of 10 kW is transferred here from liquid nitrogen to the object being cooled.

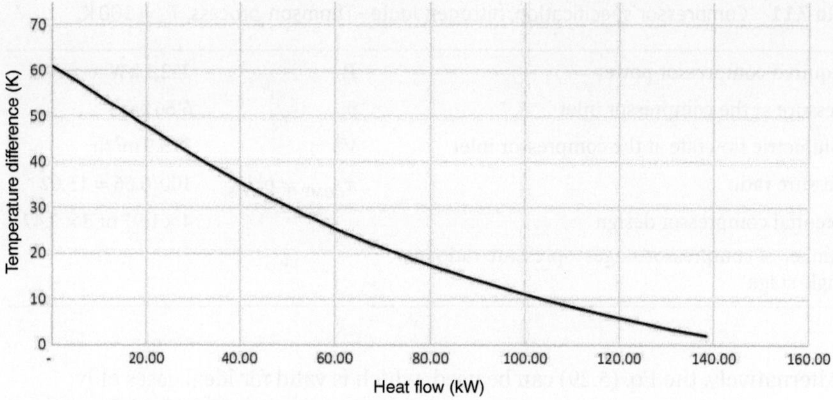

Figure 7.16 Temperature difference profile (ΔT,Q-diagram) in heat exchanger.

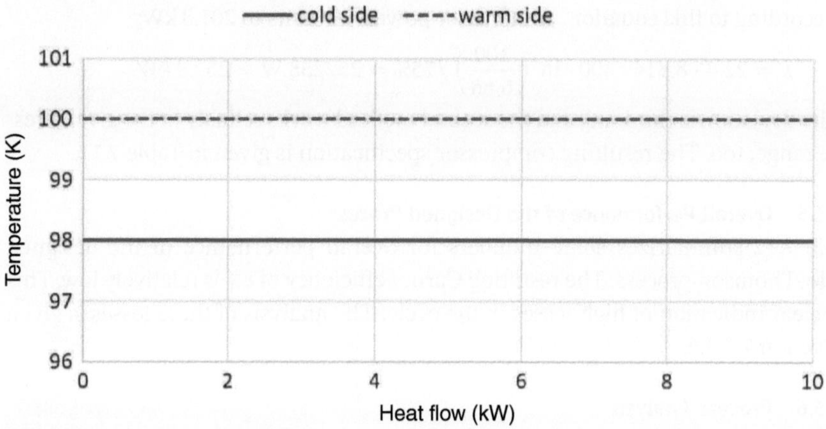

Figure 7.17 *T,Q*-diagram, evaporator, nitrogen Joule–Thomson process, $T_o = 100$ K.

7.2.3.4.3 Expansion in Expansion Valve (see Table 7.10)

Table 7.10 Joule–Thomson expansion.

Operating pressures (inlet/outlet)	99 bar / 6.76 bar
Vapor fraction at the outlet	90.4%

7.2.3.4.4 Compressor

The required compressor power can be calculated using the Eq. (5.30), which is valid for all gases without limitation:

$$P = \frac{P_{rev}}{\eta_T} = \frac{\dot{m} \cdot (e_1 - e_5)}{\eta_T}$$

According to this equation, compressor power amounts to 252.1 kW:

$$P = \frac{628\ \text{g/s} \cdot (406.6 - 165.8)\ \text{J/g}}{60\%} = 252\,082\ \text{W} \approx 252.1\ kW$$

Table 7.11 Compressor specification, nitrogen Joule–Thomson-process, $T_o = 100\,\text{K}$.

Required compressor power	$\boldsymbol{P}$	252.1 kW
Pressure at the compressor inlet	p_5	6.66 bar
Volumetric flowrate at the compressor inlet	V	315.9 m^3/h
Pressure ratio	$\pi_{COMP} = p^1/p_5$	100/6.66 = 15.02
Potential compressor design		4 × 1.97 or 3 × 2.47
number of compressor stages × pressure ratio in a single stage		

Alternatively, the Eq. (5.29) can be used, which is valid for ideal gases only:

$$P = \frac{1}{\eta_T} \dot{N} \cdot R_m \cdot T_{amb} \cdot \ln\left[\frac{p_1}{p_5}\right]$$

According to this equation, compressor power amounts to 201.8 kW:

$$P = 22.4 \cdot 8.314 \cdot 300 \cdot \ln\left(\frac{100}{6.66}\right) /75\% = 252258\,\text{W} \approx 252.3\,kW$$

The deviation is minimal, and the second method works reliably in the given pressure range, too. The resulting compressor specification is given in Table 7.11.

7.2.3.5 Overall Performance of the Designed Process

Table 7.12 summarizes some numbers for overall performance of the designed Joule–Thomson process. The resulting Carnot efficiency of 8% is relatively low. This is a clear indication of high losses in the cycle. The analysis of these losses is given in Section 7.2.3.6.

7.2.3.6 Process Analysis

This section consists of calculation of exergy loss in all process units as well as an analysis of these losses, whereby the target is to understand the cause and effect of the losses on process performance as well as how these losses can be reduced.

The procedure for calculation of exergy losses in a process unit is described in Section 4.3.3.3. It is based on an exergy balance sheet (see Eq. (4.1)) across the process unit boundary. Therefore, the exergy values for all material streams and energy

Table 7.12 Performance of the designed nitrogen Joule–Thomson process, $T_o = 100\,\text{K}$.

Cooling capacity	$\dot{Q}_o = 10\,\text{kW}$
Compressor power	$P = 252.1\,\text{kW}$
Waste heat	$\dot{Q}_{amb} = \dot{Q}_o + P = 10\,\text{kW} + 252.1\,\text{kW} = 262.1\,\text{kW}$
COP	$\varepsilon = \dot{Q}_o/P = 10/252.1 = 3.97\%$
Min required power for a reversible refrigerator (acc. to Carnot equation)	$P_{rev} = \dot{Q}_o\,(T_{amb} - T_o)/T_o = 10\,\text{kW}\,(300\,\text{K} - 100\,\text{K})/100\,\text{K} = 20\,\text{kW}$
COP of a reversible refrigerator	$\varepsilon_{CARNOT} = \dot{Q}_o/P_{rev} = 10\,\text{kW}/20\,\text{kW} = 50\%$
Efficiency	$\eta_e = \varepsilon\,/\,\varepsilon_{CARNOT} = 3.97\%\,/50\% = 0.0793 \approx 7.9\%$

Table 7.13 Exergy losses in process units of nitrogen Joule–Thomson-process, $T_o = 100$ K.

Loss related to	Exergy balance	Calculation	Result (kW)
Compressor $\dot{D}^{COMP}$	$\dot{E}_5 + P = \dot{E}_1 + \dot{D}^{COMP} \rightarrow$ $\dot{D}^{COMP} = P - \dot{E}_1 + \dot{E}_5$	$\dot{D}^{COMP} = 252.08 - 255.41 + 104.16$	100.8
Heat exchanger $\dot{D}^{HEX}$	$\dot{E}_1 + \dot{E}_4 = \dot{E}_2 + \dot{E}_5 + \dot{D}^{HEX} \rightarrow$ $\dot{D}^{HEX} = \dot{E}_1 + \dot{E}_4 - \dot{E}_2 - \dot{E}_5$	$\dot{D}^{HEX} = 255.4 + 202.1 - 312.4 - 104.2$	40.9
Joule–Thomson-valve $\dot{D}^{JT}$	$\dot{E}_2 = \dot{E}_3 + \dot{D}^{JT} \rightarrow \dot{D}_{JT} = \dot{E}_2 - \dot{E}_3$	$\dot{D}^{JT} = 312.4 - 222.9$	89.5
Evaporator $\dot{D}^{EVAP}$	$\dot{E}_3 = \dot{E}_4 + \dot{E}(\dot{Q}_o) + \dot{D}^{EVAP} \rightarrow$ $\dot{D}^{EVAP} = \dot{E}_3 - \dot{E}_4 - \dot{E}(\dot{Q}_o)$, whereas $\dot{E}(\dot{Q}_o)$ – exergy of the heat flow $\dot{Q}_o$	$\dot{E}(\dot{Q}_o) = \dot{Q}_o \cdot (T_{amb} - T_o)/T_o = 10\,\text{kW}\,(300\text{–}100\,\text{K})/100\,\text{K} = 20\,\text{kW}$ $\dot{D}^{EVAP} = 222.9 - 202.1 - 20$	0.8
Sum	—	—	232

streams entering or leaving the examined process unit are required for performing this procedure. These values were already defined during the process simulation; see Table 7.8. The procedure itself is performed by means of Table 7.13. Figure 7.18 visualizes results of calculation.

7.2.3.6.1 Loss in Expansion Valve

The largest loss in the cold part of the process (90 kW of total 232 kW), which means roughly 40% of all losses, relates to the expansion in the valve.

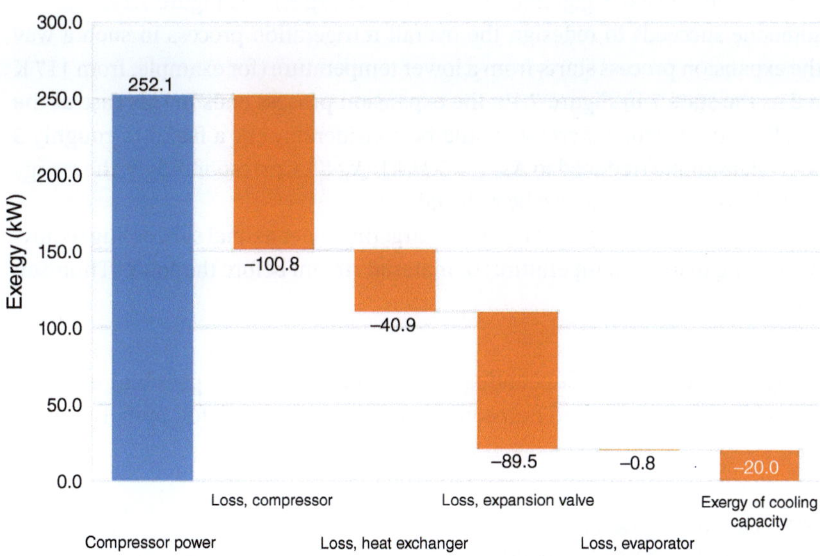

Figure 7.18 Distribution of exergy losses in given nitrogen Joule–Thomson-process for $T_o = 100$ K.

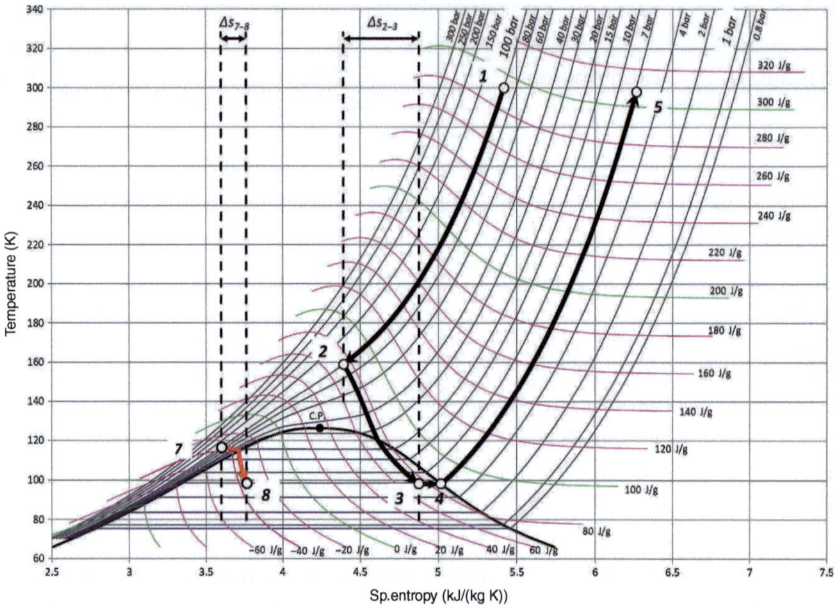

Figure 7.19 Joule–Thomson expansion in *T,s*-diagram.

A qualitative evaluation of the Joule–Thomson expansion can be performed with simple tools like a temperature–entropy diagram (the thermodynamic background is given in Section 5.1.6): a minor $\mathbf{\Delta s}$ means that the expansion process is designed well, a large increase in entropy indicates a large loss. The entropy increase for the given expansion process is gigantic: $\Delta s_{2\text{-}3} = 0.49$ kJ/(kg K), see Figure 7.19.

If someone succeeds to redesign the overall refrigeration process in such a way that the expansion process starts from a lower temperature (for example, from 117 K marked as the state 7 in Figure 7.19, the expansion process ends in this case at the state 8), then the entropy increase would be considerably (by a factor of roughly 3 for the given example) reduced to $\Delta s_{7\text{-}8} = 0.16$ kJ/(kg K). Correspondingly, the exergy loss in this process unit will also be reduced.

That is why process designers are always targeting the maximal subcooling (equivalently moving to lower temperature) of material stream before the Joule–Thomson expansion.

Learning 7.5 A maximal subcooling (as low temperature as possible) of material stream before the Joule–Thomson expansion is required for proper process design.

7.2.3.6.2 Loss in Compressor

The largest loss (101 kW of total 232 kW) relates to the compressor.

This is primarily because of the efficiency of the given machine, $\eta_T = 60\%$. This number means that 60% of the compressor power is properly converted into the compression of nitrogen; the remaining 40% of power is converted into other kinds of energy, as in the case of waste heat.[11]

It is interesting to note that the efficiency of the overall process η_e depends directly and linearly on the compressor efficiency η_T:

$$\eta_e = \frac{\dot{Q}_o}{P} = \frac{\dot{Q}_o}{P_{rev}/\eta_T} = \eta_T \cdot \frac{\dot{Q}_o}{P_{rev}} \tag{7.22}$$

This equation is valuable for process design in terms of calculation efforts. Usually, at the beginning of the conceptual design phase, the exact number for compressor efficiency η_T is not known. Nevertheless, the process designer may start the process development, perform the process calculations based on an assumed virtual number for compressor efficiency η_{T1}, and estimate the draft number for process efficiency η_{e1}. If later on an exact number for the efficiency of a real appropriate machine becomes available, the overall performance of the refrigeration process design can be simply adjusted **without performing** the whole process calculation again as

$$\eta_e = \eta_{e1} \cdot \frac{\eta_T}{\eta_{T1}} \tag{7.23}$$

Example: If the efficiency of the real compressor η_T is 5% better than originally assumed efficiency η_{T1}, then the overall efficiency of the process based on this compressor η_e is 5% better than originally calculated η_{e1}.

This correlation (Eq. (7.23)) is quite unique. The connection between the overall efficiency of the process η_e and efficiencies of other process units (heat exchanger, expansion, valve, evaporator or expander, pumps, columns, etc. for other cryogenic processes) is far from simple. If the efficiency of these components changes, then a full-size new process calculation is required for estimation of the overall process performance.

Learning 7.6 The efficiency of the overall process η_e depends directly and linearly on the compressor efficiency η_T.

7.2.3.6.3 Loss in Heat Exchanger

The loss in the heat exchanger (41 kW of total 182 kW) is quite large, too; it is roughly one-quarter of all losses.

***Explanation for high loss in the heat exchanger of a nitrogen Joule–Thomson process*:**

The high loss in the heat exchanger can be explained:

- **Partially** by the fact that the mean temperature difference in given case MTD = 11.9 K is relatively high. By comparison, an MTD of lower than 6 K is usually targeted in cryogenic processes.
- **However, the main problem** in this case is the extremely disadvantageous temperature profile (visualized by means of Table 7.14 based on Table 5.4): a huge temperature difference $\Delta T \approx 60$ K occurs here at the cold end of the heat exchanger

11 Theoretically, this waste heat may be used for heating purposes (integrated into another process) or converted into mechanical/electrical power.

Table 7.14 Impact of temperature difference in heat exchanger on exergy loss.

	Small temperature difference, less than 0.02 T	Medium temperature difference, (0.02–0.1) T	Large temperature difference, more than 0.1 T	Very large temperature difference
T = 100 K	<2 K Small loss	2–10 K moderate loss	>10 K large loss	>20 K very large loss
T = 200 K	<4 K Small loss	4–20 K moderate loss	>20 K large loss	>50 K very large loss

(on the temperature level of ≈100 K), it is a clear indication for bulky loss in heat exchanger;

- Both these unfortunate circumstances – relatively high MTD and disadvantageous temperature profile – lead to a massive loss in the heat exchanger.

What is the reason for this disadvantageous temperature profile?

As previously stated, the main problem of the classical Joule–Thomson process is its extremely disadvantageous temperature profile: a high-temperature difference at the cold end is the worst-case scenario possible.

The *T*,*h*-diagram is a convenient tool for studying heat capacity and effects caused by it, since the declination of isobars in *T*,*h*-diagram characterizes the specific heat capacity of the substance, as discussed in Section 4.2.4. Fundamentally, by nature, the specific heat capacity of every gas depends on pressure. This property is barely visible at subcritical pressure and relatively high temperature,[12] but it manages to maintain some visibility at low pressures, too. At the supercritical pressure, the impact of pressure on heat capacity becomes stronger. This is recognizable in the *T*,*h*-diagram because the high-pressure isobars are flatter than low-pressure isobars, meaning the heat capacity of a gas at a higher pressure is larger than the heat capacity of the same gas at a lower pressure.

In the case of nitrogen, this behavior is also given. The high-pressure stream on the warm side of the heat exchanger has a considerably higher heat capacity than the low-pressure stream on the cold side. Figure 7.20 illustrates the behavior in simplified form, which at the end leads to the temperature profile shown in Figure 7.15; the cold stream is warmed from temperature of 98 K up to 298 K; this cold is used for cooling a high-pressure warm stream. But this cold is not sufficient to cool the high-pressure stream down to a very low temperature. At the end, it is cooled to approximately 160 K only due to the large heat capacity of this material stream.

7.2.3.6.4 Loss in Evaporator

The loss in evaporator (less than 1 kW) is small. The cooling capacity generated in the refrigerator at 98 K is transferred to the cooling object, being kept at the

12 Therefore, a gas at these conditions is often considered an ideal gas with constant heat capacity.

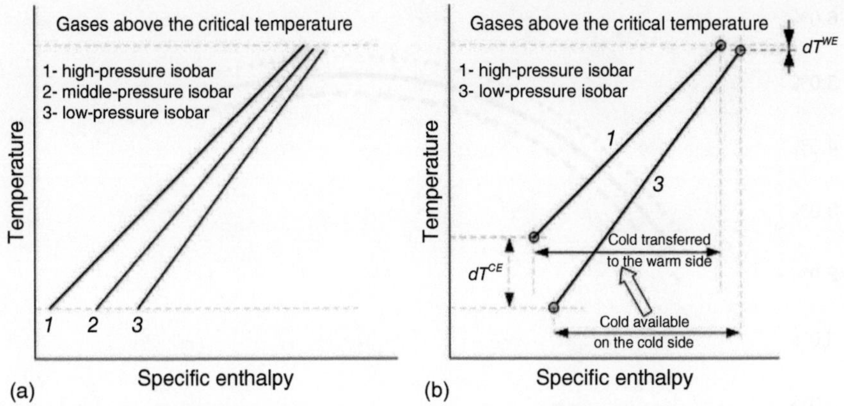

Figure 7.20 a) Typical isobars in the gaseous area of *T,h*-diagram, b) consequences for the Joule-Thomson process.

$T_o = 100\,\text{K}$. The temperature difference $\Delta T = 2\,\text{K}$ is small, and the transferred portion of heat $\dot{Q}_o = 10\,\text{kW}$ is relatively small as well at least in comparison to the main heat exchanger. Therefore, the corresponding heat transfer loss is minor.

Theoretically, this loss can be further reduced by using lower $\Delta T = 1\text{–}1.5\,\text{K}$ (the evaporator will require more surface for heat transfer at these conditions). However, the loss in the evaporator is so small in comparison to the other losses that it is not beneficial to work on improving it.

7.2.3.6.5 Explanation for Low Efficiency of the Classical Joule–Thomson Process

The following causal chain explains the low efficiency of the classical Joule–Thomson process:

1. The available refrigerants – nitrogen, argon, oxygen, and air – have a relatively low Joule–Thomson coefficients. At pressures below the critical pressure, it is close to zero, only at pressures above the critical pressure the Joule-Thomson coefficient becomes tangible. For example, the mean Joule–Thomson coefficient of nitrogen in pressure range 1–100 bar is only $\overline{\mu} \approx 0.25\,\text{K/bar}$; it becomes higher at higher pressures.
2. Therefore, a high pressure (supercritical) on the high-pressure side is required to generate a reasonable specific cooling capacity and efficiency (see Figure 7.21).
3. On the other hand, a relatively low pressure on the low-pressure process side is required to reach cryogenic cooling temperatures below 120 K.
4. The high pressure on the high-pressure side and low pressure on the low-pressure side mean a high difference in the heat capacities of the corresponding material streams.
5. This difference in the heat capacities leads to a large temperature difference at the cold end of the heat exchanger.
6. The large temperature difference at the cold end causes:
   (a) large exergy loss during the isenthalpic expansion, as well as
   (b) large exergy loss in the heat exchanger.
7. These large losses lead to low overall efficiency of the classical Joule–Thomson process.

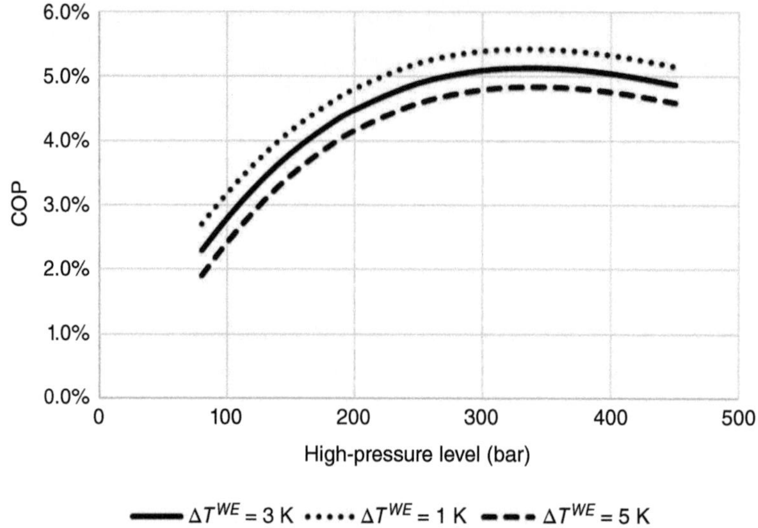

Figure 7.21 Coefficient of performance of a nitrogen-Joule–Thomson process, ΔT^{WE} – temperature difference at the warm end of the heat exchanger.

7.2.4 Process Optimization

The process calculation in Section 7.2.3 was performed for a single set of assumed parameters. Optimization is the next step required to find out the best combination. Usually, it is a two-step procedure, including: (i) the process optimization targeting the best performance in terms of power consumption, followed by (ii) the cost optimization targeting the best techno-economics. This section is focused on two corresponding examples.

7.2.4.1 Variables

The classical Joule–Thomson process is a simple process. There are only a few options that allow optimization. At a given cooling temperature T_o, there are three more or less relevant degrees of freedom:

(1) refrigerant;
(2) pressure downstream of the compressor (high-pressure level) p_1;
(3) temperature difference at the warm end of the heat exchanger ΔT^{WE}; as well as three parameters with minor impact:
(4) pressure loss at the high-pressure side of the heat exchanger $\Delta p_{1\to 2}$;
(5) pressure loss at the low-pressure side of the heat exchanger $\Delta p_{4\to 5}$;
(6) temperature difference in evaporator ΔT^{EVAP} between the cooling object and boiling refrigerant.

The pressure upstream of the compressor (low-pressure level) p_5 is not a real variable. This parameter is fixed for the given combination of required cooling temperature and selected refrigerant since:

(a) the temperature of the refrigerant boiling in evaporator T_4 is quite close to the cooling temperature:
$T_4 = T_o - \Delta T^{EVAP}$
(b) the pressure in evaporator p_4 corresponds to the boiling pressure of refrigerant at the temperature
$T_4: p_4 = p_s(T_4) = p_s(T_o - \Delta T^{EVAP})$

Therefore, the pressure on the compressor inlet is predefined by T_o, ΔT^{EVAP}, $\Delta p_{4\to5}$ as $p_5 = p_4 - \Delta p_{4\to5} = p_s(T_o - \Delta T^{EVAP}) - \Delta p_{4\to5}$.

7.2.4.2 Example of Process Optimization

An example of an optimization study is presented in this section. This study was performed for a nitrogen Joule–Thomson process with a cooling capacity of $\dot{Q}_o = 10$ kW.

Four parameters were fixed:

- $T_{amb} = 300$ K
- $\Delta p_{1\to2} = 0.1$ bar
- $\Delta p_{4\to5} = 0.1$ bar
- $T_3 = T_4 = 80$ K

Two parameters were varied:

- pressure p_1
- temperature difference at the warm end of the heat exchanger ΔT^{WE}.

It was assumed that the compression process could be performed with the same degree of efficiency and at the same cost as the entire pressure range investigated. This is a critical assumption, given that the costs of a nitrogen Joule–Thomson refrigerator depend considerably on the cost of the compressor, and the cost of the compressor depends on the number of compression stages.

Two following functions were examined during the study:

(1) the COP as a performance indicator,
(2) the Q/MTD value of the heat exchanger as an indicator for the heat exchanger costs.

The main findings for COP (shown in Figure 7.21) are:

(a) The optimum for the COP locates always in the pressure range of 330–350 bar independent of the ΔT^{WE}.
(b) The COP depends on the ΔT^{WE}. A reduction of the temperature difference at the warm end by 2 K (from 5 to 3 K or from 3 to 1 K) means a COP improvement of roughly 0.25%.

The temperature difference at the hot end of the heat exchanger impacts the mean temperature difference in the heat exchanger MTD and therefore the required heat exchange area (and heat exchanger costs). This connection is shown in the Figure 7.22.

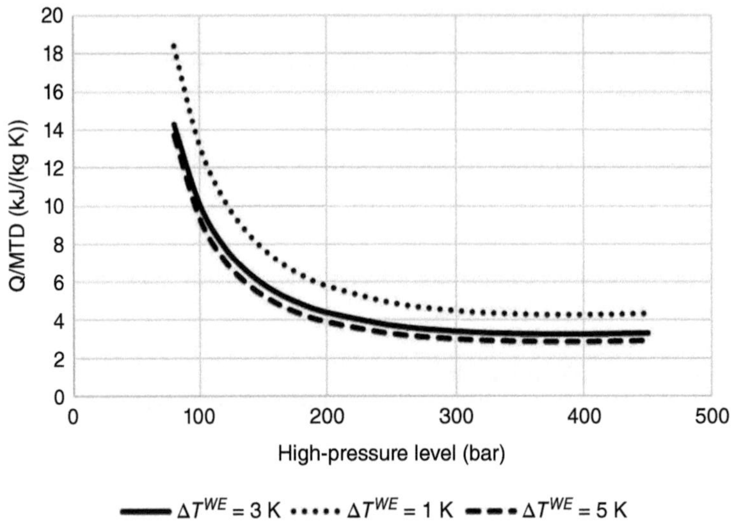

Figure 7.22 Q/MTD value for heat exchanger.

(a) The optimum is in the pressure range above 250 bar;
(b) The process design for pressure below 150 bar requires a considerably larger heat exchanger than the process design for pressures above 150 bar;
(c) The reduction of the temperature difference ΔT^{WE} from 5 to 3 K has a little impact on the Q/MTD value (and heat exchanger surface);
(d) A further reduction of the ΔT^{WE} from 3 to 1 K means a noticeably greater effect, such as a process requiring roughly 20–30% more heat exchanger surface.

7.2.4.3 In Summary

(a) Both pictures indicate that the overall optimum in pressure range is above 200 bar rather than below this value;
(b) Since both these curves are relatively flat in this pressure range, and therefore the pressure level does not impact the efficiency and the heat exchanger cost so much, then the other criteria may become important for a decision;
(c) For example, the fact that the high pressure is dangerous means additional safety measures and corresponding safety equipment are required, as well as a greater need to gain permission to execute this. Therefore, it can be assumed that a technical solution based on a moderate pressure close to 250 bar (or lower) will be better than a solution based on a high pressure of 400 bar, although the efficiencies of both processes are roughly equal.

A further optimization, which considers the characteristics and costs of the compressor, is recommended here. An example is described in Section 9.1.1.

7.2.5 Control of a Joule–Thomson Refrigerator

The typical control task for a refrigeration system is described in Section 2.4: the main target is to keep the cooled object at the specified constant cooling temperature. For this purpose, the cooling capacity of a refrigerator must be equivalent to the heat flow generated by the cooling object. The typical control loop for a refrigerator looks as follows: the control system

- monitors the temperature of the object being cooled;
- compares the temperature value with specified setpoint value;
- increases the cooling capacity of the refrigerator if the cooled object becomes warmer than the setpoint; or
- reduces the cooling capacity of the refrigerator if the cooled object becomes colder than the setpoint.

The Eq. (7.20) $\dot{Q}_o = \dot{m}\, c_p\, [\overline{\mu}\,(p_1 - p_5) - \Delta T^{WE}]$ (see Section 7.2.2.3) summarizes the variables that have an impact on the cooling capacity of a Joule–Thomson refrigerator. Not all of them may be used for control purposes, but some are more suitable than others, for example: (i) refrigerant flowrate $\dot{m}$ or (ii) pressures p_1.

Since the cycle is driven by a compressor and restricted by the Joule–Thomson valve, the refrigerant flowrate can be governed by controlling these devices:

1. The control of flowrate by the Joule–Thomson valve is a self-explaining option (see Figure 7.23a). In the simplest case, it can be done manually;
2. For the control on the compressor side, different options may be used:
   (a) The most energy-efficient option means controlling the rotation speed of the driving motor (see Figure 7.23b), by using an appropriate frequency

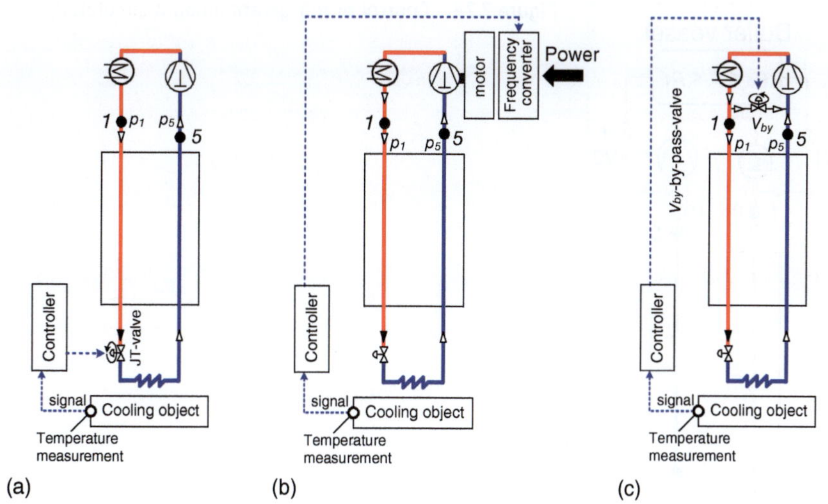

Figure 7.23 Flowrate control options: a) by means of control valve, b) by means of motor, c) by means of bypass valve.

converter. This way the machine compresses exactly the required amount of refrigerant from required low pressure to required high pressure. However, this option is relatively expensive since the frequency converter (especially for high power) and corresponding electrics are expensive pieces of equipment;

(b) The next option – the so-called bypass valve between the high-pressure and low-pressure levels, as shown in Figure 7.23c, whereby a fraction of refrigerant compressed in the compressor flows through the cold part of the process, and another fraction is provided back to the compressor inlet through the valve V_{by} – is not considered the best in terms of power since compressor drives more refrigerant than required by the refrigeration process. Still, this is an easy option in terms of control strategy;

(c) Two further options – an expansion valve on the compressor inlet and inlet guide vane control (IGV) used for control of centrifugal compressors – are described in Section 6.5.2.

An elegant control option sometimes used in cryogenic refrigeration is shown in Figure 7.24. Basically, it controls the inventory of refrigerant in the cold part of the process and therefore, the pressure levels. It is based on a buffer vessel installed parallel to the compressor by means of two valves, $V1$ and $V2$. The vessel is dimensioned in this way so that the pressure inside this vessel is maintained between the low pressure p_5 and the high pressure p_1:

- Usually, both valves $V1$ and $V2$ are closed.
- If the valve $V1$ is opened for a short time, the compressed refrigerant after the compressor goes partially into the cold part of the process and partially into the

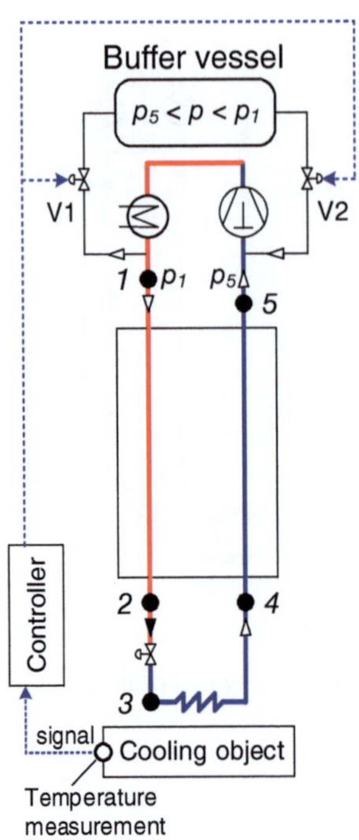

Figure 7.24 Control of refrigerant amount circulated.

buffer vessel; therefore, the amount of refrigerant circulated in the cycle as well as the pressure on high-pressure side are reduced.
- If the valve $V2$ is opened for a short time, the compressed refrigerant from the buffer vessel is filled into the cold part of the process; therefore, the amount of refrigerant circulated in the cycle as well as the pressure on high-pressure side will be slightly increased.

7.2.6 Assessment of the Joule–Thomson Process

The main advantage of the Joule–Thomson process/refrigerator is that it **is simple and highly reliable** since it does not consist of moving parts in the cold part. If the heat exchanger is properly designed and the compressor runs well, then this refrigerator safely generates the required cooling capacity.

Liquefied refrigerant: The cooling capacity is generated in the form of a liquid refrigerant. The liquid cryogen has several advantages for refrigeration consumers:
- it boils at a constant temperature,
- it has a higher density than cold gas, therefore it requires only small cooling channels,
- the heat transfer between the liquid and the cooling object is better, which leads to small temperature difference in evaporator and/or compact evaporator surfaces,
- it can be stored for back-up purposes, increasing the safety and reliability of the cooled application.

The **cooling temperature** that may be achieved by means of Joule–Thomson with pure refrigerants is limited. Theoretically, the lowest temperature is given by the triple point of oxygen of 54.4 K. Since the preferred refrigerant is the nitrogen, the triple point of nitrogen of 63.3 K is considered as a sort of natural limit.

In terms of hardware, the high-pressure compressor is a **potential weakness,** causing **high initial expenditure** as well as further expenditure for maintenance.

In comparison to other cryogenic processes, **the efficiency** of the Joule–Thomson process is **low** (less than 10%), which results in a high power bill and operation fees at the end.

Both the high initial cost and high operation cost explain why a commercial system based on the classical Joule–Thomson process/refrigerator is an exception rather than the rule. The techno-economics for a Joule–Thomson refrigerator may work in special cases only, if at all. It is presumable that the Joule–Thomson refrigerator will work in the case of single small-scale installations with relatively short operation periods, since this would mean minimal operational expenditures.

Several examples of commercial Joule–Thomson refrigerators are described in Ref. [2]. This book includes a comprehensive bibliography about Joule–Thomson refrigeration.

7.3 Brayton Process

The reversed Brayton process (referred to here as a Brayton process) is the next basic cryogenic process. The term “reversed Brayton process” indicates a connection to the original non-reversed Brayton process. It is used in power generation for conversion of combustion heat into the mechanical (and further, into the electrical) power.

7.3.1 Main Elements

Figure 7.25 shows the flow diagram. The morphology of the reversed Brayton process is similar to the morphology of the Joule–Thomson process. It consists of five process components such as the compressor, heat exchanger, expansion device, cold gas heat exchanger, and thermal insulation. They are connected in the same sequence. There are, however, two differences:

1. **Expander instead of the expansion valve:** The most obvious difference is that an expander (machine) is used here instead of the expansion valve. This device fulfills the same function – the working fluid is expanded here from the high-pressure level to the low-pressure level. The expansion in an expander has some important features already discussed in Section 5.2 that have an impact on process design:
   - It causes a temperature drop regardless of the initial state and working fluid being used;
   - At the same inlet and outlet pressure, the temperature drop is considerably higher compared to the Joule–Thomson expansion;
   - The pressure ratio achieved in a single expander stage is limited, the limitation depends on the fluid and temperature range;
   - At the outlet of the expander, the fluid is usually not liquefied (to avoid any damage to expander); it is a cold gas.
2. **Cold gas heat exchanger:** The second difference is the so-called "cold gas heat exchanger" instead of the evaporator. That is the consequence of the change

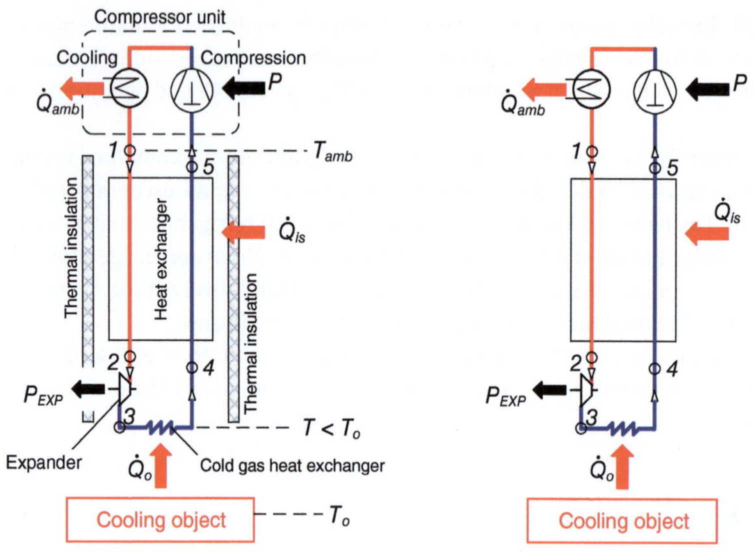

Figure 7.25 Brayton process, basic process flow diagram; left - thermal insulation is shown, right - thermal insulation is removed.

mentioned above (expander instead of the expansion valve). At the outlet of the expander, the working fluid is gaseous. It enters this heat exchanger as a cold gas. No evaporation occurs inside of this device; the sensible heat (heat capacity of the fluid) is used for cooling purposes only. Therefore, the wording "evaporator" is misleading and incorrect. Functionally, this device is a replacement for the evaporator: the heat from the object being cooled is transferred in this device to the cold gas flowing inside. In terms of mechanical design, the cold gas heat exchanger is very similar to the evaporator.

7.3.1.1 Comment About Compression

The reversed Brayton process described in most classic thermodynamics textbooks presumes an adiabatic = near-isentropic compression. But the availability of the corresponding hardware component – an adiabatic compressor – in required capacity range is quite limited. Usually, the adiabatic compressor is only available in the single-stage version. The small-/medium-scale multi-stage adiabatic version does not even exist as a product. Since the single-stage adiabatic compressor is often unable to generate a pressure ratio required for the efficient process design, isothermal compressor devices such as screw compressors or multi-stage centrifugal compressors are used instead. Therefore, the compression step in the cryogenic Brayton process is often realized as an isothermal process.

7.3.2 Process and Process Features

7.3.2.1 Process Description

Figure 7.28 shows the process in *T,s*-diagram. The sequence is as follows:

(a) The refrigerant is first compressed from the low-pressure level p_{LP} to the high-pressure level p_{HP} by means of a single-stage or multi-stage (with intermediate coolers) compressor and cooled to the approximate ambient temperature; this state is marked in BPFD as state 1;
(b) This compressed refrigerant is cooled in the countercurrent heat exchanger by the cold low-pressure stream, the state of the refrigerant at the cold end of the heat exchanger is marked as state 2, and the temperature is usually close to the required cooling temperature $\boldsymbol{T_o}$;
(c) This cold compressed fluid is expanded to the low-pressure level in the expander. During this expansion operation, a mechanical power $\boldsymbol{P_{EXP}}$ is generated, and the fluid is cooled to a lower temperature $\boldsymbol{T_3}$. At the end of the expansion, the refrigerant is still gaseous. The state at the outlet of the expansion valve is marked as state 3;
(d) This very cold refrigerant flows through the cold gas heat exchanger, cooling the object. At the outlet of the cold gas heat exchanger (state 4), the temperature of this gas is still a bit lower than the cooling temperature $\boldsymbol{T_o}$;
(e) This cold fluid flows through the heat exchanger to the compressor. In the heat exchanger, the heat is transferred from the "warm" high-pressure stream to the cold low-pressure stream, so that at the end, the low-pressure fluid is warmed

up to a temperature T_5 (state 5), which is close to the ambient temperature but still a little bit colder than the refrigerant at state 1.

(f) At the outlet of the heat exchanger, the refrigerant is in a gaseous state. It is provided to the compressor inlet.

7.3.2.2 Phase State Conditions

The classic version of the reversed Brayton process is designed to ensure the working fluid always stays in one-phase: gas or/and vapor, since the two-phase conditions in the expander may damage the machine.

In the meantime, original equipment manufacturers offer gas expanders able to manage a little liquid fraction at the outlet (usually it is specified to be lower than 5–8%) and process design engineers try to implement this degree of freedom into the process design, as already mentioned in Section 6.3.2.2.

This book presents a conservative process design approach, meaning no phase change exists inside the machine.

7.3.2.3 Working Fluids

In opposition to the Joule–Thomson effect, which may be positive or negative depending on the fluid, temperature, and pressure, the expansion of every fluid in an expander inevitably leads to a reduction in temperature.

Therefore, the reversed Brayton process has no limitation concerning the refrigerant choice. Cryogenic gases like helium, hydrogen, and neon may be employed as refrigerants. A key takeaway is this: a very low temperature-level can be achieved by using these gases, given that they have a low boiling temperature. This is an important feature of the reversed Brayton process to note.

Hydrogen: Hydrogen is an inexpensive fluid available globally. These properties make hydrogen attractive. But as a burnable substance, is considered a dangerous fluid. Therefore, other refrigerants like helium and neon are the preferred fluids for cooling temperatures below 100 K.

Neon: The neon-based version can be built based on centrifugal compressors, since molecular weight of neon is relatively high. Such a version of the Brayton process is referred to as the reversed Turbo-Brayton process. But this fluid is a) relative expensive and b) the prices are highly volatile due to limited availability. Therefore, the use of neon as refrigerant is still considered as risky for commercial applications.

Helium: A centrifugal compressor for compression of helium is not available, meaning this process can be realized only by means of other kinds of compressors. The advantage of helium is that this refrigerant is easy in terms of safety requirements, less expensive than neon (by a factor of 10) and available globally.

Air and Air Gases: For temperatures higher than 100 K, the air gases (nitrogen, oxygen, argon, and air itself) can be used as refrigerants. The most used fluid is the air, especially if the Brayton process is integrated into an air separation plant.

Ideal Gas Approach for Process Calculation: The low-temperature gases – helium, hydrogen, and neon – may be considered as ideal gases at temperatures close to the ambient temperature and moderate pressures. They are useful, for they can simplify process design considerations.

7.3.2.4 Cooling Temperature

The achievable cooling temperature depends on the working fluid. The helium-based Brayton process can reach a very low temperatures, such as 10 K.

The reversed Brayton process cannot deliver cooling capacity at a constant cooling temperature. The gas enters the cold gas heat exchanger with the temperature T_3 and leaves with the temperature T_4. The following are some points to note:

- For some applications – for example, for cooling of single-phase material streams (gases or liquids) – this temperature glide is advantageous since it leads to reduction of the loss in the cold gas heat exchanger;
- For other applications requiring a constant cooling temperature T_o, the cooling by means of cold gas may be challenging. Special measures are required to keep the whole application stable at a constant temperature if the temperature of cooling fluid varies;
- Most cryogenic applications do not require a strongly constant cooling temperature, for they can operate successfully in a temperature range as long as the temperature of the cooling object is not higher than a specified value T_o^{max}. In this case, the Brayton process is applicable. It must be designed in such a way that the temperature at the outlet of cold gas heat exchanger T_4 is still colder than the temperature T_o^{max}: $T_4 < T_o^{max}$.

7.3.2.5 Typical Process Pressure Levels

Since the expansion in an expander leads to a high-temperature drop, it is no longer necessary to employ the high pressure for driving the process. The high-pressure level is usually less than 20 bar. For the nitrogen-based process, this pressure level is subcritical. For helium-based cycles, this pressure level is already supercritical, but it is still considered relatively low. Therefore, **the Brayton process is considered a low-pressure process.**

The low-pressure level is usually in the range of 1–5 bar.

7.3.2.6 Cooling Capacity

In this section, the cooling capacity of the Brayton process is considered by means of energy balance for the cold part of the process (heat exchanger, expander, and cold gas heat exchanger). Figure 7.26 demonstrates how five energy streams are crossing the system boundary:

$\dot{Q}_o$ – heat flow from the object being cooled
$\dot{Q}_{is}$ – heat flow from the environment into the coldbox through nonideal thermal insulation
P_{EXP} – power generated by expander
$\dot{H}_1$ – enthalpy flow (in form of material flow) on the warm end of the heat exchanger, high-pressure stream, inlet
$\dot{H}_5$ – enthalpy flow (in form of material flow) on the warm end of the heat exchanger, low-pressure stream, outlet

The energy balance equation looks as follows:

$$\dot{Q}_o^{max} = \dot{Q}_o + \dot{Q}_{is} = (\dot{H}_5 - \dot{H}_1) + P_{EXP} \tag{7.24}$$

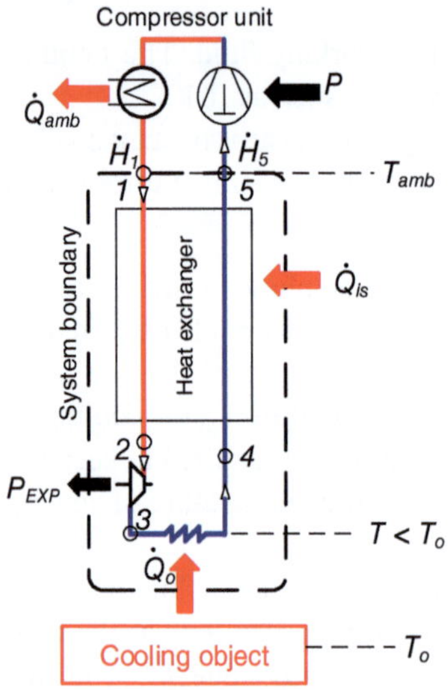

Figure 7.26 Brayton process, system boundary for cold part.

Assumption: For simplification purposes, it is assumed that the cold part of the process is thermally well-insulated, therefore the heat flow through insulation is negligibly small ($\dot{Q}_{is} \to 0$). Consequently, the energy balance equation can be simplified. It looks as follows:

$$\dot{Q}_o = (\dot{H}_5 - \dot{H}_1) + P_{EXP} \tag{7.25}$$

According to this equation, the cooling capacity of the Brayton process $\dot{Q}_o$ depends on expander power P_{exp} and the enthalpy difference at the warm end of the heat exchanger $\Delta\dot{H}^{WE} = (\dot{H}_5 - \dot{H}_1)$.

7.3.2.6.1 Ideal Gas, Infinitely Large Heat Exchanger

For an ideal gas $\Delta\dot{H}_T{}^{WE} = 0$ for infinitely large heat exchanger $\Delta T^{WE} = 0$ (see Section 7.2.2.3.1), therefore for this case, the equation can be further simplified to

$$\dot{Q}_o = P_{EXP} \tag{7.26}$$

This equation expresses that the cooling capacity of a Brayton process with an ideal gas as its working fluid is generated only by the expander.

Since the typical cryogenic gases (nitrogen and helium) can be considered as ideal gases at ambient temperature and relatively low pressure, this statement indicates cooling capacity of every Brayton process quite well.

Worth noting is that this equation can be interpreted in an opposing way as well: if the required cooling capacity is defined, the power capacity of the expander and therefore the rough dimension of expander(s) can be directly derived from this value.

7.3.2.6.2 Further Considerations

Equation (5.10) describes the expansion of an ideal gas in expander as:

$$P_{EXP} = \eta_s \cdot \dot{m}\frac{R}{M_r} \cdot T_{in} \cdot \frac{\kappa}{\kappa - 1}\left[1 - \left(\frac{p_{in}}{p_{out}}\right)^{(1-\kappa)/\kappa}\right]$$

Therefore, the cooling capacity of the Brayton cycle is as follows:

$$\dot{Q}_o \approx \eta_s \cdot \dot{m}\frac{R}{M_r} \cdot T_{in} \cdot \frac{\kappa}{\kappa - 1}\left[1 - \left(\frac{p_{in}}{p_{out}}\right)^{(1-\kappa)/\kappa}\right] \tag{7.27}$$

It is interesting to note that the cooling capacity of the Brayton process depends on the temperature on the expander inlet $T_{in} = T_2$, which is linked to the cooling temperature (which is only a bit higher than T_o). Therefore, at the same flowrate and pressure level, the cooling capacity of a refrigerator reduces in line with the cooling temperature.

Learning 7.7 The cooling capacity of the Brayton process depends on the cooling temperature. This function is a linear one. At a constant refrigerant flowrate, the colder the required cooling temperature, the smaller the cooling capacity.

7.3.2.6.3 How the Cooling Capacity of Brayton Process may be Increased

Another observation is that the cooling capacity of the Brayton process depends on refrigerant flowrate, pressure ratio p_{in}/p_{out}, and thermodynamic properties like kappa, according to Eq. (7.27). Therefore, the cooling capacity of a Brayton process can be increased by:

- increasing refrigerant flowrate,
- increasing the pressure ratio in the compressor (and therefore in expander) or
- using a refrigerant with a higher kappa

7.3.2.7 Reversible Brayton Process

A version of the Brayton process is a special one based on:

- reversible isothermal compressor,
- reversible adiabatic expander,
- heat exchanger with infinitely large heat exchanger (see Section 7.2.2.3.1) with infinitely low-pressure loss on both sides of the heat exchanger,
- a thermally well-insulated cold part, with the heat flow through insulation being negligibly small ($\dot{Q}_{is} \rightarrow 0$),
- In the cold gas heat exchanger, the heat transfer (from the fluid to the object being cooled) is realized at an infinitely low-temperature difference,
- With ideal gas as the working fluid, whereby the most important property is that the heat capacity of the material stream on the "warm" hot-pressure side of the heat exchanger equals the heat capacity of the material stream on the "cold" low-pressure side.

This version of the Brayton process is considered a fully reversible cycle. The compressor power is converted there into the cold without any loss. The Carnot equation $P = \dot{Q}_o\,(T_o/(T_{amb} - T_o))$ is valid for this cycle. The matter is only that the cooling capacity is generated in this process not at a constant temperature level, but in a temperature range $T_3 \ldots T_4$, therefore this equation must be adapted (see Section 8.1.2).

The reversible version of the Brayton process may take over the function of the Carnot process as a benchmark for purposes of comparison. It has the advantage that it relates to a realistic process and realistic refrigerator, whereas the Carnot process is only a virtual process; there are not any other cooling devices based on Carnot cycle available.

7.3.3 Example, Process Calculation, Helium Brayton Process

In terms of availability and cost, helium is one of the most appealing cryogenic fluids. It is also appealing in terms of its achievable cooling temperature. In this section, the helium Brayton process is covered in detail. This section considers an example of process calculations for a helium Brayton process.

7.3.3.1 Task Definition

A helium Brayton process for cooling capacity of $\dot{Q}_0 = 10\,\text{kW}$ must be designed. It is assumed that the applications do not require a constant cooling temperature and that they can operate successfully in a temperature range below $T_0^{max} = 80\,\text{K}$. The ambient conditions are as defined in Table 7.5. The goal is to provide:

1. Process parameters (flowrate; temperature; pressure; phase state; composition; enthalpy; entropy; exergy) at all relevant process points;
2. Total power consumption, COP, and efficiency, as well as exergy loss in every process component;
3. Figures required for designing heat exchangers and compressors.

7.3.3.2 Properties Data

The fluid property software "miniREFPROP" is used for calculating thermodynamic properties.

7.3.3.3 Process and Hardware Concept

The process concept and the hardware concept must be developed simultaneously. The main idea in this particular case is to develop a Brayton process based on aluminum-brazed plate-fin heat exchanger. This concept includes:

(a) Single-stage oil-lubricated screw compressor unit is used (see the Section 6.5.3 for more details);
(b) The evaporator is designed in the form of a tube coil wound around the object being cooled (as shown in Figure 7.27);
(c) Cold part is thermally insulated by an efficient multilayer insulation.
(d) An efficient turboexpander with electric generator as a brake-device;
(e) an efficient aluminum-brazed plate-fin heat exchanger;

Figure 7.27 Concept of helium Brayton refrigerator.

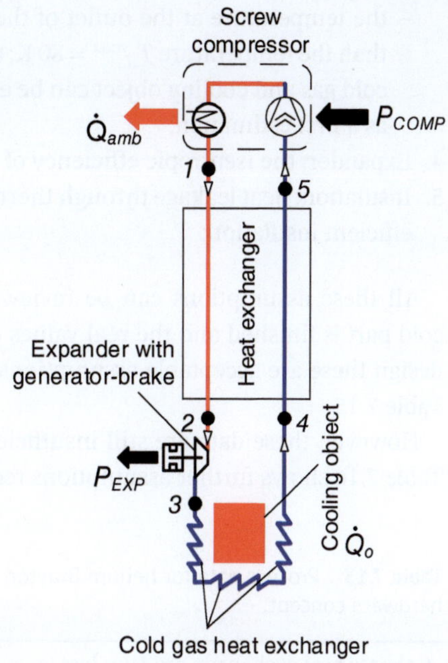

Some data required for process calculation can be directly derived from this process idea description:

1. Compressor:
   - since the compressor is oil-flooded, and therefore a continuous cooling by means of compressor oil during the compression process is applied, the compression operation can be considered and calculated as a quasi-isothermal compression; data for isothermal efficiency of an appropriate machine are necessary;
   - an oil-removal system is required since the compressor uses oil lubrication;
   - temperature T_1 is close to the ambient temperature ($T_1 \approx T_{amb} = 300\,\text{K}$) due to the efficient cooling after compression;
2. Heat exchanger:
   - for the first process design, minor pressure loss on the high-pressure side $\Delta p_{1\rightarrow 2} = 0.1\,\text{bar}$, as well as minor pressure loss on the low-pressure side $\Delta p_{4\rightarrow 5} = 0.1\,\text{bar}$ may be assumed;
   - for the temperature difference at the warm end, the value of $\Delta T^{WE} = 2\,\text{K}$ may be assumed;
3. Cold gas heat exchanger:
   - the pressure loss in the cold gas heat exchanger is not negligibly small, but rather must be considered during the process calculation where a value of $\Delta p_{3\rightarrow 4} = 0.1\,\text{bar}$ can be assumed;

- the temperature at the outlet of the cold gas heat exchanger T_4 is still colder than the temperature $T_o^{max} = 80$ K; a small temperature difference between the cold gas and cooling object can be expected there; for example $(T_4 - T_o) = 2$ K as a first estimation;

4. Expander: the isentropic efficiency of 85% can be assumed;
5. Insulation: heat leakage through thermal insulation is negligibly small due to the efficient insulation;

All these assumptions can be reviewed later when the detailed design of the cold part is finished and the real values are available, however for the first process design these are acceptable assumptions. These considerations are summarized in Table 7.15.

However, these data are still insufficient for the process simulation/calculation. Table 7.16 shows further assumptions required for the process simulations.

Table 7.15 Process data for helium Brayton process derived from the process and hardware concept.

Cold gas heat exchanger, pressure loss in	$\Delta p_{3\rightarrow4} = 0.1$ bar
Cold gas heat exchanger, min temperature difference	$T_o - T_4 = 2$ K $T_o = 80\text{ K} \rightarrow T_4 = 78$ K
Thermal insulation, heat leakage	$\dot{Q}_{is} = 0$
Heat exchanger, temperature difference at the warm end	$\Delta T^{WE} = 2$ K
Temperature at the outlet of compressor (=inlet heat exchanger)	$T_1 = 300$ K
Pressure loss in heat exchanger, high-pressure side	$\Delta p_{1\rightarrow2} = 0.1$ bar
Pressure loss in heat exchanger, low-pressure side	$\Delta p_{4\rightarrow5} = 0.1$ bar
Compression	Quasi-isothermal
The compressor efficiency can be derived from previous experiences or discussions with equipment manufacturers. For example, the value of 60% is usually assumed for this class of compressors	$\eta_T = 60\%$
Expander, isentropic efficiency	$\eta_S = 85\%$

Table 7.16 Further assumptions for process calculations.

High-pressure level: pressure p_1	$p_1 = 9$ bar
Low-pressure level: pressure p_5	$p_5 = 1.5$ bar

7.3.3.4 Process Simulation

The process calculation starts with the calculation of pressures at all relevant process points.

Pressure table:

p_1	this pressure is given: $p_1 = 9$ bar
p_2	this pressure is lower than p_1 by pressure loss in the heat exchanger, high-pressure side: $p_2 = p_1 - \Delta p_{1\to2} = 9\,\text{bar} - 0.1\,\text{bar} = 8.9\,\text{bar}$
p_5	this pressure is given: $p_5 = 1.5$ bar
p_4	this pressure is higher than p_5 by pressure loss in the heat exchanger, low-pressure side: $p_4 = p_5 + \Delta p_{4\to5} = 1.5\,\text{bar} + 0.1\,\text{bar} = 1.6\,\text{bar}$
p_3	since the pressure loss in the cold gas heat exchanger is given $\Delta p_{3\to4} = 0.1$ bar, the pressure p_3 can be shown as $p_3 = p_4 + \Delta p_{3\to4} = 1.6\,\text{bar} + 0.1\,\text{bar} = 1.7\,\text{bar}$

Simulation sequence

The following sequence can be used as an example:

Step 1: temperature T_1 is given, temperature T_5 is lower than T_1 by $\Delta T^{WE} = 2$ K and can be simply found

Step 2: temperature T_4 is given, temperature T_2 can be found by means of enthalpy balance around the heat exchanger

Step 3: calculation of the expansion process, since inlet temperature, inlet, and outlet pressures, as well as isentropic efficiency, are given, at the end, the temperature T_3 will be found

The results are shown in Table 7.17. The specific exergy as well as absolute exergy are calculated by using Eq. (4.10); the reference state is referred to as $p = 1.013$ bar and $T = 300$ K. Figure 7.28 shows the process in the T,s-diagram. The temperature of the high-pressure helium before the expansion is approximately 80 K. After the expansion, it is lower than 50 K, and in the cold gas heat exchanger, the cold low-pressure gas is heated up to 78 K before it is led into the main heat exchanger.

Table 7.17 Process data, helium Brayton process, $T_o = 80$ K.

State	Temperature (K)	Pressure (bar)	Specific enthalpy (kJ/kg)	Specific entropy (kJ/(kg K))	Flowrate (g/s)	Specific exergy (kJ/kg)	Exergy (kW)
1	300	9	1565.78	23.4419	62.46	1380.9	86.2
2	80.20	8.9	423.20	16.6043	62.46	2289.6	143.0
3	47.22	1.7	250.26	17.2944	62.46	1909.6	119.3
4	78	1.6	410.37	20.0318	62.46	1248.5	78.0
5	298	1.5	1552.95	27.1280	62.46	262.2	16.4
Reference state	300	1.013	1563.35	28.0367	1.00	0.0	0.00

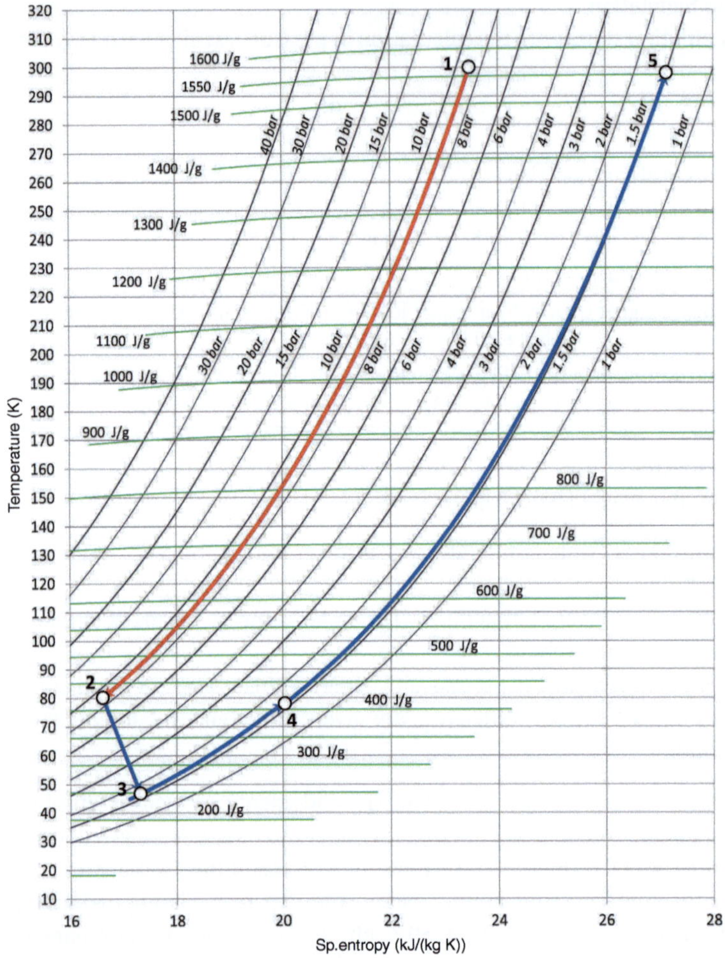

Figure 7.28 Helium Brayton process in *T,s*-diagram.

7.3.3.5 Process Units

7.3.3.5.1 Heat Exchanger

The table below (Table 7.18) summarizes the data related to the heat exchanger.

Figure 7.29 shows the corresponding *T,Q*-diagram. The temperature difference is very small, with the difference between two lines there being unrecognizable. Therefore, it is better to use the diagram given in Figure 7.30, which shows the temperature difference in the heat exchanger. The temperature difference remains constant; at the warm end and in the middle, it remains close to 2 K, while at the cold end, it increases slightly to 2.2 K.

7.3.3.5.2 Cold Gas Heat Exchanger

Figure 7.31 shows a virtual *T,Q*-diagram of the cold gas heat exchanger. The line for the cold gas corresponds to the real process data. For the temperature of the object

Table 7.18 Heat exchanger, helium Brayton process, $T_o = 80\,\text{K}$.

Heat exchanger duty	$\dot{Q}^{HEX}$	71.36 KW
Temperature difference at the warm end	$\Delta T^{WE} = T_1 - T_5 =$	$\Delta T^{WE} = T_1 - T_5 = 2\,\text{K}$
Temperature difference at the cold end	$\Delta T^{CE} = T_2 - T_4 =$	$\Delta T^{CE} = T_2 - T_4 = 80.2 - 78 = 2.2\,\text{K}$
Maximum temperature difference	$\Delta T_{MAX} =$	2.2 K
Mean temperature difference MTD	$MTD =$	$MTD = 2.05\,\text{K}$
Q/MTD value	$\dot{Q}^{HEX}/MTD =$	$Q/MTD = 34.86\,\text{kW/K}$

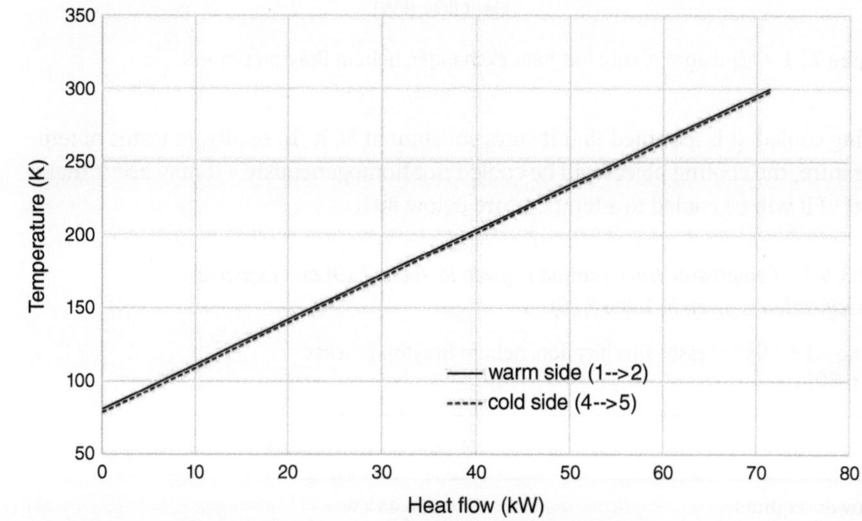

Figure 7.29 *T,Q*-diagram, heat exchanger, helium Brayton process.

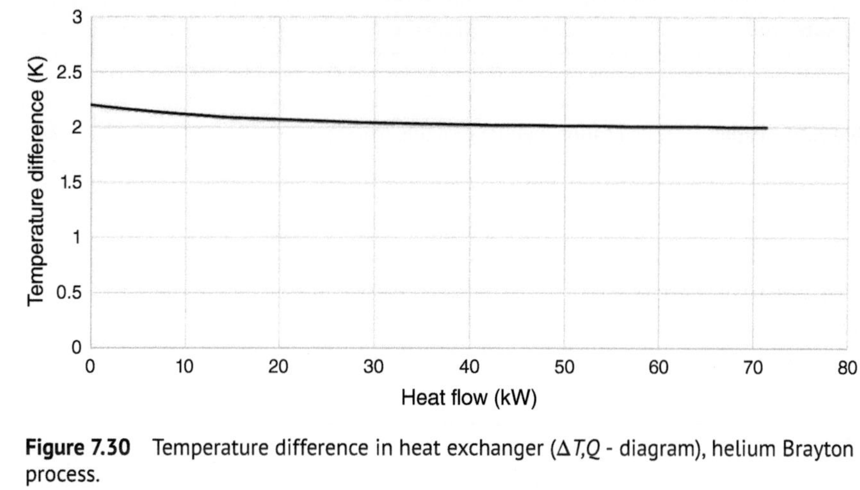

Figure 7.30 Temperature difference in heat exchanger ($\Delta T,Q$ - diagram), helium Brayton process.

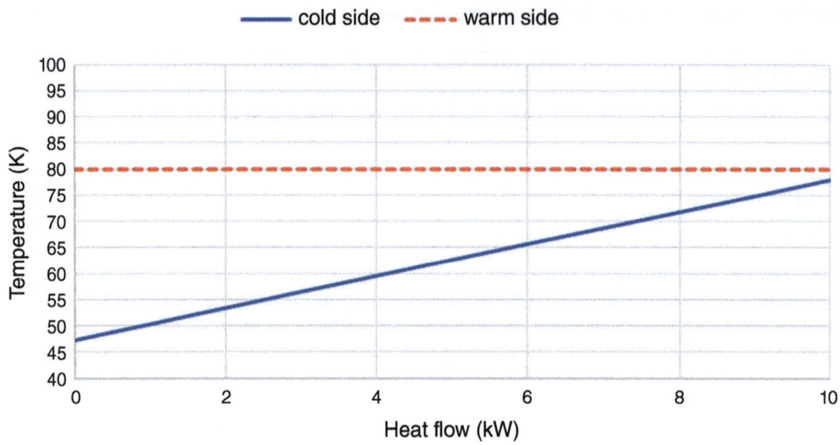

Figure 7.31 *T,Q*-diagram, cold has heat exchanger, helium Brayton process.

being cooled, it is assumed that it stays constant at 80 K. In reality, in terms of temperature, the cooling object will be cooled not homogeneously – it may occur that a part of it will be cooled to a temperature below 80 K.

7.3.3.5.3 Compressor (specification is given in Table 7.19) and Expander (specification is given in Table 7.20)

Table 7.19 Compressor specification, helium Brayton process, $T_o = 80$ K.

	Helium Brayton process
Power required	$P = 116.45$ kW
Inlet pressure	1.5 bar
Volumetric flowrate at the inlet	916 m^3/h
Pressure ratio	$9/1.5 = 6$

Table 7.20 Expander specification, helium Brayton process, $T_o = 80$ K.

	Helium Brayton process
Power generated in expander	$P_{EXP} = 10.8$ kW
Pressures at inlet and outlet	8.9 bar/1.7 bar
Volumetric flowrate at the outlet	131.1 m^3/h
Pressure ratio	5.4

Table 7.21 Performance of the helium Brayton process, $T_o = 80$ K.

Cooling capacity	$\dot{Q}_o = 10\,\text{kW}$
Total power required for the process	$P = P_{COMP} - P_{EXP} = 116.45 - 10.8 = 105.65\,\text{kW}$
COP	$\varepsilon = \dot{Q}_o\,/P = 10/105.65 = 0.0947 = 9.5\%$
Minimum power required for a reversible process (according to Carnot equation)	$P_{rev} = \dot{Q}_o\,(T_{amb} - T_o)/T_o = 10\,\text{kW}$ $(300 - 80)/80 = 27.5\,\text{kW}$
COP of the reversible process	$\varepsilon_{CARNOT} = \dot{Q}_o/P_{rev} = 10\,\text{kW}/27.5\,\text{kW} = 0.364 = 36.4\%$
Efficiency	$\eta_e = \varepsilon/\varepsilon_{CARNOT} = 0.095/0.364 \approx 26\%$

7.3.3.6 Overall Performance of the Process

Table 7.21 summarizes the performance of the helium Brayton process. It will be considered that the power generated in the expander is fully utilized. Therefore, the total power required for driving the process corresponds to the compressor power minus expander power. The efficiency of 26% is of utmost importance considering the relatively low efficiency of the screw compressor.

7.3.3.7 Process Analysis

This section consists of data for exergy loss in all process units as well as an analysis of these losses based on values given in Table 7.17. Table 7.22 and Figure 7.32 summarize the losses.

7.3.3.7.1 Loss in Compressor

The largest loss, 46.6 kW, is associated with the compressor, mainly due to its low efficiency of 60%.

7.3.3.7.2 Loss in Cold Gas Heat Exchanger

The second-largest loss – 13.8 kW – is associated with the cold gas heat exchanger. This can be explained by examining Table 7.23, based on Table 5.4. The maximum temperature difference in this heat exchanger may achieve a very high value of 33 K.

Table 7.22 Exergy loss, helium Brayton process, $T_o = 80$ K.

Element	Helium Brayton process
Compressor	46.6 kW
Heat exchanger	4.8 kW
Expander	12.9 kW
Cold gas heat exchanger	13.8 kW
Total	78.2 kW

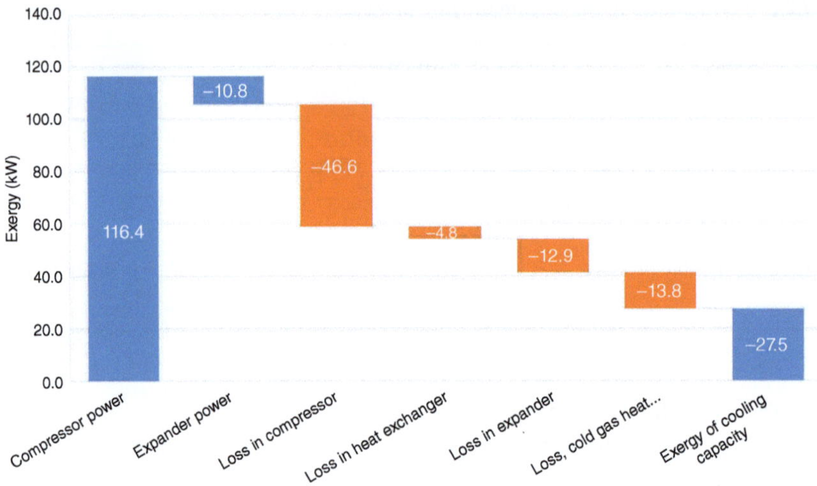

Figure 7.32 Exergy losses in the helium Brayton process.

Table 7.23 Impact of temperature difference in heat exchanger on exergy loss.

	Small temperature difference, less than 0.02 T	Medium temperature difference, (0.02–0.1) T	Large temperature difference, more than 0.1 T	Very large temperature difference
T = 100 K	<2 K Small loss	2–10 K moderate loss	>10 K large loss	>20 K very large loss
T = 200 K	<4 K Small loss	4–20 K moderate loss	>20 K large loss	>50 K very large loss

Main heat exchanger — Cold gas heat exchanger

In combination with the low-temperature level of below 80 K, it results in a large loss that exceeds the loss in the main heat exchanger (roughly 4.8 kW).

7.3.3.7.3 Loss in Main Heat Exchanger

This low loss in the main heat exchanger can be explained by means of the same Table 7.23. The temperature difference in the main heat exchanger is minimal: it amounts to 2 K at the warm end and remains low (approximately 2 K) in the whole temperature range; at the cold end it is 2.2 K (the background is that the helium isobars run almost parallel in this temperature and pressure range). Therefore, in terms of efficiency, the heat transfer in the main heat exchanger is quite optimal. The only disadvantage of this design is that the heat transfer requires an extremely large heat exchanger surface. Therefore, the heat exchanger becomes large and expensive.

7.3.3.7.4 Loss in Expander

The expander loss of 12.5 kW is considerably less compared to the expansion loss in the nitrogen Joule–Thomson process (approximately 90 kW, see Section 7.2.3.4.3).

The expander loss is predefined by the given isentropic efficiency of this device of $\eta_s = 85\%$. Using a reversible expander ($\eta_s = 100\%$), this loss could be eliminated and the required driving power for the whole process would be considerably improved. The improvement would be at least by 12.5 kW, which is more than 10% of required compressor power. That is an indication of a strong impact of the expander's efficiency on the overall process performance. It seems to be a key parameter for the overall performance of the reversed Brayton process.

Learning 7.8 In Brayton process, expander efficiency has a large impact on overall efficiency of the process.

7.3.3.7.5 Analysis, Summary

The largest loss is associated with the compressor. It can be reduced by using a more efficient compression device. This is the main and most promising direction for further improvement.

The second-largest loss is associated with the cold gas heat exchanger. An approach for reducing these losses based on the so-called isothermization method, whereby the single "expander and cold gas heat exchanger" is replaced by a sequence of multiple "expander and cold gas heat exchanger" units as shown in Figure 7.33. That is very similar to the process unit "multi-stage compressor" described in Section 5.5.

With an efficiency of 26%, the Brayton process is considerably better than a comparable nitrogen Joule–Thomson process. The following observations may help to explain this:

1. The thermodynamic properties of helium are advantageous. The heat capacity of this refrigerant does not depend on pressure in given temperature range. Therefore, the high-pressure isobars run parallel to the low-pressure isobars. It is the main reason for an advantageous temperature profile in the heat exchanger with a low-temperature difference in the whole heat exchanger. This temperature profile causes a minimal loss in the heat exchanger;
2. The expansion process of the expander can be realized in a very efficient way by means of near-isentropic expander;

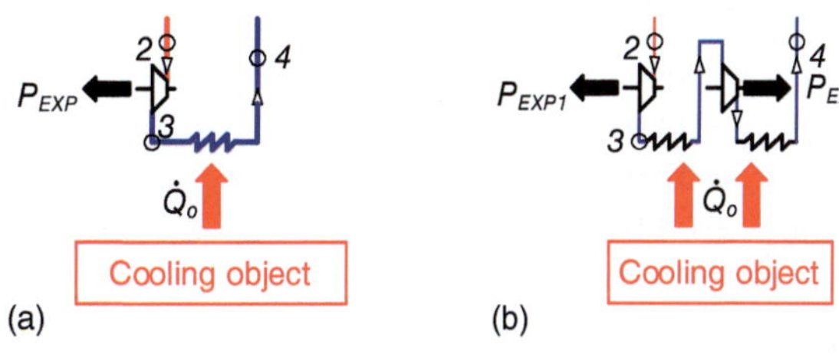

Figure 7.33 Isothermization approach (a) single expander-cold gas heat exchanger, (b) multiple "expander-cold gas heat exchanger."

3. The heat transfer in the cold gas heat exchanger causes no minor losses due to the high-temperature difference there. However, since the amount of heat transferred in this heat exchanger is relatively small, the loss is rather moderate than high;

It results in high efficiency of the overall process.

7.3.4 Process Optimization

The classical Brayton process, similar to the Joule–Thomson process, is a simple process with only a few variables that allow optimization.

7.3.4.1 Variables

Basically, there are four relevant degrees of freedom:

(1) Refrigerant (helium, neon, hydrogen, air gases, or a mixture of these gases)
(2) High-pressure level
(3) Low-pressure level
(4) Temperature difference at the warm end of the heat exchanger

There are also four parameters with minor impact:

(5) Pressure loss on the high-pressure side of the heat exchanger
(6) Pressure loss on the low-pressure side of the heat exchanger
(7) Pressure loss in the cold gas heat exchanger
(8) Minimum temperature difference in the cold gas heat exchanger

In contrast to the Joule–Thomson process, the low-pressure level is a real variable that can be selected freely.

7.3.4.2 Constraints

The constraints are as follows:

- Expander characteristics (for example, maximum pressure ratio)
- Compressor characteristics (for example, maximum pressure ratio)
- Heat exchanger design (for example, minimum temperature difference in the heat exchanger or maximum Q/MTD-value)

7.3.4.3 Example of Process Optimization

An example of an optimization study is presented in this section. This study was performed for a helium-based Brayton process with a cooling capacity of $\dot{Q}_0 = 10\,\text{kW}$.

There are six fixed parameters:

- $T_{amb} = 300\,\text{K}$
- $\Delta p_{1\to 2} = 0.1\,\text{bar}$
- $\Delta p_{3\to 4} = 0\,\text{bar}$
- $\Delta p_{4\to 5} = 0.1\,\text{bar}$
- $T_4 = 80\,\text{K}$
- $p_5 = 1.5\,\text{bar}$

Two parameters were varied:

- pressure p_1
- temperature difference at the warm end of the heat exchanger ΔT^{WE}.

It was assumed that the compression process could be performed with the same degree of efficiency and at the same cost in the whole pressure range investigated here.

Two following functions were examined during the study:

(1) The COP as a performance indicator
(2) The Q/MTD value of the heat exchanger as an indicator for the heat exchanger costs

The main findings for the COP (shown in Figure 7.34) are:

(a) The optimal range for the COP is always in the pressure range of 4–5 bar, independent of the ΔT^{WE}.
(b) The COP depends on the ΔT^{WE}. A reduction of the temperature difference at the warm end by 1 K (from 3 to 2 K or from 2 to 1 K) means a COP improvement of roughly 0.5%.
(c) The COP curves are flat if right to the optimum pressure, on the other side – left to the optimum – the efficiency (COP) drops quickly. This information is important for the choice of the right high-pressure level for the process. Usually, the system operates not only in optimal conditions, but in underload or overload, too. Therefore, the process must be optimized not for the single operation point but for a wider pressure operation range. In the given case, it would be better to set up the normal operation pressure a bit higher than the optimum. This way,

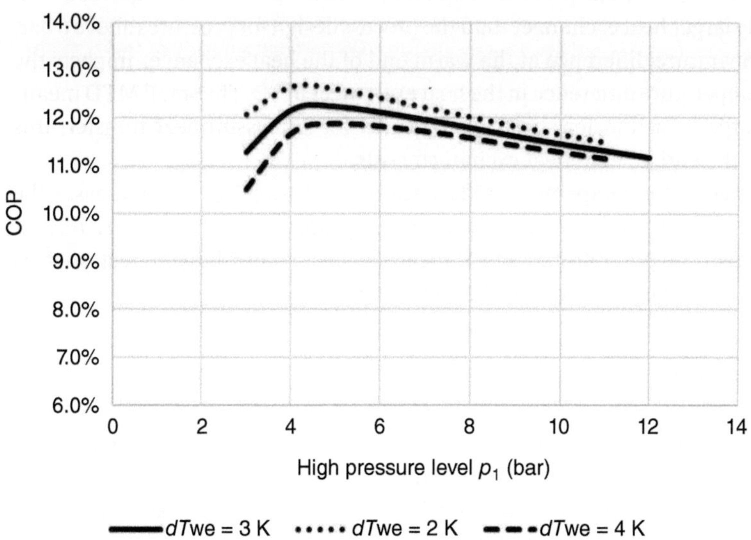

Figure 7.34 Coefficient of performance of a helium Brayton process; $dTwe = \Delta T^{WE}$ – temperature difference at the warm end of the main heat exchanger.

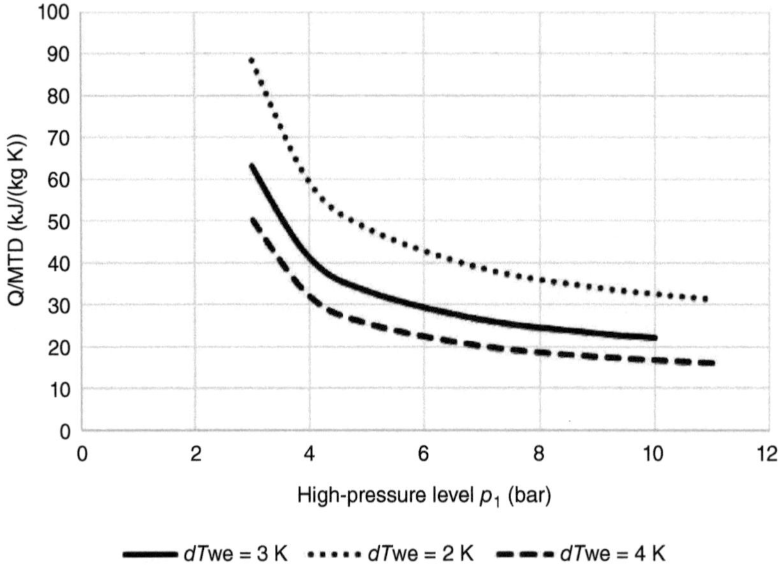

Figure 7.35 Q/MTD of the main heat exchanger, helium Brayton process
dTwe = ΔT^{WE} – temperature difference at the warm end of the main heat exchanger.

it will be secured that the process performs well even if the pressure is less than in a normal operation.

The main findings for Q/MTD (Figure 7.35) are:

(a) The optimum (minimum heat exchanger area) here is in the range of higher pressures (>10 bar). The process design for pressures below 4 bar requires a considerably larger heat exchanger than the process design for pressures above 6 bar;
(b) The temperature difference at the warm end of the heat exchanger impacts the mean temperature difference in the heat exchanger MTD. The small MTD means that a large heat transfer surface is required for successful heat transfer; this leads, at the end, to high heat exchanger costs.
The change in the temperature difference by 1 K (from 3 to 2 K) increases the requirements on the heat transfer surface by approximately 25%. A further reduction of the ΔT^{WE} from 2 to 1 K means a considerably larger effect, such as a process requiring 1/3 more heat exchanger surface.

7.3.4.4 In Summary

(a) It is difficult to make a decision for the most economic process design: the high-efficiency process requires a pressure of below 6bar, whereas the low-cost process (in terms of heat exchanger) requires a pressure that is higher, at 6bar.
(b) For a process with high efficiency (or COP > 12%), an extremely large and therefore expensive heat exchanger is needed.
(c) A further optimization, which considers the characteristics and costs of the compressor, is recommended here.

7.3.5 Control of a Brayton Refrigerator

The typical control task for a refrigeration system is described in Section 2.4: the main target is to keep the cooled object at the specified cooling temperature. For this purpose, the cooling capacity of a refrigerator must be adjusted to be equivalent to the heat flow generated by the cooling object.

Equation (7.27) summarizes the parameters and characteristics that have an impact on the cooling capacity of a Brayton refrigerator. Not all of them may be used for control purposes, but some of them are suitable – for example: (i) refrigerant flowrate $\dot{m}$ or (ii) pressure levels: high-pressure level p_1 as well as low-pressure level p_5.

Since the cycle is driven by a compressor and restricted by the expander, the refrigerant flowrate can be governed by controlling these devices:

1. The control of flowrate by the expander is a potentially viable option. The challenge is that the expansion turbine is a less robust device (compared to a Joule–Thomson valve) and must be guided much more precisely. Start-up and shutdown procedures are also usually more complicated in order to avoid damage of expander. An additional control valve on the expander inlet may be used for control purposes;
2. For the control on the compressor side, all options described in the section for Joule–Thomson process (Section 7.2.5) are applicable:
   (a) controlling the rotation speed of the driving motor by using an appropriate frequency converter,
   (b) bypass valve between the high-pressure and low-pressure levels,
   (c) expansion valve on the compressor inlet,
   (d) IGV control used for control of centrifugal compressors (described in Section 6.5.2.5). Similar to the operation of the expander, the operation of the centrifugal compressor requires more precision during start-up and shutdown in order not to damage the compressor.

The control option described in Section 7.2.5 (Figure 7.24) – based on control of the inventory of the refrigerant in the cold part of the process by means of an intermediate buffer vessel installed parallel to the compressor – is applicable to a refrigerator based on the Brayton process, too.

7.3.6 Hardware Components

7.3.6.1 Heat Exchanger

The requirements on the heat exchanger for the Brayton process are high: a large heat transfer surface, low-pressure drop, and compactness at low cost. These requirements may be more or less satisfied by brazed aluminum plate-fin heat exchangers (described in Section 6.2.1) and similar designs. But even the aluminum plate-fin heat exchanger, which is considered the most compact design, leads to a relatively large footprint of the whole system. There is a clear need for further development of heat exchangers toward higher compactness.

7.3.6.2 Compressor

The Brayton process is a low-pressure process, and it does not require high-pressure ratio in the compressor. An efficient process design may be developed based on a compressor with few compression stages.

The compressors used are either oil-lubricated screw compressors (see Section 6.5.3) or centrifugal machines (see Section 6.5.2). Generally, the screw compressor is used for small-scale applications, while the turbocompressor is targeted for middle- and large-scale applications.

Since the centrifugal compressors have a higher efficiency and require less maintenance, the general trend is to apply the turbocompressors even for small-scale applications. The interesting fact is that the scale-down of turbomachinery is a much more challenging endeavor than the scale-up, given that new concepts for bearing and drivers must be developed and implemented:

- the large centrifugal machines are based on oil-bearings, whereas the smaller machines use the magnetic or gas-bearings,
- high-frequency electric motors with permanent magnets and integrated frequency converters for control purposes are preferred as drivers for small turbocompressors, whereas large machines use conventional electric motors containing electromagnets.

7.3.6.3 Expander

The turboexpanders are the preferred kind of machines used for expanding gas in the Brayton process for large-scale, small-scale, and medium-scale applications.

Rules for the mechanical design of bearings are very similar to the rules for the compressor: the large centrifugal machines are based on oil-bearings, whereas the smaller machines use the magnetic or gas-bearings.

Another aspect relates to the loading device used for subtracting the mechanical power generated during the expansion process. Three loading devices for turboexpanders have previously been described in Section 5.2.5: oil/gas-brake, electric generator, and centrifugal compressor (booster compressor):

- the centrifugal compressor (booster compressor) is the most popular device used in large-scale systems like air separation plants,
- in small-scale cryogenics, however, the generated power is often not recycled but converted into the heat power and dissipated into the environment by means of an oil-/gas-brake,
- the option with the electric generator (based on conventional electric generators) is already practiced in large scale-systems. It has the potential to be extended to the medium/small-scale range, especially if high-frequency generators based on machines with permanent magnets are applied.

The small-scale turboexpanders are less efficient than larger machines. The isentropic efficiency of small devices is usually less than 80%, whereas the isentropic efficiency of large machines usually exceeds 80%. Very large devices can achieve an isentropic efficiency of 90%.

7.3.7 Applications

7.3.7.1 Air Separation

The largest application of the Brayton process is in the air separation; the Brayton process is intelligently integrated into the air separation plant, and causes only a minor additional cost, since it uses the existing main air compressor of the air separation unit (ASU) and the existing main heat exchanger of the ASU, so that a separate compressor and separate heat exchanger for the Brayton process are not needed. This results in an efficient and cost-efficient product.

7.3.7.2 Hydrogen Liquefaction

The helium Brayton cycle is used as an external cycle for liquefaction of hydrogen in small-scale plants.

7.3.7.3 Turbo-Brayton Refrigerator

The so-called Turbo-Brayton refrigerators, with a cooling capacity of 2–100 kW and cooling temperatures of 60–120 K, have recently been developed. The term "Turbo-Brayton" describes this system well since it is based on a combination of turbocompressor and turboexpander, usually combined in a single machine. For low cooling temperatures (<80 K), the neon or neon-based mixtures as refrigerant are preferred. A version based on air gases is preferable for cooling temperatures around 100 K. The main advantages of a Turbo-Brayton refrigerator are its high efficiency and the low maintenance it requires.

The relatively high cost of a single unit remains an issue. The hope is that the cost of a single unit will be reduced in the future if a reasonable number of units are produced and commercialized. The targeted application (and market) is the re-liquefaction of the boil-off gas, which results from evaporation (caused by heat leak through the nonideal thermal insulation) of liquefied natural gas or liquified hydrogen during the storage or transport on ships/carriers.

7.3.8 Assessment of the Brayton Process

The Brayton process has two very attractive features: (i) it is an efficient process, and (ii) low cooling temperatures of below 65 K can be reached by means of Brayton process. The second feature is neither available using the nitrogen Joule–Thomson processe, nor the mixed-fluid Joule–Thomson process (see section 7.5). It is a unique purpose of the Brayton process. Therefore, the Brayton process is considered always to have a low cooling temperature, be of high efficiency, and consequently require low operation expenditures (OPEX).

The downside of the Brayton process relates to the consequences of using an expander:

(1) **Process design issues:** according to the conservative approach, the turboexpander does not tolerate liquid at the outlet; rather, it produces mainly the cold gas. The advantages of using liquid cryogen for cooling purposes (see Section 7.2.6) are not available if the Brayton process is used;

(2) **Maintenance and reliability issues:** since the expander is a more sophisticated device in terms of mechanical design (at least in comparison to simpler expansion device like an expansion valve used in Joule–Thomson process), it is less reliable and requires more maintenance. It does not mean, however, that an expander is unreliable. On the contrary, the reliability is high, with numerous expanders being able to operate for years and decades. Generally, however, the expander's reliability is considered less than that of the valve;
(3) **Cost issues:** the cryogenic expander is a cost-intensive hardware component.

Generally, the cost of the refrigeration system based on the Brayton process is considered rather high, since it is based on relatively expensive hardware components: not only is the expander a costly device, the Brayton process requires a large heat exchanger with corresponding costs, and the compressor is a cost-intensive hardware component. That may explain why the pure Brayton process is not often implemented and materialized as a stand-alone cryogenic system.

7.4 Claude Process

Strictly speaking, the Claude process does not represent a separate sort of cryogenic process, given that it is a combination of a Joule–Thomson and Brayton process. The combination of these two processes is especially noteworthy, as it offers significant advantages in comparison to the pure Joule–Thomson process or pure Brayton process. Ultimately, almost all modern cryogenic plants are now based on the Claude process. By contrast, plants based on the pure Joule–Thomson process or the pure Brayton process are a rarity. Therefore, it makes sense to consider the Claude process separately.

7.4.1 Basic Claude Process, Main Elements

The Claude process was invented, developed, and commercialized by French innovator Georges Claude. Figure 7.36 shows a version of this process, referred to as "basic Claude process," adapted for refrigeration purposes. It differs slightly from the process introduced originally for the air liquefaction purposes in 1902, see [3]. This "basic Claude process" is a sort of virtual process existing in textbooks only, since the morphology of real nitrogen Claude process and real helium Claude process differs from "basic Claude process." Nevertheless, from didactic perspective, it is better to start first consideration using basic Claude process, and switch later to real Claude processes.

The process includes the following elements:

1. Compressor unit, which is usually a multi-stage isothermal compressor
2. Three heat exchangers HEX1, HEX2, and HEX3
3. Expander
4. Expansion valve
5. Evaporator
6. Thermal insulation

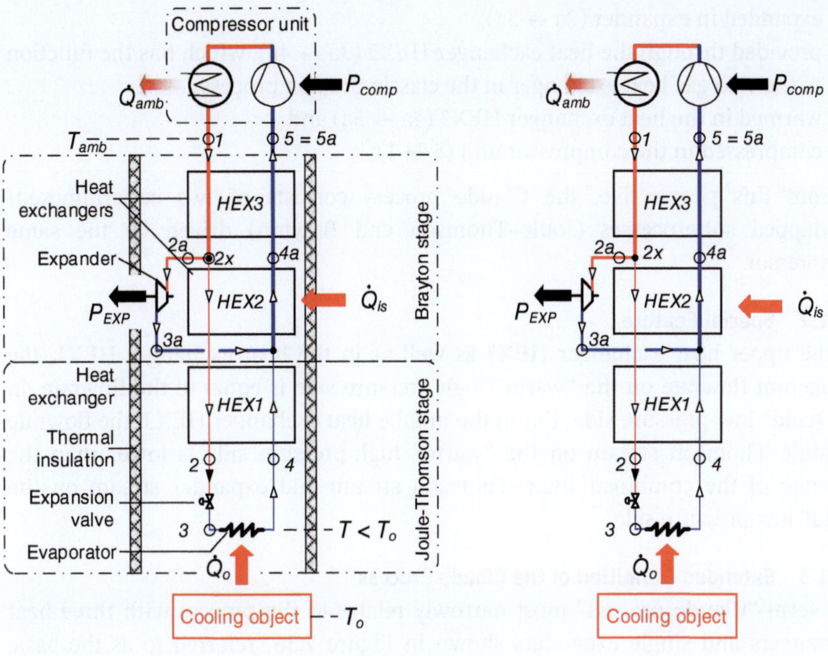

Figure 7.36 Basic Claude process, process flow diagram. Left - the thermal insulation is shown, right - the thermal insulation is removed.

All these process units and their functions have been previously described in Section 7.2 (nitrogen Joule–Thomson process) and Section 7.3 (Brayton process).

The most salient details here are:

- This process includes two expansion devices – an expansion valve and an expander;
- There are three separate heat exchangers – HEX1, HEX2, and HEX3 – used in this process instead of a single (main) heat exchanger.

7.4.1.1 Morphology

Morphologically, the basic Claude process is a combination of a classic Joule–Thomson process and a Brayton process:

1. The first – the Joule–Thomson process – can be recognized in Figure 7.36 as the sequence marked as 1 → 2 → 3 → 4 → 5, whereas the compressed fluid is
   - cooled in heat exchangers HEX3, HEX2, HEX1 (1 → 2),
   - expanded in expansion valve (2 → 3),
   - provided through the evaporator (3 → 4),
   - warmed in heat exchangers HEX1, HEX2, HEX3 (4 → 5) and
   - compressed in the compressor unit (5 → 1);
2. The second – the Brayton process – can be recognized in Figure 7.36 as the sequence marked as 1 → 2a → 3a → 4a →5a, whereas the compressed fluid is
   - cooled in heat exchanger HEX3 (1 → 2a),

- expanded in expander (2a → 3a),
- provided through the heat exchanger HEX2 (3a → 4a), which has the function of the cold gas heat exchanger in the classic Brayton process,
- warmed in the heat exchanger HEX3 (4a → 5a) and
- compressed in the compressor unit (5 → 1);

From this perspective, the Claude process consists of two superimposed/overlapped subprocesses (Joule–Thomson and Brayton) driven by the same compressor.

7.4.1.2 Special Feature

In the upper heat exchanger HEX3 as well as in the heat exchanger HEX1, the refrigerant flowrate on the "warm" high-pressure side is equal to the flowrate on the "cold" low-pressure side. But in the middle heat exchanger HEX2, the flowrate of Joule–Thomson stream on the "warm" high-pressure side is lower than the flowrate of the combined Joule–Thomson stream and expander stream on the "cold" low-pressure side.

7.4.1.3 Extended Definition of the Claude Process

The term "Claude process" most narrowly relates to the process with three heat exchangers and single expanders shown in Figure 7.36 (referred to as the basic Claude process). In a broader sense, however, Georges Claude invented not only the process but a universal process design principle – that Joule–Thomson stages and Brayton stages can be combined. This is applicable to the design of cryogenic processes.

Therefore, the term "Claude process" is not merely used for the single "basic Claude process," but for the large family of similar processes based on a combination of Joule–Thomson stages and Brayton stages. The number of variations is quite unlimited, given that the number of Brayton stages and Joule–Thomson stages is unlimited. Moreover, there are numerous further degrees of freedom. For example:

- The Brayton stages can be ordered with or without an overlap (in terms of temperature level),
- Every Brayton stage may include single expander or several expanders connected in series or in parallel,
- The expanders can be designed as reciprocal or as centrifugal machines,
- The mechanical power generated in the expander can be utilized by means of an electric generator, compressor, or other devices.

All these combinations are considered as the Claude process.

7.4.2 Process and Process Features

7.4.2.1 Main Idea of the Claude Process

In terms of hierarchy, the Joule–Thomson stage is the most important part of the Claude process since the required cooling temperature and cooling capacity is generated there, making it the "superior" process. The role of the Brayton stage is basically to ensure that the Joule–Thomson process performs well. It is a sort of "subordinate" process. In terms of thermodynamics, the main purpose of the Brayton stage is to minimize irreversibilities in the Joule–Thomson stage. The main source of irreversibility in the Joule–Thomson process is the expansion operation. Therefore, the Brayton stage must be integrated into the Claude process in such a way that the temperature of the Joule–Thomson stream before the expansion would be reduced to a minimum doable value.

In order to achieve this goal, the temperature at the expander inlet as well as the flowrate of the expander stream must be optimized:

- If the temperature at the inlet is too high →the temperature at the expander outlet will be high, and the subcooling of the Joule–Thomson stream will be limited by this temperature;
- If the flowrate is too small → the Brayton stage will not generate enough cooling capacity to cool the Joule–Thomson stream to the lowest temperature.

Both situations will result in reduced liquid fraction at the expansion valve outlet (high flash-gas generation) and correspondingly reduced cooling capacity of the process.

7.4.2.2 Basic Claude Process, Process Description

The sequence is the following:

(a) The refrigerant is first compressed from the low-pressure level p_{LP} to the high-pressure level p_{HP} (supercritical pressure) by means of a compressor and cooled to the approximate ambient temperature; this state is marked in Figure 7.36 as state 1;
(b) This compressed refrigerant is cooled in the countercurrent heat exchanger HEX3 by the cold low-pressure stream, and the state of refrigerant at the cold end of this heat exchanger is marked as state 2x;
(c) At this point, the refrigerant flow is divided into two streams.
The first stream is usually referred to as the Joule–Thomson stream, and the second stream is referred to as the expander stream (or turbine stream). The flowrate of the expander stream is usually higher (by a factor of more than 2) than the flowrate of the Joule–Thomson stream;
(d) Expander stream:
  - The expander stream is expanded in the expander from the high-pressure level p_{HP} to the low-pressure level p_{LP}. The temperature of this stream reduces correspondingly. The state on the expander outlet is marked in Figure 7.36 as state 3a; this stream is gaseous,
  - This cold gas from the expander outlet is provided through the heat exchanger HEX2. The cold of the cold expander stream is applied there for cooling the Joule–Thomson stream. Simultaneously, the cold gas is warmed up (process step 3a → 4a);
  - It is provided further to the heat exchanger HEX3. This cold fluid flows through the heat exchanger HEX3 to the compressor. In the heat exchanger,

the heat is transferred from the "warm" high-pressure stream to the cold low-pressure stream. The low-pressure fluid is warmed up to a temperature T_5 (state 5), which is close to the ambient temperature, but still a little bit colder than the refrigerant at state 1. At the outlet of the heat exchanger, the refrigerant is in a gaseous state.

(e) Joule–Thomson stream:
   - The Joule–Thomson stream flows through the heat exchanger HEX2, where it is cooled by the cold expander stream in a countercurrent direction. It enters the Joule–Thomson stage, where:
   - It is cooled to the temperature T_2 (state 2, the state at this point corresponds to the supercritical liquid) in heat exchanger HEX1;
   - It is expanded in the expansion valve to the low-pressure level p_{LP}, and this process (2 → 3) is a typical Joule–Thomson expansion, whereby the Joule–Thomson stream partially liquefies and a part of refrigerant evaporates generating flash-gas. This two-phase fluid flows through the evaporator and adsorbs the heat power from the object being cooled. At the outlet of the evaporator (state 4), the liquid is evaporated and the state corresponds to the dew point,
   - This cold vapor is provided through the heat exchanger HEX1, where it warmed up to a temperature close to the temperature T_{3a} cooling the material stream on the warm side of this heat exchanger;
   - At the outlet of the heat exchanger HEX1, the Joule–Thomson stream is mixed with the expander stream; it flows through the heat exchangers HEX2 and HEX3 to the compressor inlet.

7.4.2.2.1 Staging

The process design engineers use to structure the basic Claude process in special way. The process is divided into two stages:

(a) Low-temperature stage, which includes heat exchanger HEX1, expansion valve, and evaporator. This stage is referred to as a Joule–Thomson stage, as shown in Figure 7.36 (left).This stage is indeed identical to the classic Joule–Thomson process;
(b) The stage referred to as the Brayton-stage, which includes heat exchanger HEX3, expander and heat exchanger HEX2, as shown in Figure 7.36 left.

7.4.2.3 Typical Process Pressure Levels

7.4.2.3.1 High-Pressure Level and Pressure Dilemma

The Claude process is a combination of a Joule–Thomson and Brayton stage. Both these loops are driven by the same compressor. The dilemma is that the requirements on the process pressure from the Joule–Thomson process's perspective differs much from the requirements on the process pressure from the Brayton process:

- The Joule–Thomson process is a high-pressure process requiring supercritical pressure on the high-pressure side of the process;
- On the contrary, the Brayton process is a low-pressure process usually requiring subcritical pressure level.

Therefore, a compromise always determines the choice of the right process pressure. The usual procedure is to select the high-pressure level that is (i) higher as critical pressure but (ii) not extremely high as it is usually required by the pure Joule–Thomson process. Therefore, commonly, the high-pressure level of 40–80 bar is practiced in the nitrogen/air-based Claude process, and 8–16 bar in helium-based processes. Other approaches to manage the pressure dilemma are discussed in Section 9.1.4.

7.4.2.3.2 Low-Pressure Level

***Pressure ratio*:** Some further restrictions caused by the use of centrifugal expanders must be taken into account. It is salient to note that the limitation on maximum pressure ratio in a single expander stage has great impact on process design.

Helium: For the helium-based Claude process, the low-pressure level of 1.1–1.6 bar is typical. In combination with the high-pressure level of 8–16 bar, this means a high-pressure ratio that cannot be performed by means of a single stage helium turboexpander. Therefore, the process design for helium system is more complex than the basic Claude process shown in Figure 7.36.

Air gases: The low-pressure level of 5–7 bar is typically used in the nitrogen-based or air-based Claude process. The background is that the nitrogen Claude systems are usually connected to or even integrated into an air separation plant (for more information, see Section 8.3.1). The pressure level of 5-7 bar is a kind of standard pressure level used in air separation. Therefore, the all connected process units usually take over this pressure level to achieve process compatibility. In combination with the high-pressure level of 40–80 bar, it means a high pressure-ratio in the expander of 8–10. This is manageable by a single-stage expander at temperatures below 200 K. However, since it is very close to the upper limit, the process design engineers are always looking for an appropriate solution for this problem.

7.4.2.4 Cooling Capacity

The approach previously used in Section 7.3.2.6 may be applied to consider the cooling capacity of the Claude process with a system boundary, as shown in Figure 7.37.

The corresponding energy balance equation is:

$$\dot{Q}_o + \dot{Q}_{is} = (\dot{H}_5 - \dot{H}_1) + P_{EXP} \tag{7.28}$$

If the heat flow through the thermal insulation is negligibly small ($\dot{Q}_{is} \rightarrow 0$), the equation can be rewritten as:

$$\dot{Q}_o = (\dot{H}_5 - \dot{H}_1) + P_{EXP} \tag{7.29}$$

Formally, this equation is identical to the energy balance of the Brayton process (see Eq. (7.25)). However, it is a good practice to regard the specifics of the considered process and discuss what this specific may mean for the key learnings about the cooling capacity.

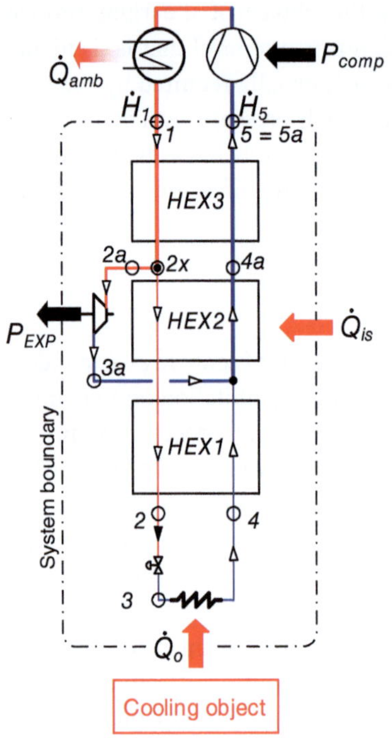

Figure 7.37 Claude process, system boundary for cold part.

Since the Claude process is a high-pressure process, the ideal gas-assumption (used in Section 7.3.2.6 for Brayton process) is not always applicable to Claude process, and the real properties of gases (nitrogen, helium) must be considered.

7.4.2.4.1 Process with Infinitely Large Heat Exchanger

For the case of infinitely large heat exchanger, the Eq. (7.29) can be further simplified to:

$$\dot{Q}_o = \Delta \dot{H}_T^{\ WE} + P_{EXP} \tag{7.30}$$

A deep dive into the properties of cryogenic substances shows that the enthalpy difference $\Delta \dot{H}_T^{\ WE}$ for the nitrogen in the typical pressure range is a slightly positive value, whereas the enthalpy difference $\Delta \dot{H}_T^{\ WE}$ for helium is quite close to zero or slightly negative. Therefore:

for nitrogen process $\dot{Q}_o > P_{EXP}$

for helium process $\dot{Q}_o < P_{EXP}$

7.4.2.4.2 Process with a Real Heat Exchanger

Usually, in terms of numbers, the contribution of the Joule–Thomson effect is less than contribution of expander. However, in the case of the high-pressure Claude processes with nonideal gas (for example nitrogen), the contribution of the Joule–Thomson effect becomes tangible. Consequently, this effect must be considered.

The following equations can be generated by combining the Eqs. (7.18) and (7.30):

$$\dot{Q}_o = \left(\Delta\dot{H}_T^{WE} - \dot{m}\cdot c_p\,\Delta T^{WE}\right) + P_{EXP} \tag{7.31}$$

or extended even further thanks to using Eq. (7.27) for a near-isentropic expansion to:

$$\dot{Q}_o = \dot{m}c_p\,\left[\overline{\mu}\,(p_1 - p_5) - \Delta T^{WE}\right] + \eta_s\cdot\dot{m}\frac{R}{M_r}\cdot T_{2a}\frac{\kappa}{\kappa-1}\left[1-\left(\frac{p_{2a}}{p_{3a}}\right)^{\frac{(1-\kappa)}{\kappa}}\right] \tag{7.32}$$

where

the first summand	$\dot{m}\,c_p\,\left[\overline{\mu}\,(p_1 - p_5) - \Delta T^{WE}\right]$	describes the impact of the Joule–Thomson effect
the second summand	$\eta_s\cdot\dot{m}\frac{R}{M_r}\cdot T_{2a}\frac{\kappa}{\kappa-1}\left[1-\left(\frac{p_{2a}}{p_{3a}}\right)^{\frac{(1-\kappa)}{\kappa}}\right]$	describes the contribution of the expander.

According to this equation, the cooling capacity of the corresponding Claude process can be increased by means applicable to the Joule–Thomson process (see Section 7.2.2.3.3) as well as by means applicable to the Brayton process (see Section 7.3.2.6.3).

7.4.3 Nitrogen Claude Process

Figure 7.38 shows the nitrogen Claude process. The special feature of this process is that the coldest (lower) heat exchanger HEX1 is not required for this process and is removed.[13] Therefore, this process consists of only two heat exchangers: HEX2 and HEX3.

7.4.3.1 Process Description

Figure 7.39 shows an example of a nitrogen Claude process, described further in Section 7.4.4 in detail in the *T,s*-diagram. The process sequence is as follows:

(a) The nitrogen is compressed from the low-pressure level $p_{LP} = p_5$ (in this example close to 7 bar) to the high-pressure level $p_{HP} = p_1$ (60 bar in this example) by means of a multi-stage compressor and cooled to the approximately ambient temperature. This state is marked as state 1;

(b) This compressed refrigerant is cooled in the countercurrent heat exchanger HEX3 by the cold low-pressure stream to the temperature of roughly 150–200 K (in the given example $T_{2a} = 178$ K);

13 A process referred to as Kapitsa process may be found in the classic cryogenic literature; see [4]. It is interesting to note that the morphology of the Kapitsa process (developed for liquefaction purposes) is quite similar to the morphology of the nitrogen Claude process. However, that is a pure coincidence. The thermodynamics of the low-pressure Kapitsa process differs principally from the thermodynamics of the high-pressure nitrogen Claude process, see section 9.2.2.

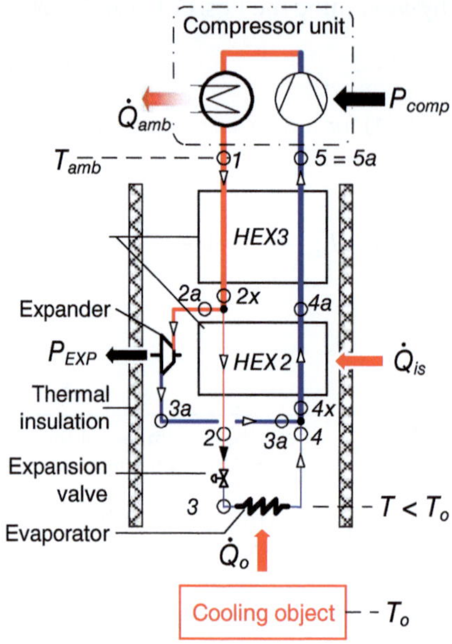

Figure 7.38 Nitrogen Claude process. Left - the thermal insulation is shown, right - the thermal insulation is removed.

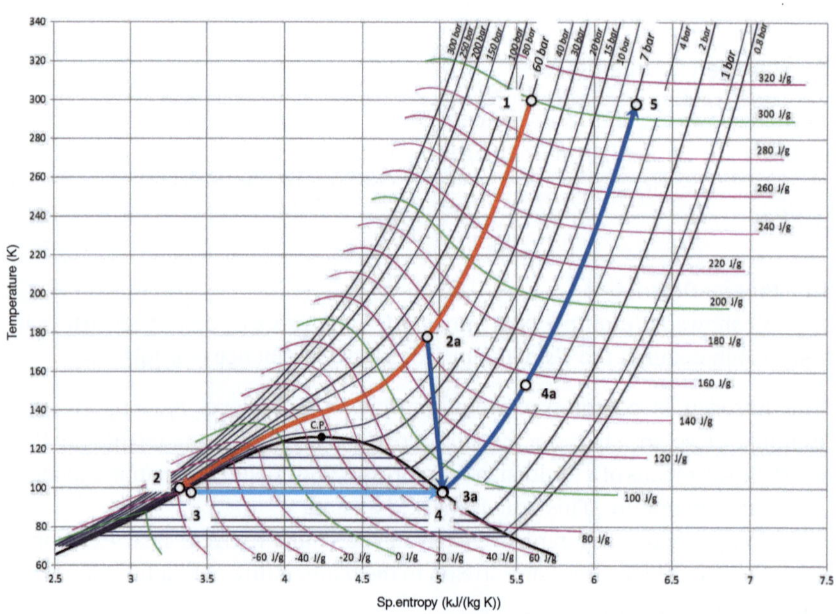

Figure 7.39 Nitrogen Claude process in *T,s*-diagram.

(c) At this point, the refrigerant flow is divided into two streams: the Joule–Thomson stream and the expander stream. The flowrate of the expander stream is usually higher than the flowrate of the Joule–Thomson stream;
(d) The expander stream is expanded from the high-pressure level p_{HP} to the low-pressure level p_{LP}. The temperature of this stream reduces correspondingly. The targeted state on the expander outlet is the dew point (or slightly above). In the given example, $T_{3a} \approx 98$ K.
(e) The expander stream is still gaseous; it is mixed with the cold gas from the evaporator, and this combined stream is provided through the heat exchanger HEX2.
(f) In the heat exchanger HEX2, the Joule–Thomson stream is cooled to the temperature T_2. The target is to achieve the lowest possible temperature. In the given example, the cold gas at the expander outlet is very cold ($T_{3a} = 98$ K), and the flowrate of the expander stream is adjusted in such a way so that the Joule–Thomson stream is cooled to the temperature of approximately 100 K.
(g) The Joule–Thomson stream is expanded in the expansion valve to the low-pressure level p_{LP}. This process ($2 \rightarrow 3$) is a typical Joule–Thomson expansion whereby the Joule–Thomson stream liquefies, but a part of the refrigerant (flash-gas) evaporates.
(h) This two-phase fluid flows through the evaporator and absorbs the heat flow from the object being cooled; at the outlet of evaporator (state 4), the liquid is evaporated.
(i) This cold vapor is mixed with the cold expander stream, and this combined stream flows through the heat exchangers HEX2 and later HEX3 to the compressor inlet. In the HEX2, it is heated to $T_{4a} \approx 150$ K and in HEX3 to $T_5 \approx 297$ K.

7.4.4 Example, Process Calculation, Nitrogen Claude Process

An example of a process calculation for a nitrogen Claude process is considered in this section.

7.4.4.1 Task Definition

A nitrogen Claude process for cooling capacity $\dot{Q}_o = 10$ kW for constant cooling temperature $T_o = 100$ K must be designed. The ambient conditions are given in Table 7.5. The target is to provide:

1. Process parameters (flowrate; temperature; pressure; phase state; composition; enthalpy; entropy; exergy) at all relevant process points;
2. Total power consumption, COP, and efficiency, as well as exergy loss in every process component;
3. Figures required for designing heat exchangers and compressors.

7.4.4.2 Process and Hardware Concept

The process concept is based on application of a bath evaporator (see Figure 7.40), similar to the design used in the nitrogen Joule–Thomson process (see Section 7.2.3).

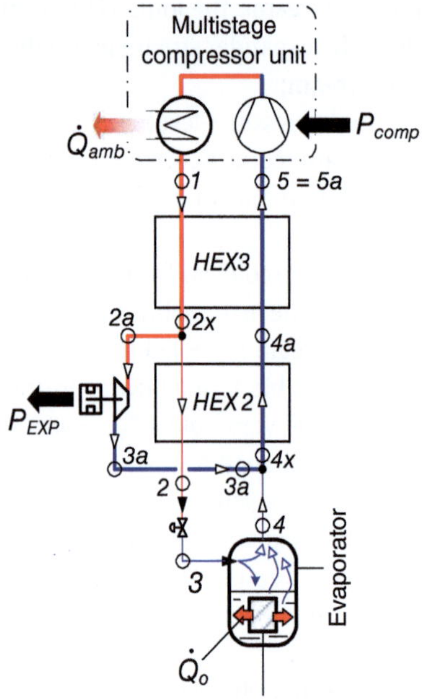

Figure 7.40 Concept of the nitrogen Claude process.

The cooling object is submerged in liquid nitrogen. This ensures safe and efficient cooling.

The other process features are:

(a) an efficient aluminum-brazed plate-fin heat exchanger;
(b) an oil-free multi-stage centrifugal compressor;
(c) an efficient multilayer insulation,
(d) an efficient turboexpander with an electric generator as the brake device, where the generated power is utilized;

Further process design considerations are as follows:

1. Compressor:
   - since a multi-stage centrifugal compressor is applied, the compression operation can be considered a quasi-isothermal compression;
   - it will be assumed that the high-pressure level corresponds to the $p_1 = 60$ bar;
   - data for compressor efficiency can be derived from previous experiences or discussions with equipment manufacturers. In the given example, it is assumed to be 70% (typical value for small centrifugal compressors);
   - an oil-removal system is not required since the compressor used is oil-free;
   - temperature T_1 is close to the ambient temperature ($T_1 \approx T_{amb} = 300$ K) due to the efficient cooling after compression;
2. Heat exchangers:
   - for the first process design, minor pressure loss on the high-pressure side $\Delta p_{1\rightarrow 2x} = \Delta p_{2x\rightarrow 2} = 0.05$ bar

as well as minor pressure loss on the low-pressure side
$\Delta p_{4x \to 4a} = \Delta p_{4a \to 5a} = 0.05\,\text{bar}$
may be assumed for all heat exchangers;
- for the temperature difference at the warm end the value of $\Delta T^{WE} = 2\,\text{K}$ may be assumed;

3. Evaporator:
   - the pressure loss in the evaporator is negligibly small $\Delta p_{3 \to 4} = 0\,\text{bar}$;
   - a small temperature difference between the nitrogen in evaporator and the cooling object can be expected there; for example $(T_4 - T_o) = 2\,\text{K}$ as a first estimation;
4. Expander: the isentropic efficiency of 85% can be assumed;
5. Insulation: heat leakage through thermal insulation is negligibly small due to the efficient insulation;

All these assumptions can be reviewed later, when the detailed design of the cold part is finished and the real values are available; however, for the first process design these are acceptable assumptions. These considerations are summarized in Table 7.24.

7.4.4.3 Properties Data

The fluid property software "miniREFPROP" is used for calculating thermodynamic properties.

7.4.4.4 Process Simulation

The process calculation starts with calculating pressures at all relevant process points.

Table 7.24 Input data for nitrogen Claude process derived from the hardware concept.

Thermal insulation, heat leakage	$\dot{Q}_{is} = 0$
Compression	*Quasi-isothermal*
The compressor efficiency	$\eta_T = 70\%$
High-pressure level	$p_1 = 60\,\text{bar}$
Pressure loss in heat exchangers, high-pressure side	$\Delta p_{1 \to 2x} = \Delta p_{2x \to 2} = 0.05\,\text{bar}$
Pressure loss in heat exchangers, low-pressure side	$\Delta p_{4x \to 4a} = \Delta p_{4a \to 5a} = 0.05\,\text{bar}$
Pressure loss in evaporator	$\Delta p_{3 \to 4} = 0\,\text{bar}$
Temperature at the outlet of compressor (= inlet heat exchanger HEX3)	$T_1 = 300\,\text{K}$
Heat exchanger, min temperature difference	$\Delta T^{min} = 2\,\text{K}$
Minimum temperature difference in evaporator	$T_o - T_4 = 2\,\text{K}$ $T_o = 100\,\text{K} \to T_4 = 98\,\text{K}$
Expander, isentropic efficiency	$\eta_S = 85\%$

7.4.4.4.1 Pressure Table

p_1	$p_1 = 60$ bar as given in Table 7.24
$p_{2x} = p_{2a}$	this pressure is lower than p_1 by pressure loss in the heat exchanger HEX3, high-pressure side: $p_{2a} = p_{2x} = p_1 - \Delta p_{1\to 2x} = 60\,\text{bar} - 0.05\,\text{bar} = 59.95\,\text{bar}$
p_2	this pressure is lower than p_{2x} by pressure loss in the heat exchanger HEX2, high-pressure side: $p_2 = p_{2x} - \Delta p_{2x\to 2} = 59.95\,\text{bar} - 0.05\,\text{bar} = 59.9\,\text{bar}$
p_3	pressure p_3 can be found as the boiling pressure that corresponds to given boiling temperature T_3 by means of "miniREFPROP" using temperature as input parameter. The temperature T_3 amounts to $T_3 = T_4 = 98$ K, (see Table 7.24). Consequently, $p_3 = 6.76$bar
$p_4 = p_{4x} = p_{3a}$	This pressure is equal to the pressure p_3, ($\Delta p_{3\to 4} = 0$, according to Table 7.24). Therefore, $p_4 = 6.76$ bar
p_{4a}	This pressure is lower than p_4 by pressure loss in the heat exchanger HEX2, low-pressure side: $p_{4a} = p_{4x} - \Delta p_{4x\to 4a} = 6.76\,\text{bar} - 0.05\,\text{bar} = 6.71\,\text{bar}$
p_5	This pressure is lower than p_{4a} by pressure loss in the heat exchanger HEX3, low-pressure side: $p_5 = p_{4a} - \Delta p_{4a\to 5} = 6.71\,\text{bar} - 0.05\,\text{bar} = 6.66\,\text{bar}$

7.4.4.4.2 Simulation Sequence

The following sequence can be used as an example:

Step 1:	**Obtaining the temperature at the expander inlet T_{2a}:** • An appropriate assumption for the temperature T_{2a} must be made, for example $T_{2a} = 180$ K; • Calculation of the expansion process in the expander must be performed, and the outlet temperature T_{3a} defined; • The temperature T_{2a} must be iteratively adjusted targeting the temperature at the expander outlet being as close as possible to the temperature at the evaporator outlet $T_{3a} \approx T_4 = 98$ K.
Step 2:	**Obtaining the flowrate of the Joule–Thomson stream $\dot{m}_{JT} = \dot{m}_2 = \dot{m}_3$:** An appropriate assumption for a temperature upstream expansion valve T_2 must be made. A good starting point is the assumption that the minimum temperature difference $\Delta T^{min} = 2$ K can be achieved at the cold end of the heat exchanger HEX2. This would mean $T_2 = T_4 + \Delta T^{min} = 98\,\text{K} + 2\,\text{K} = 100\,\text{K}$; An appropriate assumption for the flowrate of the Joule–Thomson stream must be made; for example, $\dot{m}_{JT} = 60$ g/s; Calculation of the expansion process in the Joule–Thomson valve as well as the calculation of the process in evaporator must be performed, and the generated cooling capacity must be defined; The flowrate $\dot{m}_{JT}$ must be iteratively adjusted, targeting the generated cooling capacity being equal to the required cooling capacity of the process $\dot{Q}_o = 10$ kW.

Step 3:	**Obtaining the flowrate of the expander stream $\dot{m}_{EXP}$:** • An appropriate assumption for the flowrate of the expander stream must be made, for example $\dot{m}_{EXP} = 200$ g/s. The total mass flowrate can be calculated as sum of the Joule–Thomson stream and the expander stream: $\dot{m}_1 = \dot{m}_{EXP} + \dot{m}_{JT}$; • Calculation of the whole process must be performed; • The flowrate $\dot{m}_{EXP}$ must be iteratively adjusted, targeting the suitable temperature profile in both heat exchangers HEX2 and HEX3 with local temperature difference ΔT being higher or equal to the minimum temperature difference required $\Delta T \geq \Delta T^{min} = 2$ K. Optimally, the temperature difference at the warm end of the heat exchanger HEX3 corresponds to the required minimal temperature difference $\Delta T^{WE} = \Delta T^{min} = 2$ K.
Step 4	Tuning these three parameters (temperature at the expander inlet, flowrate of the Joule–Thomson stream, and flowrate of the expander stream), targeting the local temperature difference ΔT being higher or equal to the minimum temperature difference required $\Delta T \geq \Delta T^{min} = 2$ K in the whole process.

The tricky part is the right process design in the heat exchanger HEX2. The slope of the isobar p_{HP} in temperature range below 200 K (which corresponds to the cooling of the Joule–Thomson stream in HEX2, process step 2a → 3), is strongly nonlinear. Consequently, the pinch points may occur there. Therefore, the temperature profile in the heat exchangers (*T,Q*-diagram of the heat exchanger) must be checked after every process calculation, and the pinch points must be identified and analyzed. In case of minimal temperature difference is lower than specified minimal value $\Delta T < \Delta T^{min}$, the flowrate of the expander stream must be adjusted correspondingly.

Table 7.25 shows the results. Figure 7.39 shows the process in *T,s*-diagram. The specific exergy as well as absolute exergy are calculated by using Eq. (4.5); the reference state is referred to as $p = 1.013$ bar and $T = 300$ K.

7.4.4.5 Process Units

Only main process units are considered in this section. The mixing of two streams (material streams 3a and 4) with different temperatures is not discussed.

7.4.4.5.1 Heat Exchangers

Table 7.26 summarizes the data related to heat exchangers. The corresponding temperature profiles and temperature difference profiles are shown in Figures 7.41–7.44.

HEX3: In the upper heat exchanger HEX3, the temperature profile looks very similar to the profile of the classic Joule–Thomson process: at the warm end, the lines get closer, while at the cold end, the temperature difference increases to the maximum value, which is approximately 25 K in the given case. This temperature profile is typical for high-pressure processes; it is due to the difference in heat capacities between the supercritical warm stream and the subcritical cold material stream at the same flowrate.

HEX2: However, the temperature profile in the heat exchanger HEX2 looks different in comparison to the lower part of the classic Joule–Thomson process: the temperature difference at the cold end is reduced to the minimum.

Table 7.25 Output process data, nitrogen Claude process, $T_o = 100$ K.

State	Temperature (K)	Pressure (bar)	Vapor fraction (mol/mol)	Specific enthalpy (J/g)	Specific entropy (J/(g K))	Mass flowrate (g/s)	Specific exergy (J/g)	Exergy (kW)
1	300	60	1	299.026	5.5919	208.00	362.83	75.47
2	100	59.9	0.0	−71.764	3.3166	62.78	674.65	42.36
3	98	6.76	0.037	−71.764	3.3929	62.78	651.76	40.92
4	98.00	6.76	1.0	87.516	5.0182	62.78	323.45	20.31
5	297.99	6.66	1.0	307.852	6.2722	208.00	167.56	34.85
2a	177.30	59.95	1.0	143.760	4.9172	145.22	409.99	59.54
3a	98.02	6.76	1.0	87.540	5.0184	145.22	323.40	46.96
4a	152.36	6.71	1.0	152.586	5.5530	208.00	228.06	47.44
2x	177.30	59.95	1.00	143.760	4.9172	208.00	409.99	85.28
4x	98.01	6.76	1	87.533	5.0183	208.00	323.41	67.27
Reference state	300	1.013	1	311.199	6.8419	1.00	0.00	0.00

Table 7.26 Heat exchanger data, nitrogen Claude process.

		HEX3	HEX2	HEX2 + HEX3
Heat exchanger duty	$\dot{Q}^{HEX}$	32.3 KW	13.53 kW	45.83 KW
Temperature difference at the warm end	ΔT^{WE}	2.01 K	24.94 K	—
Temperature difference at the cold end	ΔT^{KE}	24.94 K	1.99	—
Max temperature difference	ΔT^{MAX}	24.94 K	24.94 K	24.94 K
Min temperature difference	ΔT^{MIN}	2.01 K	1.99 K	1.99 K
Mean temperature difference MTD	MTD	6.95 K	11.03 K	7.80 K
$\dot{Q}_{HEX}/MTD$	$\dot{Q}^{HEX}/MTD$	4.65 kW/K	1.23 kW/K	5.88 kW/K

The heat flow transferred in the heat exchanger HEX2 is relatively small, since only the Joule–Thomson stream is cooled here. In the HEX3 both material streams, however – the Joule–Thomson stream and the expander stream – are cooled.

A very special shape of the line on the "warm" side of this heat exchanger with different curvatures above and below the critical temperature can be observed. This sort of shape is typical for supercritical material streams. In given case, this shape looks well (temperature difference between warm and cold streams is positive). But generally speaking, from the process design perspective, this shape is a bit risky, since it can lead to unexpected pinches in the heat exchanger if the flowrates of material streams are not well adjusted.

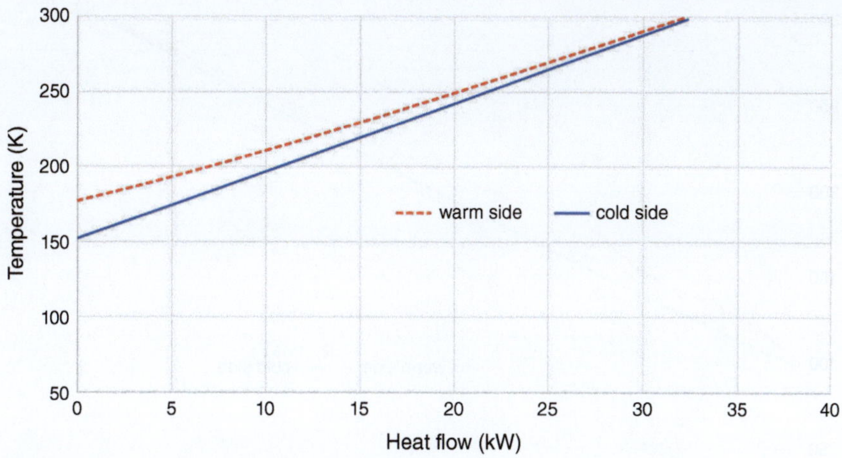

Figure 7.41 *T,Q*-diagram, heat exchanger HEX3.

Figure 7.42 *T,Q*-diagram, heat exchanger HEX2.

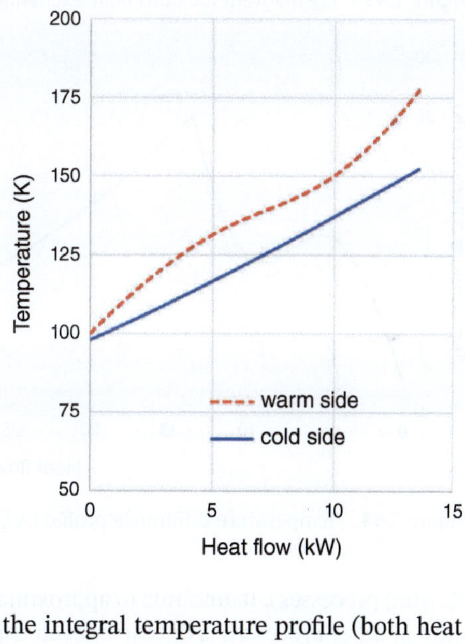

HEX2 + HEX3: It is useful to consider the integral temperature profile (both heat exchangers HEX2 and HEX3 together), as shown in Figure 7.43, for it characterizes the overall process.

The minimum temperature difference $\Delta T^{min} = 2\,\mathrm{K}$ is reached on the warm end of the heat exchanger HEX3 as well as at the cold end of the heat exchanger HEX2. The maximum temperature difference of approximately 25 K is reached at the temperature level $T = 150$–$175\,\mathrm{K}$.

What is special about this process is that the heat flow being transferred from the warm to the cold side in the heat exchanger is considerably lower in comparison to other previously considered cryogenic processes (such as Joule–Thomson and

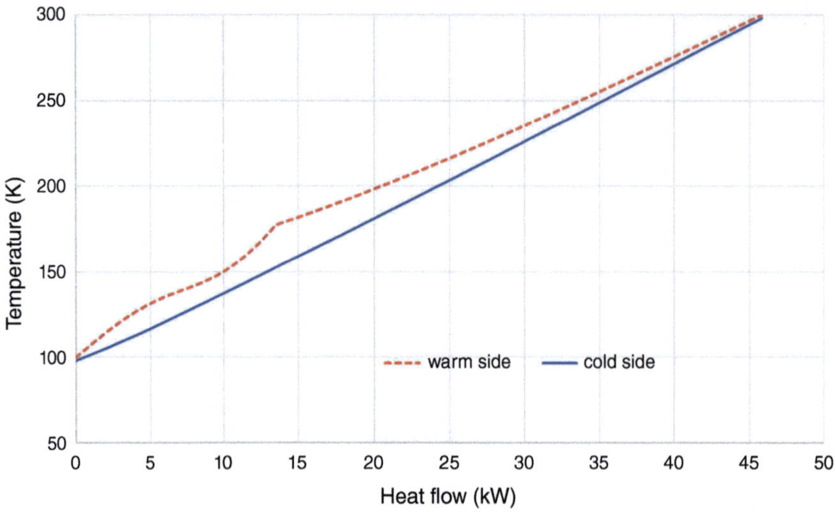

Figure 7.43 *T,Q*-diagram for both heat exchangers HEX2 and HEX3 together.

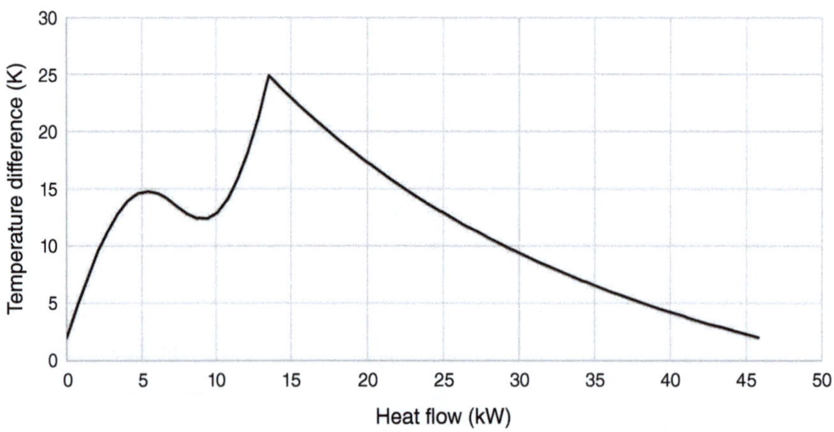

Figure 7.44 Temperature difference profile ($\Delta T,Q$-diagram) for both heat exchangers HEX2 and HEX3 together.

Brayton processes). It amounts to approximately 46 kW, whereas the heat exchanger duty in other processes is double or even triple this value.

Sizing: the volume of the heat exchanger can be estimated roughly using Eq. (5.17). According to this equation, the volume amounts to less than 0.2 m^3.

$$V = \frac{\dot{Q}^{HEX}/MTD}{C} = \frac{5.9\,\text{kW/K}}{30\ldots40\frac{\text{kW}}{\text{K}\cdot\text{m}^3}} = 0.15 - 0.2\,\text{m}^3$$

7.4.4.5.2 Evaporator

Figure 7.45 shows the temperature profile in the evaporator. There, the nitrogen boils at a constant temperature of 98 K, which keeps the cooling object at a constant temperature $T_o = 100$ K.

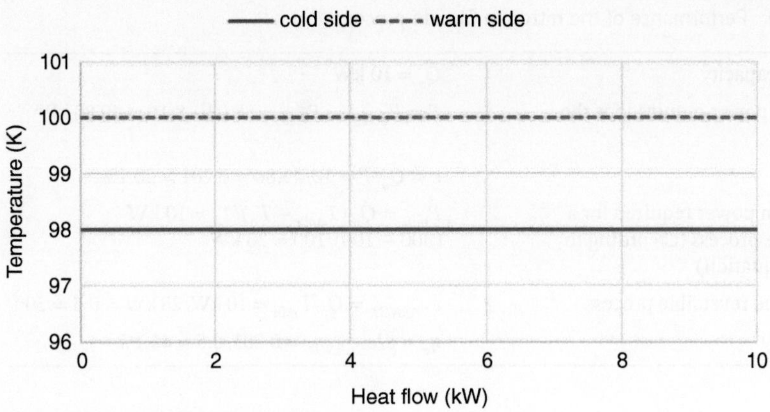

Figure 7.45 *T,Q*-diagram for evaporator.

7.4.4.5.3 Compressor (specification is given in Table 7.27) and Expander (specification is given in Table 7.28)

Table 7.27 Compressor specification, nitrogen Claude process, $T_o = 100$ K.

Compressor power	58.02 kW
Operating pressures (inlet/outlet)	6.71 bar/60 bar
Pressure ratio	$60/6.71 = 8.94$
Compressor concept	4×1.73
Number of compression stages × pressure ratio per stage	
Volumetric flowrate at the inlet	99.37 m^3/h

Table 7.28 Expander specification.

Generated power	8.16 kW
Operating pressures (inlet/outlet)	59.95 bar/6.76 bar
Total pressure ratio	8.87
Volumetric flowrate in the outlet nozzle	18.85 m^3/h

7.4.4.5.4 Expansion Valve

The process of isenthalpic expansion is designed in an excellent way, since the vapor fraction at the outlet amounts to only 3.7% mol, see Table 7.29.

Table 7.29 Joule–Thomson expansion.

Operating pressures (inlet/outlet)	59.9 bar/6.76 bar
Vapor fraction at the outlet	3.7 mol%

Table 7.30 Performance of the nitrogen Claude process.

Cooling capacity	$\dot{Q}_o = 10$ kW
Total net power required for the process	$P = P_{COMP} - P_{EXP} = 58.02 - 8.16 = 49.86$ kW
COP	$\varepsilon = \dot{Q}_0/P = 10/49.86 = 0.201 = 20.1\%$
Minimum power required for a reversible process (according to Carnot equation)	$P_{min} = \dot{Q}_o\,(T_{amb} - T_o)/T_o = 10$ kW $(300 - 100)/100 = 20$ kW
COP of the reversible process	$\varepsilon_{CARNOT} = \dot{Q}_o/P_{min} = 10\,\text{kW}/20\,\text{kW} = 0.5 = 50\%$
Efficiency	$\eta_e = \varepsilon/\varepsilon_{CARNOT} = 0.201/0.5 = \mathbf{40.1\%}$

7.4.4.6 Overall Performance

Table 7.30 summarizes the performance of the nitrogen Claude process. It is considered that the power generated in the expander is utilized, therefore the total net power required for the process is calculated as compressor power minus expander power. The efficiency, 40.1% is an excellent result.

7.4.4.7 Process Analysis

This section consists of data for exergy loss in all process units as well as an analysis of these losses based on values given in Table 7.25. The losses and loss distribution are shown in Table 7.31 and Figure 7.46.

7.4.4.7.1 Loss in Compressor

The largest loss – 17.4 kW – is associated with the compressor. This value is relatively low in comparison to other processes due to:

(a) relatively efficient compressor with $\eta_T = 70\%$,
(b) relatively low nitrogen flowrate being compressed in this process: 208 g/s in comparison to 628 g/s in nitrogen Joule–Thomson process.

7.4.4.7.2 Loss in Heat Exchanger

The second largest loss – 6 kW – is associated with the heat exchanger. That is a very low value. There are two reasons for this great result: (i) favorable temperature profile and (ii) reduced heat exchanger duty (heat transferred in the heat exchanger).

Table 7.31 Exergy loss, nitrogen Claude process.

Compressor loss	17.41 kW
Loss HEX3	2.78 kW
Loss HEX2	3.22 kW
Loss expander	4.41 kW
Loss expansion valve	1.44 kW
Loss evaporator	0.61 kW
Loss total	29.86 kW

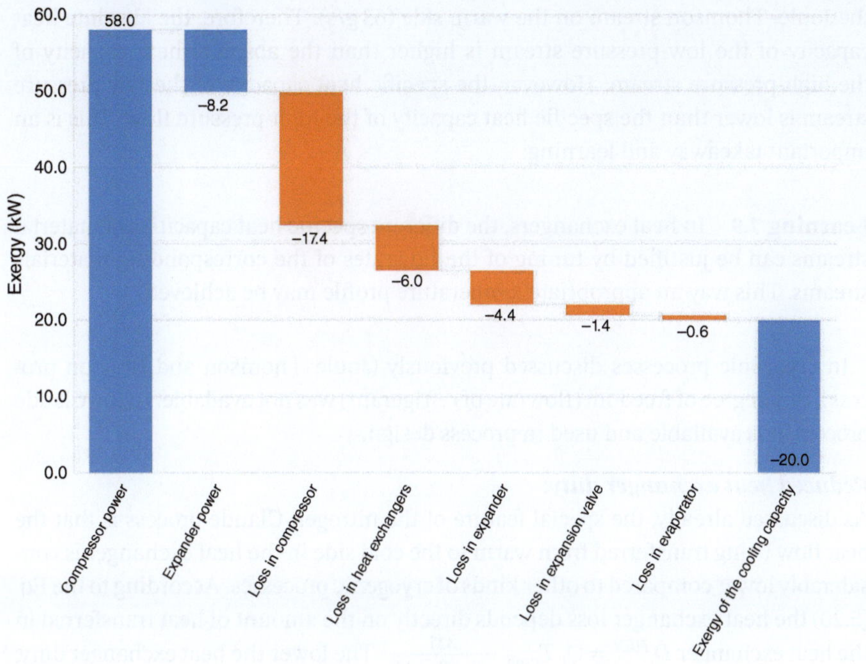

Figure 7.46 Exergy loss in the nitrogen Claud process.

Table 7.32 Impact of temperature difference in heat exchanger on exergy loss.

	Small temperature difference, less than 0.02 T	Medium temperature difference, (0.02–0.1) T	Large temperature difference, more than 0.1 T	Very large temperature difference
T = 100 K	<2 K Small loss	2–10 K moderate loss	>10 K large loss	>20 K very large loss
T = 200 K	<4 K Small loss	4–20 K moderate loss	>20 K large loss	>50 K very large loss

(Hand-drawn annotations on the table: HEX2, HEX3)

***Favorable temperature profile*:**
The low loss in the heat exchanger may be explained by means of Table 7.32 based on Table 5.4. According to this table, the loss is either minimal or small. Only in the middle (150–175 K), where the temperature difference reaches 25 K, is the loss increased.

The key factor ensuring optimal performance of the heat exchanger is the favorable temperature profile in the heat exchanger HEX2: the temperature difference at the cold end is reduced to the minimum. This is because the flowrate of nitrogen on the cold side (208 g/s) is considerably higher than the flowrate of the Joule–Thomson stream on the warm side (63 g/s). Therefore, the absolute heat capacity of the low-pressure stream is higher than the absolute heat capacity of the high-pressure stream. However, the specific heat capacity of the low-pressure stream is lower than the specific heat capacity of the high-pressure flow. This is an important takeaway and learning:

Learning 7.9 In heat exchangers, the different specific heat capacities of material streams can be justified by tuning of the flowrates of the corresponding materials streams. This way an appropriate temperature profile may be achieved.

In cryogenic processes discussed previously (Joule–Thomson and Brayton process), this degree of freedom (flowrate of refrigerant) was not available. In the Claude process, it is available and used in process design.

Reduced heat exchanger duty:
As discussed already, the special feature of the nitrogen Claude process is that the heat flow being transferred from warm to the cold side in the heat exchanger is considerably lower compared to other kinds of cryogenic processes. According to the Eq. (5.20) the heat exchanger loss depends directly on the amount of heat transferred in the heat exchanger $\dot{D}_i^{HEX} = \dot{Q}_i\, T_{amb}\, \frac{\Delta Ti}{T_{c_i}\,(T_{c_i}+\Delta T)}$. The lower the heat exchanger duty, the lower the loss. That is a fundamental learning:

Learning 7.10 The loss in heat exchanger can be considerably decreased by reducing the heat exchanger duty. That is a thermodynamic version of the popular wisdom: "the fewer things you own the less you can lose." Instinctively, the process engineers work hard optimizing the temperature profile in heat exchanger(s). However, the greatest improvement can be achieved by reducing the working fluid flowrates.

7.4.4.7.3 Loss in Expander

The expander loss (4.4 kW in given example) is relatively small due to the high isentropic efficiency of the applied machine (85%). Consequently, the entropy increase is small, and the corresponding line in the *T,s*-diagram is quite vertical.

The expander loss is three times larger than the loss in the Joule–Thomson valve, which is a bit unexpected. The background is that both basic ingredients – the specific entropy rise (0.1 J/(g K)) and the expander flowrate – are higher than the flowrate and specific entropy increase of the Joule-Thomson stream by a factor of 2.3.

7.4.4.7.4 Loss in Expansion Valve

The loss in the Joule–Thomson valve (approximately 1.4 kW) is extremely small since the working fluid is precooled to a very cold temperature before expansion. The entropy increase (with less than 0.08 J/(g K)) is practically unrecognizable in the *T,s*-diagram, Figure 7.39.

The low irreversibility of this expansion process is shown by the fact that the liquid fraction after the expansion is extremely high; the flash-gas amounts to only 3.7%, and the 96.7% of the compressed Joule–Thomson stream is liquefied during the expansion.

7.4.4.7.5 Loss in Evaporator

The small loss in the evaporator (less than 1 kW) indicates that the valuable 98 K cold is efficiently used here for maintaining the cooling object at 100 K.

7.4.4.7.6 Analysis, Summary

The high efficiency of the nitrogen Claude process can be explained as follows:

1. Due to a smart integration of the Brayton process, the losses in the Joule–Thomson stage of the process are considerably reduced:
   - in the Joule–Thomson valve: by moving the isenthalpic expansion to extremely low temperatures;
   - in the heat exchanger HEX2: by the favorable temperature profile and reduced amount of heat being exchanged there. It leads to the reduced flowrate of the working fluid in the Joule–Thomson stage;
2. The losses in the Brayton stage of the process itself are relatively low since this process is based on a highly efficient expander. In the Claude combination the losses are further reduced due to the favorable temperature profile and reduced amount of heat being exchanged in both heat exchangers HEX3 and HEX2;

To summarize, in the Claude process the disadvantages of the single Joule–Thomson and the single Brayton process are eliminated, whereas the advantages of the single Joule–Thomson and the single Brayton process are united. Therefore, the combination delivers a promising, highly efficient process.

7.4.5 Helium Claude Process

Figure 7.47a shows the typical helium Claude process. The general structure corresponds to the structure of the basic Claude process; see Figure 7.36: it consists of two overlapped subprocesses – the Joule–Thomson process (marked as 1 → 2 → 3 → 4 → 5) and the Brayton process (marked as 1 → 2a → 3a → 4a →5) driven by the same compressor unit.

The Joule–Thomson part of the process remains the same. However, the Brayton part looks a bit different. The most salient details are:

1. **Two expanders and the connection of these:**
   - The process includes two expanders;
   - Basically, these expanders are connected serially – the expander 2 after the expander 1;
   - This is not a simple connection, for it includes three process steps:
     (a) the expander stream is expanded in the expander 1 from the high-pressure level to an intermediate-pressure level;

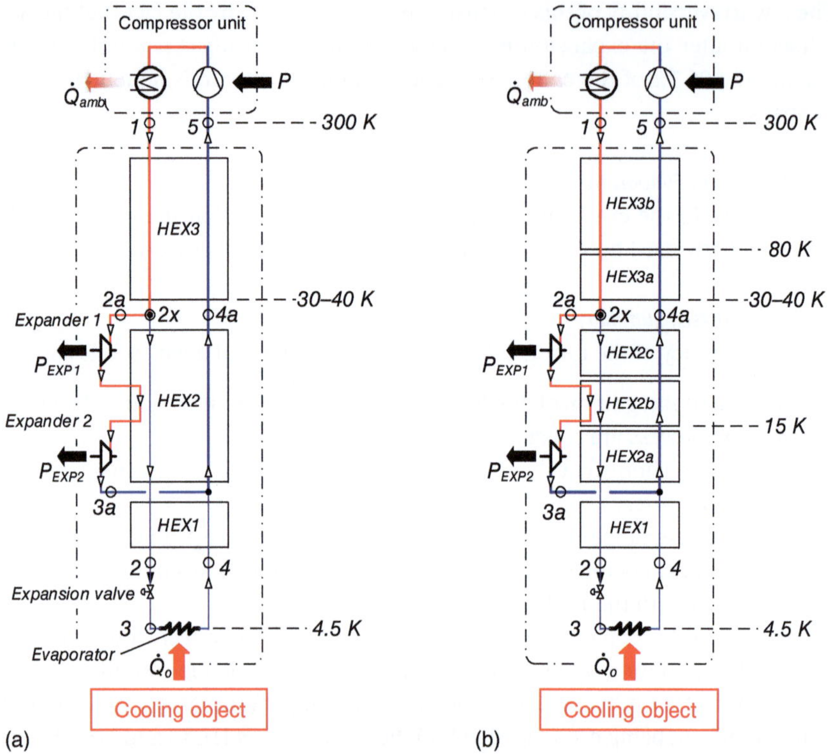

Figure 7.47 Helium Claude process with two turboexpanders (a) simplified view, (b) including details concerning heat exchangers.

(b) it is provided through the heat exchanger HEX2, where it is further cooled;
(c) before it is expanded in the expander 2 from the intermediate-pressure level to the low-pressure level.

2. **Mechanical design of the heat exchanger process unit HEX2:** The process sequence with two expanders and intermediate cooling has consequences for the mechanical design of the heat exchanger HEX2. Usually, this process unit is divided into three separate heat exchangers – HEX2a, HEX2b, and HEX2c, as shown in Figure 7.47b. The background is that the flowrate of the expander stream is considerably higher than the flowrate of the Joule–Thomson stream, and the mechanical design of the heat exchanger HEX2b (with expander stream) differs much from the mechanical design of the HEX2a and HEX2c.
3. **Mechanical design of the heat exchanger process unit HEX3:** Similarly, the mechanical design of the heat exchanger HEX3 includes two separate heat exchangers – HEX3a and HEX3b, as shown in Figure 7.47b. It relates to operational items. The performance of helium refrigerators/liquefiers can be improved by integration of liquid nitrogen if it is available on-site. In this case, the liquid nitrogen will be integrated into the heat exchanger HEX3b. Therefore, HEX3b is often designed for three material streams in advance; this feature is not shown in Figure 7.47 but in Figure 9.9. The HEX3a remains a conventional two-stream heat exchanger.

7.4.5.1 Process Description

The sequence of the helium Claude process is as follows:

(a) The helium is first compressed from the low-pressure level p_{LP} to the high-pressure level p_{HP} by means of a compressor and cooled to the approximately ambient temperature. This state is marked in Figure 7.47b as state 1. The typical pressure levels are given in Section 7.4.2.3;

(b) This compressed helium is cooled;
- in the countercurrent heat exchanger HEX3b to a temperature of approximately 80 K (liquid nitrogen temperature level), and further
- in the countercurrent heat exchanger HEX3a to a temperature of 30–40 K by the cold low-pressure stream. The state of refrigerant at the cold end of this heat exchanger is marked in Figure 7.47b as the state 2x;

(c) At this point, the refrigerant flow is divided into two sub-streams, the Joule–Thomson stream and the expander stream. The flowrate of the expander stream is higher than the flowrate of the Joule–Thomson stream;

(d) the expander stream is:
- expanded in the expander 1 from the high-pressure level to an intermediate-pressure level;
- provided through the heat exchanger HEX2b, where it is further cooled to approximately 15 K;
- before it is expanded in the expander 2 from the intermediate-pressure level to the low-pressure level. The temperature of this stream reduces correspondingly. The state on the outlet of expander 2 is marked in Figure 7.47b as the state 3a; this stream is gaseous;
- this cold gas is provided through the heat exchangers HEX2a, HEX2b, and HEX2c. The cold of the cold expander stream is applied there for cooling of the Joule–Thomson stream. Simultaneously, the expander stream is warmed up (process step 3a → 4a).
- provided further through the heat exchangers HEX3a and HEX3b to the compressor inlet, where it is warmed up to the temperature T_5 which is close to the ambient temperature.

(e) The Joule–Thomson stream is:
- provided through the heat exchangers HEX2c, HEX2b, and HEX2a, where it is cooled by the cold expander stream in a countercurrent direction, and enters the Joule–Thomson stage, where it is:
- cooled in the heat exchanger HEX1 to the temperature T_2 (state 2, the state at this point corresponds to the supercritical liquid) and
- expanded in the expansion valve to the low-pressure level p_{LP}. This process (2 → 3) is a typical Joule–Thomson expansion, whereby the Joule–Thomson stream partially liquefies and a part of refrigerant evaporates, generating flash-gas;
- this two-phase helium flows through the evaporator and absorbs the heat flow from the object being cooled. At the outlet of evaporator (state 4), the liquid is evaporated; the state corresponds to the dew point;

- this cold vapor is provided through the heat exchanger HEX1, where it warmed to a temperature close to the temperature T_{3a}, cooling the Joule–Thomson stream on the warm side of this heat exchanger;
- at the outlet of the heat exchanger HEX1, the Joule–Thomson stream is mixed with the expander stream, and it flows through the heat exchangers HEX2abc and HEX3ab to the compressor inlet.

7.4.5.2 Special Feature

In the heat exchangers HEX3ab, as well as in the heat exchangers HEX1 and HEX2b, the refrigerant flowrate on the "warm" high-pressure side is equal to the flowrate on the "cold" low-pressure side.

In the heat exchangers HEX2a and HEX2c, the flowrate of Joule–Thomson stream on the "warm" high-pressure side is lower than the flowrate of the combined Joule–Thomson stream and expander stream on the "cold" low-pressure side.

7.4.5.3 Staging

The process design engineers used to structure the helium Claude process in

- The precooling stage, where the helium is cooled to liquid nitrogen temperature level,
- The core Claude process

as shown in Figure 7.48.

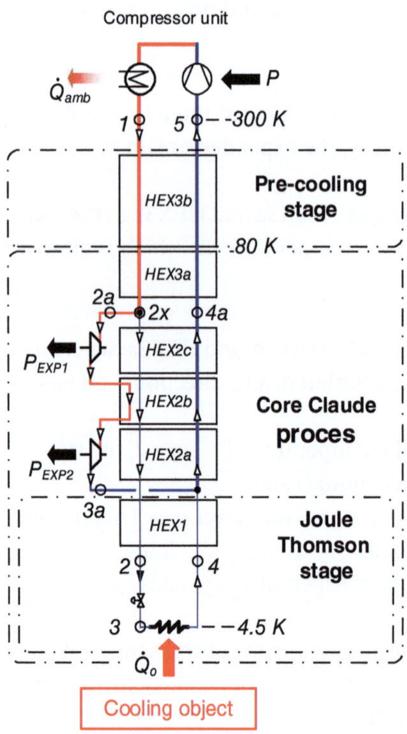

Figure 7.48 Helium Claude process, typical staging.

The core process is usually divided into

- The Joule–Thomson stage, which includes HEX 1, expansion valve and evaporator, and
- The Brayton stage

In terms of hierarchy, the Joule–Thomson stage remains the "superior" part of the process. The roles of the precooling stage and the Brayton stage are essential to ensure that the Joule–Thomson process performs well. They are "subordinate" processes.

7.4.6 Example, Process Calculation, Helium Claude Process

This section considers an example of process calculation for a helium Claude process.

7.4.6.1 Task Definition

A helium Claude process for a cooling capacity $\dot{Q}_0 = 100\,\text{W}$ for constant cooling temperature $T_0 = 4.6\,\text{K}$ must be designed.

The ambient conditions are given in Table 7.5. The target is to provide:

1. Process parameters (flowrate; temperature; pressure; phase state; composition; enthalpy; entropy; exergy) at all relevant process points;
2. Total power consumption, COP, and efficiency, as well as exergy loss in every process component;
3. Figures required for designing heat exchangers and compressors.

7.4.6.2 Process and Hardware Concept

Figure 7.49 shows the process flow diagram. The process concept is based on hardware components used for small-size laboratory-type helium liquefiers employed in academia research. This is based on:

(a) The oil-lubricated screw compressor (see the Section 6.5.3 for more details);
(b) Efficient aluminum-brazed plate-fin heat exchangers for all heat exchangers including evaporator;
(c) Efficient turboexpanders with gas-brakes, where the generated power is dissipated into the environment;
(d) Efficient multilayer thermal insulation;

Further process design considerations are as follows:

1. Compressor:
   - since an oil-lubricated screw compressor is applied, the compression operation can be considered as a quasi-isothermal compression;
   - it is assumed that the high-pressure level corresponds to the $p_1 = 12\,\text{bar}$;
   - compressor efficiency is assumed to be 60%,
   - an oil-removal system is required,

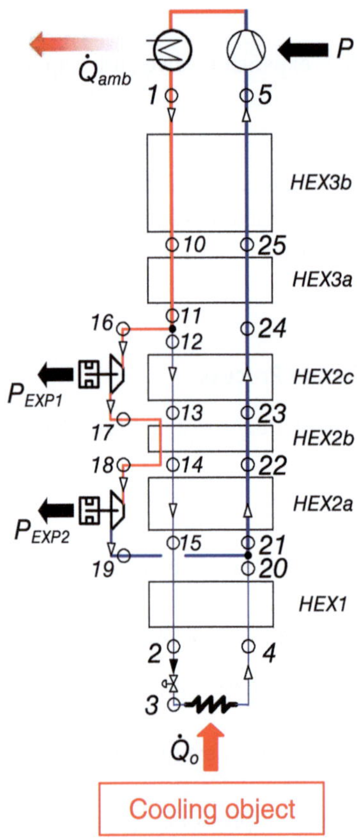

Figure 7.49 Helium Claude process, PFD for process calculation.

- temperature T_1 is close to the ambient temperature ($T_1 \approx T_{amb} = 300$ K) due to the efficient cooling after compression;

2. Expander:
   - since the expanders used here are small-size machines, a low isentropic efficiency of 62.5% is assumed for both expanders,
   - the intermediate-pressure level is assumed to be 4 bar;
3. Heat exchangers:
   - for the first process design, minor pressure loss on both sides – the high-pressure side, as well as the low-pressure side – are assumed; the values are given in Table 7.33,
   - minimum temperature difference of $\Delta T^{MIN} = 0.5$ K is targeted;
4. Evaporator:
   - the pressure loss in the evaporator is negligibly small $\Delta p_{3\rightarrow4} = 0$ bar;
   - a small temperature difference between the nitrogen in the evaporator and the cooling object can be expected there; for example $(T_4 - T_o) = 0.1$ K as a first estimation;
5. Insulation: The multilayer insulation is a very efficient kind of thermal insulation. But since the cooling temperature is very low, some heat will leak into the system and reduce the effective cooling capacity of the refrigerator.

Table 7.33 Input data for process calculation, helium Claude process, $T_o = 4.6$ K.

Cooling capacity	$\dot{Q}_o = 100$ W
Heat leak through thermal insulation	$\dot{Q}_{is} = 100$ W
Cooling temperature	$T_o = 4.6$ K
Temperature at the evaporator outlet	$T_4 = 4.5$ K
High-pressure level	$p_1 = 12$ bar
Intermediate-pressure level (for expander calculation)	$P_{17} = 4$ bar
Low-pressure level	$p_3 = p_4 = 1.25$ bar
Expanders, Isentropic efficiency	$\eta_{S\_EXP\_1} = \eta_{S\_EXP\_2} = 62.5\%$
Pressure drop in passages of heat exchangers HEX3a and HEX3b	0.05 bar
Pressure drop in passages of heat exchangers HEX1 and HEX2	0.025 bar
Compressor, Isothermic efficiency	$\eta_T = 60\%$

For simplification purposes, here it is assumed that a) the maximum heat leak to the cooling object /and evaporator amounts to $\dot{Q}_{is} = 100$ W, b) the whole heat leak is provided to the evaporator directly. Therefore, the cooling capacity generated in the evaporator must be $\dot{Q}_{evap} = \dot{Q}_o + \dot{Q}_{is} = 100\,\text{W} + 100\,\text{W} = 200\,\text{W}$.
That is a **very critical assumption** (see Section 7.4.6.8), since it has a huge impact on flowrates, dimensions, and costs of the entire system. This assumption must be reviewed when the detailed mechanical design of the cold part is finished and the real values are available.

7.4.6.2.1 Comment About Thermal Insulation

In previous process design examples, performed for liquid nitrogen temperature level, it was always assumed that the heat leak through thermal insulation is minor and can be neglected. For liquid helium temperature level, this approach does not work well.

The first reason is that

(a) the cold part of a refrigeration process for liquid helium temperature consists of much more hardware components than a nitrogen process,
(b) these components for helium processes (mainly heat exchangers) are relatively large devices, even for small-scale refrigerator. It leads to a voluminous coldbox with relatively large outer surface and, at the end, an increased heat leak even through a very efficient multilayer thermal insulation.

The second reason is that the cold generation at liquid helium temperature level requires substantial amount of power, by one order of magnitude more in comparison to systems for liquid nitrogen temperature level. Therefore, even a small inaccuracy in specification of required cooling capacity will lead to a massive underestimation of flowrate, required power, size of hardware components, and, at the end, cost of entire system. Consequently, for helium refrigeration plants, the heat leak through thermal insulation must be considered even in very early conceptual phase.

All these considerations are summarized in Table 7.33.

The process is complex. For simplification purposes, some values, which are usually to be found during the process optimization procedure, will be simply predefined given here:

- Total mass flowrate of gas $\dot{m}_1 = 50\,\text{g/s}$
- In the upper heat exchanger (HEX3b) the high-pressure flow is cooled to $T_{10} = 82$ K,
- In the second heat exchanger (HEX3a) to $T_{11} = 38$ K,
- In the heat exchanger HEX2b the helium is cooled further to the temperature $T_{14} = T_{18} = 13$ K,
- The temperature on the inlet expansion valve is $T_2 = 5.4$ K

With these data, which are summarized in Table 7.34, the process is well-defined. Only the flowrate of the Joule–Thomson stream, as well as the flowrate of the expander stream, is unknown. Both these parameters must be found during the process calculation procedure.

Table 7.34 Further Input data for process calculation, helium Claude process, $T_o = 4.6$ K.

Temperature T_{10}	$T_{10} = 82$ K
Temperature T_{11}	$T_{11} = 38$ K
Temperature $T_{14} = T_{18}$	$T_{14} = 13$ K
Temperature T_2	$T_2 = 5.4$ K
Total mass flowrate $\dot{m}_1$	$\dot{m}_1 = 50$ g/s

7.4.6.3 Properties Data

The fluid property software "miniREFPROP" is used for calculating thermodynamic properties.

7.4.6.4 Process Simulation

The results are shown in Table 7.35.

The specific exergy as well as absolute exergy are calculated by using Eq. (4.5); the reference state is referred to as $p = 1.013$ bar and $T = 300$ K. Figure 7.50 shows the process in *T,s*-diagram.

7.4.6.5 Process Units

7.4.6.5.1 Heat Exchanger

Table 7.36 summarizes the data related to heat exchangers. Figures 7.51 and 7.52 show the corresponding temperature profiles and temperature difference profiles.

HEX3a and HEX3b: In the upper heat exchangers HEX3, the high-pressure helium is cooled to the temperature of 82 K and further to 38 K (see Figure 7.51). Since helium behaves as an ideal gas at these conditions, the heat capacities on the

Table 7.35 Output process data, helium Claude process, $T_o = 4.6$ K.

	Temperature (K)	Pressure (bar)	Vapor fraction (mol/mol)	Specific enthalpy (J/g)	Specific entropy (J/(g K))	Flowrate (g/s)	Specific exergy (J/g)	Exergy (kW)
1	300.00	12.0	1	1566.76	22.8447	50	1569.1	78.45
2	5.40	11.8	1	8.71	0.1461	17.35	6820.6	118.34
3	4.50	1.3	0.381	8.71	1.8927	17.35	6296.6	109.25
4	4.50	1.3	1	20.24	4.4547	17.35	5539.5	96.11
5	294.08	1.1	1	1532.46	27.7034	50	77.1	3.86
10	82.00	11.95	1	433.23	16.1054	50	2457.3	122.87
11	38.00	11.9	1	201.59	12.0596	50	3439.4	171.97
12	38.00	11.9	1	201.59	12.0596	17.35	3439.4	59.67
13	29.53	11.875	1	155.91	10.7033	17.35	3800.6	65.94
14	13.00	11.85	1	60.12	5.8930	17.35	5147.9	89.32
15	10.40	11.825	1	42.01	4.3396	17.35	5595.8	97.09
16	38.00	11.9	1	201.59	12.0596	32.65	3439.4	112.30
17	29.53	4.	1	157.44	13.0427	32.65	3100.3	101.23
18	13.00	3.975	1	68.04	8.5959	32.65	4345.0	141.86
19	9.75	1.275	1	53.54	9.5683	32.65	4038.7	131.86
20	9.75	1.275	1	53.54	9.5683	17.35	4038.7	70.07
21	9.75	1.275	1	53.54	9.5683	50	4038.7	201.94
22	10.90	1.25	1	59.82	10.2173	50	3850.3	192.52
23	28.26	1.225	1	151.44	15.3010	50	2416.8	120.84
24	31.29	1.2	1	167.29	15.8768	50	2260.0	113.00
25	75.81	1.15	1	398.93	20.5706	50	1083.5	54.17
Ref. state	300.00	1.0	1	1563.35	28.0635	1	0.0	0.00

high-pressure and low-pressure sides do not differ so much. The temperature profile looks very similar to the profile of the classic Brayton process, and the temperature difference is constant: at the warm end and in the middle, it remains close to 6 K, while it increases slightly at the cold end to 6.7 K.

Figure 7.52 shows the combined diagram for heat exchangers HEX2 (HEX2c, HEX2b, and HEX2a) and HEX1. In this part of the process, the Joule–Thomson stream is cooled from $T_{12} = 38$ K further to ≈30 K, respectively 13 K, ≈10 K and ≈5.4 K. The whole heat flow transferred in this part of the process is 6.3 kW. It is approximately 50% of the duty of the heat exchanger HEX 3b and approximately 10% of the heat exchanger HEX3a. If this graph were to be shown in the same scale as the graph in Figure 7.51, it would be a tiny, narrow picture. Paradoxically, this is the most important part of the process.

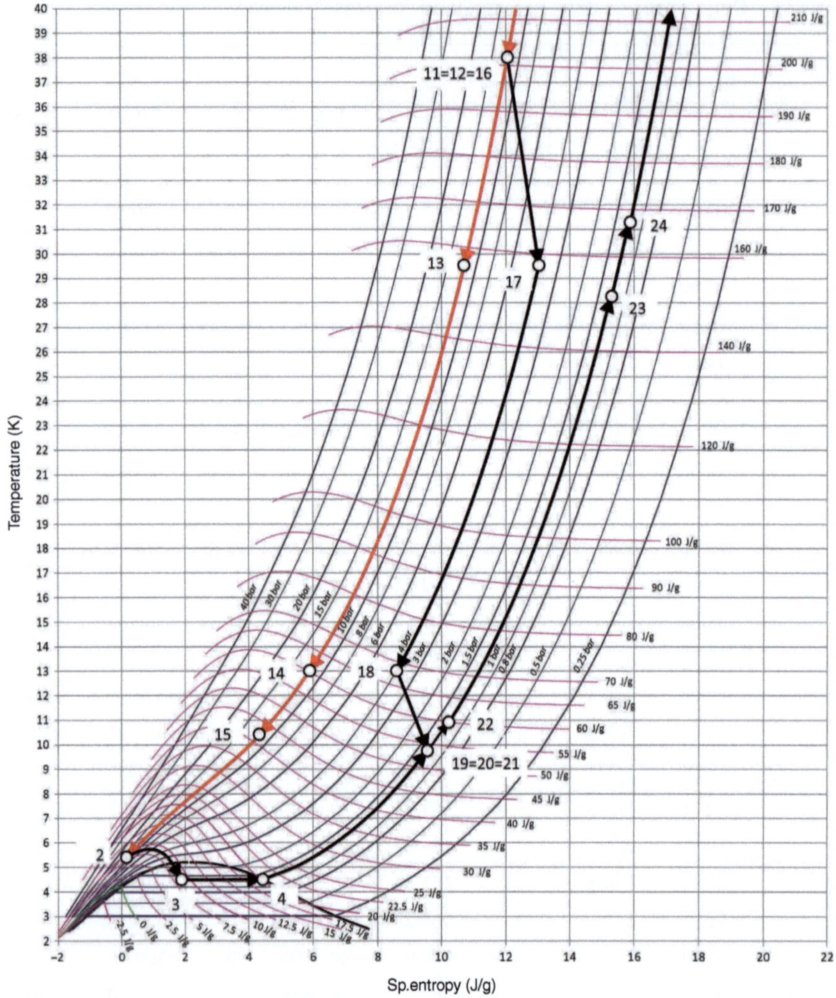

Figure 7.50 Helium Claude process in *T,s*-diagram.

HEX2c: In the heat exchanger HEX2c, the relatively small Joule–Thomson stream (17.35 g/s) is cooled by the cold stream with the maximum flowrate (50 g/s). Therefore, the temperature of this stream drops steeply from 38 to 30 K.

HEX2b: In the heat exchanger HEX2b, the flowrate of both "warm" streams is equal to the flowrate of the cold stream. In this sense, this heat exchanger is very similar to the HEX3ab. But the nonideal gas behavior of helium becomes slightly visible at these very low temperatures: the temperature difference at the cold end (2.1 K) is a bit higher than the temperature difference at the warm end (1.3 K).

HEX2a: The conditions in the heat exchanger HEX2a are similar to the conditions in the heat exchanger HEX2c: the relatively small Joule–Thomson stream is cooled there by the cold stream with the max flowrate. Therefore, the temperature of this

Table 7.36 Heat exchanger data, helium Claude process.

		HEX3b	HEX3a	HEX2c	HEX2b	HEX2a	HEX1
Duty $\dot{Q}^{HEX}$	kW	56.68	11.58	0.79	4.58	0.314	0.58
MTD	K	5.98	6.38	3.26	1.55	1.21	1.19
Q/*MTD* (*UA*)	kW/K	9.47	1.82	0.24	2.95	0.26	0.49
ΔT^{min}	K	5.92	6.19	1.27	1.27	0.65	0.65
ΔT^{max}	K	6.19	6.71	6.71	2.1	2.1	1.61

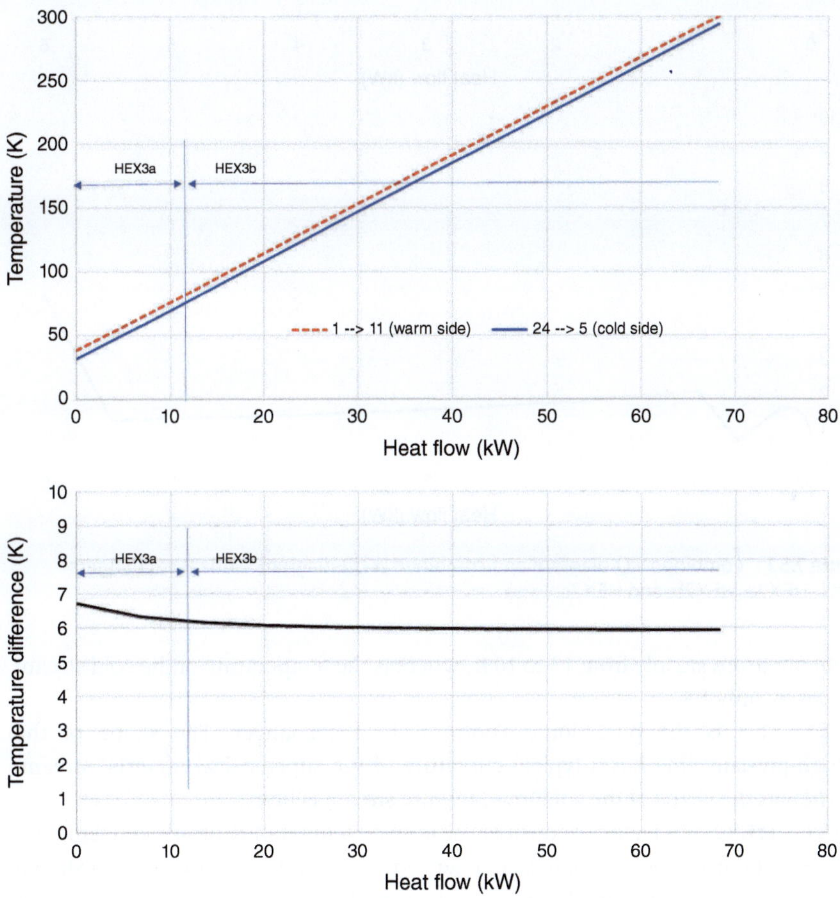

Figure 7.51 Combined *T*,*Q*-diagram and Δ*T*,*Q*-diagram for heat exchangers HEX3a and HEX3b.

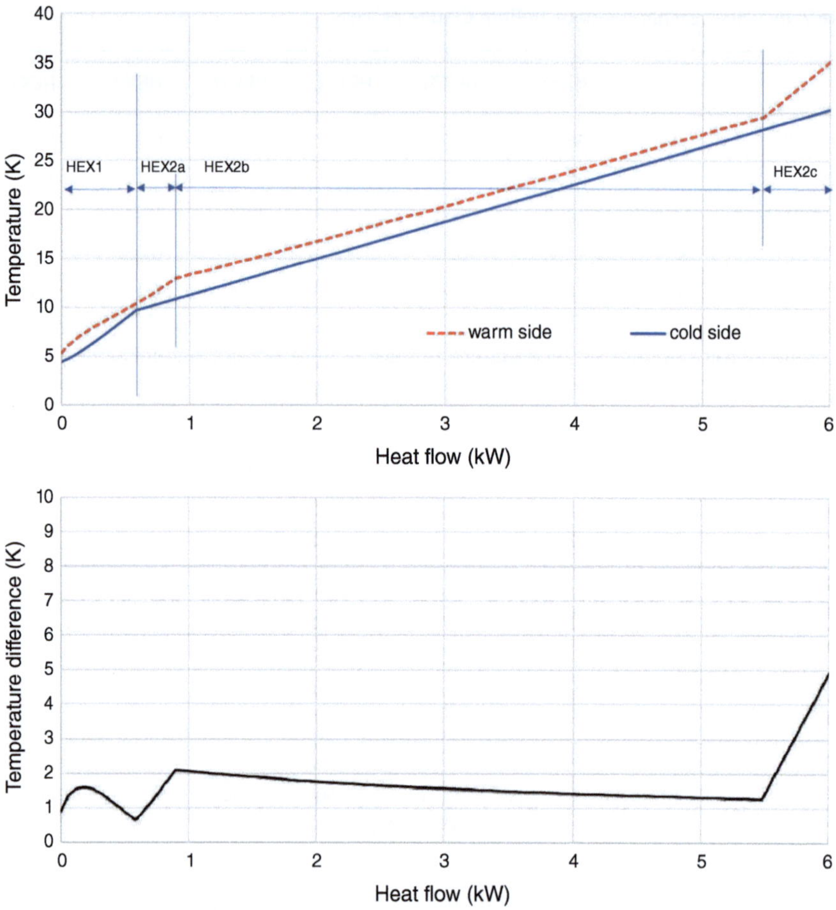

Figure 7.52 Combined *T,Q*-diagram and combined Δ*T,Q*-diagram for heat exchangers HEX1, HEX2a, HEX2b, and HEX2c.

stream drops steeply from 13 to 10 K, whereas the temperature of the cold stream rises by only 1 K.

HEX1: This is the pure Joule–Thomson heat exchanger. The shape of the high-pressure line has a typical curvature of the supercritical material stream, whereas the shape of the cold low-pressure stream is linear.

HEX1 + HEX2: It is useful to consider the integral temperature profile and temperature difference profile in heat exchangers HEX2abc and HEX1 together, as shown in Figure 7.52, since they characterize the overall process: Some observations are:

- the temperature difference remains lower than 2.5 K in the whole process below 30 K, which is a very positive message since it means low exergy loss;
- the profile is complex; it shows some potential pinches at the
  - cold end of HEX2c,
  - cold end of HEX2a – warm end of HEX1,
  - cold end of HEX1

- The minimum temperature difference $\Delta T^{min} = 0.65$ K is reached at the inlet into the Joule–Thomson stage;
- The maximum temperature difference of approximately 6.7 K is reached at the warm end of HEX3c;
- At the warm end of HEX2a, the temperature difference of approximately 2.1 K is reached.

Size:
The heat exchanger HEX3a is the largest one, mainly due to the fact that heat flow transferred there (from the warm side to the cold side) and Q/MTD value is high. The second-largest heat exchanger is the HEX2b, with Q/MTD = 2.95 kW/K due to a low temperature difference there, followed by HEX3a (Q/MTD = 1.82 kW/K). All other heat exchangers are very small.

7.4.6.5.2 Evaporator

The temperature profile in the evaporator is shown in the Figure 7.53. The helium boils there at a constant temperature of 4.5 K, keeping the cooling object at $T_0 = 4.6$ K. Q/MTD = 100 W/0.1 K = 1000 W/K = 1 kW/K.

7.4.6.5.3 Compressor (specification is given in Table 7.37)

Table 7.37 Compressor specification for helium Claude process.

Compressor power	124.3 kW
Pressures (inlet/outlet)	1.1 bar/12 bar
Pressure ratio:	10.9
Volumetric flowrate at the inlet	1000 m^3/h

7.4.6.5.4 Expander

The specification for expander is given in Table 7.38. It is interesting to note that the pressure ratio, as well as the volumetric flowrate at the outlet, is the same for both

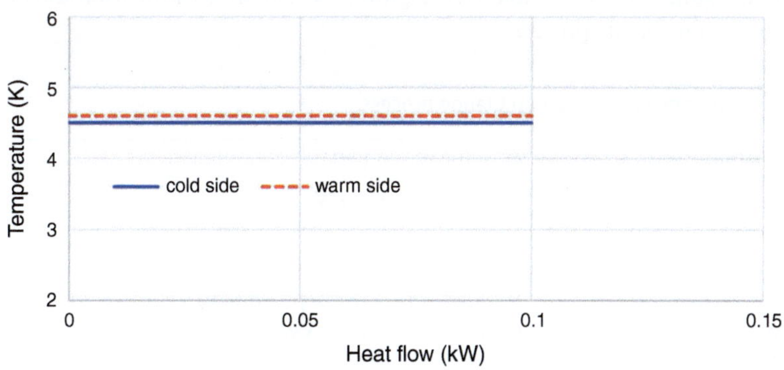

Figure 7.53 *T,Q*-diagram for evaporator.

Table 7.38 Expander specification for helium Claude process.

	EXP1	EXP2
Generated power	1.44 kW	0.47 kW
Operating pressures (inlet/outlet)	11.9/4 bar	3.975/1.275 bar
Pressure ratio:	2.98	3.12
Volumetric flowrate in the outlet nozzle	18.1 m^3/h	18 m^3/h

expanders. This means that the mechanical design of both machines (expander 1 and expander 2) is similar. Two identical expanders may be applied in this process, which helps considerably with engineering efforts.

7.4.6.5.5 *Expansion Valve (specification is given in Table 7.39)*

Table 7.39 Joule–Thomson expansion in helium Claude process.

Operating pressures (inlet/outlet)	11.8/1.3 bar
Vapor fraction at the outlet	0.38 mol/mol

7.4.6.6 Overall Performance

Table 7.40 summarizes the performance of the helium Claude process. It is considered that the power generated in the expander is not utilized. Therefore, the total net power required for the process corresponds to the compressor power. The resulting process efficiency of 5.2% corresponds to a midrange level. It indicates that the process design is targeting cost efficiency rather than power efficiency.

7.4.6.7 Process Analysis

This section consists of data for exergy loss (based on values given in Table 7.35) in all process units as well as analysis of these losses.

The losses are summarized in Table 7.41. Figure 7.54 visualizes the distribution of losses in the helium Claude process.

Table 7.40 Performance of the helium Claude process.

Cooling capacity	$\dot{Q}_o = 100\,\text{W} = 0.1\,\text{kW}$
Total net power required for the process	$P = P_{COMP} = 124.3\,\text{kW}$
COP	$\varepsilon = \dot{Q}_o/P = 0.1/124.3 = 0.0008 = 0.08\%$
Minimum power required for a reversible process (according to the Carnot equation)	$P_{min} = \dot{Q}_o\,(T_{amb} - T_o)/T_o = 0.1\,\text{kW}\,(300 - 4.6)/4.6 = 6.42\,\text{kW}$
COP of the reversible process	$\varepsilon_{CARNOT} = \dot{Q}_o/P_{min} = 0.1\,\text{kW}/6.42\,\text{kW} = 0.0155 = 1.55\%$
Overall process efficiency	$\eta_e = \varepsilon/\varepsilon_{CARNOT} = 0.08\%/1.55\% \approx \mathbf{0.052 = 5.2\%}$

Table 7.41 Helium Claude process.

Loss compressor	49.73 kW
Loss heat exchanger HEX3b	5.90 kW
Loss heat exchanger HEX3a	9.72 kW
Loss heat exchanger HEX2c	1.58 kW
Loss heat exchanger HEX2b	7.66 kW
Loss heat exchanger HEX2a	1.65 kW
Loss heat exchanger HEX1	4.79 kW
Loss expander 1	9.63 kW
Loss expander 2	9.53 kW
Loss expansion valve	9.09 kW
Loss evaporator	0.29 kW
Loss heat leak through insulation	6.42 kW
Total loss:	115.99 kW

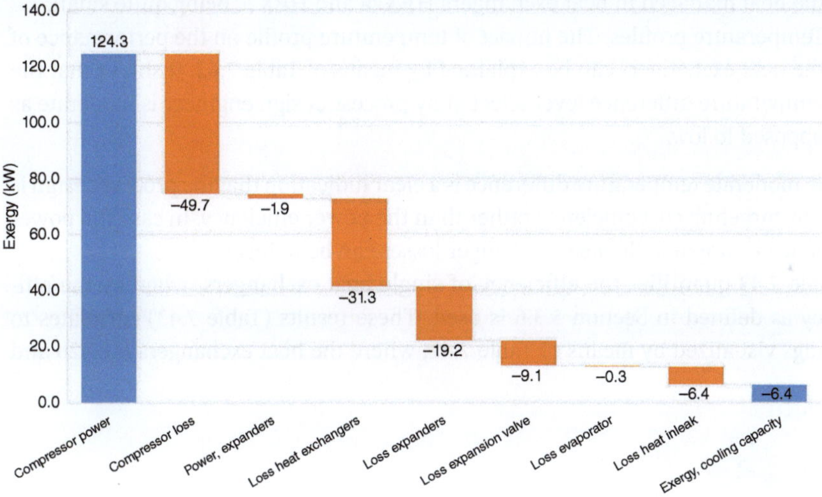

Figure 7.54 Distribution of losses in helium Claude process.

7.4.6.7.1 Loss in Compressor

The largest loss – 49.7 kW – is associated with the compressor. This value is high due to relatively low compressor efficiency $\eta_T = 60\%$. All comments given in Section 7.2.3.6.2 (compressor, Joule–Thomson process) are valid for the helium Claude process, too.

7.4.6.7.2 Loss in Heat Exchanger

The loss in heat exchangers is the second largest. The distribution between the single heat exchangers is shown in Figure 7.55, indicating that the contribution of heat exchangers HEX3a, HEX2b, and HEX1 is the largest. This is because of:

Table 7.42 Impact of temperature difference in heat exchangers on specific exergy loss.

	Small temperature difference, less than 0.02 T	Medium temperature difference, (0.02–0.1) T	Large temperature difference, more than 0.1 T	Very large temperature difference
$T < 5$ K	<0.1 K Small loss	0.1–0.5 K moderate loss	>0.5 K large loss HEX1	
$T = 20$ K	<0.4 K Small loss	0.4–2 K moderate loss HEX2b	HEX2a HEX2c >2 K large loss HEX3a	
$T = 100$ K	<2 K Small loss	2–10 K moderate loss HEX3b	>10 K large loss	
$T = 200$ K	<4 K Small loss	4–20 K moderate loss	>20 K large loss	

(a) Relatively large amounts of heat being transferred in these heat exchangers, with the heat managed in heat exchangers HEX2a and HEX2c being quite small,
(b) Temperature profiles. The impact of temperature profile on the performance of the heat exchangers can be explained by means of Table 7.42. It shows that the temperature difference level selected by process design engineer is moderate as opposed to low.

The moderate temperature difference is a clear indication that the process design is done by targeting cost efficiency rather than the power efficiency. In case the power efficiency is targeted, the heat exchanger losses can be reduced.

Table 7.43 quantifies the efficiency of single heat exchangers, whereby the efficiency as defined in Section 5.3.6 is used. These results (Table 7.43) correlates to findings visualized by means of Table 7.42, where the heat exchangers HEX2b and

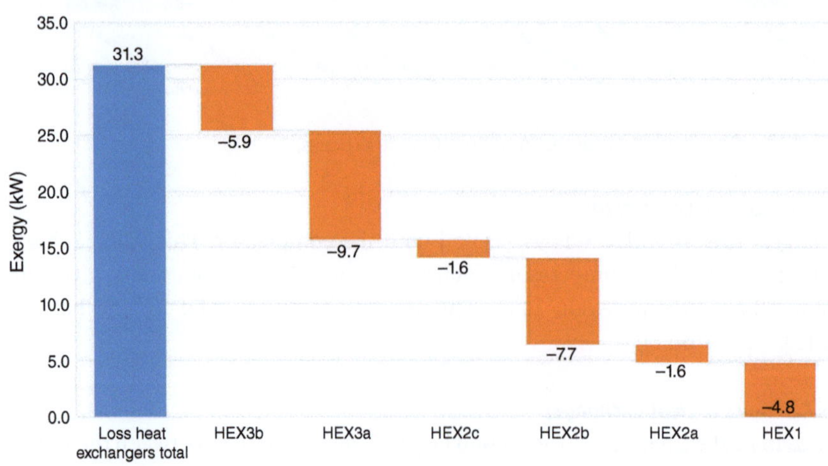

Figure 7.55 Distribution of heat exchanger losses in helium Claude process.

Table 7.43 Exergy efficiency of single heat exchangers.

	Loss in heat exchanger (kW)	Exergy difference on the cold side (kW)	Exergy difference on the warm side (kW)	Efficiency (%)
HEX3b	5.90	50.3	44.41	88.3
HEX3a	9.72	58.8	49.11	83.5
HEX2c	1.58	7.8	6.27	79.9
HEX2b	7.66	71.7	64.01	89.3
HEX2a	1.65	9.4	7.77	82.5
HEX1	4.79	26.0	21.25	81.6

HEX3b are identified as the best, whereas process design for heat exchangers HEX1, HEX2c, HEX2a, and HEX 3a is less efficient.

These efficiency numbers must be considered as extension to the Table 7.43. For example,

- the exergy efficiency of the heat exchanger HEX2c with 79.9% is the lowest, but the exergy loss in this heat exchanger is not so high, since this heat exchanger is small (Q/MTD = 0.24 kW/K) and only small amount of heat is transferred inside of this heat exchanger.
- the exergy efficiency of the heat exchanger HEX2b with 89.3% is the highest, but the exergy loss in this heat exchanger is quite high, since this heat exchanger is quite large (Q/MTD = 2.95 kW/K) and a lot of heat is transferred inside of this heat exchanger.

7.4.6.7.3 Loss in Expander

There are two kinds of losses associated with the expander:

(a) The power generated by expanders (1.9 kW) is not used in the process and is "lost" since it is dissipated into the environment,

(b) The exergy loss of 19.2 kW in expanders itself is due to irreversibilities caused by given mechanical design.

The impact of the loss of expander power is very minor compared to the other loss in the process as a whole.[14] By contrast, the impact of direct expander loss is 10 times higher. But even this relatively high number does not truly reflect the negative contribution of the irreversibilities in the expander. The negative contribution is much higher. The argumentation goes as follows:

Generally, the expander loss is predefined by the given low isentropic efficiency of this device of $\eta_s = 62.5\%$. If a reversible expander is used, the flowrate of the expander stream may be reduced by a factor of 1.6 (100%/62.5%) for the same expander power (and therefore the cooling capacity of the Brayton process): to

14 Therefore, from a techno-economics perspective, the integration of the expander power into the process makes no sense.

20.4 g/s instead of 32.65 g/s, and the total helium flowrate compressed in the process would be reduced to approximately 38 g/s instead of 50 g/s. Correspondingly, the compressor's power would be reduced by ≈25%. This would mean a considerable improvement of the entire process.

In reality, the overall process becomes even better. Since a lower temperature may be achieved at the outlet of the expander EXP2, the Joule–Thomson stream will be cooled to a lower temperature before the Joule–Thomson expansion.

Here is a clear message about a strong impact of the expander's efficiency on the overall process performance. Every seemingly small improvement in the expander efficiency has a tangible effect on the process efficiency.

7.4.6.7.4 Loss in Expansion Valve

The loss in the Joule–Thomson valve (9.1 kW) is medium. The moderate irreversibility of this expansion process is expressed in the fact that the liquid fraction after the expansion is in the middle, the flash-gas amounts to 38%, and 62% of the compressed Joule–Thomson stream is liquefied during the expansion. That is considerably better than the nitrogen Joule–Thomson process, but it is worse in comparison to the nitrogen Claude process.

7.4.6.7.5 Loss in Evaporator

The small loss in the evaporator (0.3 kW) is an indication that the valuable 4.5 K-cold is efficiently used here for maintaining the cooling object at 4.6 K.

7.4.6.7.6 Loss Caused by Heat Leak Through Thermal Insulation

The "apparently" low loss related to the heat leak to the evaporator through the non-ideal thermal insulation (6.4 kW in terms of exergy) can be interpreted that this loss has a minor impact on the efficiency of the entire process. But this interpretation is wrong. The correct message is as follows:

> the process itself is able to generate a twice cooling capacity (200 W instead of 100 W), but the thermal insulation is so bad, that only half of this capacity is available for cooling the cooling object, whereas half of the generated cooling capacity is "destroyed" by this heat leak. If the heat leak is avoided, the refrigerator would require half as much in terms of power and equipment size.

This is a clear message about the major impact of the heat leak on the overall process performance. Every improvement in the mechanical design and thermal insulation, which reduces the heat leak has a considerable effect on the process efficiency. It makes always sense to invest in an efficient thermal insulation.

Learning 7.11 It is always better to avoid the heat leak than to generate a very costly cooling capacity required to cover the loss caused by the heat leak.

7.4.6.7.7 Analysis Summary

Potentially, the helium Claude process may achieve high efficiency, if an efficient compressor, efficient expanders, and large heat exchangers are used. The process under consideration is based on commercially available hardware components like oil-lubricated screw compressors and gas-beared expanders with gas-brakes. The heat exchangers, especially HEX1, HEX2b, and HEX3a, are undersized. This results in there being moderate efficiency in combination with moderate cost. This is an appealing combination.

The performance of the expanders (isentropic efficiency) has a tangible impact on the overall performance of the helium Claude refrigerator.

7.4.6.8 Review of Simulation Procedure

The process design is an iterative procedure. Usually, after the first iteration, the engineer will develop a very rough idea about power requirements, hardware components (size, number), and how the system may look like. That is a good time to review assumptions made at the beginning about pressure drops, temperature differences, efficiencies of rotating equipment, and similar parameters. Particularly, assumptions rated as critical assumptions must be reviewed.

7.4.6.8.1 Heat Leak

In given case, the assumption about thermal insulation was recognized as very critical (see Section 7.4.6.2), since it has a huge impact on the flowrates, dimensions, power, and costs of the entire system. It must be reviewed and either confirmed or tuned. In given case, the argumentation goes as follows:

Originally, the value of parasitic heat $\dot{Q}_{is} = 100\,\text{W}$ was allocated only to evaporator. But the cold part of the process (located in the insulated coldbox) consists of seven heat exchangers (evaporator plus six process heat exchangers). Three of them are relatively small: HEX2c with Q/MTD = 0.24 kW/K, HEX2a with Q/MTD = 0.26 kW/K and HEX1 with Q/MTD = 0.49 KW/K. But the three other process heat exchangers are relatively large: HEX3b with Q/MTD = 9.47 kW/K, HEX2b with Q/MTD = 2.95 kW/K, and HEX3a with Q/MTD = 1.82 KW/K, see Table 7.36. They are considerably larger than evaporators with Q/MTD = 1 KW/K in terms of inner heat transfer surface, but in terms of outer surface and volume, too. In reality, the heat flow through nonideal isothermal insulation will be distributed among all these heat exchangers. For the correct process design, the parasitic heat must be allocated to all heat exchangers. Table 7.44 illustrates an approach, whereby (i) the parasitic heat is allocated to heat exchangers according to their Q/MTD-values, (ii) it is allocated to the four largest apparatuses only, this way the process simulation of small heat exchangers HEX1, HEX2a, and HEX2c is simplified.

7.4.6.8.2 Pressure Drop in Heat Exchangers

The assumptions concerning pressure drop (especially on the low-pressure side) may be considered as critical. For example, originally, for heat exchangers HEX3a and HEX3b, the pressure drop was assumed as equal. After the first iteration, this assumption can be reviewed targeting a more realistic distribution of pressure drop.

Table 7.44 Example of allocation of parasitic heat to heat exchangers.

Heat exchanger	Q/MTD (kW/K)	Percentage of total Q/MTD	Proposal for allocation of heat inleak to heat exchangers (W)
HEX3b	9.47	58%	58
HEX3a	1.82	11%	11
HEX 2c	0.24	1%	0
HEX2b	2.95	18%	22
HEX2a	0.26	2%	0
HEX1	0.49	3%	0
Evaporator	1	6%	9
Total Q/MTD	16.23	—	—

7.4.7 Process Optimization

In the classic Joule–Thomson process, as well as in the pure Brayton process, the process design engineer has no possibilities to actively design the temperature profile in the heat exchanger. In this sense, the Claude process is more flexible, and the temperature profile can be impacted by setting the expander inlet temperature and expander flowrate. The optimization procedure is generally focused on the search for the optimal arrangement of the Joule–Thomson and the Brayton stages.

Variables: Since the process is more complex, the number of optimization variables is a bit higher than in a single Joule–Thomson or single Brayton process. For example, six parameters may be identified as relevant variables for the simple basic Claude process with predefined working fluid:

(1) High-pressure level
(2) Flowrate of the expander stream
(3) Flowrate of the Joule–Thomson stream
(4) Temperature at the expander inlet
(5) Temperature at the inlet of the expansion valve
(6) Temperature difference at the warm end of the heat exchanger

Despite several additional variables with minor impacts like pressure losses in heat exchangers and the evaporator, there is a minimum temperature difference in the evaporator. The number of variables for optimizing a more complex helium Claude process is even higher. Therefore, an optimization procedure is necessary when designing the Claude process.

Constraints: The Claude process represents a combination of the Joule–Thomson stage and one (or more) Brayton stages. Consequently, all constraints valid for the Joule–Thomson process and Brayton process are valid for the Claude process, too. The most important constraints are preset by the Brayton process. These constraints are:

- Expander characteristics, for example, maximum pressure ratio
- Heat exchanger design, for example, minimum temperature difference in heat exchanger or maximum Q/MTD-value.

7.4.8 Control of a Claude Refrigerator

The typical control task for a refrigeration system is described in Section 2.4: the main target is to keep the cooled object at the specified cooling temperature. For this purpose, the cooling capacity of a refrigerator must be adjusted to equal the heat flow generated by the cooling object.

Since the Claude process is a combination of a Joule–Thomson process and a Brayton process, control of it as a process is a more complex task: both these subprocesses must be controlled properly. Equation (7.32) summarizes the parameters and characteristics that have an impact on the cooling capacity of the Claude refrigerator. There are many such parameters. Therefore, the number of theoretically available options for control strategies is quite high. The control approaches can be divided into two groups, depending on which subprocess leads and which follows:

1. The first approach considers the Joule–Thomson stage as a "superior" process since the required cooling temperature and cooling capacity is generated therein. In framework of this approach, the main control task relates to controlling of the Joule–Thomson stage, for example, by means described in Section 7.2.5. The Brayton stage is a sort of "subordinate" process. It must just follow the changes on the Joule–Thomson-stage, for example, using the control options described in Section 7.2.5.
2. Another approach is based on an opposing strategy: the cooling capacity of the Brayton stage is controlled mainly, while the Joule–Thomson stage follows (in this case automatically). This approach is based on the learning that the performance of the Joule–Thomson stage depends on the temperature level at the warm end of the Joule–Thomson heat exchanger HEX1: if this temperature level rises, the cooling capacity of the Joule–Thomson stage reduces. Therefore, the changes in the Brayton part of the process may lead to changes in the Joule–Thomson stage.

Both these approaches are practiced, with the final decision for one or the other strategy being dependent on the actual process design and given circumstances.

7.4.9 Hardware Components

7.4.9.1 Compressor

The Claude process is a high-pressure process, requiring high-pressure ratio in the compressor. The nitrogen-based Claude process uses multi-stage isothermal centrifugal machines - four to six stages are usually required for an efficient process design. The helium-based version uses an oil-lubricated screw compressor, since no centrifugal compressors for helium compression are commercially available.

7.4.9.2 Heat Exchanger

The requirements on the heat exchanger for the Claude process may be satisfied by the brazed aluminum plate-fin heat exchangers. Other kinds of heat exchangers are usually not considered.

Different approaches for mechanical design of the heat exchangers – and correspondingly of the coldbox – are pursued:

- Since the heat exchangers for a nitrogen Claude process are rather short and thick, engineers practice combining both heat exchangers HEX1 and HEX2 into a single heat exchanger core/unit with corresponding nozzles at the warm end, cold end, and in the middle, where the expander stream is withdrawn.
- The heat exchangers for the helium Claude process are rather long and thin. Therefore, engineers employ an opposite approach for the helium process: the heat exchangers are not combined to a single core, but rather divided into a number of single heat exchangers that can be folded down to a very compact structure.

7.4.9.3 Expander

The turboexpanders are the preferred kinds of machines for the expansion of gas in the Claude process for both versions – nitrogen and helium. The large centrifugal machines for nitrogen process use oil-bearings, whereas the smaller helium machines apply gas-bearings.

Another aspect relates to the loading device used for removal of the mechanical power generated during the expansion process. Three loading devices for turboexpanders were described already in Section 5.2.5: oil/gas-brake, electric generator, and centrifugal compressor (booster compressor):

- the last is the most popular device used in large-scale systems like air separation plants and nitrogen liquefiers,
- whereas in helium cryogenics the generated power is often not recycled but converted into the heat and dissipated into the environment by means of oil/gas-brake,
- the option with electric generator (based on conventional electric generators) is practiced in large-scale systems only. It has the potential to be extended to the medium/small-scale range, especially if high-frequency generators based on machines with permanent magnets are to be applied.

The small-scale turboexpanders are less efficient than larger machines. The isentropic efficiency of small devices is usually less than 80%, whereas the isentropic efficiency of large machines is usually higher than 82%. Very large devices can achieve an isentropic efficiency of 90%.

7.4.9.4 Commercial Availability

All hardware components for the large-scale nitrogen-based Claude process are commercially available. The number of suppliers is relatively high, and market works well in this segment.

The situation in the helium refrigeration is a bit different. The compressors and heat exchangers are available here, too. But the number of expander manufacturers, which are the key elements of this process, is very limited.

7.4.10 Applications

The Claude process is used almost everywhere in cryogenics:

Air separation: The largest application of the Claude process is the air separation, whereby the integrated Claude process is applied for following use cases:

1. Air separation process with so-called internal compression. This process is most suitable for integration of the Claude process since it already consists of a high-pressure compressor and an appropriate heat exchanger. Therefore, the integration of a Claude process causes only minor additional costs.
2. Plants for liquid air products (liquid oxygen, liquid nitrogen), whereby the Claude process may be implemented as a separate stand-alone nitrogen liquefier unit or as an integral part of the main air separation process.

Helium liquefaction: The Claude process is the standard process for helium liquefaction. There are no alternative solutions available today.

Hydrogen liquefaction: The state-of-the-art technology for middle-size and large-size hydrogen liquefaction is based on the Claude process.

Liquefaction of natural gas: Middle-size and small-size natural gas liquefaction are often based on the Claude process, although the general trend is to use the mixed-fluid process there.

7.4.11 Assessment of the Claude Process

The Claude process has several attractive features:

- **No limitation concerning cooling temperature:** every temperature level can be achieved with an appropriate refrigerant (helium, neon, hydrogen, nitrogen, etc.).
- **Efficiency:** the Claude process is potentially very efficient. The large-scale nitrogen-based version may achieve an excellent efficiency in the range above 30%.
- **Liquefied refrigerant:** the cooling capacity is generated in the form of a liquid refrigerant. The liquid cryogen has several advantages for refrigeration consumers, see Section 7.2.6.

The weakness of the Claude process stems from:

- Consequences of using an expander, mainly regarding maintenance and reliability issues,
- Cost issues: the cost of refrigeration systems based on Claude process is considered rather high, since it is based on relatively expensive hardware components like a high-pressure compressor and expander.

7.5 Mixed-Fluid Joule–Thomson Process

The classical Joule–Thomson process described in Section 7.2 is usually realized with nitrogen as working fluid. Other gases like argon, oxygen, or air may be used as refrigerants for this process too. And of course, all imaginable mixtures of these substances, even with addition of some helium, neon, or even hydrogen, are virtually applicable; these refrigerant options are sometimes considered for special single applications. However, from the thermodynamic perspective, the overall process does not change so much by using these gases; the pressure range and typical phase states remain the same. All learnings, equations, statements, and comments given for the nitrogen Joule–Thomson process in Section 7.2 are still valid for processes based on mixtures of gases mentioned above. Therefore, all these processes are considered as classical Joule–Thomson processes.

However, the thermodynamics of the Joule–Thomson process change dramatically if a special sort of mixture is used: the mixtures of nitrogen[15] and hydrocarbons such as methane, ethane, ethylene, propane, butane, and/or their halogenated derivatives. The nitrogen fraction in these mixtures is usually less than ⅓, with the remaining ⅔ being hydrocarbon-based substances. Therefore, hydrocarbons dominate the properties of these kinds of multicomponent fluids. That is the main reason why the thermodynamics of the overall process change so much (Table 7.49 gives an impressive summary of the differences), that the new process – the mixed-fluid Joule–Thomson process – is considered as a separate class of refrigeration processes.

The term "mixed-fluid Joule–Thomson process" is not fully established in the cryogenic nomenclature. Other terms like mixed refrigerant, mixed gas, and mixed fluid refrigerant (MCR) are used synonymously with the term mixed fluid. Correspondingly, expressions like the mixed-gas process/cycle, the mixed-refrigerant process/cycle, the mixed-fluid process/cycle refer to the same kind of processes.

A useful source of information about mixed-fluid Joule–Thomson refrigeration is the book listed as Ref. [4], it includes many process simulation examples. Another source of information is Ref. [2] (Section 8), especially for its comprehensive bibliography. Each book gives an encyclopedic view on mixed-fluid Joule–Thomson refrigeration.

7.5.1 Main Elements

Morphologically the mixed fluid Joule–Thomson process is identical to the nitrogen Joule–Thomso process. The main elements are:

1. Compressor unit,
2. Counter-current heat exchanger,
3. Expansion valve, often referred to as a Joule-Thomson valve,
4. Evaporator,
5. Thermal insulation.

7.5.2 Mixed Fluid as Refrigerant, Properties

This section deals with thermodynamic properties of mixed fluid refrigerants, as the thermodynamics of mixed-fluid Joule–Thomson processes is determined by them.

15 It is also possible to use other air gases (argon, oxygen, etc.) or mixtures of these gases instead of pure nitrogen.

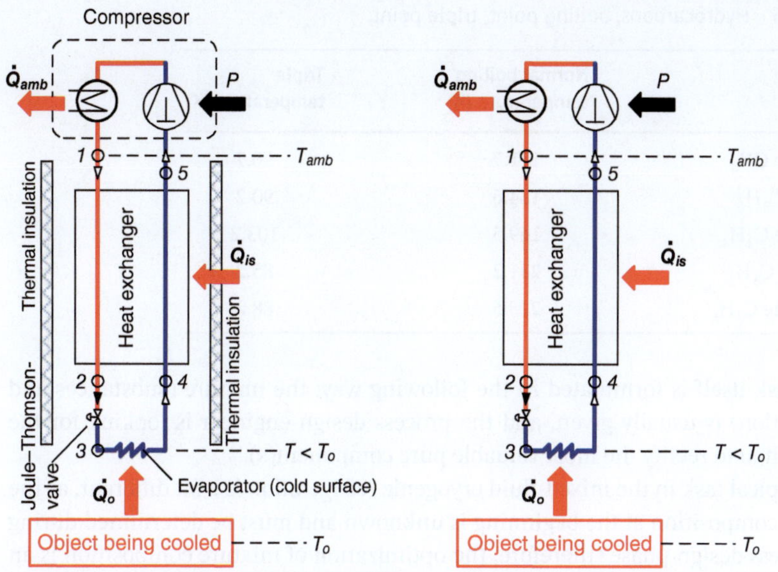

Figure 7.56 Joule–Thomson process, basic process flow diagram left: with thermal insulation, right: without thermal insulation.

7.5.2.1 Main Components of Mixed Fluid

In this book only one mixed-fluid family is considered, including:

- nitrogen,
- alkanes (methane, ethane, propane, and butanes),
- alkenes (ethylene, propylene) and
- helium

The learnings and statements are certainly applicable to other mixed-fluid families based on other substances.

If they are liquified and cooled further, a special feature of the hydrocarbons used for mixed-fluid Joule–Thomson processes is that they remain liquid down to very low temperatures; the solidification, characterized by triple point, occurs at a temperature below 110 K. In the case of propane or propylene, the solidification occurs below 90 K (see Table 7.45).

7.5.2.2 Thermodynamic Fundamentals for Mixed Fluids

The fundamentals of thermodynamics for gas mixtures are well described in textbooks for chemical process engineering, as shown in [5, 6]. Most of these texts focus on problems around the separation of mixtures by means of rectification, whereby these processes are usually characterized by:

(a) the boiling temperatures of gases being separated do not differ so much,
(b) the whole separation process occurs in a two-phase area in relatively narrow range of temperatures.

Table 7.45 Hydrocarbons, boiling point, triple point.

	Normal boiling temperature (K)	Triple temperature (K)
Methane CH_4	111.7	90.7
Ethane C_2H_6	184.6	90.2
Ethylene C_2H_4	169.5	103.2
Propane C_3H_8	231.2	85.2
Propylene C_3H_6	225.5	88.2

The task itself is formulated in the following way: the mixture (substances and composition) is usually given, and the process design engineer is looking for the best method to rectify the most valuable pure component(s).

The typical task in the mixed-fluid cryogenic refrigeration is a bit different, as the mixture composition at the beginning is unknown and must be determined during the process design phase (therefore, the optimization of mixture composition is an important part of the process design). The following details complete the picture:

(a) the mixed fluid consists of a number (usually more than four) of substances with very different boiling temperatures, distributed quite uniformly. There are no "most valuable" components; alternatively, all substances are important.
(b) during the process, the mixture goes through three or even four different phase state areas (vapor, two-phase state, liquid state), and the thermodynamic properties of working fluid in every phase state are necessary for the process calculation.

In this sense, the thermodynamics for mixed-fluid Joule–Thomson refrigeration is potentially more challenging than the phase equilibria data required for pure separation processes. Poling et al. [7] helps one develop a better understanding of the thermodynamics of mixtures.

7.5.2.3 *T,s*-Diagram for Typical Mixed Fluid

The *T,s*-diagram for a MCR with composition as given in Table 7.46 is shown in Figure 7.57, and a zoom for the temperature range 70–150 K is presented separately in Figure 7.58. Both pictures were created by means of a common Soave–Redlich–Kwong (SRK) equation of state (EOS), with the binary interaction parameters as

Table 7.46 Example of mixed fluid refrigerant composition.

Nitrogen	Methane	Ethylene	Propane	i-butane
30 mol %	23 mol %	17.mol %	10 mol %	20 mol %

They are no any special reasons why this mixture composition was chosen as an example, only that this mixture was already described and considered in [8]. It is used in this book for illustrative purposes.

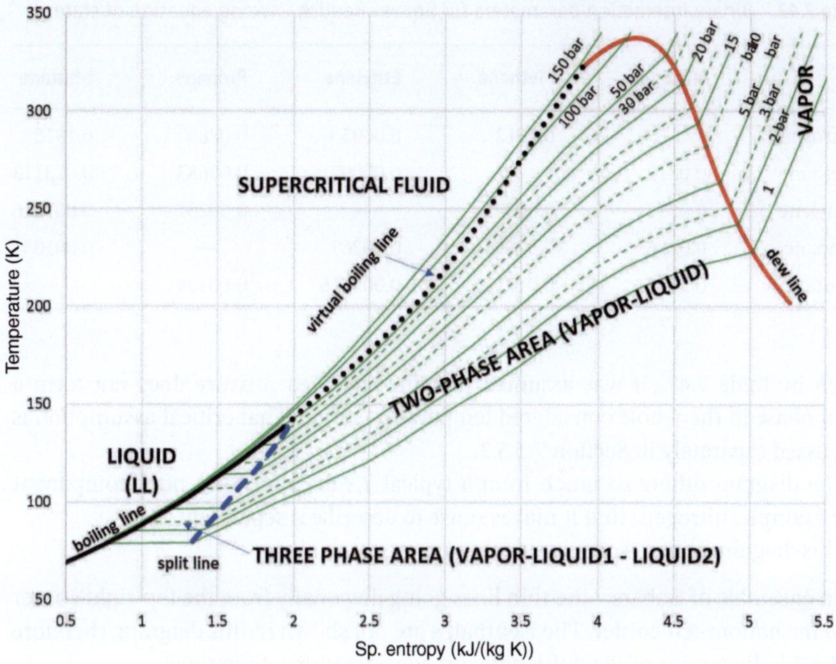

Figure 7.57 *T,s*-diagram for a typical nitrogen–hydrocarbon mixture.

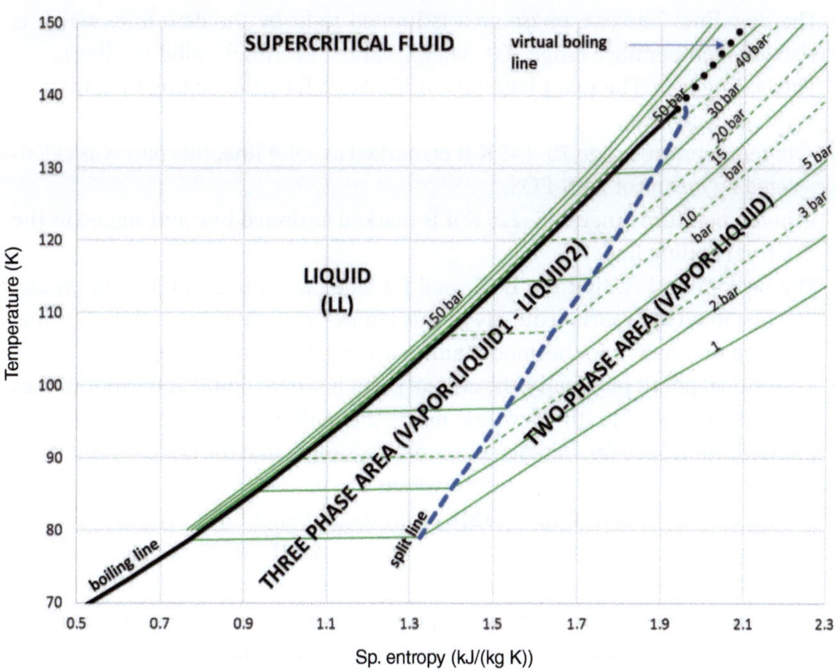

Figure 7.58 *T,s*-diagram for a typical nitrogen–hydrocarbon mixture (three-phase area).

Table 7.47 Binary interaction parameters for Soave–Redlich–Kwong equation of state.

	Nitrogen	Methane	Ethylene	Propane	i-butane
Nitrogen	—	0.0312	0.0695	0.0886	0.1315
Methane	0.0312	—	0.00097	0.00683	0.013113
Ethylene	0.0695	0.00097	—	0.00267	0.007016
Propane	0.0886	0.00683	0.00267	—	0.001041
i-butane	0.1315	0.013113	0.007016	0.001041	—

given in Table 7.47. It was assumed that the specified mixture does not form a solid phase in the whole considered temperature range. That critical assumption is discussed separately in Section 7.5.5.3.

The diagram differs so much from a typical *T,s*-diagram for a pure component (for example, nitrogen) that it makes sense to describe it separately.

This diagram consists of:

- an ensemble of isobars – the thin lines going diagonally from the top-right corner to the bottom-left corner. The isenthalps are not shown in this diagram, therefore the *T,h*-diagram is used additionally for considerations if required.
- three phase transition lines – the dew line, the boiling line, and the split line – marking the borders between different phase areas:
  - The dew line: The two-phase area is limited right by the dew line, which is visible in temperature range 200–340 K (marked as a thick, solid red line);
  - The boiling line: The two-phase area is limited left by the boiling line, marked as thick black line.
    - in temperature range 70–145 K it is marked as solid line, this part was calculated by means of RKS-EOS,
    - in temperature range 145–225 K it is marked as dotted line and signed as the virtual boiling line.

    The background is, that the EOS used for calculation was not able to predict this line in an appropriate quality in this temperature range, therefore this line is drawn manually (no thermodynamics behind it, it is pure art);
  - The critical point: physically, the critical point is where liquid and vapor phases are indistinguishable (the dew line meets the boiling line). Another more formal definition used for a single-component system is that the critical point is the highest pressure and temperature at which two phases can exist. However, this second definition is **not applicable to** multicomponent fluid systems, for this point neither represents the maximum pressure nor the maximum temperature for vapor-liquid coexistence because of the so-called retrograde phenomenon.[16] The critical point for given MCR composition locates somewhere on the virtual boiling line, in the area where this line meets the dew line;

16 More information about it you can find here: https://www.e-education.psu.edu/png520/m4_p3.html.

- The split line is a new kind of line, not appearing at all in *T,s*-diagrams for pure substances; it may occur only in multicomponent fluid systems.

At temperatures below 145 K and pressures below 42 bar, the given mixture forms three phases, a vapor phase, and two liquid phases: one liquid consists mainly of liquid nitrogen, whereas the other liquid contains less nitrogen and more liquid hydrocarbons. These two liquids are immiscible - the behavior is well-known from the fluid system mineral oil–water (see Figure 7.59), where the oil droplets swim in the water and/or form a separate liquid layer.[17] In simplified terms, the thermodynamics behind this state is that the molecules of a first liquid differ so much from molecules of the second liquid that there is no way to mix the two. Two liquid phases (the nitrogen-based and the hydrocarbon-based) are also immiscible.

The split line is a border between the temperature–pressure range, where the two liquid phases coexist (left to the split line), and temperature–pressure range, where the liquid is homogeneous (right to the split line) and does not form two separate liquids;

- In the liquid area, these two liquids still coexist. Therefore, the liquid must be considered as a mixture of two liquids (liquid1 and liquid2), and is therefore referred to as the LL-area;
- The behavior mentioned above (two liquid phases) may or may not occur - it depends on the composition (substances and fraction of these substances in the mixture) and temperature–pressure conditions. For example, see the mixture

Figure 7.59 Mineral oil–water mixture as a liquid–liquid (LL) fluid system. Source: pat_hastings/Adobe Stock.

17 "Science expert Christopher Grayce compares the interaction between oil and water molecules to how boys interact with girls at recess. Grayce says, "Oil molecules like to stick to other oil molecules more than they like to stick to water molecules." Unless the oil molecules have a way of joining with water molecules, the two will not mix. Mineral oil has nonpolar molecules, and water has polar ones." see https://sciencing.com/mineral-oil-water-dont-mix-6107887.html.

shown in Table 7.46, and consider how with ethane instead of ethylene it does not form two liquids;

- These three phase transition lines (the dew line, the boiling line, and the split line) mark the borders between the following phase areas:
  - vapor area: right to the dew-line;
  - liquid area (or liquid–liquid area): left to the boiling line;
  - two-phase area (vapor–liquid area or VL-area): between the split and the dew line;
  - three-phase area (vapor–liquid1–liquid2 area) between the boiling line and the split line, often referred to as VLL-area;
  - the area to the left of the virtual boiling line is considered in this book as supercritical state;[18]
- The isobars have a remarkable shape for the following reasons:
  - in the vapor area the shape is quite linear;
  - in the three-phase area, it is rather horizontal;
  - in the two-phase area the curvature is complex; see Section 7.5.2.5;
  - at the supercritical conditions and in the liquid there are no anomalies, making it similar to the behavior of pure substances.
- All subcritical isobars have a knee if they cross the phase transition lines.

7.5.2.4 *T,h*-Diagram for a Typical Mixed Fluid

The *T,h*-diagram for a mixed fluid with composition as given in Table 7.46 is shown in Figure 7.61. Figure 7.62 shows a zoom for the temperature range 70–150 K.

In two-phase area, the isobars have an interesting pattern: they run almost parallel, and at lower temperatures the distance between these lines is small, whereas at

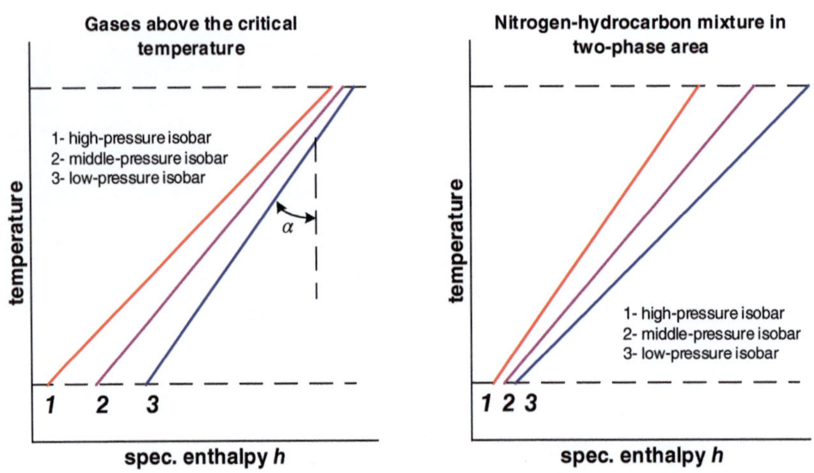

Figure 7.60 Typical isobars pattern left – nitrogen; right – mixture in two-phase area.

18 The discussion about all these definitions and phenomena is very interesting, but it is beyond the scope of this book.

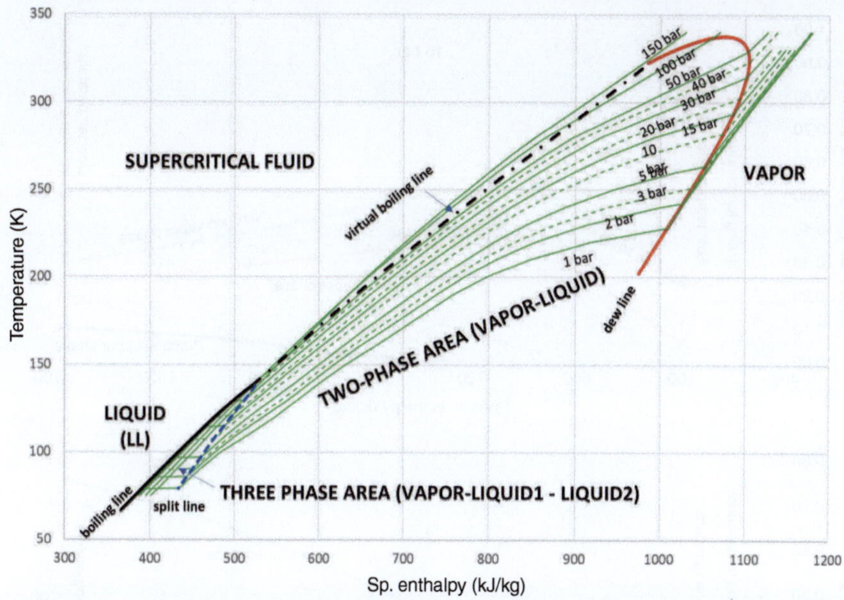

Figure 7.61 *T,h*-diagram for a typical nitrogen–hydrocarbon mixture.

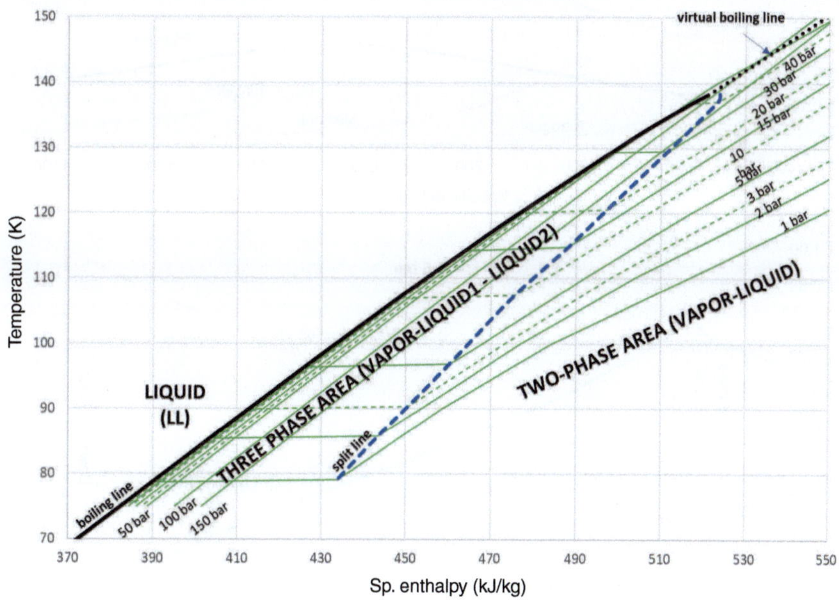

Figure 7.62 *T,h*-diagram for a typical nitrogen–hydrocarbon mixture (three-phase area).

higher temperatures the interspace becomes bigger so that a fan-like pattern forms. This behavior is opposite to the behavior of the conventional pure substances, where the isobars are aligned more closely with higher temperatures. Figure 7.60 (right) shows this behavior in simplified form;

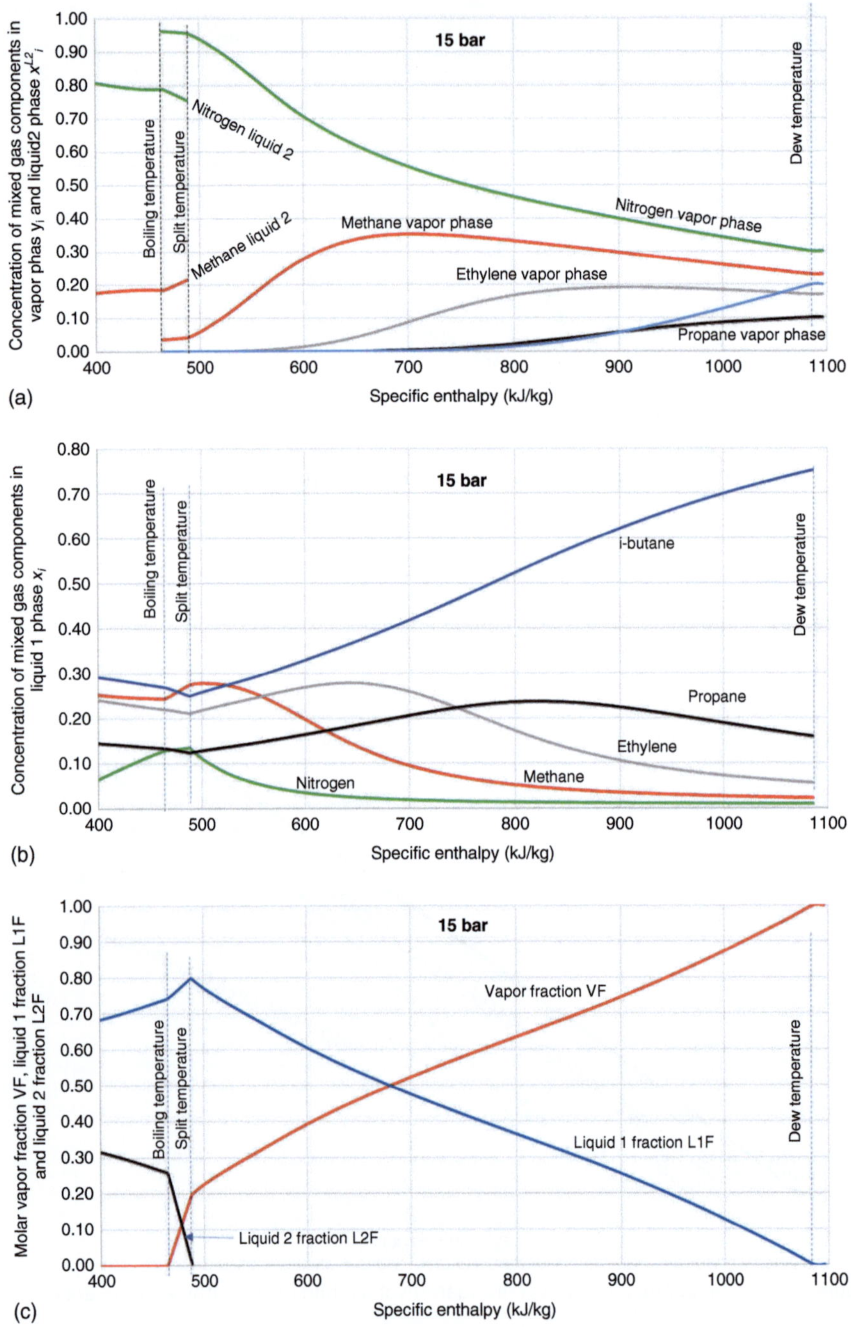

Figure 7.63 Composition and fractions of vapor phase, liquid phase, and liquid 2 phase.

This opposite pattern is only typical for the two-phase area; in the vapor area, gas area, as well as in the liquid area, the mixed fluid behaves similarly to conventional pure substances; the pattern is like that shown in Figure 7.60 left.

7.5.2.5 Isobaric Cooling/Heating Of Mixed Fluid

In both the *T,s*- and *T,h*-diagrams, the isobars run diagonally, crossing different phase areas, from the top-right corner to the bottom-left corner. For example:

- the 15 bar-isobar starts in the vapor area. In the vapor area no anomalies occur, with the behavior of mixed fluid there being similar to the behavior of pure substances;
- The dew temperature for this pressure is approximately 294 K. Below this temperature, in the two-phase area, the shape of isobar is slightly accentuated.
- At the temperature of ca. 115 K, this isobar crosses the split line and enters the three-phase area. In this area, the isobar runs horizontally until it meets the boiling line (roughly 114 K) and enters the liquid (liquid–liquid) area.

In the two-phase area, the shape of mixed-fluid isobars is notable, being a result of processes/phenomena, occurring during the isobaric condensation (or evaporation) of mixed fluid. It depends on the mixed-fluid composition; the vapor fraction as well as composition of vapor and liquid fractions change considerably during this process. Figure 7.63 gives an example of a typical mixture (composition as given in Table 7.46, pressure 15 bar), presenting:

(a) molar concentration of mixed-fluid components in vapor phase y_i and partially in the liquid2 phase x_i^{L2};
(b) molar concentration of mixed-fluid components in the liquid1 phase x_i, as well as;
(c) vapor phase fraction VF, fraction of first liquid L1F, and fraction of second liquid phase L2F.

The enthalpy of 1100 kJ/kg (right side of the scale) corresponds to approximately ambient temperature, and the enthalpy of 400 kJ/kg (left side of the scale) corresponds to a temperature of ca. 80 K. Therefore, for consideration of a cooling (condensation) process the diagram must be analyzed from right to left.

These figures deliver a lot of information, with some of the most apparent observations being:

- the vapor composition at the dew temperature corresponds to the overall mixed-fluid composition. The dew liquid consists mainly of butane and propane. The concentration of ethylene, methane, and nitrogen in liquid at this temperature is minor;
- with lower temperatures, the components migrate from the vapor phase into the liquid phase starting with propane, later ethylene, methane, and finally nitrogen. Consequently, the butane in the liquid phase becomes more and more diluted during this process;
- Correspondingly, the vapor phase enriches on nitrogen. At the split point, the vapor fraction consists mainly of nitrogen;

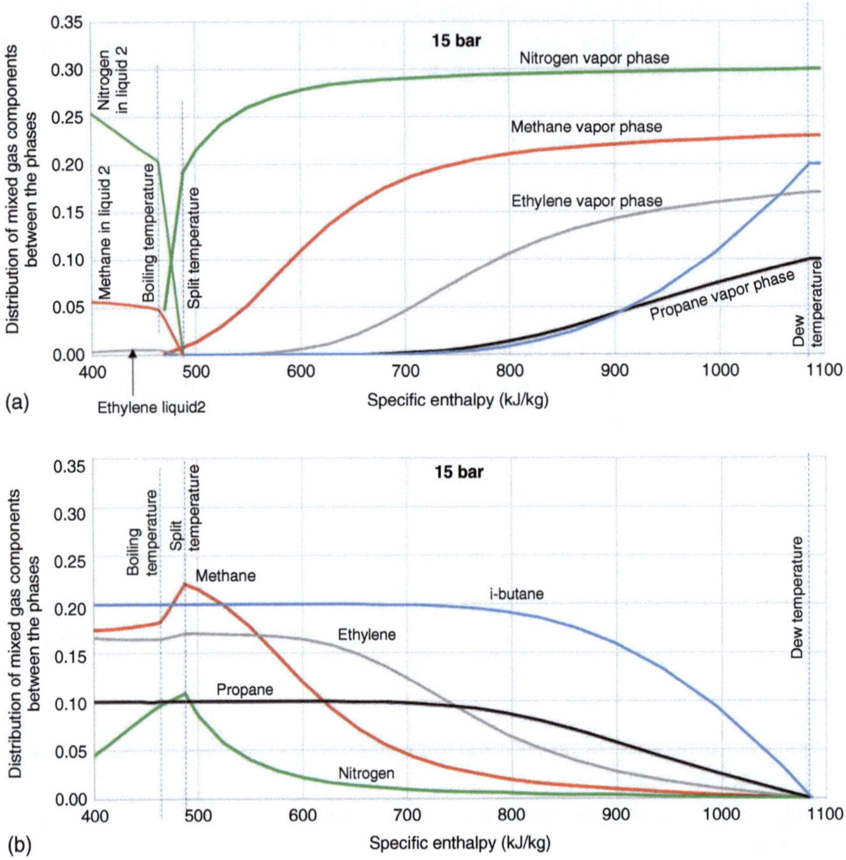

Figure 7.64 Distribution of mixed-fluid components between the phases $VF{\cdot}y_i$, $L1F{\cdot}x_i$, $L2F{\cdot}x_i^{L2}$.

- In VLL area the nitrogen condenses from vapor phase into the liquid2 phase, and a part of methane migrates from liquid1 phase into the liquid2 phase;
- During the further cooling below boiling point, the nitrogen contained in liquid1 migrates to liquid2, with other components concentrated in the liquid 1 phase.

Figure 7.64 shows the distribution of mixed-fluid components between these three phases; the concentrations of components are multiplied here by fraction of corresponding phase, and these numbers – $VF{\cdot}y_i$, $L1F{\cdot}x_i$, $L2F{\cdot}x_i^{L2}$ – are plotted in these two diagrams. This is useful additional information. For example, one can see that:

- nitrogen does not really progress into the liquid1 phase, but instead concentrates in the vapor phase and later liquefies into the liquid2 phase directly with only a small portion of nitrogen entering the liquid1;
- i-butane, propane moves from the vapor phase into the liquid1 only, with no other movements;
- ethylene behaves similarly, condensing from vapor into liquid1; later (at cold temperatures), a small portion of ethylene diffuses into the liquid2;

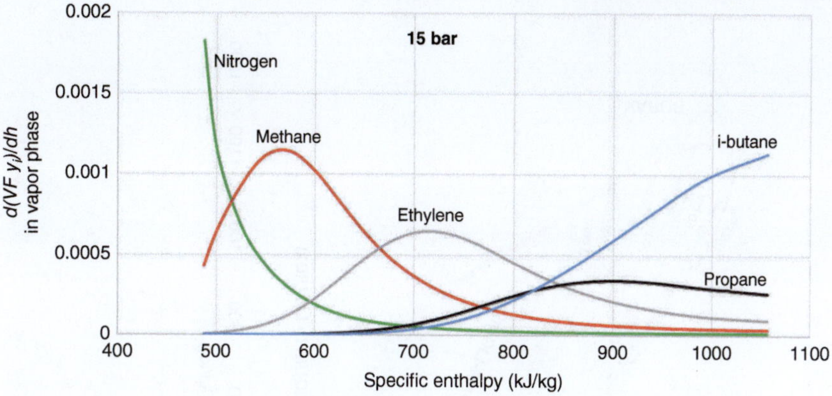

Figure 7.65 Intensity of phase transition, vapor phase.

- the behavior of methane is individualized: at the beginning of the cooling process, it moves from the vapor phase into the liquid1 phase completely. But later (at very cold temperatures,) a tangible portion moves from liquid1 phase into the liquid2.

Figure 7.65 shows the derivation $d(VF{\cdot}y_i)/dh$ for the vapor phase. This parameter characterizes the intensity of phase transition. It is possible to recognize the most intensive condensation of mixed-fluid components according to enthalpy (and therefore temperature) range:

- the most intensive condensation of propane occurs at roughly 900 kJ/kg (≈260 K)
- the most intensive condensation of ethylene occurs at 720 kJ/kg (≈210 K)
- the methane condensation is most intensive at 560 kJ/kg (≈150 K)
- the nitrogen condensation occurs below 500 kJ/kg (<125 K)

7.5.2.6 Comparison Mixed Fluid Versus Nitrogen

7.5.2.6.1 Two-Phase Area

Figure 7.66 shows *T,h*-diagrams for a typical mixed fluid and nitrogen. Both these pictures are already known, but in Figure 7.66 they are shown in approximately the same scale (temperature as well as specific enthalpy) and with a similar pressure range (1–150 bar). This representation helps with comparison purposes.

These diagrams differ greatly. The most visible difference is that in both dimensions, the two-phase area of the mixed fluid is considerably larger than the two-phase area of nitrogen, both in temperature range and specific enthalpy range. Both these observations are explainable by mixed-fluid properties:

- the large temperature range is due to the fact that a typical mixed fluid consists of not low-boiling substances such as nitrogen and methane, but high-boiling substances such as propane and butane, too.
  The high-boiling substances like butanes have a normal boiling temperature quite close to the ambient temperature. For example, propane liquifies at pressure of roughly 8 bar at the ambient temperature. Therefore, the presence of these

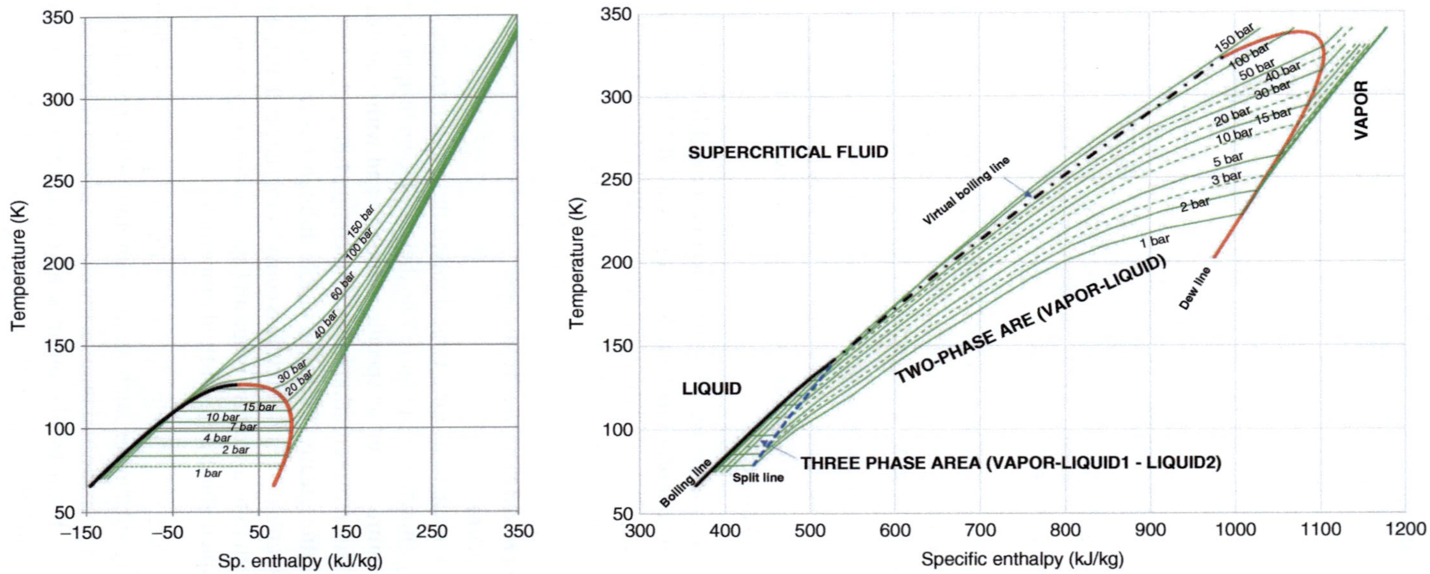

Figure 7.66 Comparison: T,h-diagram for a typical nitrogen–hydrocarbon mixture versus nitrogen.

substances (with a relatively high concentration of more than 20%) in the mixed fluid extends the two-phase area to a level of typical ambient temperature;

- The large width of the two-phase area (in comparison to the nitrogen) arises from the point that high-hydrocarbons generally have a larger enthalpy of vaporization (methane 511 kJ/kg, ethylene 482 kJ/kg, propane 430 kJ/kg, and i-butane 367 kJ/kg) in comparison to nitrogen (199 kJ/kg).

7.5.2.6.2 Three-Phase Area

Some similarities exist between the two-phase area of pure nitrogen and the three-phase area for the mixed fluid since it consists of mainly nitrogen. Both areas may be compared in two dimensions: in temperature range as well as in specific enthalpy range. The enthalpy of vaporization of the liquids2 phase at 1 bar amounts to approximately 45 kJ/kg; however, the concentration of the nitrogen in the mixture amounts to 30 mol% and only 85% is collected in liquid 2. The nitrogen's enthalpy of vaporization (199 kJ/kg) multiplied by these factors yields a number roughly around 50 kJ/kg, somewhat close to 45 kJ/kg.

7.5.2.6.3 Isothermal Enthalpy Difference at the Ambient Temperature (warm end of heat exchanger)

Figure 7.67 is derived from Figure 7.61. It shows the isothermal enthalpy difference (difference between the specific enthalpy at 300 K and 1 bar and the specific enthalpy for other pressures at 300 K) for two substances, nitrogen and mixed fluid. The line for nitrogen is more or less linear. The line for mixed fluid is more complex because at pressures below the dew point (approximately 19 bar) this line is linear as well. However, at the dew point, the slope changes disruptively, and the Δh increases considerably. At pressures above 50 bar, the line becomes accentuated; at pressures above 100 bar, it becomes quite linear again.

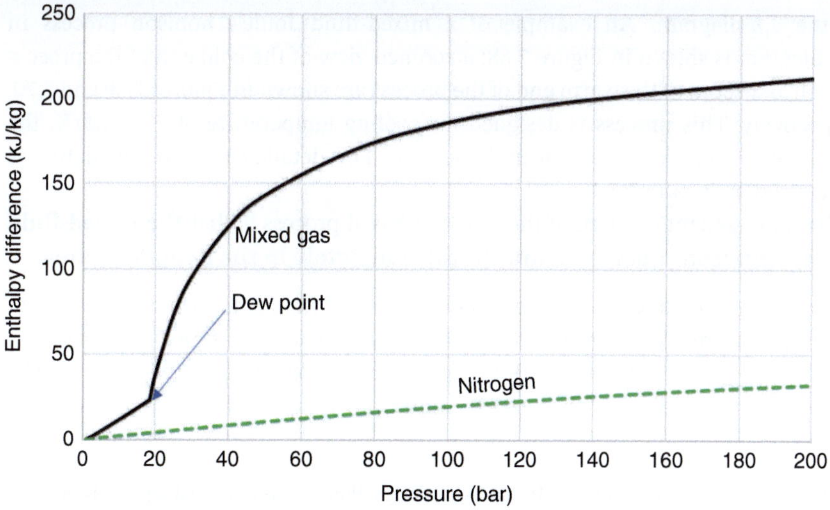

Figure 7.67 Difference between the specific enthalpy at 300 K and 1 bar and the specific enthalpy at 300 K and other pressures, for nitrogen and mixed fluid.

In the whole range of pressures, the enthalpy difference for mixed fluid is considerably larger than that of nitrogen, by a factor of 5–15.

7.5.3 Process Features

Figure 7.56 shows the BPFD for mixed-fluid Joule–Thomson process. Morphologically, there are no differences between the mixed-fluid Joule–Thomson and the classical nitrogen Joule–Thomson processes. Both processes consist of the same five main hardware components – **compressor, heat exchanger, expansion valve, evaporator, and thermal insulation.**

In the steady-state operation, the process sequence also appears identical:

- the refrigerant is compressed in a compressor and then cooled in the aftercooler to the approximately ambient temperature;
- the compressed refrigerant is cooled in the countercurrent heat exchanger;
- the refrigerant is expanded to the low-pressure level in a corresponding device for an isenthalpic expansion in two-phase area;
- this two-phase refrigerant flows through the evaporator and cools the object;

Then, the cold refrigerant is transmitted through the heat exchanger where it warms to a temperature close to the ambient temperature, then returns to the compressor inlet.

It is important to note that this is only an overview of the process. When considering the process in detail, many differences become apparent.

7.5.3.1 Phase State Conditions

There are many differences to note upon closer examination of the process. The *T,h*-diagram is a convenient tool for thermodynamic considerations since the Joule–Thomson expansion is an isenthalpic process visualized as a vertical line in the *T,h*-diagram. An example of a mixed-fluid Joule–Thomson process in *T,h*-diagram is shown in Figure 7.68; a zoomed view of the cold end of the process as well as a zoom of the warm end of the process are shown in Figures 7.69 and 7.70, respectively. This process is designed for cooling temperature of $T_0 = 100\,\text{K}$, the mixed-fluid composition given in Table 7.46. The detailed process calculation is performed in Section 7.5.4.

The most salient element of the nitrogen-based process is that **the mixed-fluid Joule–Thomson process occurs almost completely in the two-phase area**.

(a) In this particular example, the pressurized mixed fluid enters the heat exchanger (state 1) already in two-phase condition since the temperature T_1 is lower than the dew point[19] (where the HP-isobar crosses the dew line); see Figure 7.70. The two-phase condition at the heat exchanger inlet is not at all a strong requirement.

19 It means some complication for the hardware design; this issue is discussed separately in Section 8.4.7.

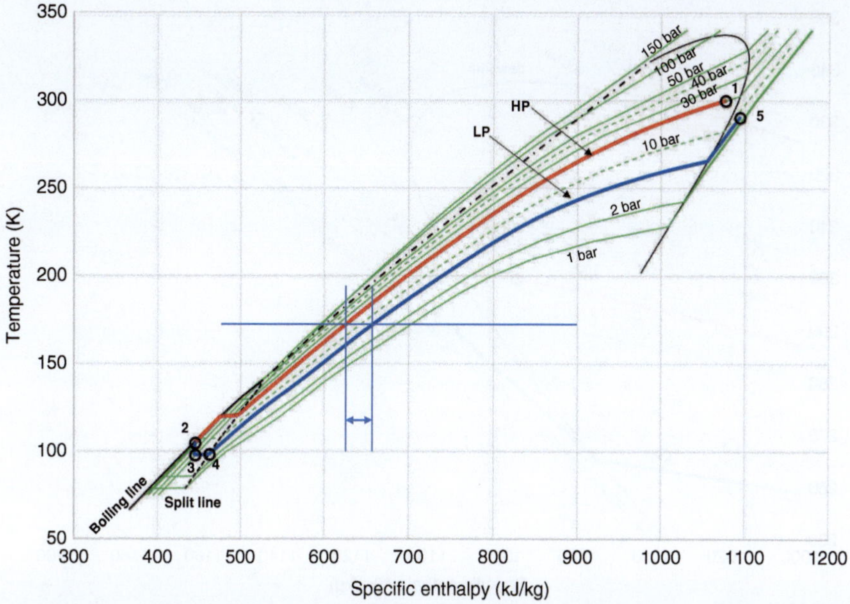

Figure 7.68 Typical mixed-fluid Joule–Thomson process in *T,h*-diagram, HP-high-pressure side, LP-low-pressure side of the process.

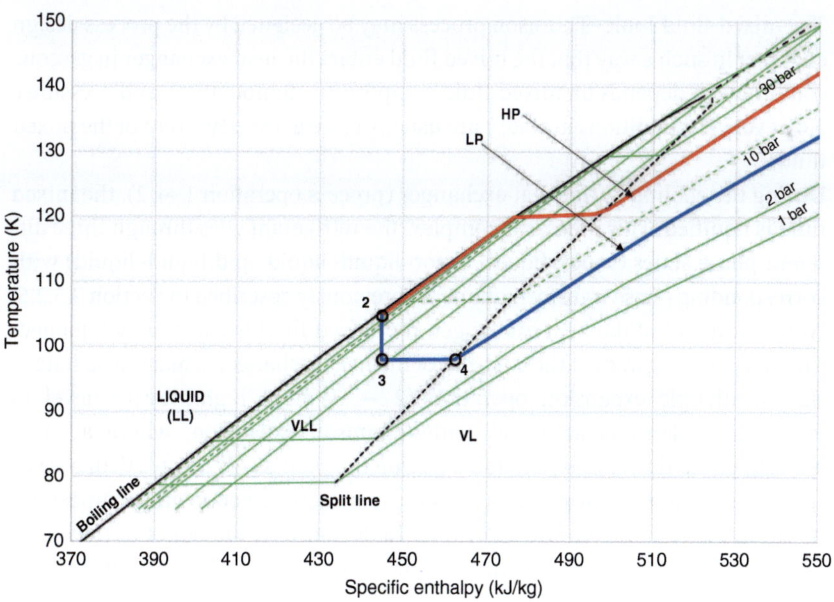

Figure 7.69 Typical mixed-fluid Joule–Thomson process in *T,h*-diagram, cold end HP-high-pressure side, LP-low-pressure side of the process.

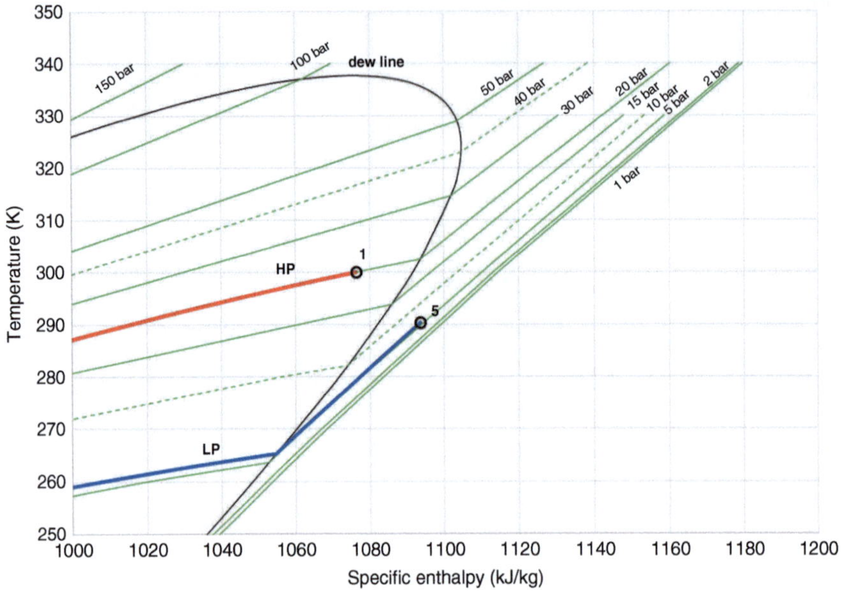

Figure 7.70 Typical mixed-fluid Joule–Thomson process in *T,h*-diagram, warm end HP-high-pressure side, LP-low-pressure side of the process.

The mixed-fluid Joule–Thomson process may be designed by the process design engineer in such a way that the mixed fluid enters the heat exchanger in gaseous conditions (it depends on mixed-fluid composition, temperature, and pressure). However, the conditions in state 1 are usually close to the dew point of the mixed fluid;

(b) During the cooling in the heat exchanger (process operation 1 → 2), the mixed fluid is liquified. This process is complex, the refrigerant goes through three different phase states (vapor–liquid, vapor–liquid–liquid, and liquid–liquid) with corresponding phase transformations, as previously described in Section 7.5.2.5. At the cold end of the heat exchanger, the mixed fluid is completely liquefied. The state of refrigerant at the cold end of the heat exchanger is marked as state 2;

(c) The isenthalpic expansion operation (2 → 3) from high-pressure level to low-pressure level is a typical Joule–Thomson expansion, where a small amount of the flash-gas is generated during the process. In this particular case, mainly the liquid2 (nitrogen-rich liquid) expands, the fraction and composition of liquid phase 1 remain the same (see Table 7.48). At the outlet of the valve, the vapor fraction is small and the flow looks more like liquid with some minor vapor fraction;

(d) This refrigerant flows through the evaporator, where the liquid is partially evaporated (3 → 4). In this case, the liquid2 evaporates completely (see Table 7.48). Consequently, the vapor fraction increases. The liquid1 is still available, meaning the fluid is in two-phase conditions (VL);

Table 7.48 Change of phase compositions during the isenthalpic expansion process.

		Phase fraction mol/mol	Concentrations				
			Nitrogen mol/mol	Methane mol/mol	Ethylene mol/mol	Propane mol/mol	i-butane mol/mol
State before expansion	Vapor	0	—	—	—	—	—
	Liquid1	0.724	0.114	0.246	0.226	0.137	0.276
	Liquid2	0.276	0.788	0.187	0.022	0.002	0.001
State after expansion	Vapor	0.115	0.981	0.019	0.000	0.000	0.000
	Liquid1	0.738	0.100	0.268	0.226	0.135	0.271
	Liquid2	0.146	0.774	0.203	0.021	0.002	0.001
State after evaporator	Vapor	0.225	0.979	0.021	0.000	0.000	0.000
	Liquid1	0.775	0.103	0.291	0.219	0.129	0.258
	Liquid2	0.000	—	—	—	—	—

(e) This two-phase fluid enters the heat exchanger on the cold end. There it is heated and the liquid1 evaporates. This process is quite the opposite to the cooling = condensation process on the warm side, where mixed-fluid components migrate from the liquid phase into the vapor phase starting with nitrogen, then followed by methane, ethylene, propane, and lastly i-butane. The dew point is in given case at approximately 265 K. At this temperature the liquid has completely evaporated. This gaseous mixed fluid is heated up further to a temperature T_5, which is quite close to the ambient temperature (operation 4 → 5);

(f) It is given to the compressor inlet where it is compressed to high-pressure level and cooled in an aftercooler to temperature T_1 (operation 5 → 1)

7.5.3.2 Difference Between the Mixed-Fluid-Based and Nitrogen-Based Joule–Thomson Process

See Table 7.49

7.5.3.3 Temperature Profile in Heat Exchanger

The processes in the heat exchanger (condensation on warm side and evaporation on the cold side) result in a relatively complex temperature profile. Figure 7.71 shows an example of the process considered in Sections 7.5.4 and 7.5.3.1. The abscissa here is the specific heat transferred in the heat exchanger. At temperatures below 150 K, the temperature difference between the warm and the cold streams is relatively small; it is difficult to recognize the value there. Thus, the temperature difference diagram is shown separately in Figure 7.72. In this case, the temperature difference at the warm end is ca. 10 K and at the cold end it is approximately 7 K. The minimum value $\Delta T^{min} \approx 1$ K is achieved quite close to the cold end, more precisely at the split point of high-pressure stream.

Table 7.49 Difference between the mixed-fluid Joule–Thomson process and the nitrogen Joule–Thomson process.

	Nitrogen Joule–Thomson process **Supercritical process**	**Mixed-fluid Joule–Thomson process** **Subcritical process**
Process units		
Compressor	– Compression to supercritical pressures (100–200 bar), – more than three compressor stages	– Compression to subcritical pressures (less than 50 bar), – 1–2-compressor stages
Heat exchanger	– Temperature difference at the cold end: up to 60–70 K – Temperature difference at the warm end: 3–5 K – Mean temperature difference: >10 K – The pinch locates always at the warm end of the heat exchanger	– Temperature difference at cold end: small, usually less than 5 K – Temperature difference at the warm end: 3–5 K – Temperature difference in the heat exchanger is relatively small; the mean temperature difference is less than 10 K – The pinch may occur at the warm end, cold end, or in the middle of the heat exchanger, and it may be a local pinch or a distributed pinch area
Expansion (Joule–Thomson) valve	– Large temperature difference between the inlet and the outlet, – Low liquid fraction at the outlet (<15%)	– Small temperature difference between the inlet and the outlet – Liquid fraction at the outlet >80%
Evaporator	Vapor–liquid equilibrium	Vapor–liquid–liquid2 equilibrium
Characteristic process point		
High-pressure stream at the warm end of the heat exchanger	Gaseous	Gaseous or two-phase (partially liquefied)
Warm side of the heat exchanger	Pressurized gas (single phase) is cooled down	– Two-phase fluid – the mixed fluid is liquified during the cooling process, – the composition of the liquid phase differs from the composition of the vapor phase, – Both these compositions (vapor and liquid) as well as vapor fraction (distribution between vapor and liquid fraction) continuously change with the temperature, – At low temperatures, even three-phase conditions may occur before the fluid becomes liquid

Table 7.49 (Continued)

	Nitrogen Joule–Thomson process Supercritical process	**Mixed-fluid Joule–Thomson process Subcritical process**
In front of the expansion valve	Supercritical fluid	- Liquid
Downstream the expansion valve	- Two-phase flow, - low liquid fraction, (more vapor than liquid)	- Two-phase flow (or even three-phase flow), - small vapor fraction (more liquid than vapor)
Outlet evaporator	Vapor, single-phase flow	- Two-phase flow, small vapor fraction (more liquid than vapor)
Cold side of the heat exchanger	Single phase, gas is warmed up	- Two-phase fluid - the mixed fluid is evaporated, - the composition of the liquid phase differs from the composition of the vapor phase, - Both these compositions (vapor and liquid) as well as vapor fraction (distribution between vapor and liquid fraction) continuously change with the temperature, - At the warm end, the refrigerant is completely evaporated
Low-pressure stream at the warm end of the heat exchanger	Gaseous	Vapor

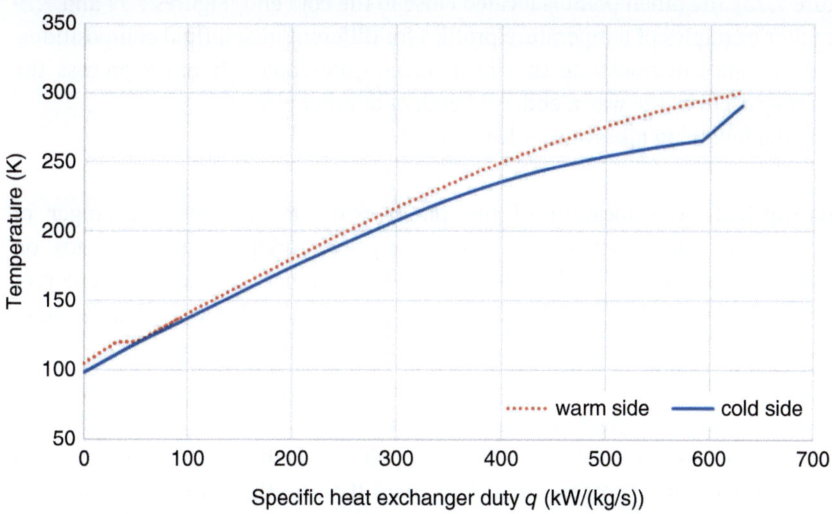

Figure 7.71 *T,Q*-profile in the heat exchanger.

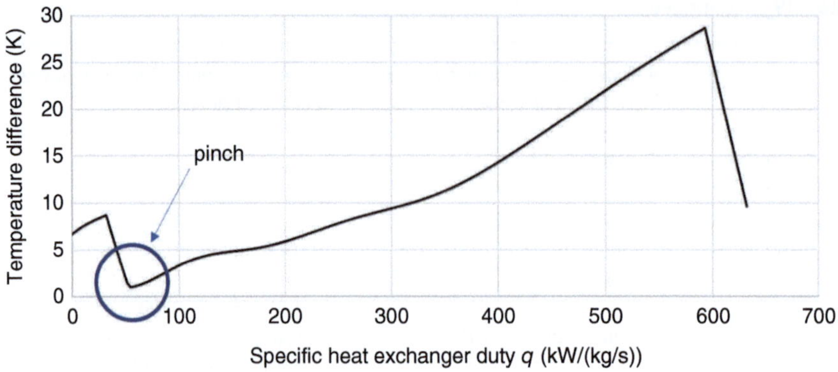

Figure 7.72 Temperature difference profile (ΔT,Q-diagram) in heat exchanger.

7.5.3.4 Location of Pinch Point(S)

First, let us define the pinch point:

Important definition	**The point (or area) with minimal temperature difference is referred to as a pinch point (or pinch area).**

Since the driving force (ΔT) at the pinch point is minimal, the heat transfer works sluggishly there and requires a lot of surface. It may considerably impact the heat exchanger cost. Consequently, the pinch points are highly relevant for process design considerations.

The next aspect relates to the location of the pinch point in heat exchanger because it impacts the process simulation sequence. In the example given in (Figure 7.72), the pinch point is located close to the cold end. Figures 7.77 and 7.79 give other examples of temperature profiles for different mixed-fluid compositions. These examples demonstrate that in a mixed-fluid Joule–Thomson process the pinch may occur at the warm end, cold end, or at other places.

Note the following takeaway, or learning:

Learning 7.12 The location of the pinch point in the heat exchanger of a mixed-fluid Joule–Thomson process is a priori unknown and depends on mixed-fluid composition and pressure levels. The pinch can occur at different places in the heat exchange, for example, at the warm end, cold end, at the split point, etc.

7.5.3.4.1 Consequences for Process Simulations Procedure

The statement that the location of the pinch point in the heat exchanger is a priori unknown is the bane of process design engineers. It means that the conventional procedure based on the definition of temperature difference on the warm end of the heat exchanger ΔT^{WE} (as it is used, for example, for definition of the classical nitrogen

Joule–Thomson process) is insufficient for determining mixed-fluid Joule–Thomson processes. In the worst-case scenario, it may be that:

- the temperature difference at the warm end of the heat exchanger is still positive $\Delta T^{WE} > 0$, while
- the minimal temperature difference (on the cold end and/or in the middle of the heat exchanger) is close to zero (or even negative: $\Delta T^{min} \leq 0\,\text{K}$)

The last situation would indicate that the corresponding process is not at all feasible since the heat cannot be transferred from a colder material stream to a warmer material stream by nature. To avoid such upsets, special measures must be introduced.

The Learning 7.12 has consequences for the process simulation procedure:

1. The definition of the temperature difference at the warm end ΔT^{WE} is insufficient for the proper process simulations; this parameter is an output value rather than an input parameter;
2. The most important requirement (and consequently an important input parameter) is the specification of the minimal temperature difference in the heat exchanger ΔT^{min};
3. During the process design phase, an additional procedure must be used that guarantees the temperature difference in every cross-section of the heat exchanger is $\Delta T \geq \Delta T^{min}$ at least by visual check of the temperature profile or temperature difference profile.

7.5.3.5 Cooling Capacity of a Mixed-Fluid Joule–Thomson Process

The Eq. (7.11), originally introduced for cooling capacity of the nitrogen Joule–Thomson process, is fully applicable to the mixed-fluid Joule–Thomson refrigerator, too. According to this equation, the cooling capacity of the mixed-fluid Joule–Thomson process depends only on the enthalpy difference at the warm end of the heat exchanger $\Delta\dot{H}^{WE}$, (assuming $\dot{Q}_{is} = 0$):

$$\dot{Q}_o = \Delta\dot{H}^{WE} = \dot{H}_5 - \dot{H}_1 = \dot{m} \cdot (h_5 - h_1)$$

Equation (7.33) represents an extension of Eq. (7.9). It postulates that the cooling capacity of a Joule–Thomson process is defined by the enthalpy difference in heat exchanger at every cross-section, as shown in Figure 7.73:

$$\dot{Q}_o = \dot{H}_{5x} - \dot{H}_{1x} = \dot{m} \cdot (h_{5x} - h_{1x}) \tag{7.33}$$

where $\dot{H}_{5x}$ and $\dot{H}_{1x}$ (respectively, h_{5x} and h_{1x}) – are the enthalpies of high-pressure and low-pressure streams for every cross-section of the heat exchanger.

Since the mixed-fluid Joule–Thomson process is determined by minimal temperature difference in the heat exchanger ΔT^{min}, it makes sense to define the cooling capacity of mixed-fluid Joule–Thomson refrigerator in terms of ΔT^{min}. The challenge is that the location of the pinch point in the heat exchanger is a priori unknown. Therefore, the procedure is a bit complex:

1. it is presumed that the pinch point (and therefore the minimum temperature difference at this point) may occur on every location $T_{1x}\ \varepsilon\ [T_2 \ldots T_1]$,

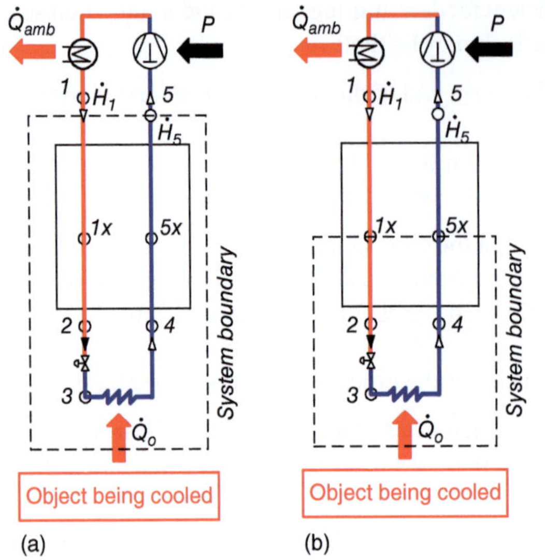

Figure 7.73 Mixed-fluid Joule–Thomson process, system boundary: a) for the whole cold part of the process, b) for a part of the process below the cross-cut.

2. the enthalpy difference in every cross-section of the heat exchanger must be checked:
   $\Delta h_{1x} = h\,(p_{LP}, T_{1x} - \Delta T^{min}) - h\,(p_{HP}, T_{1x})$,
   where
   $h\,(p_{HP}, T_{1x})$ – specific enthalpy for given high-pressure level and temperature T_{1x}
   $h\,(p_{LP}, T_{1x} - \Delta T^{min})$ – specific enthalpy for given low-pressure level and temperature $T = T_{1x} - \Delta T^{min}$
3. the minimum value of these enthalpy differences will occur at the real pinch point, and it corresponds to the cooling capacity of the cycle.

The corresponding equation, which describes the procedure mentioned above, is as follows:

$$\dot{Q}_o = min\,[\dot{m} \cdot (h\,(p_{LP}, T_{1x} - \Delta T^{min}) - h\,(p_{HP}, T_{1x}))] \quad \text{for every } T_{1x} \in [T_2 \ldots T_1] \tag{7.34}$$

7.5.3.5.1 Process with Infinitely Large Heat Exchanger

The Eq. (7.34) is not really transparent due to min-function used there. The approach used in Section 7.2.2.3.1 (cooling capacity of classic Joule–Thomson process) makes this equation more tangible. This approach is based on consideration of a process with an infinitely large heat transfer surface. In this case, the heat transfer may be realized at an infinitesimal temperature difference and therefore $\Delta T^{min} \to 0$. Equation (7.34) can be simplified to:

$$\dot{Q}_o = \min\,[\dot{m} \cdot (h\,(p_{LP}, T_{1x}) - h\,(p_{HP}, T_{1x}))] \text{ for every } T \in [T_2 \ldots T_1]$$

$$\dot{Q}_o = \min\,[\dot{m} \cdot \Delta h_T] = \dot{m} \cdot \min\,[\Delta h_T] \text{ for every } T \in [T_2 \ldots T_1] \tag{7.35}$$

where:

$\Delta h_T = h\,(p_{LP}, T) - h\,(p_{HP}, T)$ – specific isothermal enthalpy difference between the high-pressure level and the low-pressure level.

In principle, Equation (7.35) expresses that the cooling capacity of a mixed-fluid Joule–Thomson process (with infinitely large heat exchanger) corresponds to the isothermal enthalpy difference at the most narrow location in the *T,h*-diagram (pinch point) for given high-pressure level and low-pressure level. That is an important learning and simultaneously a practical advice, on how to find out the specific cooling capacity of a mixed-fluid Joule–Thomson process:

(a) for given mixed-fluid composition, both isobars (high-pressure and low-pressure isobar) are calculated and considered in *T,h*-diagram (see for example Figure 7.68). Specifically, the enthalpy difference at the same temperature (isothermal enthalpy difference) is analyzed;
(b) the cooling capacity of the process corresponds to the enthalpy difference at the most narrow place. For example, in Figure 7.68 (better Figure 7.69), it is at 120 K.

7.5.3.5.2 How the Cooling Capacity of the Process May Be Increased

Since the enthalpy difference depends on the pressure difference ($p_1 - p_5$) and the temperature difference at the pinch point corresponds to the minimal temperature difference ΔT^{min}, the following sentence may be formulated:

Learning 7.13 The cooling capacity of a mixed-fluid Joule–Thomson process can be increased by the following measures:

- increase of refrigerant flowrate $\dot{m}$ (trivial solution),
- increase of process pressure on the high-pressure side $p_{HP} = p_1$,
- use of other mixed fluid (substances and fractions) with larger Δh_T (isothermal enthalpy difference).

The minimal temperature difference in the heat exchanger ΔT^{min} has a negative impact on the cooling capacity of the Joule–Thomson process. Therefore, the minimization of ΔT^{min} must be targeted during the process design phase.

Learning 7.13 is formulated in more generically than Learnings 7.3 and 7.4. It is applicable to all kinds of Joule–Thomson refrigerators including the classical nitrogen Joule–Thomson process.

7.5.3.6 Typical Process Pressure Levels

7.5.3.6.1 High-Pressure Level

Since the enthalpy difference at the warm end of the heat exchanger for the mixed process is considerably higher than it is for the nitrogen process (see Section 7.5.2.6), the extremely high pressure is no longer required for this process. That is a great advantage from engineering perspective. The most obvious benefit is that the mixed-fluid Joule–Thomson process may be realized by means of a compressor with less stages: just one (single stage compressor) or two (dual stage compressor). Such a process would be a cost-efficient solution, at least in comparison to the nitrogen Joule–Thomson process, which requires an expensive multi-stage machine.

Another advantage relates to safety, especially regarding permission required for operation of cryogenic plants. The corresponding regulations differ from country to country, but the general rule is that the lower the pressure level the easier and cheaper is the approval procedure.

In summary, in mixed-fluid-based cryogenics processes, the high-pressure level p_{HP} may be reduced to the values below the critical pressure of the mixed fluid, usually lower than 50 bar. Therefore, in terms of pressure, the typical mixed-fluid Joule–Thomson processes for cooling purposes are designed often as subcritical processes.[20]

7.5.3.6.2 Low-Pressure Level

In the nitrogen Joule–Thomson process, the low-pressure level p_{LP} is obviously predefined by required cooling temperature T_0, since the cooling temperature T_0 is quite close to the boiling temperature of the liquid T_b, which evaporates in the evaporator at given pressure p_{LP} according to Eq. (7.21): $T_0 = T_b\,(p_{LP}) + \Delta T^{EVAP}$.

This equation is also applicable to some limited classes of mixed fluids, such as for mixtures characterized by two liquid phases at low temperatures (see the mixture (Table 7.46) used in Section 7.5.2.3 for illustration purposes). Since the second liquid phase consists of mainly nitrogen, it evaporates at a more or less constant temperature like pure nitrogen.

Commonly, however, the Eq. (7.21) is not applicable to the mixed fluid refrigerant; this is so because:

(a) mixed fluid usually evaporates at a gliding temperature (the temperature changes during the evaporation of a mixed fluid), not at the constant temperature,
(b) the evaporation temperature depends not only on the pressure but on the composition (substances as well as their fractions) of the mixed fluid.

The general principle – low pressure means a lower boiling temperature – remains the same. However, the procedure for finding the right value is more complex; since the low-pressure level is a real degree of freedom, it should be found during the overall process optimization.

The pressure level of 1–6 bar is used often, whereby the subatmospheric conditions – pressure lower than the ambient pressure – should be avoided (see Section 7.2.2.4 for arguments that explain this).

7.5.4 Example Process Calculation, Mixed-Fluid Joule–Thomson Process

An example of process calculations for a mixed-fluid Joule–Thomson process is considered in this section.

---

20 Multicomponent refrigerant processes for liquefaction of natural gas are designed often as supercritical processes, similar to the classic nitrogen Joule–Thomson process.

7.5.4.1 Task Definition

A mixed-fluid Joule process for cooling capacity of $\dot{Q}_0 = 10\,\text{kW}$ at the cooling temperature $T_0 = 100\,\text{K}$ must be designed; ambient conditions are as defined in Table 7.5. The target is to provide:

1. process parameters (flowrate; temperature; pressure; phase state; composition; enthalpy; entropy; exergy) at all relevant process points;
2. total power consumption, COP, and efficiency, as well as exergy loss in every process component;
3. figures required for designing heat exchangers and compressors.

7.5.4.2 Properties Data

A common Soave–Redlich–Kwong (SRK) equation of state (EOS) is used for calculating data properties; the corresponding binary interaction parameters are given in Table 7.47. It is assumed that the specified mixed fluid does not form a solid phase in the whole considered temperature range.

7.5.4.3 Process and Hardware Concept

The process concept and the hardware concept must be developed simultaneously. The main idea in this particular case is to develop a mixed-fluid Joule–Thomson process based on:

(a) an efficient aluminum-brazed plate-fin heat exchanger (see Section 6.2.1 for more details). This kind of heat exchanger is characterized by: (i) high specific volumetric surface (m^2/m^3), therefore a process with minimal temperature difference of 1 K or even lower may be designed; (ii) pressure loss on the high-pressure and low-pressure sides, where it is usually small (≈0.1 bar);
(b) single-stage oil-lubricated screw compressor unit is used (see the Section 6.5.3 for more details). This machine is characterized by a moderate pressure ratio of 3–5 achievable in one compressor stage. The compressor unit includes an aftercooler able to cool the compressed mixed fluid to ambient temperature T_{amb} and an oil-removal system;
(c) the evaporator is designed in the form of a tube coil wounded around the object being cooled (as shown in Figure 7.74);
(d) cold part is thermally insulated by an efficient multilayer insulation.

Some data required for process calculation can be directly derived from this process idea description. For example:

1. Since the compressor is oil-flooded, and therefore a continuous cooling by means of compressor oil during the compression process is applied, the compression operation can be considered and calculated as a quasi-isothermal compression; data for isothermal efficiency of an appropriate machine are necessary;
2. An oil-removal system is required since the compressor uses oil lubrication;
3. Temperature T_1 is close to the ambient temperature ($T_1 \approx T_{amb} = 300\,\text{K}$) due to the efficient cooling after compression;

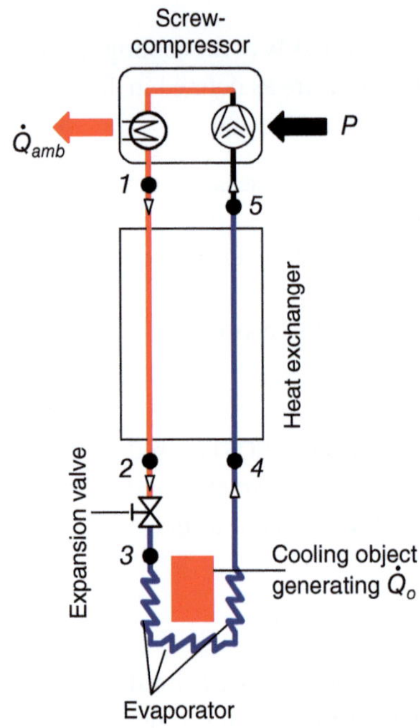

Figure 7.74 Mixed-fluid Joule–Thomson refrigerator for process calculations.

4. Heat exchanger: for the first process design:
   a small min temperature difference in the heat exchanger $\Delta T^{min} = 1\,\text{K}$ may be assumed;
   minor pressure loss on the high-pressure side $\Delta p_{1\rightarrow 2} = 0.1$ bar as well as minor pressure loss on the low-pressure side $\Delta p_{4\rightarrow 5} = 0.1$ bar may be assumed;
5. the pressure loss in the evaporator is not negligibly small, it must be considered during the process calculation where a value of $\Delta p_{3\rightarrow 4} = 0.1$ bar can be assumed;
6. it may be assumed that the cooling object is cooled in an efficient way because of direct contact between the object and liquid nitrogen, therefore a small temperature difference between the liquid and cooling object can be expected; for example $(T_o - T_3) = 2\,\text{K}$ as a first estimation;
7. heat leakage through thermal insulation is negligibly small due to the efficient insulation;

All these assumptions can be reviewed later, when the detailed design of the cold part is finished and the real values are available; however, for the first process design these are acceptable assumptions. These considerations are summarized in Table 7.50.

However, these data are still insufficient for the process simulation/calculation. Further assumptions required for the process simulations are given in Table 7.51.

In this form, the task is well specified, and corresponding process simulation can be performed.

Table 7.50 Input process data derived from the hardware concept.

Pressure loss in evaporator	$\Delta p_{3\to4} = 0.1$ bar
Min temperature difference in evaporator (at the outlet of evaporator)	$T_o - T_4 = 2$ K $T_o = 100\text{ K} \to T_4 = 98$ K
Heat leakage through thermal insulation	$\dot{Q}_{is} = 0$
Min Temperature difference in the heat exchanger	$\Delta T^{min} = 1$ K
Temperature at the outlet of compressor (=inlet heat exchanger)	$T_1 = 300$ K
Pressure loss in heat exchanger, high-pressure side	$\Delta p_{1\to2} = 0.1$bar
Pressure loss in heat exchanger, low-pressure side	$\Delta p_{4\to5} = 0.1$ bar
Pressure loss in evaporator	$\Delta p_{3\to4} = 0.1$ bar
Compression	Quasi-isothermal

Table 7.51 Further assumptions for process calculations.

The compressor efficiency can be derived from previous experiences or discussions with equipment manufacturers. For example, the value of 60% is usually assumed for this class of compressors	$\eta_T = 60\%$
The mixed-fluid composition from Table 7.46 can be used for the first calculations	See Table 7.46
High-pressure level: pressure p_1	$p_1 = 20$ bar

7.5.4.4 Process Simulation

The process simulation may be executed with every available process simulator.

The process calculation starts with calculation of pressures at all relevant process points:

7.5.4.4.1 Pressure Table

p_1	This pressure is given: $p_1 = 20$ bar
p_2	This pressure is lower than p_1 by pressure loss in the heat exchanger, high-pressure side: $p_2 = p_1 - \Delta p_{1\to2} = 200\text{ bar} - 0.1\text{ bar} = 19.9$ bar
p_4	The state at the outlet of evaporator (4) is characterized by temperature $T_4 = 98$ K; another fact is that this point locates on the split line (the second liquid fraction is completely evaporating L2F = 0). These inputs are sufficient for clear definition of the state and the corresponding pressure is $p_4 = 5.42$ bar
p_5	This pressure is lower than p_4 by pressure loss in the heat exchanger, low-pressure side: $p_5 = p_4 - \Delta p_{4\to5} = 5.42\text{ bar} - 0.1\text{ bar} = 5.32$ bar
p_3	Since the pressure loss in evaporator is given $\Delta p_{3\to4} = 0.1$ bar the pressure p_3 can be found as $p_3 = p_4 + \Delta p_{3\to4} = 5.42\text{ bar} + 0.1\text{ bar} = 5.52$ bar

7.5.4.4.2 Simulation Sequence

It is an iterative procedure. The conventional simulation sequence is:

Step 1:	Make appropriate assumption(s) for starting value(s)
Step 2:	Perform a process simulation
Step 3:	Check the results
Step 4:	Adjust the starting values until the correct solution is be found

In the case of the mixed-fluid Joule–Thomson process, you may use this sequence as an example:

Step 1:	Assumptions: - assumption for state 2 (before expansion), for example, $T_2 = 110\,\text{K}$ - assumption for flowrate, for example, 1000 kg/s
Step 2:	Process calculation: Since the pressure after expansion p3 is known → the calculation of the isenthalpic expansion process 2 → 3 is performed; Followed by calculation of the process in evaporator; Since states 1 and state 4 are already defined (and state 2 is assumed), it is possible to find the state 5 by means of enthalpy balance for the heat exchanger At the end of this procedure all states are known;
Step 3:	Check results: - Check the min temperature difference in heat exchanger ΔT^{min} (advice: maximize the number of zones in critical areas of the heat exchanger to clearly identify the pinch points) - Check the cooling capacity $\dot{Q}_0$
Step 4:	Adjust: - Increase T_2 if calculated ΔT^{min} is lower than required $\Delta T^{min} = 1\,\text{K}$ (or decrease if it is higher) - Increase flowrate if calculated cooling capacity is lower than required $\dot{Q}_0 = 10\,\text{kW}$ (or decrease if it is higher) - Go to step 2

The process simulation was performed by means of process simulator UNISIM. The results are shown in Table 7.52. The specific exergy as well as absolute exergy are calculated by using Eq. (4.10); the reference state is referred to as $p = 1.013$ bar and $T = 300$ K. Figure 7.68 shows the process in T,h-diagram.

7.5.4.5 Process Units

Some data presented in this section (mainly related to the heat exchanger) are generated by the process simulator and unfortunately there is no easier way to derive them.

7.5.4.5.1 Heat Exchanger

Table 7.53 summarizes the process data for the heat exchanger.

Table 7.52 Output process data, mixed-fluid Joule–Thomson process, $T_o = 100$ K.

	Temperature (K)	Pressure (bar)	Vapor fraction (mol/mol)	Specific enthalpy (kJ/kg)	Specific entropy kJ/(kg K)	Flowrate (g/s)	Flowrate (mol/s)	Specific exergy (kJ/kg)	Exergy (kW)
1	300.00	20	0.972	1076.3	4.5582	578.34	17.58	218.89	126.59
2	104.74	19.9	0	444.9	1.3404	578.34	17.58	552.84	319.73
3	97.95	5.52	0.107	444.9	1.3678	578.34	17.58	544.61	314.97
4	98.00	5.42	0.225	462.2	1.5452	578.34	17.58	508.68	294.19
5	290.36	5.32	1.000	1093.6	4.9309	578.34	17.58	124.39	71.94
Reference state	300.00	1.01	1	1114.4	5.4147	578.34	17.58	0.00	0.00

Table 7.53 Heat exchanger data, mixed-fluid Joule–Thomson process, $T_o = 100$ K.

The heat transferred in the heat exchanger (heat exchanger duty)	$q^{HEX} = h_5 - h_4$ $\dot{Q}^{HEX} = \dot{m} \cdot q_{HEX}$	$q_{HEX} = 1093.6 - 462.2 = 631.4$ kJ/kg 0.578 kg/s $\cdot$ 631.4 kJ/kg $= 364.9$ kW
Temperature difference at the warm end	$\Delta T^{WE} = T_1 - T_5$	$300 - 290.36 = 9.64$ K
Temperature difference at the cold end	$\Delta T^{CE} = T_2 - T_4$	$104.74 - 98 = 6.74$ K
Max temperature difference	ΔT^{max}	28.61 K
Min temperature difference	ΔT^{min}	1 K
Mean temperature difference MTD	MTD	6.88 K
Q/MTD	$\dot{Q}^{HEX}/MTD$	364.9 kW/6.88 K = 53.04 kW/K

Figure 7.71 shows the temperature profile in the heat exchanger. Figure 7.72 shows the temperature difference profile in the heat exchanger. The pinch point locates in the lower 1/6th of the heat exchanger.

7.5.4.5.2 Expansion in Expansion Valve (is specified in Table 7.54)

Table 7.54 Joule–Thomson expansion.

Operating pressures (inlet/outlet)	19.9 bar/5.52 bar
Vapor fraction at the outlet	10.7%

Table 7.55 Compressor specification, mixed Joule–Thomson-process, $T_o = 100$ K.

Required compressor power	P_{COMP}	91.1 kW
Pressure at the compressor inlet	p_5	5.32 bar
Volumetric flowrate at the compressor inlet	V	278.7 m³/h
Pressure ratio	$\pi_{COMP} = p_1/p_5$	20/5.32 = 3.76
Potential compressor design: number of compressor stages × pressure ratio in a single stage	—	1 × 3.76 or 2 × 1.94

7.5.4.5.3 Compressor

The compressor specification is given in Table 7.55. The required compressor power can be calculated using the Eq. (5.31), which is valid for all gases without limitation:

$$P_{COMP} = \frac{P_{rev}}{\eta_T} = \frac{(\dot{E}_1 - \dot{E}_5)}{\eta_T}$$

According to this equation, compressor power amounts to 91.1 kW:

$$P_{COMP} = \frac{(126.59 - 71.94)}{60\%} = 91.1\ kW$$

7.5.4.5.4 Evaporator

The *T-Q* diagram for the evaporator is shown in Figure 7.75. The mixed fluid boils in the evaporator at a nearly constant temperature between 97.9 and 98 K. It is because of special behavior related to two liquid phases, described in Section 7.5.2.3, where the liquid phase 2 consists of mainly pure nitrogen and therefore boils at nearly constant temperature. That is not always the case. For other families of mixed fluids, this temperature glide may be more pronounced.

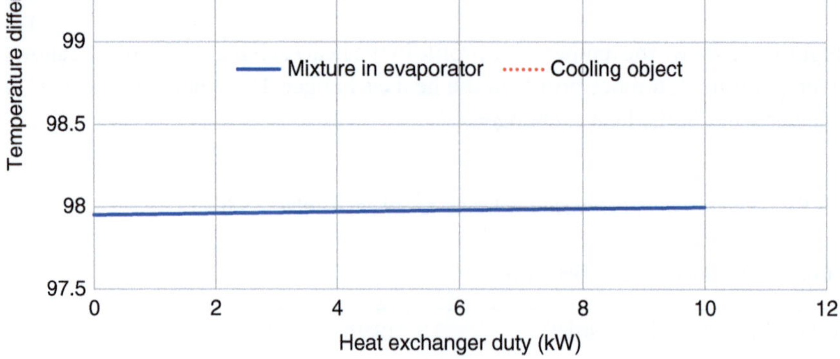

Figure 7.75 Temperatures in the evaporator, mixed Joule–Thomson process, $T_o = 100$ K.

7.5.4.6 Overall Performance of the Designed Process

Table 7.56 summarizes some numbers for overall performance of the designed mixed-fluid Joule–Thomson process.

With an efficiency of 22%, the mixed-fluid Joule–Thomson process is remarkably more efficient than a comparable nitrogen Joule–Thomson process. Thus, the thermodynamic losses in this process are small, and the drive power of the compressor is better converted into cold.

Table 7.56 Performance of the designed mixed-fluid Joule–Thomson process, T_o =100 K.

Cooling capacity	$\dot{Q}_o = 10\,\text{kW}$
Compressor power	$P_{COMP} = 91.1\,\text{kW}$
Waste heat	$\dot{Q}_{amb} = \dot{Q}_o + P = 10\,\text{kW} + 91.1\,\text{kW} = 101.1\,\text{kW}$
COP	$\varepsilon = \dot{Q}_o/P = 10/91.1 = 10.98\%$
Min required power for a reversible refrigerator (acc. to Carnot equation)	$P_{rev} = \dot{Q}_o\,(T_{amb} - T_o)/T_o = 10\,\text{kW}$ $(300\,\text{K} - 100\,\text{K})/100\,\text{K} = 20\,\text{kW}$
COP of a reversible refrigerator	$\varepsilon_{CARNOT} = \dot{Q}_o/P_{rev} = 10\,\text{kW}/20\,\text{kW} = 50\%$
Efficiency	$\eta_e = \varepsilon/\varepsilon_{CARNOT} = 10.98\%/50\% = 0.2195 \approx 22\%$

7.5.4.7 Process Analysis

This section consists of the calculation of exergy loss in all process units as well as an analysis of these losses, whereby the goal is to understand the cause and effect of the losses on process performance as well as how these losses can be reduced.

The procedure for calculation of exergy losses in a process unit is described in Section 4.3.3.3. The results are shown in Table 7.57 and Figure 7.76.

Table 7.57 Exergy losses in process units of mixed-fluid Joule–Thomson-process, $T_o = 100\,\text{K}$.

Loss related to	Exergy balance	Mixed Joule–Thomson process, 100 K	Nitrogen Joule–Thomson process, 100 K (for comparison)
Compressor $\dot{D}^{COMP}$	$\dot{E}_5 + P = \dot{E}_1 + \dot{D}^{COMP}$ $\rightarrow \dot{D}^{COMP} = P - \dot{E}_1 + \dot{E}_5$	$\dot{D}^{COMP} = 36.4\,\text{kW}$	100.8 kW
Heat exchanger $\dot{D}^{HEX}$	$\dot{E}_1 + \dot{E}_4 = \dot{E}_2 + \dot{E}_5 + \dot{D}^{HEX}$ $\rightarrow \dot{D}^{HEX} = \dot{E}_1 + \dot{E}_4 - \dot{E}_2 - \dot{E}_5$	$\dot{D}^{HEX} = 29.1\,\text{kW}$	40.9 kW
Joule–Thomson-valve $\dot{D}^{JT}$	$\dot{E}_2 = \dot{E}_3 + \dot{D}^{JT} \rightarrow \dot{D}^{JT} = \dot{E}_2 - \dot{E}_3$	$\dot{D}^{JT} = 4.8\,\text{kW}$	89.5 kW
Evaporator $\dot{D}^{EVAP}$	$\dot{E}_3 = \dot{E}_4 + \dot{E}\,(\dot{Q}_o) + \dot{D}^{EVAP}$ $\rightarrow \dot{D}^{EVAP} = \dot{E}_3 - \dot{E}_4 - \dot{E}\,(\dot{Q}_o)$	$\dot{D}^{EVAP} = 0.8\,\text{kW}$	0.8 kW
Sum	—	70.7	232 kW

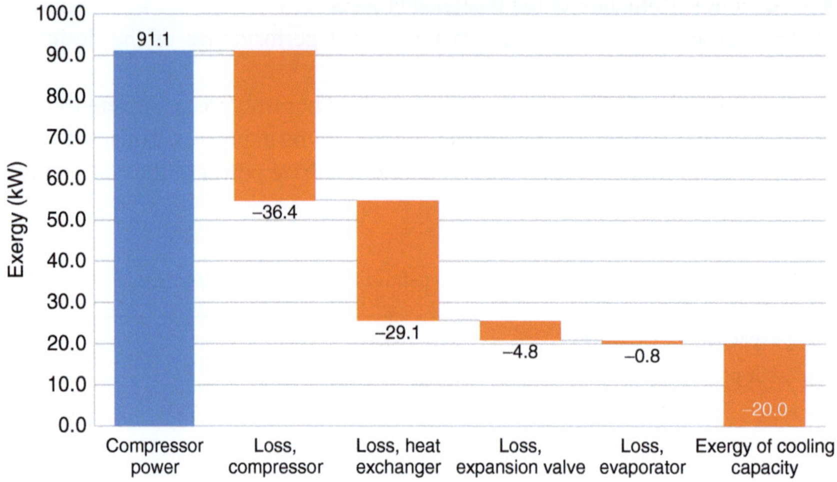

Figure 7.76 Distribution of losses, mixed-fluid Joule–Thomson process, $T_o = 100$ K.

7.5.4.7.1 Loss in Heat Exchanger

In contrast to the nitrogen-based process

- the heat transferred in the mixed-fluid heat exchanger $\dot{Q}^{HEX}$ is considerably larger: 365 kW versus 138.4 kW;
- The mean temperature difference is considerably smaller MTD = 6.88 K versus 11.9 K;

Therefore, it is possible that the mixed-fluid heat exchanger requires a larger heat transfer surface.

Unfortunately, a direct comparison of heat exchangers (volumes and surfaces) based on the numbers given above is not possible given that:

- The heat exchanger in the nitrogen process is a gas–gas heat exchanger; consequently, the heat transfer coefficient realized there remains in the order of magnitude of 100 W/(m^2 K), whereas the heat transfer coefficient in the mixed-fluid Joule–Thomson process is in the order of magnitude of 1000 W/(m^2 K), since it is driven mainly by condensation and evaporation processes;
- Different kinds of heat exchangers are used in process calculations: plate-fin heat exchangers for mixed-fluid Joule–Thomson processes versus spiral-wound heat exchangers for nitrogen-based processes.

Although the heat transferred in the mixed fluid heat exchanger is by a factor of roughly 2.5 larger than in the nitrogen-based process, the heat exchanger loss in the mixed process is considerably lower: 29.1 kW versus 41 kW. Section 5.3.5 explains the approach used in analyzing the loss in a heat exchanger. According to the Section 5.3.5, the loss in the heat exchanger depends on the temperature difference and temperature level. Table 7.58 shows this principle.

Table 7.58 Impact of temperature difference in heat exchanger on exergy loss.

	Small temperature difference, less than 0.02 T	Medium temperature difference, (0.02–0.1) T	Large temperature difference, more than 0.1 T	Very large temperature difference
$T = 100$ K	<2 K Small loss	2–10 K moderate loss	>10 K large loss	>20 K very large loss
$T = 200$ K	<4 K Small loss	4–20 K moderate loss	>20 K large loss	>50 K very large loss

According to this approach, the temperature profile in a mixed Joule–Thomson process is favorable in terms of loss:

- at temperatures below 150 K, the temperature difference is small, meaning the loss in this part of the heat exchanger is minimal,
- in the middle the temperature difference is moderate, and therefore the loss there is moderate,
- it is only in the upper part (250–300 K) where the temperature difference increases up to 25 K, but since this temperature level is close to the ambient temperature, the loss is not that great.

Ultimately, this favorable temperature profile leads to low heat exchanger loss.

7.5.4.7.2 Loss in Compressor

The largest loss is associated with the compression of the mixed fluid. This loss is relatively high since a machine with a relatively low efficiency of 60% is used. If used, a more efficient compressor would translate to a more efficient process.

In terms of absolute values (36.4 kW) this loss is considerably lower than the corresponding loss in the nitrogen-based process (101 kW), although the compressor efficiencies for both these processes are equal.[21]

The process parameters indicate that this compressor may be realized as a cost-efficient device:

1. the compressor may be designed as a single-stage machine;
2. the volumetric flowrate at the compressor inlet is approximately 280 m^3/h; it is roughly 15% less in comparison to the volumetric flowrate of the comparable nitrogen Joule–Thomson process (320 m^3/h).
   Therefore, the compressor itself as well as piping work are more compact;
3. the required driving power is low, therefore the whole electric part (motor, power supply device/transformator, etc.) is designed for lower power – meaning there is a clear cost advantage;

21 Problems to solve: find an explanation for it.

4. the amount of the waste heat to be rejected into the environment by means of cooling fluid (air or water) is low; therefore, the whole cooling part (aftercooler, driver, piping, etc.) is more compact and less costly.

7.5.4.7.3 Loss in Expansion Valve

The loss associated with the expansion process is reduced to 4.8 kW. It is a remarkable improvement in comparison to the nitrogen-based process, where the corresponding loss is roughly 90 kW. The background is that:

- the liquid in front of the Joule–Thomson valve is cooled to a very cold temperature, with the difference between this temperature and the cooling temperature amounting to $T_2 - T_o = 104.74\,\text{K} - 100\,\text{K} = 4.74\,\text{K}$. At the end, the expansion process generates a small temperature drop of 6.74 K only;
- the entropy increase is minor: $s_3 - s_2 = 1.37 - 1.34 = 0.03\,\text{kJ/(kg K)}$, meaning that the irreversibility is substantially reduced.

The Learning 7.5 is perfectly implemented in the mixed-fluid Joule–Thomson process.

7.5.4.7.4 Loss in Evaporator

In comparison to other hardware components, the loss in the evaporator is only minor due to a small temperature difference in the evaporator (which is close to 2 K).

7.5.4.7.5 Explanation for High Efficiency of the Mixed-Fluid Joule–Thomson Process

With efficiency of 22%, the mixed-fluid Joule–Thomson process is roughly three times better than the comparable nitrogen Joule–Thomson process. The following causal chain helps to explain this fact:

1. The thermodynamic properties of mixed refrigerants are advantageous. The first advantage relates to large enthalpy difference at the warm end, which directs to a relatively high specific cooling capacity even at moderate pressures. Therefore, the process does not require compression to high-pressure level. It may be realized at subcritical pressures;
2. The second advantage relates to a favorable shape of mixed-fluid isobars in subcritical (two-phase) area. It is a reason for an optimal temperature profile in the heat exchanger with low-temperature difference in the middle and at the cold end and moderate temperature difference in the upper fourth of the heat exchanger. This temperature profile causes a moderate loss in the heat exchanger only;
3. Due to small temperature difference at the cold end the expansion process generates minor exergy loss only;
4. The reduced loss in heat exchanger and in Joule–Thomson valve means that the overall process becomes less irreversible, which results in relatively high efficiency.

7.5.4.7.6 What Can be Further Improved

The temperature profile in the heat exchanger is favorable. However, there is still potential for further improvement:

1. Of course the temperature difference at the cold end $\Delta T^{CE} = 6.74\,\text{K}$ can be reduced to $\Delta T^{min} = 1\,\text{K}$, which would mean:
   - decrease of loss in the heat exchanger, as well as
   - decrease of loss associated with isenthalpic expansion because the entropy generation during the expansion process will be further reduced
2. Similarly, the temperature difference at the warm end $\Delta T^{WE} = 9.64\,\text{K}$ can be further reduced to ΔT^{min}. The temperature difference at the warm end is not so critical, causing only a minor loss. Nevertheless, it generates some loss, which can be potentially reduced.

Therefore, an optimization of mixed-fluid composition would be useful in said case. According to [8], the fully optimized mixture composition (in terms of exergy loss) is characterized by the three coexisting pinch points, usually located:

1. at the cold end: $\Delta T^{CE} = \Delta T^{min}$,
2. at the warm end: $\Delta T^{WE} = \Delta T^{min}$,
3. in the lower third of the heat exchanger, usually around or close to the split point (high-pressure side):

   $\Delta T^{MIDDLE} = \Delta T^{min}$,

   where the min temperature difference is synchronously achieved.

7.5.5 Design of the Mixed Fluid

The optimization of mixed-fluid composition is the easiest way to improve the process performance since it requires neither changes on the hardware components nor changes in process morphology. Therefore, the design of the mixture is an important part of the overall process design procedure.

The tasks pursued by process design engineer may differ, since this depends on numerous factors like availability of substances, availability of process design tools, availability of properties data, targeted level of mixture complexity (four components, five components, or even more), and so on and so forth.

7.5.5.1 Mixed-Fluid Composition

A typical task often pursued by engineers can be defined as follows:

- a set of substances is fixed (for example, nitrogen/methane/ethylene/propane/i-butane),
- the high-pressure level is fixed,
- the low-pressure level may be slightly varied to achieve the required cooling temperature,

Optimization of concentrations/fractions of all given components must be performed.

The optimization is an iterative procedure:

1. the process calculation/simulation is performed for an initial mixed-fluid composition,
2. the temperature profile in the heat exchanger is studied, the pinches/pinch areas are recognized and analyzed (location and impact on process performance),
3. the mixed-fluid composition (fractions of mixture components) is adjusted,
4. the process calculation is performed again, until the optimal composition (three coexisting pinches) is identified.

Clearly, the adjustment of the mixed-fluid composition is the key operation. The principle is:

(a) identify the tightest pinch and the corresponding temperature range,
(b) identify the substance(s) boiling/condensing intensively in this temperature area, (the corresponding procedure is described in Section 7.5.2.5 and illustrated by means of Figure 7.65),
(c) increase the fraction of this substance(s) in the mixed fluid targeting de-pinching.

The corresponding adjustment can be performed intuitively/manually or automatically by means of integrated algorithm, since this procedure can be easily mathematically formalized. Alexeev [8] describes an example of an appropriate formal procedure. In terms of mathematics, this task is relatively simple. It can be easily managed by modern powerful optimizers. The manual/intuitive adjustment procedure is often used, too.

7.5.5.1.1 An Example of a Typical Sequence for Adjustment of the Mixed-Fluid Composition

(a) **Initial Mixed-fluid Composition:** The initial mixed fluid consists of 0.3 mol of nitrogen, 0.23 mol of methane, 0.17 mol of ethylene, 0.1 mol of propane, and 0.2 mol of i-butane. This mixed fluid is referred to as "mix 1" in Table 7.59. The resulting mol fractions are given in Table 7.59, row 3. The corresponding process calculation was already performed in Section 7.5.4;

Table 7.59 Adjustment of mixed-fluid composition.

	Mix 1		Mix 2	
	mol	mol %	mol	mol %
Nitrogen	0.3	30%	0.3	28.6%
Methane	0.23	23%	0.28	26.7%
Ethylene	0.17	17%	0.17	16.2%
Propane	0.1	10%	0.1	9.5%
i-butane	0.2	20%	0.2	19.0%

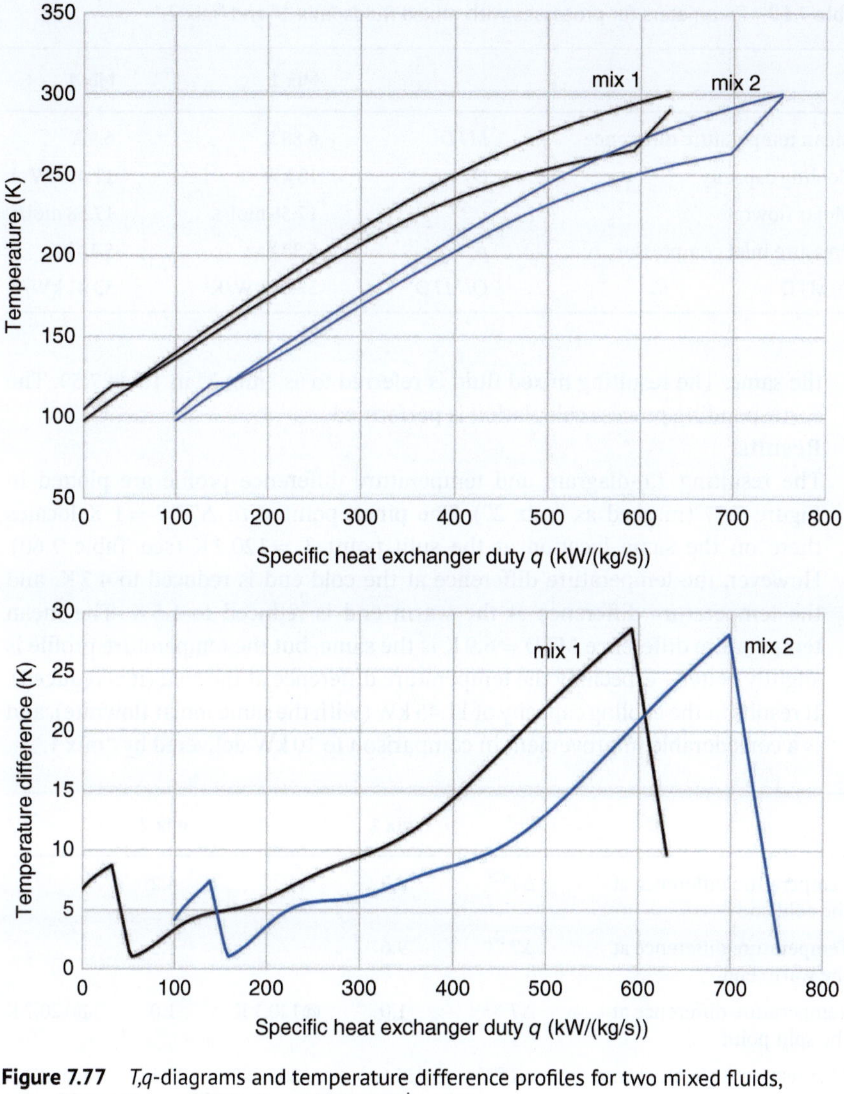

Figure 7.77 *T,q*-diagrams and temperature difference profiles for two mixed fluids, $p_{HP} = 20$ bar, $T_{amb} = 300$ K, $T_o = 100$ K, $\Delta T^{min} = 1$ K.

Results:

The *T,q*-diagram and temperature difference profile are plotted in Figure 7.77 (marked as "mix 1"). The temperature difference at the warm end is 9.6 K, and at the cold end is 6.74 K. The mean temperature difference in the heat exchanger is 6.88 K. The pinch point with $\Delta T^{min} = 1$ K locates there at the split point $T = 120.1$ K (see Table 7.60).

(b) **Adjustment of the Mixed-fluid Composition:** There are two substances, nitrogen, and methane, that are intensely boiling/condensing in this temperature range. The process design engineer increased the amount of the methane in the mixed fluid by 0.05 mol, and the amount of other components remained

Table 7.60 Parameters for processes with mixed fluids "mix 1" and "mix 2."

		Mix 1	Mix 1
Mean temperature difference	MTD	6.88 K	6.9 K
Cooling capacity	$\dot{Q}_0$	10 kW	11.45 kW
Molar flowrate		17.58 mol/s	17.58 mol/s
Pressure inlet compressor	p_5	5.32 bar	5.14 bar
Q/MTD	Q/MTD	53.07 kW/K	53.41 kW/K

the same. The resulting mixed fluid is referred to as "mix 2" in Table 7.59. The corresponding process calculation is performed.

Results:

The resulting T,q-diagram and temperature difference profile are plotted in Figure 7.77 (marked as "mix 2"). The pinch point with $\Delta T^{min} = 1$ K locates there on the same location at the split point $T = 120.7$ K (see Table 7.60). However, the temperature difference at the cold end is reduced to 4.2 K, and the temperature difference at the warm end is reduced to 1.5 K. The mean temperature difference MTD = 6.9 K is the same, but the temperature profile is slightly better – especially the temperature difference at the cold (it is reduced). It results in the cooling capacity of 11.45 kW (with the same molar flowrate), and is a considerable improvement in comparison to 10 kW delivered by "mix 1."

		mix 1		mix 2	
Temperature difference at the cold end	ΔT^{CE}	6.7		4.2	
Temperature difference at the warm end	ΔT^{WE}	9.6		1.5	
Temperature difference at the split point	ΔT^{SPLIT}	1.0	@120.1 K	1.0	@120.7 K
Max temperature difference	ΔT^{max}	28.6	@265.3 K	28.1	@263.1 K

From the example given, it is important to note the following:

Learning 7.14

1. The temperature profile in the heat exchanger is sensitive to the mixed-fluid composition; on one hand, this is advantageous; the performance of the process may be significantly improved by smooth tuning of mixed-fluid composition; on the other hand, this sensitivity is disadvantageous, since the performance of the process may decrease if the actual mixed-fluid composition deviates from the optimal one.
2. The influence of mixed-fluid composition on the process efficiency is complex, and a detailed analysis is required to find the optimum.

7.5.5.2 Ballast Gas Addition

An addition of a non-condensable gas (usually helium or neon) to the mixed fluid is used for reducing cooling temperature without reduction of the low-pressure level. Alternatively, the low-pressure level may be increased while cooling temperature remains the same. This trick may also be applied to the conventional nitrogen Joule–Thomson process.

The thermodynamic background is as follows: the non-condensable gas always remains in the vapor phase, where it reduces the partial pressure of all other mixed-fluid components. In the evaporator, it reduces the partial pressure of the liquid-phase components that are being evaporated there. Consequently, the liquid boils at a lower temperature.

7.5.5.2.1 Use Cases

- The addition of a non-condensable gas helps avoid vacuum conditions in the process if a cooling temperature of below 80 K is required;
- Another useful effect relates to the compressor sizing: since the higher pressure on the compressor inlet means a higher mass flowrate at the same volumetric flowrate, it is possible to create a refrigerator for higher cooling capacity (at the same cooling temperature) based on the same compressor by increasing the low pressure level in the process (and addition of ballast gas). It does not mean, however, that the efficiency of the cycle remains the same. Usually, the efficiency will decline, since the non-condensable gas does not generate any cooling capacity, but rather requires some power for compression in the compressor.[22] However, this trick helps optimize the cost of the compressor (and therefore of refrigerator).

7.5.5.2.2 Fluid Phase Behavior

The fluid system with non-condensable gas itself is an interesting object for investigation. This mixed fluid has no clear boiling line/boiling point. The presence of the non-condensable gas means that the liquid phase(s) is (are) always coexisting with a gaseous phase, even at very low temperatures. At higher pressures, the fluid system becomes supercritical (one phase) without a visible phase change.

7.5.5.2.3 An Example for Ballast Gas Addition

The example considered in this section is based on the same mixed fluid "mix 1" used in all previous examples. The helium-containing mixed fluid is referred to as "mix 3"; see Table 7.61. This mixed fluid is produced from the mixed fluid "mix 1" by adding 0.05 mol of helium.

Figure 7.78 shows the corresponding processes in temperature–enthalpy diagram; the pressure on the low-pressure side was maintained at 1.2 bar for both processes. In the process with "mix 1," the mixed fluid boils in evaporator at a constant temperature of approximately 80 K; in the process with helium-contained mixed fluid, the boiling temperature varies from 76 to 78 K. It means the cooling temperature in the process is reduced by 2 K.

---

22 Therefore, it is referred to as "ballast gas."

Table 7.61 Mixed-fluid compositions, mix 3.

	Nitrogen	Methane	Ethylene	Propane	i-butane	Helium
mol	0.3	0.23	0.17	0.1	0.2	0.05
mol (%)	27.27	25.45	15.45	9.09	18.18	4.55

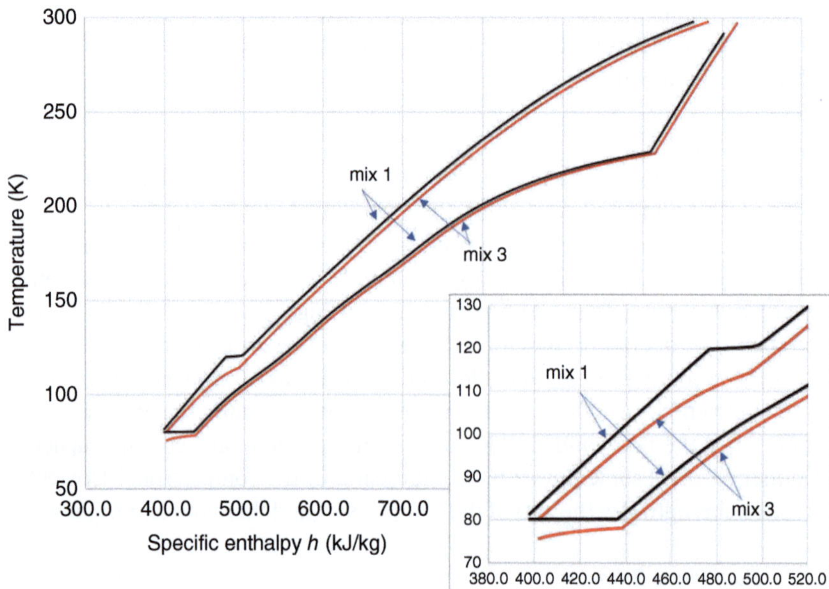

Figure 7.78 Process in T,h-diagram for two mixed fluids, $p_{HP} = 20$ bar, $p_{LP} = 1.2$ bar, $T_{amb} = 298$ K, $\Delta T^{min} = 1$ K.

Other observations are as follows:

- There are some differences in the shape of the high-pressure isobar in VLLE area: the mix 1- isobar runs horizontally there, whereas the course of mix 3 – isobar below the split point is declining;
- The temperature difference profiles are shown in Figure 7.79: in the upper part the profiles are quite similar; at the warm end as well as at the cold end, the mix 3 – profile is a bit better because of only a small temperature difference.

7.5.5.3 Solidification

One important requirement for failure-free operation of a refrigerator based on mixed-fluid Joule–Thomson process is that the applied mixed fluid does not form a solid phase. The solid particles can cause clogging, which disturbs the normal circulation of the fluid in the refrigerator. In this sense, the most critical locations are the cold end of the heat exchanger, expansion valve, and the evaporator, since the lowest temperature is achieved there.

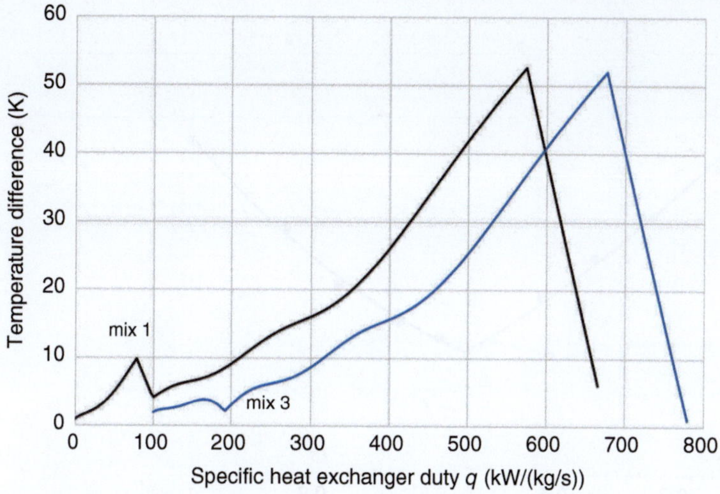

Figure 7.79 Temperature difference profiles for two mixed fluids, $p_{HP} = 20$ bar, $p_{LP} = 1.2$ bar, $T_{amb} = 298$ K, $\Delta T^{min} = 1$ K.

The solidification of main pure substances used for mixed-fluid design occurs at temperatures 85–100 K (see Table 7.45). But, surprisingly, the mixed-fluid Joule–Thomson refrigerators can operate successfully down to temperatures of 70 K [10]. This behavior can be explained (i) partially by solid–liquid equilibria in multicomponent fluids, and (ii) partially by phenomena related to the so-called metastability.[23]

7.5.5.3.1 Solid–Liquid Equilibria

The solidification temperature of hydrocarbon-based mixtures (liquidus temperature) may be lower than the solidification temperature of pure hydrocarbons. Figure 7.80 illustrates this behavior for methane–propane mixture. This mixture forms a eutectic at the propane concentration of ≈0.55 mol/mol, the solidification temperature is reduced to lower than 70 K. This temperature is considerably lower than the solidification temperature of pure propane (85.2 K) and pure methane (90.7 K). Other binary hydrocarbon mixtures behave in a similar way.

Mogorychny [9] showed that the addition of a third mixture component led to a further reduction of the liquidus temperature as a rule. Potentially, the ternary mixtures may remain liquid at even lower temperatures.

An accurate prediction of solidification temperature (liquidus temperature) is a complicated issue. The conventional sequence for description of solid–liquid equilibria consists of two steps:

1. the first, the kind of equilibria must be identified: (i) number of coexisting phases as well as (ii) internal behaviors between the phases (soluble/insoluble, etc.);
2. second, the corresponding phase transition lines are calculated and fitted to available experimental data by means of appropriate tuning coefficients.

23 https://en.wikipedia.org/wiki/Metastability.

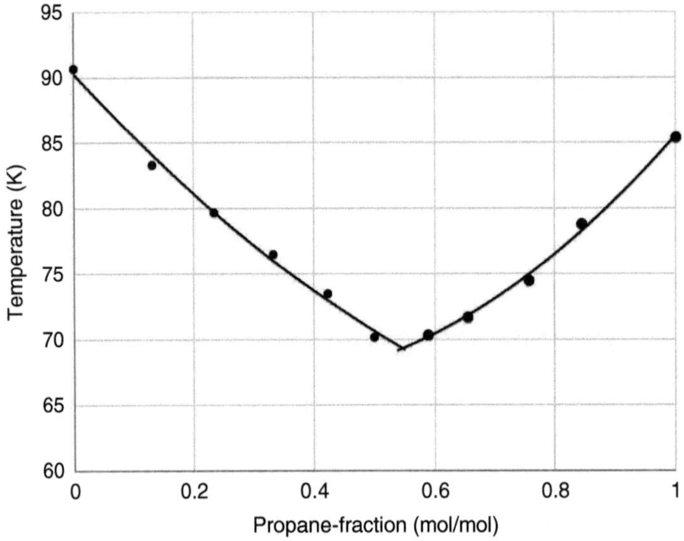

Figure 7.80 Liquidus-line (solidification) for methane–propane mixture according to Ref. [12].

The thermodynamic theory for description of solid–liquid equilibria for binary mixtures at low temperatures works well if it is combined with some experimental data. But the extension of this theory to multi-component mixtures is a complicated issue, as it struggles even with ternary mixtures:

- The prediction of the number of coexisting phases is challenging since the number of potential options increases;
- Even if the number of phases is known, the phase behavior (individual substances can be soluble, insoluble, or partially soluble in this mixture) is often unclear;
- There are some rules available, but they are not universally applicable, and every mixture must be characterized individually (the best by means of an experiment).

7.5.5.3.2 Metastability

It is unclear what kind of solid phase occurs at these temperatures. The classic thermodynamic model considers the solid phase as a kind of crystal-like structure, whereas the existing operational experiences indicate that it is rather a paste-like highly viscous substance.

Potentially, this behavior can be explained by the so-called metastability phenomenon. According to this phenomenon, the (supercooled) liquid mixtures may remain liquid even at temperatures below the solidification point until some disturbance like applied mechanical vibrations or something working as condensing seed doping initiates crystallization centers, whereby a solidification will suddenly occur. The subcooling below the solidification temperature may reach 10–20 K in mixtures of propane–propylene, methane–propane, and *n*-butane–*i*-butane as reported in [9].

It is still unclear if this is a “real” metastability (meaning a stable state), which may exist for months, years, decades, or longer, for the fluid system located in a local minimum in terms of potential energy. Or it may only be a kind of "temporary" state that occurs due to extremely slow dynamics of phase transition at low temperatures in highly viscous liquid phase. In this last case, the equilibrium (solid phase) will be achieved at a later point.

7.5.5.3.3 Summary

In terms of practical applications, all phenomena related to metastability are useful since they help to avoid solidification and related problems in the operation of mixed-fluid Joule–Thomson refrigerators.

There are a number of open issues concerning the solidification of multicomponent mixtures at low temperatures: an accurate and secure method for prediction of equilibria is still not available as well as the model for metastability and/or for thermal hysteresis. However, it is positive that new approaches based on molecular modeling[24] introduced in the fields of computational chemistry, drug design, etc. may be applied to cryogenic substances (which are relatively simple molecules) and used for prediction of liquid–solid equilibria in the future.

7.5.6 Optimization of a Mixed-Fluid Joule–Thomson Process

The overall process optimization procedure is similar to the procedure described in Section 7.2.4 for the nitrogen Joule–Thomson process. However, in contrast to the nitrogen process with three relevant degrees of freedom, the mixed-fluid-based process offers considerably more variables. They are:

(1) min temperature difference in the heat exchanger ΔT^{min} which is a substitute for temperature difference at the warm end ΔT^{WE};
(2) pressure downstream of the compressor (high-pressure level) p_{HP};
(3) pressure upstream of the compressor (low-pressure level) p_{LP}; which is a real variable if a ballast gas addition is practiced;
(4) molar fraction of ballast gas,
(5) molar fractions of all mixture components, usually more than four,

as well as a couple of variables that have minor impacts, like pressure drops in hardware components or the temperature differences possible in the evaporator and similar.

Therefore, the whole process optimization procedure – which **inevitably** includes the optimization of the mixed-fluid composition – requires more effort and investment in the development of appropriate (software) tools as well as in the necessary education of process design engineers.

7.5.7 Control of a Mixed-Fluid Joule–Thomson Refrigerator

All the approaches developed for controlling nitrogen Joule–Thomson process (described in Section 7.2.5) are also applicable to the mixed-fluid Joule–Thomson process.

---

24 https://en.wikipedia.org/wiki/Molecular_modelling.

7.5.7.1 A New Option

Theoretically, it is also possible to control the system by controlling the mixed-fluid composition. However, this option requires a complex instrumentation. For example:

- the mixed-fluid composition must be monitored by means of sophisticated and expensive devices such as gas chromatographs, mass-spectrometers, or similar devices;
- a system for removal (in case of excess) or for injection (in case of lacking) of a mixed-fluid component or even several subsystems of this kind is (are) necessary.

From an economic standpoint, this option may make sense in large-scale applications only, where even small efficiency improvements (and corresponding power savings) achieved by the tuning of mixed-fluid composition in real time can justify the investment into the corresponding instrumentation.

For middle-scale systems, a simplified approach is considered. It includes a regular check of the mixed-fluid composition (every three or six months depending on the application) by taking and analyzing mixed-fluid samples in an external laboratory. The mixed fluid is replaced completely if a critical deviation from the original composition is recognized.

For small-scale applications, this procedure is even simpler – the mixed fluid is replaced completely if the system is not able to deliver the specified cooling capacity.

7.5.8 Hardware Components

Although there exist many differences in thermodynamics, there are not any differences in the morphology between the mixed-fluid Joule–Thomson and the classical nitrogen Joule–Thomson processes. Both these processes consist of the same five main hardware components such as compressor, heat exchanger, expansion valve, evaporator, and thermal insulation.

Question:
It is assumed that a laboratory has an existing nitrogen Joule–Thomson refrigerator. The laboratory would like to improve performance of this system. It is known that the efficiency can be improved by using a mixed fluid instead of nitrogen. A mixed fluid (composition, see Table 7.46) was purchased, the system is filled with this refrigerant, and is ready to begin. Will this approach work successfully?

Answer:
The answer is: "rather will not" since the hardware components originally developed for the nitrogen-based process are not ideally suited to an operation with a mixed fluid. They are very similar, but are either oversized or undersized:

- Heat exchanger:
  First, the amount of heat flow to be transferred in the heat exchanger in a mixed-fluid Joule–Thomson process is considerably larger (by a factor of 2–2.5); the mean temperature difference is considerably smaller (by a factor of 2–3) in comparison to the nitrogen-based process; Second, the density of mixed fluid in the lower part of the heat exchanger (in terms of temperature) is higher since a

fraction of the fluid is liquefied there, meaning the mixed-fluid Joule–Thomson process requires a lower hydraulic diameter of the channels; In summary, the heat exchanger required for mixed-fluid Joule–Thomson process is rather slim and long, whereas the heat exchanger for nitrogen process is rather thick and short; Therefore, the heat exchanger dimensioned for nitrogen is probably unable to perform in the mixed-fluid-based process.

- Expansion valve:
  The nitrogen is gaseous and relatively warm in front of the Joule–Thomson valve and the mixed fluid is liquid and cold (high density), therefore the volumetric flowrate is considerably smaller. Downstream of the valve, the difference in the volumetric flowrate is even larger since the flash-gas flowrate in the nitrogen process is considerably larger than the flash-gas flowrate in the mixed-fluid Joule–Thomson process. The valve designed for the nitrogen process is too large for the mixed-fluid Joule–Thomson process. Although not the worst case, it will probably be challenging to control the valve since it is near to "close" position. However, it may still work.
- Evaporator:
  At the same flowrate, the mixed-fluid Joule–Thomson process generates a higher cooling capacity (by a factor of 3–5), and the evaporator designed for nitrogen processes is undersized for mixed-fluid operation;
- Compressor:
  - the nitrogen compressor compressed to 100 bar (or higher), for mixed-fluid Joule–Thomson process a subcritical pressure (<50 bar) would be sufficient, therefore the compressor is over-engineered for the given task;
  - a potential problem relates to the fact that the mixed fluid forms a liquid phase at pressures higher than 15 bar and temperatures close to the ambient temperature. These conditions occur at the outlet of the intermediate cooler, respectively at inlet of an intermediate compressor stage. The liquid at the compressor stage inlet means a potential damage risk: the droplets will hit the suction valves[25] and damage them. If the liquid does get through the valves in the liquid form, it may damage the cylinder and piston (may be identified by mechanical vibration) since it is not compressible fluid.

7.5.8.1 Compressor for Mixed-Fluid Refrigerator

The mixed-fluid Joule–Thomson process does not require very high pressure. An efficient process can be designed using a moderate pressure level of less than 30 bar. The portfolio of compressors available for this pressure range is wide; in principle, every appropriate compressor can be used. The decision depends more on the capacity range, as shown in the Figure 7.81.

- Centrifugal compressors (see Section 6.5.2) are used for large-scale mixed-fluid Joule–Thomson systems,
- Screw compressors (see Section 6.5.3) are used for medium-size plants,
- Hermetic compressors are the basis for small-scale systems.

25 If a piston compressor used.

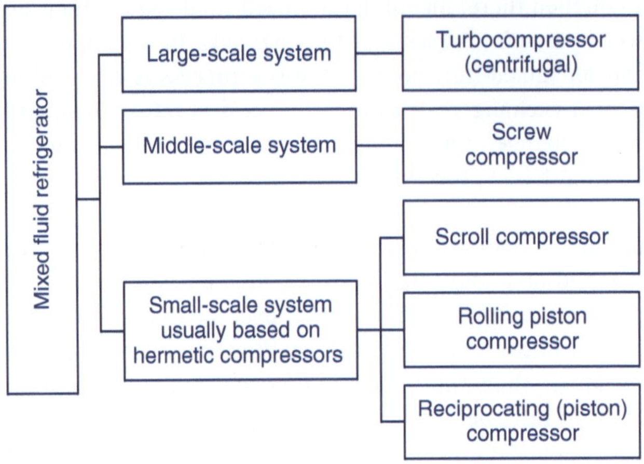

Figure 7.81 Overview, compressor for mixed-fluid Joule–Thomson process.

7.5.8.1.1 Number of Stages and Efficiency

- **Multiple stages:** The centrifugal compressor is often a multi-stage machine and therefore a relatively cost-intensive hardware component. The advantage is that an optimal pressure level may be realized by means of these compressors, resulting in high efficiency of the overall mixed-fluid Joule–Thomson process.
- **Single stage:** most of the compressors available on the market are single-stage machines, which offer many advantages as described below, but also one disadvantage as well: the pressure ratio they achieve is relatively small, so the efficiency of the resulting mixed-fluid Joule–Thomson process is reduced in comparison to processes based on multi-stage compressors.
- The pressure level and pressure ratios required for the mixed-fluid Joule–Thomson process can be covered well by a two-stage compressor. But the low-cost two-stage compressors are not really available on the market, they are still the exception rather than the rule;

7.5.8.1.2 Compressors for Conventional Refrigeration and/or Air Conditioning

The single stage compressors used in conventional refrigeration and air conditioning are very attractive hardware components since these units are produced in large quantities worldwide. They are robust and extremely reliable machines, and the price level is very economical due to high competition in this market segment. These compressors are the machines of choice when a low-cost, highly reliable system is needed and high efficiency is not the primary goal.

There are three options when it comes to mechanical design, in order of size from the largest to smallest:

1. Scroll compressors;
2. Rolling piston compressors;
3. Reciprocating machines;

Hermetic Compressors: All of these compressors are what are known as hermetic or hermetically sealed compressors (for further information, see Ref. [11]). With these units, the compressor and the motor driving it are integrated and operate within the pressurized gas envelope of the system. The motor is designed to be cooled by operating while surrounded by the refrigerant gas being compressed. It is not possible to repair a one-piece welded steel casing; if the hermetic fails, it is simply replaced with an entirely new unit. The primary advantage of a hermetic is that there is no route for the gas to leak out of the system.

7.5.8.1.3 Lubrication

The compressors developed for application in conventional refrigeration and air conditioning are commonly oil-lubricated machines. Therefore, a corresponding oil removal system is required after the compressor.

Centrifugal compressors do not use lubrication in the main process passage. They are considered a "dry" machines, so they require no additional process units such as oil removal systems.

7.5.8.2 Heat Exchanger for Mixed-Fluid Joule–Thomson System

There are no heat exchangers designed and developed specifically for mixed-fluid Joule–Thomson refrigerators. The conventional approach is to look at the heat exchangers available on the market, as they may meet the requirements.

In terms of mechanical design, three following options are often considered:

- The brazed aluminum plate-fine exchanger (see Section 6.2.1) is applicable for mixed-fluid Joule–Thomson processes, thanks to the moderate pressure level and appropriate temperature difference profile. This design is very compact; therefore, it is possible to realize an efficient process with low-temperature difference;
- The spiral-wound heat exchanger (see Section 6.2.2) is also considered a suitable option;
- The multitube heat exchanger can be used for small-scale mixed-fluid Joule–Thomson refrigerators, although today this design looks more like a relict;

7.5.8.2.1 Thermal Design

The design of the heat exchanger for a mixed-fluid Joule–Thomson process is one of the most complex and challenging engineering tasks since it is characterized by

(a) multiphase flow;
(b) multi-component mixture;
(c) wide temperature (and liquid fraction) range.

The process design is extremely challenging in terms of fluid mechanics and heat transfer. The performance of the heat exchanger can be affected by a range of phenomena, including flow maldistribution, phase slip, and even backflow, de-mixing, or fluid phases, et cetera – the most salient are shown in Table 7.62.

Table 7.62 Phenomena in the heat exchanger for mixed-fluid Joule–Thomson process.

Phenomena	Description
Multiple flow patterns	The liquid fraction changes from 0% at the warm end to 100% at the cold end. Therefore, the fluid mechanics are extremely complex: the multiphase flow can demonstrate all possible flow patterns (bubbly, slug, churn, annular, or droplet flow) in the same heat exchanger;
Flow maldistribution	This phenomenon relates to a non-uniform distribution of fluid among the channels of a heat exchanger. It results in a reduced effective temperature difference (the heat exchanger requires more heat transfer surface) and higher pressure drop. Even the homogeneous distribution of the single-phase fluid is a complex task. The distribution of a two-phase flow is a very challenging issue, discussed separately in Section 8.5.5 (where approaches how to manage the distribution of two-phase flow are considered). The maldistribution issue relates to all kinds of heat exchangers; specifically, the plate-fin heat exchanger's performance may be considerably affected due to the mechanical design based on multiple parallel channels on the high-pressure side as well as on the low-pressure side. The spiral-wound heat exchanger has one space (one passage) on the low-pressure side only, making the maldistribution on this side less challenging.
Phase slip (phase shift)	This effect occurs in the heat exchanger if the vapor phase and the liquid phase have different velocities due to an inappropriate geometry (diameter and form) of the channel. Usually, the liquid phase moves slower than the vapor phase. An extreme form of the phase slip is the so-called "liquid backflow," whereby the liquid moves in the opposite direction of the vapor. Such a situation may occur in vertical channels if there is a large amount of liquid and velocity momentum, and/or non-uniform distribution of the vapor phase is insufficient to carry up the liquid. The flow then becomes pulsating and unstable. The consequence of the phase slip is that the compositions of vapor and the liquid phases in a cross-section are no longer in equilibrium. This leads to parasitic evaporation/condensation processes between the coexisting phases (additionally, to the main heat transfer). It results in a reduced effective temperature difference (with the heat exchanger requiring additional heat transfer surface) and increased pressure drop. The worst-case scenario would be that the heat exchanger – and therefore the overall process – does not work at all and must be replaced.
Non-homogeneity	The process simulations assume that both (vapor and liquid phases) are homogeneously mixed in every cross-section and they have the same temperature and phase compositions corresponding to the equilibrium composition. But it is not always the case, as in the case of there being no phase slip. The consequences are similar to the consequences of the phase-slip effect.

Table 7.62 (Continued)

Phenomena	Description
De-mixing and accumulation of liquid phase	The process simulations assume that the mixed-fluid composition circulated in the heat exchanger corresponds to the composition of the mixture filled in. The mechanical design of the heat exchanger may include some details and elements (such as death ends and sampling lines) that cause a local separation and accumulation of the liquid phase. If an amount of liquid (which has a different composition in comparison to the overall mixture) is excluded from the main circulation loop, the composition of the remaining fluid differs from the optimal composition. A similar situation may occur if the heat exchanger is untightened, the leakage follows from one phase (liquid or vapor) only, and the composition of this phase differs from the overall mixed-fluid composition. In terms of solidification, it is potentially dangerous to de-mix at low temperatures: since the composition of the liquid phase in this case differs from the original composition, it may form a solid phase and cause operational problems.
Heat transfer in multicomponent condensation and evaporation	The heat transfer is dominated by the evaporation and condensation processes. It is generally assumed that the heat transfer coefficients during the evaporation/condensation of mixtures can be substantially smaller than those of the pure components of the mixture. The models developed for pure substances are not applicable to multicomponent mixtures. The heat transfer in multicomponent condensation and evaporation is a science in and of itself. Numerous studies and books around this topic were performed and later published, but the theoretical description has yet to be completed. The main challenge is that the heat transfer is linked to the mass transfer between the phases. The compositions of liquid and vapor phases as well as the thermodynamic properties of corresponding phase (heat conductivity, viscosity, heat capacity, surface tension, etc.) are changing during the heat transfer operation. A further challenge related to the mixed fluid Joule–Thomson system is that the liquid fraction – and correspondingly the flow pattern – varies widely in the heat exchanger and depends on the mixture's temperature. Since the flow mechanics has a clear impact on the heat transfer coefficient, a corresponding number of heat transfer models (describing every flow pattern) are required to describe the whole process.

7.5.9 Assessment of the Mixed-Fluid Joule–Thomson Process

The mixed-fluid Joule–Thomson process maintains all advantages of the nitrogen Joule–Thomson process: in terms of hardware it is simple and highly reliable since it does not consist of moving parts in the cold part. If the heat exchanger is properly designed and the compressor runs well, then this refrigerator safely generates the required cooling capacity.

But the mixed-fluid Joule–Thomson process overcomes the weak points of the nitrogen process, for it can be highly efficient at a low cost. Unfortunately, both these targets cannot be reached simultaneously; it is either the high-efficient process or the low-cost. Therefore, it is always a candidate for both kinds of applications:

(a) requiring low operational expenditures (OPEX);
(b) requiring low initial capital expenditures (CAPEX).

The lowest **cooling temperature** is approximately 70 K.

7.5.9.1 Challenging Issues

- In terms of process design, the overall process optimization is a complex task requiring special tools and knowledge about the mixed-fluid properties in the whole temperature range from solidification point to the dew point.
- The design of the heat exchanger is the next challenging task; see Section 7.5.8.2.
- Operationally, the management of the mixed-fluid composition in the system may be an issue.

7.5.9.2 Applications

- In a version adapted for liquefaction purposes the mixed-fluid Joule–Thomson process is used widely for natural gas liquefaction; see Section 8.5;
- The low-cost version of mixed-fluid Joule–Thomson process (based on small single-stage compressor) is commercialized by several companies; for example, by MMR Technologies as a part of a miniature nitrogen liquefier or as a small-scale refrigerator referred to as "Kleemenko-cryocooler."

7.6 Cryogenic Refrigeration Processes, Life Cycle

The product and innovation managers use the so-called S-curve for tracking products and innovations. This approach is based on the assumption that a technical system develops in a way that is analogous to a biological system: it goes through four stages comparable to infancy, growth, maturity, and decline. Applying this method for characterization of cryogenic processes, the following observation can be made:

7.6.1 Joule–Thomson Process

- The classic "stand-alone" Joule–Thomson process is at the end of its "decline phase." This process is not competitive anymore and is applied only in very special cases;

- As a part of the Claude process (Joule–Thomson stage) it is already in "maturation phase";
- However, the Joule–Thomson process experiences a kind of reincarnation in the form of the mixed-fluid Joule–Thomson process.

7.6.2 Brayton Process

- The Brayton process is still a key process in air separation (maturation phase);
- The Brayton process is part of the Claude process in the maturation phase, too;
- As a "stand-alone" system based on novel centrifugal compressor with magnetic bearings and permanent magnet motors, this process experiences the growth phase.

7.6.3 Claude Process

- The Claude process is definitely situated in the middle of the maturation phase. It is the main process for helium liquefaction, hydrogen liquefaction, nitrogen liquefaction, and air separation.
- The new development is the Claude process using the Joule–Thomson expander instead of the valve. This innovation is in the "growth phase."

7.6.4 Mixed-Fluid Joule–Thomson Process

The mixed-fluid Joule–Thomson process is in the middle of the maturation phase. Recently, It is applied widely to middle-scale and large-scale natural gas liquefaction purposes.

References

1 Bloch, H.P., *A practical guide to Compressor Technology, Section 6: Metal Diaphragm Compressors*, pp. 127–136, McGraw-Hill, 1995, ISBN 0-07-005937-3

2 Maytal, B.Z. and Pfotenhauer, J.M. (2013). *Miniature Joule Thomson Cryocooling, Principles and Practice*. Springer. ISBN: 978-1-4419-8284-1.

3 Kolbe, L. (1920). *Flüssige Luft*. Leipzig: Johann Ambrosius Barth.

4 Venkatarathnam, G. (2008). *Cryogenic Mixed Refrigerant Processes*, 40. Springer.

5 Sattler, K. and Feindt, H.J. (1995). *Thermal Separation Processes, Principles and Design*. VCH. ISBN: 3-527.28622-5.

6 Stichlmair, J.G. and Fair, J.R. (1998). *Distillation: Principles and Practice.* Wiley VCH.

7 Poling, B.E., Prausnitz, J.M., and O'Connell, J.P. (2000). *The Properties of Gases and Liquids*. McGraw-Hill Professional.

8 Alexeev, A. (1999). Untersuchung zur Weiterentwicklung einer Gemisch-Joule-Thomson-Kältemaschine auf Basis ölgeschmierter Kompressoren, in German. PhD thesis. TU Dresden. 9783932715648.

9 Mogorychny, V.I. (1986). Development of methods for prediction of liquid-solid equilibria in multicomponent solutions of cryogens, in Russian. PhD thesis. Moscow Power Engineering Institute, UDK. 621.565:621.59.

10 Khatri, A. and Boyarski, M. (1997). A throttle cycle cryocooler operating with mixed gas refrigerants in 70 K to 120 K. In: *Cryocoolers 9* (ed. R.G. Ross Jr.), 515–520. New York: Plenum Press.

11 Jungnickel, H., Agsten, R., and Kraus, W.E. (1990). Grundlagen der Kältetechnik, Chapter 4.1 Verdichter. Verlag Technik GmbH Berlin. In: *ISBN 3-341-00806-3*.

12 Gmehling, J. (1989). Solid-Liquid Equilibrium Data Collection. Vol. VIII, Part 1: Binary Systems (DECHEMA Chemistry Data Series). Von H. Knapp, M. Teller und R. Langhorst. Herausgeg. von D. Behrens und R. Eckermann. DECHEMA, Frankfurt/M. 1987. LXXVI, 474 S., zahlr. Abb. u. Tab., geb., DM 290. *Chemie Ingenieur Technik*, 61(4), 347–347. Portico. https://doi.org/10.1002/cite.330610423

8

Liquefaction of Cryogenic Gases

The current chapter focuses on design processes for liquefaction purposes. The trade in liquefied industrial gases is a large business with a worldwide footprint. The corresponding plants for liquefaction of cryogenic gases (helium, hydrogen, nitrogen, and natural gas) are in widespread use throughout the industry, much more often than the plants for pure refrigeration purposes discussed in Section 7.2, 7.3, 7.4, and 7.5.

The liquefiers for nitrogen, helium, and natural gas are the primary focus of this chapter. The learnings presented here may be applied to hydrogen liquefaction as well.

The thermodynamic basics for liquefaction processes are very similar to the thermodynamics for refrigeration processes. Therefore, it is not necessary to dive deeply into the thermodynamic analysis of these processes; the corresponding examples of liquefaction processes are removed from this section and discussed in a separate, supplementary section.

This chapter considers several examples of industrial liquefaction systems, including hardware components.

8.1 Thermodynamic Basics for Liquefaction

Liquefiers and refrigerators are, thermodynamically speaking, very similar systems, but they are not identical. This chapter discusses the special thermodynamic features related to liquefaction systems.

8.1.1 Definitions for Liquefaction Processes

8.1.1.1 Isobaric Liquefaction Process in T,s-Diagram

Figure 8.1 shows an isobaric liquefaction process in a T,s-diagram. Exemplarily, the working fluid is nitrogen.

The gaseous fluid in initial state 1 is converted here to liquid state 2. The overall process can be considered as a two-step operation: first, the fluid is cooled to its dew point $\boldsymbol{D}$ (operation $\boldsymbol{1 \rightarrow D}$) by removal of the sensible heat $\boldsymbol{Q_1 = H_1 - H_D}$; second, the fluid is condensed by removal of the latent heat $\boldsymbol{Q_2 = H_D - H_2}$ from the dew point

Process Design for Cryogenics, First Edition. Alexander Alekseev.

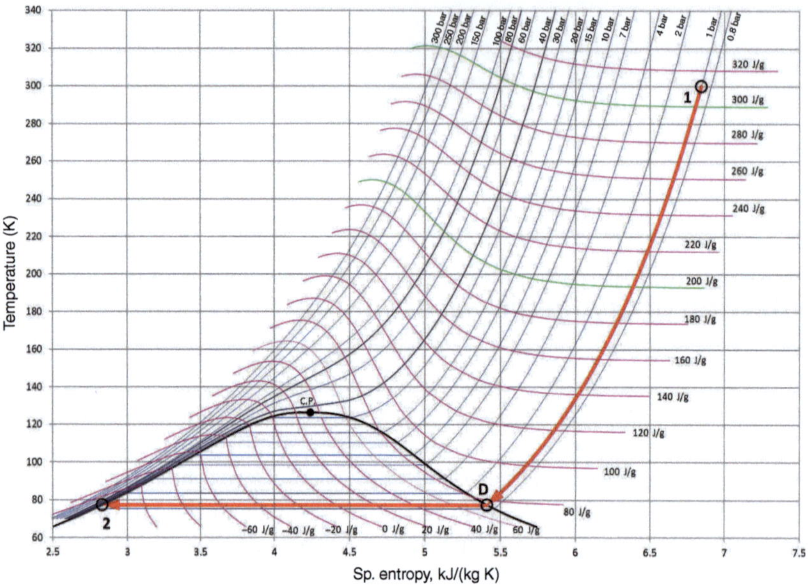

Figure 8.1 T,s-diagram of isobaric nitrogen liquefaction.

to the boiling point (operation $\boldsymbol{D \rightarrow 2}$). In the first step, the temperature of the fluid changes from $\boldsymbol{T_1}$ to $\boldsymbol{T_D} = \boldsymbol{T_2}$; during the second operation – condensation – the temperature remains constant at $\boldsymbol{T_D} = \boldsymbol{T_2}$.

8.1.1.2 Liquefier

Figure 8.2a shows – in a very simplified way – the function of a liquefier, which realizes the same operation from an external perspective: a gaseous fluid in initial state 1 (characterized by temperature T_1, pressure p_1, mass flow $\dot{m}$, composition z, vapor fraction $VF = 1$, specific enthalpy h_1, and entropy s_1) is converted here to liquid state 2 ($VF = 0$, T_2, p_2, $\dot{m}$, z, h_2, and s_2) using mechanical power $\boldsymbol{P}$. The waste heat $\dot{Q}_{amb}$ generated during this process is dissipated to the environment via cooling water or cooling air.

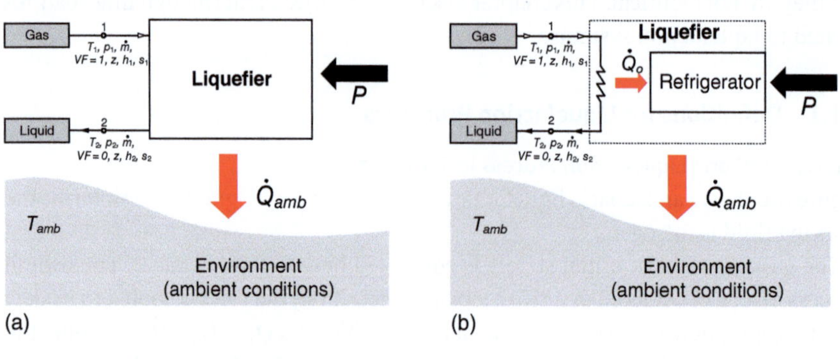

Figure 8.2 Liquefier from external perspective.

The energy conservation law for this system is as follows:

$$\dot{m}\,(h_2 - h_1) = P - \dot{Q}_{amb} \tag{8.1}$$

8.1.1.3 Liquefaction Capacity

The mass flow $\dot{m}$ of gas liquefied in a liquefier represents the liquefaction capacity of the machine. It is typically expressed in (kg/s) or (tons per day). Sometimes the volumetric units, such as (l/s) or (m^3/h), are used for the liquid state.

8.1.1.4 Connection to Refrigeration Processes

Thermodynamically, the liquefier can be considered as a sort of refrigerator that absorbs heat (heat flow) from the fluid being liquefied, as shown in Figure 8.2b. The cooling capacity of this refrigerator corresponds to the enthalpy difference of the fluid between the inlet and outlet, as stated below:

$$\dot{Q}_o = \dot{m}\,(h_1 - h_2) \tag{8.2}$$

The special feature of this refrigerator adapted for liquefaction purposes is that the cooling capacity should be provided not at a constant temperature level but at a changing temperature in the temperature range from initial temperature T_1 (usually, the ambient temperature) to T_2.

8.1.2 Minimum Liquefaction Power/Work

An important thermodynamic question related to liquefaction processes concerns the minimum power required for liquefaction of a certain flow of a given substance – known as minimum liquefaction power. It corresponds to the power of a liquefaction system based on a reversible liquefaction process.

8.1.2.1 Minimum Liquefaction Power via Carnot Equation

A similar matter was already discussed in Section 7.1.5; namely, the minimum power required for generating cooling capacity $\dot{Q}_o$ at a **constant** temperature $\boldsymbol{T_o}$. The Carnot equation (Eq. 7.4), which is valid for the reversible refrigerator, was used for this purpose.

Generally, this concept based on reversible refrigerators is also applicable to liquefaction processes. However, in the liquefaction case, the Carnot equation can only be applied directly to the second step of the liquefaction process – the condensation operation – where the heat from the fluid is removed at a constant temperature. For application to the first part of the process – the cooling of the gas to the dew point at changing temperature – the Carnot concept should be slightly adapted. The basic idea is to provide the required cooling capacity not by a single reversible refrigerator, but by a set of reversible refrigerators, each of which provides a corresponding partial cooling capacity at a different cooling temperature. This ensures coverage of the entire temperature range from $\boldsymbol{T_1}$ to $\boldsymbol{T_D}$.

Figure 8.3 illustrates this principle:

- For the cooling from the initial state 1 to the dew temperature $\boldsymbol{T_D}$, multiple reversible refrigerators (1· · ·n) are used. Each refrigerator $\boldsymbol{i}$ delivers a small

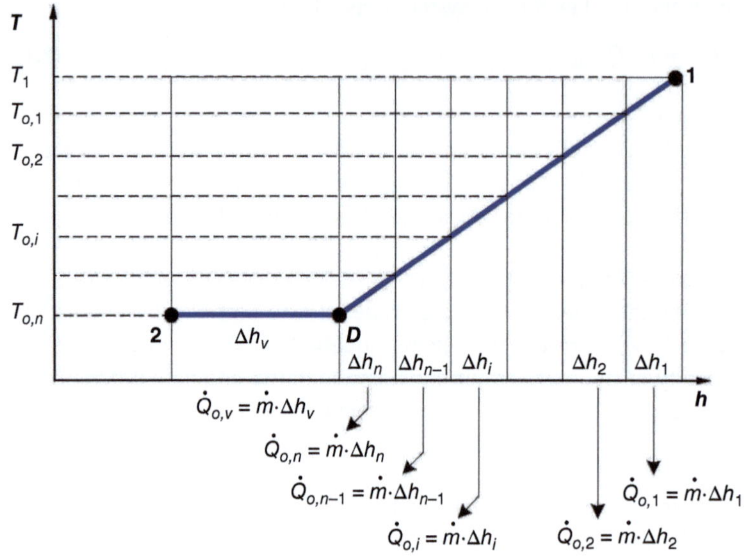

Figure 8.3 Scheme for calculating the minimum liquefaction work, Δh_V-specific latent heat.

portion of the cooling capacity $\dot{Q}_{0,i} = \dot{m}\,\Delta h_i$ (from $\dot{Q}_{0,1}$ to $\dot{Q}_{0,n}$) at the corresponding temperature level $T_{0,i}$ (from $T_{0,1}$ to $T_{0,n}$);

- The cooling capacity for the condensing part $\dot{Q}_{0,v} = \dot{m}{\cdot}\Delta h_v$ (where Δh_V is specific evaporation heat) is delivered by a single reversible refrigerator with a cooling capacity $\dot{Q}_{0,v}$ at the constant cooling temperature $T_{0,n}$.

The Carnot equation defines the minimum mechanical power of each reversible refrigerator:

$$P_{min\ i} = \dot{Q}_{0,i}\left[\frac{T_{amb} - T_{0,i}}{T_{0,i}}\right]$$

The minimum power for the entire system is calculated as the sum:

$$P_{min} = \Sigma P_i = \dot{Q}_{0,v}\left[\frac{T_{amb} - T_{0,n}}{T_{0,n}}\right] + \Sigma\left[\dot{Q}_{0,i}\left[\frac{T_{amb} - T_{0,i}}{T_{0,i}}\right]\right] \tag{8.3}$$

The larger the number of refrigerators, the more precise the calculation. In mathematics, this kind of task is solved using integral calculus, which leads to the following equation:

$$P_{min} = \dot{m}\cdot\Delta h_V\left[\frac{T_{amb} - T_{0,n}}{T_{0,n}}\right] + \dot{m}\cdot\int_{T_2}^{T_1} c_p\frac{T_{amb} - T_{0,i}}{T_{0,i}}dT \tag{8.4}$$

8.1.2.2 Minimum Liquefaction Power via Exergy

Another method for calculation of the minimum power is based on the thermodynamic function "exergy," which is described in Section 4.5. According to definition, the exergy of a material stream, a given state 1 $\dot{E}_1$ (p_1, T_1, $\dot{H}_1$, $\dot{S}_1$) describes the minimum power required to convert this material stream (in reversible fashion) from the ambient conditions to this state 1; the corresponding equation is as follows:

$$\dot{E}_1 = (\dot{H}_1 - \dot{H}_{amb}) - T_{amb}\,(\dot{S}_1 - \dot{S}_{amb})$$

Therefore, the minimum liquefaction power can be found as the difference between the exergy of the fluid in the state 1 and the exergy in the liquefied state 2:

$$P_{min} = \dot{E}_2 - \dot{E}_1 = \dot{H}_2 - \dot{H}_1 - T_{amb}\,(\dot{S}_2 - \dot{S}_1) \tag{8.5}$$

or in terms of specific values:

$$P_{\min} = \dot{m} \cdot (e_2 - e_1) = \dot{m}\,[h_2 - h_1 - T_{amb}\,(s_2 - s_1)] \tag{8.6}$$

8.1.2.3 Minimum Liquefaction Work for Typical Cryogenic Gases

Table 8.1 shows the minimum liquefaction work for typical cryogenic gases. There are three key takeaways:

(a) In terms of specific molar values (in kJ/mol), the minimum liquefaction work is quite similar for all cryogenic gases: it amounts to 22 ± 6 kJ/mol.
(b) The minimum liquefaction work in terms of specific mass values (in kJ/kg) varies greatly, from 3-digit to 5-digit values. The value for hydrogen, for example, is approximately 10 times higher than for neon. This is mainly due to the difference in molecular weight.
(c) Hydrogen requires the largest effort for liquefaction, exceeding even helium.

8.1.2.4 Key Performance Indicators for Liquefiers

Two indicators are used for characterization of liquefaction processes/plants: efficiency and specific power.

Table 8.1 Minimum liquefaction work for cryogenic gases, $T_{amb} = 300$ K, $p_{amb} = 1.0$ bar.

	Minimum liquefaction work (kJ/mol)	Minimum liquefaction work (kJ/kg)	Minimum liquefaction work (kWh/kg)
Helium	26.5	6633	1.84
e-Hydrogen, including ortho-to-para conversion	28.14	14068	3.91
n-Hydrogen, without ortho-to-para conversion	24.21	12103	3.36
Neon	31.8	1572	0.44
Nitrogen	20.5	730.5	0.20
Air	20.6	710	0.20
Oxygen	19.7	615	0.17
Methane	16.8	1050	0.29

8.1.2.4.1 Efficiency of a Liquefaction System

The "minimum liquefaction power" is used as a benchmark for evaluation of liquefaction processes in terms of energy efficiency. The procedure is similar to that used for refrigerators: the efficiency of the entire liquefaction system η_{LIQ} is defined as the quotient of power required in a reversible liquefaction process $\boldsymbol{P}_{rev}$ and actual power consumption of the given system $\boldsymbol{P}$:

$$\eta_{LIQ} = \frac{P_{rev}}{P} \tag{8.7}$$

8.1.2.4.2 Specific Liquefaction Power

The specific liquefaction power $\boldsymbol{p}_{sp}$ is defined as the actual power consumption of the liquefaction system $\boldsymbol{P}$ divided by the liquefaction capacity, usually specified as mass flowrate $\dot{\boldsymbol{m}}$:

$$p_{sp} = \frac{P}{\dot{m}} \tag{8.8}$$

This value is usually given in kW/kg, kWh/(kg/s) or similar units.

8.1.3 Specifics of Process Design for Liquefaction

8.1.3.1 Early Liquefaction Processes

Historically, liquefaction processes were introduced and commercialized before refrigeration processes, which were derived later from the liquefaction process. However, in this book, the reverse sequence is followed, hence the refrigeration process is covered first.

This section examines two previously discussed processes – the Joule–Thomson and the Claude processes – both of which were introduced roughly 100 years ago. In terms of process design, both of these processes are valuable for understanding the main principle of liquefaction and the related basic challenges.

8.1.3.1.1 Joule–Thomson Process for Liquefaction

Figure 8.4a (left) shows a nitrogen liquefaction process based on a nitrogen Joule–Thomson refrigeration process. It differs a bit from the classic Joule–Thomson process for refrigeration purposes discussed in Section 7.2. The most salient change is a phase separator installed downstream of the Joule–Thomson valve instead of by the evaporator. The main purpose of this device is to separate the liquid phase from the vapor phase. In terms of mechanical design, the phase separator represents a simple vessel. It is sizeable enough to reduce the velocity of the fluid entering the vessel, targeting the order of magnitude of 0.1 m/s. Under these conditions, the liquid droplets generated after the isenthalpic expansion will precipitate and collect in the sump of the vessel, while the vapor fraction collects above the liquid phase. This is how the phase separation process occurs. Additional internal components such as wire screens or baffles may be integrated into this vessel for improvement of separation performance.

In the Joule–Thomson liquefaction process, the vapor is withdrawn from the top of the phase separator and provided further into the main cycle. The liquid nitrogen product, LIN, is removed from the sump of the vessel. Since a fraction of the working

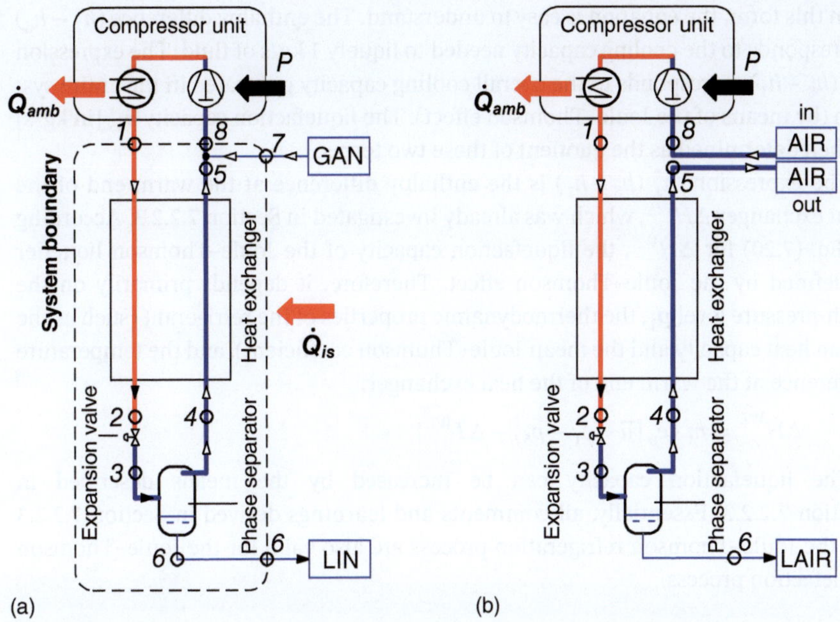

Figure 8.4 Joule–Thomson processes for liquefaction purposes: (a) Nitrogen Joule–Thomson process (GAN – gaseous nitrogen, LIN – liquid nitrogen), (b) Linde process with single expansion (AIR – air, LAIR – liquid air).

fluid is continuously withdrawn from the main process, a corresponding flowrate of the working fluid (in this case, gaseous nitrogen [GAN]) is fed into the system to maintain material balance. The gaseous fluid is usually fed upstream of the compressor. From an external perspective, the nitrogen appears to enter the system as a gas and leave it as a liquid.

In practice, the liquid can be withdrawn from the phase separator either continuously or discontinuously. In the latter case, the phase separator must be dimensioned sufficiently to guarantee corresponding storage capacity. If it is designed as a large vessel, it can assume the function of a storage vessel. In this case, the vessel has two functionalities, phase separation, and liquid storage. In terms of mechanical design, the phase separator with storage function is usually constructed as a separate cryogenic vessel positioned next to the liquefaction system and connected by corresponding insulated pipelines.

Liquefaction Capacity of the Joule–Thomson Liquefier The liquefaction capacity of a Joule–Thomson liquefier can be studied using the enthalpy balance for the cold part of the process according to system boundary as shown in Figure 8.4a, assuming that the heat leakage through the thermal insulation is negligible ($\dot{Q}_{is} \approx 0$):

$$\dot{H}_8 - \dot{H}_1 = \dot{H}_7 - \dot{H}_6 \tag{8.9}$$

Considering that $\dot{m}_6 = \dot{m}_7$, the equation can be transformed to:

$$\dot{m}_6 = \dot{m}_7 = \dot{m}_1 \frac{h_8 - h_1}{h_7 - h_6} \tag{8.10}$$

In this form, the equation is easy to understand. The enthalpy difference $(h_7 - h_6)$ corresponds to the cooling capacity needed to liquefy 1 kg/s of fluid. The expression $\dot{m}_1\ (h_8 - h_1)$ corresponds to the overall cooling capacity generated in the entire system (by means of the Joule–Thomson effect). The liquefaction capacity $\dot{m}_6$ [in kg/s] is then determined as the quotient of these two terms.

The expression $\dot{m}_1\ (h_8 - h_1)$ is the enthalpy difference at the warm end of the heat exchanger $\Delta\dot{H}^{WE}$, which was already investigated in Section 7.2.2.3. According to Eq. (7.20) for $\Delta\dot{H}^{WE}$, the liquefaction capacity of the Joule–Thomson liquefier is defined by the Joule–Thomson effect. Therefore, it depends primarily on the high-pressure level $\boldsymbol{p_1}$, the thermodynamic properties of the refrigerant (such as the mean heat capacity and the mean Joule–Thomson coefficient), and the temperature difference at the warm end of the heat exchanger:

$$\Delta\dot{H}^{WE} = \dot{m} \cdot c_p\,[\overline{\mu} \cdot (p_1 - p_5) - \Delta T^{WE}]$$

The liquefaction capacity can be increased by the means described in Section 7.2.2.3. Essentially, all comments and learnings derived in Section 7.2.2.3 for the Joule–Thomson refrigeration process are also valid for the Joule–Thomson liquefaction process.

8.1.3.1.2 Linde Process

Figure 8.4b shows the liquefaction process originally developed and realized by Carl Linde in 1895. In German-speaking countries, this process is referred to as the "Linde process with single expansion."[1] The world's first industrial plant for air liquefaction was based on this process. The ambient air is used as the working fluid, which is drawn directly from the environment by the compressor. The vapor from the top of the phase separator flows through the heat exchanger and is also exhausted into the environment. There is no direct pipework connection between the heat exchanger outlet and the compressor inlet. The cycle is closed via the ambient air. The first system ever built produced roughly 10 liters of liquid per day, which was a relatively large quantity at that time.

8.1.3.1.3 Basic Claude Process for Liquefaction

Figure 8.5 shows what is known as the "basic Claude process" for liquefaction purposes with three heat exchangers and a single expander (shown as centrifugal machine). The liquefaction process is very similar to the refrigeration process, the only difference being that the evaporator is replaced by a phase separator. Actually, the basic Claude process shown in Figure 8.5 differs a bit from the process

1 In English-speaking countries, this process is referred to as the "Linde–Hampson-process" or "Hampson–Linde-process" as a tribute to the British engineer and innovator William Hampson, who applied for a corresponding patent on 23rd May 1895. This was the first patent for air liquefaction, since Carl Linde's patent was dated 5th June 1895. Linde's patent, however, was more comprehensive (Hampson did not submit the full description until 1896). Additionally, Carl Linde, who had been studying air liquefaction for many years, had already demonstrated a full-size functioning system. The legal dispute between these two parties regarding priority and patent rights went on for several years. Ultimately, the legal issue was decided mainly in favor of Carl Linde.

Figure 8.5 Basic Claude process for nitrogen liquefier, GAN – gaseous nitrogen, LIN – liquefied nitrogen.

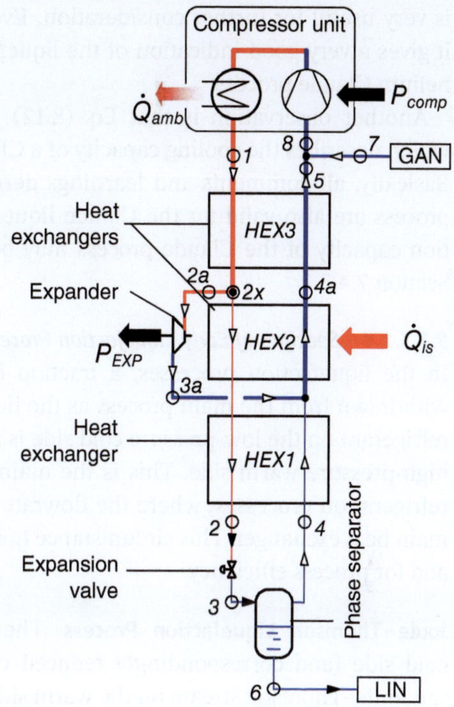

originally developed by George Claude [1]. However, since the "basic Claude process" implements all features originally introduced by George Claude but is a bit easier, this process is more suitable for explanation of Claude's idea.

Liquefaction Capacity of the Claude Liquefier The liquefaction capacity of the basic Claude process can be studied using the enthalpy balance for the cold part of the process shown in Figure 8.5, assuming that the heatage leak due to nonideal thermal insulation is negligible ($\dot{Q}_{is} \approx 0$):

$$\dot{H}_8 - \dot{H}_1 + P_{EXP} = \dot{H}_7 - \dot{H}_6 \tag{8.11}$$

This equation can be transformed to:

$$\dot{m}_6 = \dot{m}_7 = \frac{\dot{m}_1\,(h_8 - h_1) + P_{EXP}}{h_7 - h_6} \tag{8.12}$$

For a process with an ideal gas and an infinitely large heat exchanger, the enthalpy difference at the warm end is zero $(h_8 - h_1) = 0$. Therefore, this equation can be further simplified as follows:

$$\dot{m}_6 = \dot{m}_7 = \frac{P_{EXP}}{h_7 - h_6} \tag{8.13}$$

This equation expresses that the liquefaction capacity of a Claude process with an ideal gas is defined by the power produced in the expander (or multiple expanders).

Since the behavior of cryogenic gases at ambient temperature (especially that of nitrogen and helium) does not deviate much from that of ideal gas, this statement is very useful for further consideration. Even if this equation is not 100% correct, it gives a very good indication of the liquefaction capacity of a typical nitrogen or helium Claude process.

Another observation is that Eq. (8.12) is akin to Eq. (7.31) in Section 7.4.2, which describes the cooling capacity of a Claude process for refrigeration purposes. Basically, all comments and learnings derived there for the Claude refrigeration process are also valid for the Claude liquefaction process. Likewise, the liquefaction capacity of the Claude process may be increased by the means described in Section 7.4.2.

8.1.3.1.4 Specifics of Early Liquefaction Processes

In the liquefaction processes, a fraction of the working fluid is separated and withdrawn from the main process as the liquid product. Therefore, the flowrate of refrigerant on the low-pressure cold side is always smaller than the flowrate on the high-pressure warm side. This is the main difference compared to homonymous refrigeration processes, where the flowrate remains the same on both sides of the main heat exchanger. This circumstance holds several consequences for the process and for process efficiency.

Joule–Thomson Liquefaction Process The reduced refrigerant flowrate on the cold side (and correspondingly reduced cold capacity of this flow) means that the Joule–Thomson stream on the warm side cannot be cooled down as far as in the Joule–Thomson refrigeration process. The temperature difference at the cold end of the heat exchanger becomes even larger. The exergy loss in the heat exchanger as well as exergy loss in the expansion valve will increase. These losses become visible and tangible as reduced liquid fractions after the Joule–Thomson expansion. Therefore, the Joule–Thomson liquefaction process is generally less efficient than the refrigeration process.

Basic Claude Liquefaction Process In the Joule–Thomson stage of the Claude process, the consequences are the same as described above for the classic Joule–Thomson process: the temperature difference at the cold end of the heat exchanger HEX1 rises. The exergy loss in this heat exchanger as well as exergy loss in the expansion valve will also increase.

In the Brayton part of the process, principally, the reduced flowrate on the low-pressure side can be balanced by increasing the flowrate of the expander stream. This allows the temperature difference in the heat exchangers HEX2 and HEX3 to be maintained (in comparison to the refrigeration process). However, since the flowrate of the expander stream will be increased, and since the heat duty of the heat exchangers HEX2 and HEX3 will be increased, the exergy loss there will increase – even at the same temperature difference. The other losses rise as well: the losses in the expander and in the compressor rise due to increased flowrate and higher required power. Ultimately, these higher losses become apparent as reduced liquid fraction at increased compressor power.

In the version for refrigeration purposes, the target is always a high liquid fraction after the expansion valve. In the liquefier version, the high liquid fraction means a low vapor fraction and, consequently, reduced flowrate for the material stream on the cold side of the Joule–Thomson heat exchanger (HEX1), increased temperature difference at the cold end of the heat exchanger, and so on, creating a kind of vicious cycle that affects the performance of the liquefaction process.

Summary The liquefaction processes are less efficient compared to the homonymous refrigeration processes.

8.1.3.2 Process Design Approaches for Recent Liquefaction Plants

Since liquefaction of cryogenic gases requires considerable power, the power cost contributes heavily to the total cost of cryogenic liquids. Consequently, high efficiency is required to reduce the operational expenditures and achieve a positive business case.

The early processes such as the classic Joule–Thomson liquefaction process or the basic Claude liquefaction process described in Section 8.1.3.1 are not efficient enough. In recent decades, these processes, especially the Claude process, have been improved considerably in terms of efficiency to satisfy the current market needs.

As described in Section 8.1.3.1, in the case of liquefaction processes, the main challenge is the lack of cooling capacity on the cold (low-pressure) side. Therefore, the basic impetus for improvement is always the addition of cooling capacity there. Principally, there are two possibilities:

Either

(a) the additional cooling capacity is generated (actively) by means of an additional cooling process or
(b) the cooling capacity is imported (passively) from an external cold source.

Two process examples illustrating both these approaches will be considered in this section:

(a) Implementation of an additional Brayton stage into a Claude-based liquefaction process as an example of integration of an additional cooling stage, and
(b) Implementation of liquid nitrogen as an external cooling source into a process for helium liquefaction.

A further approach is based on application of an external reversed Rankine-type refrigerator. Table 8.2 provides some information. This method is considered in Sections 8.4.5.5 and 8.4.6.1 (liquefaction of natural gas) in detail.

8.1.3.2.1 Implementation of Additional Brayton Stage

The main principle can be illustrated by means of a nitrogen liquefaction process, as shown in Figure 8.6. It is derived from the nitrogen Claude process for refrigeration purposes.

This contains the following features:

- The Joule–Thomson heat exchanger HEX1 is not required and is removed;
- Phase separator is introduced instead of evaporator;

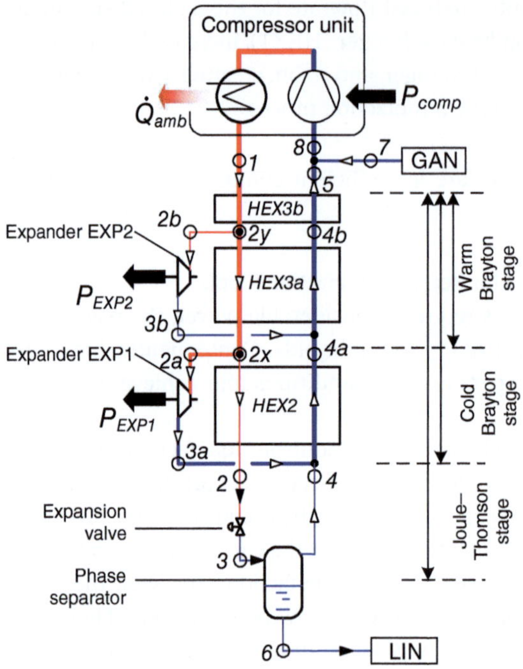

Figure 8.6 Dual-expander nitrogen Claude liquefier.

- The upper heat exchanger (former HEX3) is divided into two separate heat exchangers HEX3a and HEX3b;
- The second expander is introduced and implemented in the upper part of the process. Usually, this second expander is referred to as the "warm-expander," whereas the expander EXP1 is referred to as the "cold-expander," since the temperature at the inlet of the expander EXP1 is colder than the temperature at the inlet of the EXP2.

Since two expanders are used, this process is often referred to as the "dual-expander Claude process."

Morphologically, the process includes a new additional Brayton stage that may be recognized as the process chain $1 \rightarrow 2y \rightarrow 2b \rightarrow 3b \rightarrow 4b \rightarrow 5$. The heat exchanger HEX3b, the expander EXP2, and the heat exchanger HEX3a take over the roles of the respective main heat exchanger, expander, and cold gas heat exchanger of this integrated Brayton stage.

The entire process sequence is as follows:

(a) The nitrogen is first compressed from the low-pressure level $p_{LP} = p_8$ to the high-pressure level $p_{HP} = p_1$ (supercritical pressure) by means of a compressor and cooled to approximately ambient temperature. This state is denoted as state 1 in Figure 8.6.
(b) This compressed refrigerant is slightly cooled (by 10–40 K) in the countercurrent heat exchanger HEX3b by the cold low-pressure stream. The state of the refrigerant at the cold end of this heat exchanger is shown as State 2y.
(c) At this point, the refrigerant flow is divided into two streams. The smaller one (labeled 2b) is referred to as the warm-expander stream. The remaining

material stream consists of the Joule–Thomson stream and the cold-expander stream.

(d) The warm-expander stream is expanded in the warm-expander EXP2 from the high-pressure level p_{HP} to the low-pressure level p_{LP}. The temperature of this stream reduces correspondingly. In Figure 8.6, the state at the expander outlet is shown as State 3b, and this stream is gaseous. This cold gas is mixed with the cold gas from the lower part of the process and proceeds through heat exchangers HEX3a and HEX3b, then further to the compressor inlet;
(e) The remaining stream (consisting of the Joule–Thomson stream and the cold-expander stream) is cooled in heat exchanger HEX3a (State 2*x*) before it is divided into the Joule–Thomson stream and the cold-expander stream (State 2a);
(f) The cold-expander stream is expanded in the cold-expander EXP1 from the high-pressure level p_{HP} to the low-pressure level p_{LP}. The temperature of this stream reduces correspondingly. In Figure 8.6, the state at the expander outlet is shown as State 3a, and this stream is gaseous. This cold gas from the expander outlet is provided, through heat exchanger HEX2 and further through HEX3a and HEX3b, to the compressor inlet;
(g) The Joule–Thomson stream flows through heat exchanger HEX2, where it is cooled by the cold-expander stream in countercurrent direction to the temperature T_2 (State 2, which corresponds to supercritical liquid). It is expanded in the expansion valve to the low-pressure level p_{LP}; this process ($2 \rightarrow 3$) is a typical Joule–Thomson expansion, whereby the Joule–Thomson stream partially liquefies, with a part of refrigerant evaporating and thereby generating flashgas. The generated liquid is separated in the phase separator. The flashgas is mixed with the cold-expander stream and provided to the compressor inlet via the heat exchangers.

The introduction of the warm Brayton stage allows active control of the temperature profile in heat exchanger HEX3a, whereas the temperature profile in heat exchanger HEX2 is controlled by the cold Brayton stage. This means the optimal temperature difference in heat exchangers HEX2 and HEX3a is achievable. The temperature profile in heat exchanger HEX3b is not actively controlled, but rather it results autonomously.

The optimization of the process design involves primarily the optimal distribution of the refrigerant flow between the warm-expander, cold-expander, and Joule–Thomson valve, as well as optimal temperatures at the inlets of these devices.

The main advantage of this process, which achieves the minimal temperature difference in both cold heat exchangers plus maximum subcooling upstream of the expansion valve, is that the corresponding losses are minimized. This results in a highly efficient liquefaction process.

The disadvantages are that

(1) an additional, relatively expensive component – an expander – is required, therefore the costs are increased, and
(2) the process grows more complex and requires more effort for control (involving maintenance, engineering, etc.).

Liquefaction Capacity of the Dual-Expander Claude Process The dual-expander Claude process is used often for the design of industrial nitrogen liquefiers. The liquefaction capacity of this process can be estimated using the enthalpy balance for the cold part of the process. If the heat leak through the thermal insulation ($\dot{Q}_{is} \approx 0$) considered as negligibly small, the following equation can be used:

$$\Delta\dot{H}_{N2} = P_{EXP1} + P_{EXP2} + \Delta\dot{H}^{WE} \tag{8.14}$$

whereas

$\Delta\dot{H}_{N2} = \dot{H}_7 - \dot{H}_6 = \dot{m}_{N2}\,\Delta h_{7-6}$ – Enthalpy difference between the nitrogen feed and the liquid product, which corresponds to the cooling capacity of the liquefier

$\Delta\dot{H}^{WE} = \dot{H}_8 - \dot{H}_1$ – Enthalpy difference at the warm end of the heat exchanger.

The corresponding liquefaction capacity can be found as follows:

$$\dot{m}_{N2} = \frac{\Delta\dot{H}_{N2}}{\Delta h_{7-6}} = \frac{P_{EXP1} + P_{EXP2} + \Delta\dot{H}^{WE}}{\Delta h_{7-6}} \tag{8.15}$$

Assuming that $\Delta\dot{H}^{WE}$ is significantly smaller than the expander power (as considered in Section 7.4.2.4) this equation can be further simplified as follows:

$$\dot{m}_{N2} = \frac{\Delta\dot{H}_{N2}}{\Delta h_{7-6}} \approx \frac{P_{EXP1} + P_{EXP2}}{\Delta h_{7-6}} \tag{8.16}$$

This equation expresses that the liquefaction capacity is defined mainly by the power of both expanders. Vice versa, this equation can be used to estimate expander power for required liquefaction capacity.

Exercise:
Given: The total mechanical capacity of both expanders is given as $P_{EXP1} + P_{EXP2} = 1\,MW$. The specific enthalpy difference between the gaseous state at ambient temperature and the liquid state is given as 400 J/g.
Task: Estimate the liquefaction capacity of a dual-expander Claude process.

Solution: According to Eq. (8.16):

$$\dot{m}_{N2} \approx \frac{P_{EXP1} + P_{EXP2}}{\Delta h_{7-6}} = \frac{1\,000\,000\ \mathrm{W}}{400\ \mathrm{J/g}} = 2500\ \mathrm{g/s} = 2.5\ \mathrm{kg/s} = 9\ \mathrm{t/h}$$

8.1.3.2.2 Liquid Nitrogen Precooling

The approach described in Section 8.1.3.2.1 – integration of an additional Brayton stage[s] – is applicable to the helium liquefaction process as well. However, the most common method for improving a helium liquefaction process is based on the use of an external cold source in the form of liquid nitrogen.[2]

---

2 This is generated by an external cryogenic air separation or nitrogen liquefaction plant, for example, by means of two expander processes as described in Section 8.1.3.2.1.

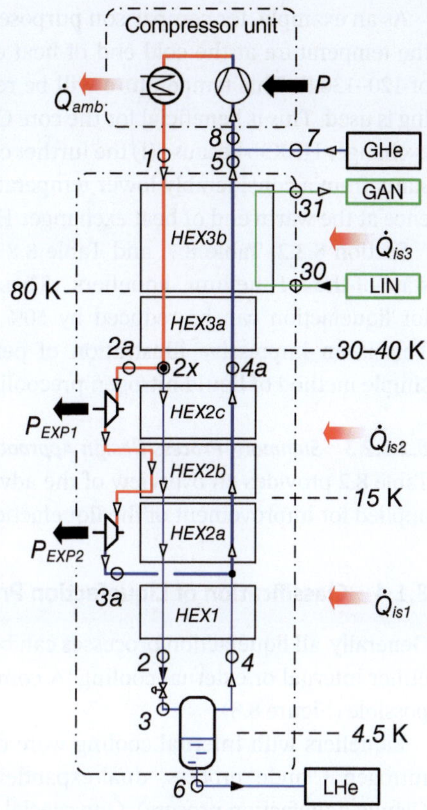

Figure 8.7 Helium Claude liquefier with liquid nitrogen precooling.

The principle is shown in Figure 8.7. The process is derived from the helium Claude process for refrigeration purposes. The main changes are:

- The phase separator instead of the evaporator;
- The upper (in terms of temperature) heat exchanger HEX3b consists of additional passages for the third material stream, nitrogen.

The liquid nitrogen (the state is denoted as 30) is fed into the cold end of the heat exchanger HEX3b. It is evaporated within the heat exchanger, warmed up to nearly ambient temperature (state is denoted as 31), and released into the environment at the warm end of this heat exchanger. This simple measure helps

(1) to cool the warm high-pressure helium flow in the heat exchanger HEX3b to the liquid nitrogen temperature level and control this temperature by managing flow of liquid nitrogen, and
(2) to balance the heat leakage through the nonideal thermal insulation in the upper part of the process (heat exchanger HEX3b), referred to as $\dot{Q}_{is3}$ in Figure 8.7. Since the heat exchanger HEX3b is the largest one, this portion of the heat leakage is also the largest.

As an example, for comparison purposes: without the liquid nitrogen precooling, the temperature at the cold end of heat exchanger HEX3b would be in the range of 120–130 K. This temperature will be reduced to 80 K if liquid nitrogen precooling is used. This is beneficial for the core Claude liquefaction process below the heat exchanger HEX3b because (i) the further cooling of the high-pressure helium stream starts from a considerably lower temperature level, and (ii) the temperature difference at the warm end of heat exchanger HEX3a can be minimized.

Section 8.3.2, Table 8.7, and Table 8.8 provide data on power consumption for state-of-the-art helium liquefiers. These data show that the specific power for liquefaction can be reduced by 50% if liquid nitrogen precooling is applied. This is an impressive illustration of performance improvement using the very simple method of liquid nitrogen precooling.

8.1.3.2.3 *Summary, Process Design Approaches for Liquefaction*

Table 8.2 provides an overview of the advantages and disadvantages of approaches applied for improvement of the liquefaction process.

8.1.4 Classification of Liquefaction Processes

Generally, all liquefaction processes can be structured in two groups, i.e. those with either internal or external cooling. A combination of both of these methods is also possible (Figure 8.8).

Liquefiers with internal cooling were discussed in Section 8.1.3 (Linde process, nitrogen Claude process, dual-expander nitrogen Claude process, and helium Claude liquefaction process). One special property of these processes is that the gas being liquefied (feed gas) and the working fluid circulating in the process are the same substance. The same compressor drives the entire process.

However, in principle, the liquefaction process can be designed such as shown in Figure 8.9, where the fluid being liquefied (feed gas) is cooled by means of an external closed-cycle refrigerator (or multiple external refrigerators) with its own working fluid. Such a liquefaction process is referred to as a **process with external cooling**.

8.1.5 Liquefaction with External Cooling

This section explains the main principle of external cooling as well as an example of a process with external cooling.

8.1.5.1 Principle of External Cooling

Figure 8.9 visualizes the general concept of a process with external cooling.

Morphologically, the entire process consists of two separate subprocesses that are thermally connected by means of a common heat exchanger:

Subprocess #1, for the Feed Gas: The subprocess for the gas being liquefied is designed as a Joule–Thomson process ($1 \rightarrow 2 \rightarrow 3 \rightarrow 4 \rightarrow 5 \rightarrow 8$) with feed stream 7 and product stream 6. The feed gas is:

Table 8.2 Process design approaches for liquefaction purposes.

	Additional Brayton stage[s]
Main application	Nitrogen liquefaction, natural gas liquefaction, can be considered for hydrogen liquefaction as well as for helium liquefaction.
Principle and effect on power efficiency	The temperature difference in every heat exchanger bypassed by every expander stream can be minimized; maximum subcooling of the Joule–Thomson stream upstream of the expansion valve is achieved. Therefore, the corresponding losses are minimized. It improves efficiency of every liquefaction process considerably.
Cost efficiency:	An additional, relatively expensive hardware component – expander – is required, which leads to higher CAPEX.
Further aspects:	The process becomes more complex and requires more effort in terms of control, maintenance, engineering, etc.
	Precooling by Liquid Nitrogen
Main application	Helium liquefaction, hydrogen liquefaction
Principle and power efficiency	This approach presumes that the cold generation in a nitrogen liquefier is more efficient than in a helium liquefier, in terms of efficiency and cost. Therefore, the import of external cold in the form of liquid nitrogen from a nitrogen liquefier to a helium liquefier improves the efficiency of the helium liquefier. The background: – Nitrogen liquefiers (or air separation plants) are based on large-scale high-efficiency centrifugal compressors (Section 6.5.2) and large turboexpanders (see Section 6.3.3), whereas – Helium liquefiers are based on medium-scale screw compressors (Section 6.5.3), which are less efficient, and small-/medium-scale expanders (Section 6.3.4), which are less efficient in comparison to the hardware used in nitrogen liquefaction.
Cost efficiency	The implementation of liquid nitrogen precooling requires only relatively simple and inexpensive measures: an additional passage in the "warm" heat exchanger and some hardware for liquid nitrogen supply. Therefore, it is a very cost-efficient operation.
Further aspects	Transport, storage, and management of liquid nitrogen generate some associated costs and liquid loss. These costs depend on the distance between the liquid nitrogen production site and the location of the helium system requiring supply.
	External Reversed Rankine-type Refrigerator
Main application	natural gas liquefaction and nitrogen liquefaction, can be considered for hydrogen liquefaction as well as for helium liquefaction.
Principle and power efficiency	The idea is to use an additional external refrigerator to cool the feed gas to an intermediate temperature level before it is fed onward to the main liquefaction process.

Table 8.2 (Continued)

	This solution is preferred if the external refrigerator performs considerably better than the main liquefaction process. Correspondingly the precooling of the feed gas in the external refrigerator requires less power than in the main process. That is usually the case if (i) a Rankine-type refrigerator is used and (ii) a temperature level of 240–270 K is targeted for precooling purposes.
Cost efficiency	An additional hardware component – external refrigerator – is required. Therefore, CAPEX usually increases.
Further aspects	The process becomes more complex and requires more effort in terms of control, maintenance, engineering, etc.

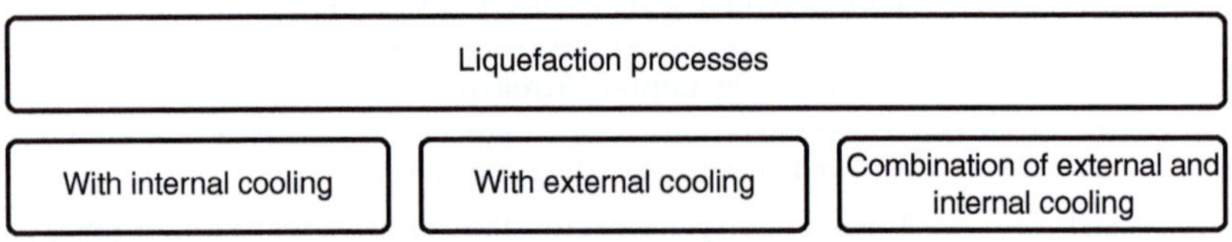

Figure 8.8 Classification of the liquefaction processes.

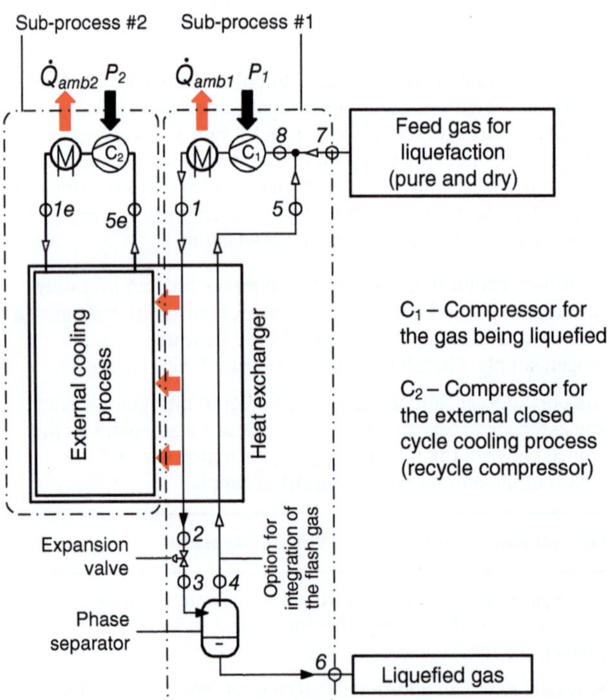

Figure 8.9 Principle of liquefaction process with external cooling.

(a) Compressed in the compressor unit C_1 to the supercritical pressure. A multi-stage compressor is usually applied;
(b) Cooled in the main heat exchanger ($1 \rightarrow 2$) to a temperature T_2. The target is always to achieve maximum subcooling (minimal temperature) before expansion. The enthalpy difference of the fluid in the heat exchanger $\Delta\dot{H} = \dot{H}_1 - \dot{H}_2$ corresponds to the required cooling capacity to be generated by subprocess #2, which is an external closed-cycle refrigeration process;
(c) The compressed and cooled fluid expanded in the expansion valve ($2 \rightarrow 3$) to the required product pressure. Depending on the process condition, the fluid is partially or fully liquefied and provided to the phase separator;
(d) The liquid product (marked as 6) is withdrawn from the phase separator;
(e) For the handling of the flashgas (marked as 4), there are different process design options available (see Section 8.1.5.2 for more information). The most obvious option is shown in Figure 8.9, where the flashgas is recirculated through the heat exchanger ($4 \rightarrow 5$) back to the inlet of the feed compressor K1; there, it is mixed with feed stream 7 to produce stream 8.

Subprocess #2, for the Main Refrigeration Process: The external refrigeration process generates the cooling capacity $\dot{Q}_o$ required for removal of $\dot{Q}_o = \Delta\dot{H} = \dot{H}_1 - \dot{H}_2$ from the material stream being liquefied.

(a) It may use a separate working fluid.
(b) A separate compressor C_2 drives the process, compressing the working fluid from the state (see Figure 8.9) 5e to the state 1e.
There are many options available for the main refrigeration process as well as for the working fluid. The most common processes are:
    - the mixed-fluid Joule–Thomson process;
    - the single/multiple Brayton stage[s] with nitrogen or helium as the working fluid;
    - the nitrogen-based or helium-based Claude process.

All these options can also be combined. Therefore, a great number of potential combinations can be created. The main goal of the process designer is to find the most economically efficient one.

8.1.5.1.1 *Example of Liquefaction Process with External Cooling*

Figure 8.10 shows a virtual liquefaction process with external cooling process derived from the classic dual-expander Claude process by physical separation of the Brayton stages from the Joule–Thomson stage.

Morphologically, it consists of two subprocesses:

1. A Joule–Thomson cycle ($1 \rightarrow 2 \rightarrow 3 \rightarrow 4 \rightarrow 5$) driven by the compressor C_1.
2. A refrigeration subprocess based on two Brayton stages, whereas
   - the first Brayton stage is represented by the process chain 1e $\rightarrow$ 2a $\rightarrow$ 3a $\rightarrow$ 4a $\rightarrow$ 5e,
   - the second Brayton stage is represented by the process chain 1e $\rightarrow$ 2b $\rightarrow$ 3b $\rightarrow$ 4b $\rightarrow$ 5e, driven by the recycle compressor C_2.

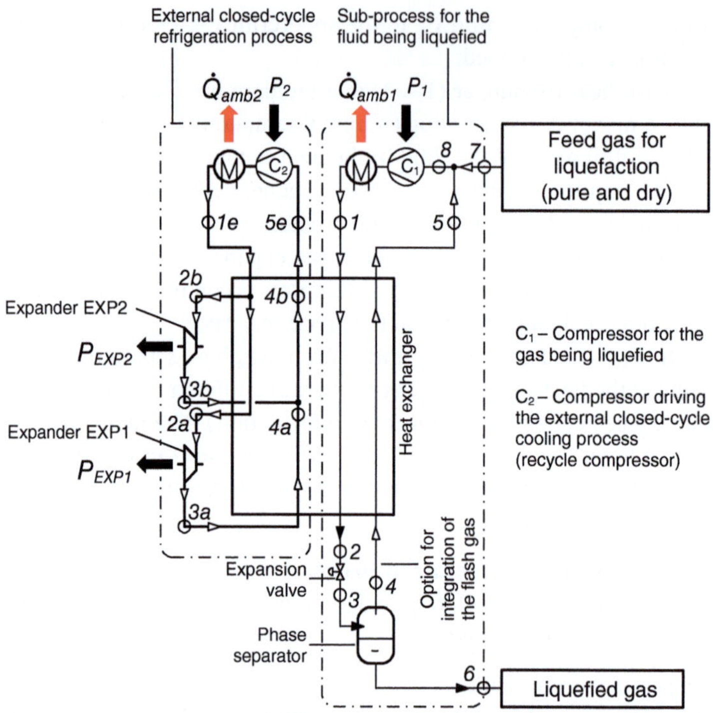

Figure 8.10 Example of a liquefaction process with external cooling, derived from the dual-expander Claude process.

The main process features are:

- Both subprocesses are thermally connected by a common heat exchanger, which consists of multiple passages for different material streams;
- The cold stream 3b from the outlet of expander EXP2 flows through the middle part of the heat exchanger (process 3b → 4b) and cools the Joule–Thomson stream as well as the high-pressure stream to the expander EXP1 there;
- The cold stream 3a from the outlet of expander EXP1 flows through the lower part of the heat exchanger (process 3a → 4a) and cools the Joule–Thomson stream;
- In the example shown, the flash-gas is provided through the heat exchanger to the inlet of compressor C_1. Since this gas provides some cooling capacity, the main refrigeration process is slightly relieved;
- The compressors C_1 and C_2 operate at different pressure levels, meaning there are additional degrees of freedom.

This kind of process is used, for example, for medium-scale natural gas liquefaction, with natural gas as the feed gas. The refrigeration process (two Brayton stages) can use nitrogen, methane, or a nitrogen-hydrocarbon mixture (for example, a nitrogen-methane mixture) as its working fluid. In the case of natural gas liquefaction, the feed gas [entering the system] is delivered at high pressure. In this case, the compressor C_1 is not required.

8.1.5.1.2 Other Process Examples

Two typical examples of a process with external cooling are:

(1) The mixed-fluid Joule–Thomson process for natural gas liquefaction, where the natural gas part is designed as a Joule–Thomson cycle and the main refrigeration cycle is based on a mixed-fluid Joule–Thomson process, see Section 8.4.5.1;
(2) The helium Brayton cycle for hydrogen liquefaction, where the hydrogen feed gas is cooled by an external refrigeration process based on a helium Brayton or helium Claude process [4, 5].

8.1.5.2 Management of Flashgas

The design of liquefaction processes with external cooling often involves the problem of the flashgas being generated after the Joule–Thomson expansion.[3] Depending on the temperature before the expansion, the flashgas fraction may amount to 10% of the total Joule–Thomson flow or even higher. In terms of relative values, this is a small fraction in comparison to other flows managed in a liquefier. However, in terms of absolute numbers, this flowrate is relatively high due to the high flowrates of gases being processed in industrial plants. Therefore, it has been managed in an appropriate way.

Table 8.3 illustrates four of the most prevalent options applied by process designers:

The first option is often favored by process design engineers. The main idea is to subcool the Joule–Thomson stream far enough so that no flashgas is generated during expansion, at least during normal operation of the plant. Since the precooling is achieved by means of an external cooling process, this option presumes that the correspondingly low-temperature level is achieved by the applied external cooling cycle. It incurs an increased power requirement for the corresponding external cooling process since generation of cooling capacity at lower temperatures requires more power according to the second law of thermodynamics. However, the performance of the overall liquefaction process usually improves given that the liquefaction rate is maximized (with no flashgas). The main disadvantage here relates to the increased capital expenditures that are necessary to realize such an efficient process. The decision for/against such a process depends on the extent to which the additional CAPEX may justify the reduced operational expenditures.

The second option is the recycling of the flashgas as shown in Figure 8.10. This is a two-step approach:

(a) first, the cold flashgas from the phase separator is routed from the cold end to the warm end via a separate passage of the main heat exchanger. For some designs of the heat exchanger – for example, for a spiral-wound heat exchanger (SWHE) – it can be challenging to integrate an additional passage on the low-pressure side. Therefore, in such cases, other process options should be considered.

3 Usually, in the process with internal cooling, the flashgas is recycled to the inlet of the main recycle compressor. Potentially, the options discussed here for processes with external cooling are also applicable for processes with internal cooling.

Table 8.3 Common process design options for flash-gas management.

	Process option	Advantages	Disadvantages	Comment
1	Extensive subcooling of the Joule–Thomson stream before the expansion, with the aim of zero flashgas	Zero flashgas, therefore, no need for further measures; high efficiency	Requires appropriate process design to realize extensive subcooling before Joule–Thomson expansion	
2	Recycling/integration of flashgas into the main liquefaction process	No emission of flashgas	Several additional hardware components may be required, at additional cost. Some issues for mechanical design of heat exchanger	
3	Use of flashgas for other purposes on site (not related to the cryogenic process)	No emission of flashgas		For example, the flash of natural gas can be used as fuel gas for gas turbines or gas motors on site
4	Release of flashgas into the environment (directly or by means of flaring for burnable substances)	Low-cost solution	Emission of flashgas	For example, nitrogen flash can be released into the environment

(b) in the second step, the flashgas is injected into the feed stream. If the pressure of the flashgas is lower than pressure of the feed gas, the flashgas has to be compressed in a separate compressor (not shown in Figure 8.10) before injection.

Both operations (heating in the heat exchanger and compression of the flashgas) cause additional capital expenditures for the corresponding hardware components. That is the main tradeoff for this process option.

Option 3 as well as option 4 are self-explanatory.

8.1.5.3 Internal Versus External Cooling

By nature, processes with external cooling are more complex and sophisticated than processes with internal cooling, given that they require

- multiple compressors,
- heat exchanger[s] with more passages for different substances,
- more auxiliary subsystems, such as inventory systems for two different working fluids, or
- additional equipment for flashgas management.

Therefore, the general expectation is that the process with external cooling is more cost-intensive. On the other hand, the process with external cooling provides more degrees of freedom for process design and optimization. Therefore, this process may perform better in terms of efficiency and operational expenditures. A general recommendation is difficult to make. The decision for or against a process with external cooling cycle depends on the application and local conditions (existing infrastructure, transport limitations, feed gas pressure, etc.).

In some applications, feed-gas for liquefaction is delivered/supplied at a relatively high pressure. Two typical examples are

(1) feed gas for natural gas liquefaction (further details can be found in Section 8.4.2), where the pressure is usually higher than 50 bar and
(2) hydrogen feed for hydrogen liquefaction produced by means of combination of steam methane reformer and a pressure swing adsorption (PSA) system, see Ref. [6], where the feed gas pressure is usually higher than 20 bar.

In this special case (feed gas at high pressure), a separate feed compressor (at corresponding cost) is not required. Therefore, a liquefaction plant with an external cooling process has better chances than one with an internal cooling process, due to comparable quantity and size of equipment, but potentially better process efficiency.

8.2 Nitrogen Liquefaction

A nitrogen liquefier (or "nitrogen liquefier unit," NLU) is almost always combined with an air separation plant. Physically, it is either directly integrated into an air separation plant or designed as a "stand-alone" unit to be placed adjacent to the air separation plant and connected by corresponding piping and instruments. The stand-alone plant is considered in this chapter.

8.2.1 General Setup, Feed Gas, and Products

An air separation unit (ASU) is usually able to supply a considerable amount of pressurized gaseous nitrogen (PGAN) at 5–6 bar and an even greater amount of pressureless gaseous nitrogen (GAN). Both of these streams can be used as feed gas for the nitrogen liquefier. The PGAN is the preferred feed, since the feed gas will be compressed further anyway to the high-pressure level inside the NLU, and because compression of PGAN requires less power and simpler equipment (a more compact compressor with less stages), as compared to compression of GAN.

The feed gas is liquefied inside the NLU. The product stream – liquified nitrogen – is usually produced at the same pressure level as the feed (at 5–6 bar); this product is often referred to as pressurized liquid nitrogen (PLIN). Alternatively, the liquid can be produced at low-pressure level (close to ambient pressure), this product is referred to as liquid nitrogen LIN.

Figure 8.11 shows the simplified schematic diagram of the overall setup of the NLU–ASU combination. The air separation plant supplies the NLU with PGAN. The PGAN is liquefied there to PLIN and/or LIN.

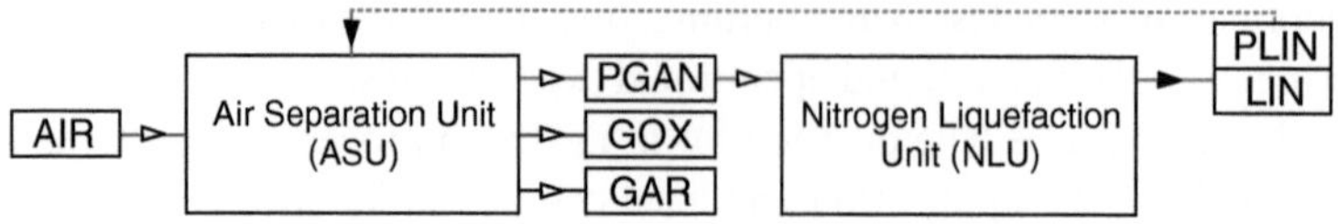

Figure 8.11 General setup for stand-alone nitrogen liquefier.

Optionally, a fraction of the LIN/PLIN can be returned to the ASU to cover, fully or partially, the refrigeration demands of the ASU. This option (referred to as LIN assist) is very useful if the cold generation function or unit inside the ASU is out of operation due to failure or maintenance. This connection allows the full functionality and availability of the ASU to be upheld even under such circumstances.

Virtually, in the case of ASU–NLU combination and LIN assist, the ASU does not need its own cold generation at all, because it can be supplied continuously with cold/cooling capacity in the form of PLIN/LIN from the nitrogen liquefier. The ASU process may be simplified, with corresponding advantages in terms of cost efficiency.

8.2.2 Power Requirements

Exact power requirements depend on the specific process used, but a very rough estimation can be made using the concept of "min liquefaction work" introduced in Section 8.1.2 and Eq. (8.7) ($\eta_{LIQ} = P_{rev}/P$).

For pressureless nitrogen GAN as feed, the min liquefaction work amounts to 771 kJ/kg. Assuming NLU efficiency of 40%, the value of 1927 kJ/kg = 535 kWh/t for required specific power can be derived. This results in power costs of 27 \$/t if a power rate of 0.05 \$/kWh is assumed.

The power requirement depends on the feed gas pressure since the min liquefaction work depends on the feed gas pressure as given in Table 8.4. If the pressurized nitrogen PGAN is used at a pressure of 6 bar, the required min liquefaction work is only 611.4 kJ/kg, i.e. 20% less than min liquefaction work for pressureless GAN (771 kJ/kg). This is the main motivation for using PGAN as feed instead of GAN, as discussed in Section 8.2.1.

8.2.3 Main Elements

The main elements of a nitrogen liquefier are the recycle compressor, heat exchanger, and turboexpander.

8.2.3.1 Compressor

The recycle compressor is the most cost-intensive component of a nitrogen liquefier. The cost of the compressor (including the motor) amounts to one-quarter to one-third of the total capital expenditures. In terms of operational expenditures, the recycle compressor is even more prominent, since the power consumption of a liquefaction plant is determined mainly by the power consumption of its recycle compressor, and the largest share of maintenance and service expenditures relates to those of the recycle compressor. All these facts reflect the extraordinary importance

Table 8.4 Min liquefaction work for nitrogen, depending on feed gas pressure, T_{amb} = 300 K.

Pressure	Min liquefaction work			
bar	kJ/kg	kJ/mol	kJ/nm^3	kWh/t
1	770.9	21.59	963.6	214.1
2	709.2	19.86	886.5	197.0
3	673.1	18.85	841.4	187.0
4	647.5	18.13	809.4	179.9
5	627.6	17.57	784.5	174.3
6	611.4	17.12	764.3	169.8

of the recycle compressor for the overall performance and technical economics of the plant.

The thermodynamics of the compressor as a process unit has already been described in Section 5.5. The specifics of centrifugal machines used for nitrogen liquefaction, as well as typical restrictions and limitations, are described in Section 6.5.2.

8.2.3.2 Heat Exchanger

The industrial nitrogen liquefier design is based on the brazed aluminum plate-fin heat exchanger (PFHE). The mechanical design is described in Section 6.2.1.

8.2.3.3 Large Expander

In nitrogen liquefaction, medium-size turboexpanders with oil bearings are used. Their power usually lies within the range of 0.5–5 MW. The same type of expander is used for air separation or for small-scale natural gas liquefaction. For large-scale natural gas liquefaction, the power requirements are higher: a capacity of 20 MW or even higher is needed.

The thermodynamics of the expander as a process unit as well as typical restrictions and limitations have already been described in Section 5.2. The loading devices that receive the mechanical power generated during expansion are described in Section 5.2.5. For nitrogen liquefaction, a single-stage centrifugal compressor (often referred to as a booster) is used as the loading device, "because of its simplicity and ease of control" according to Haselden [7].

8.2.3.4 Impact of Heat Exchanger Design on Process Design

In the case of nitrogen liquefiers, the availability of brazed aluminum PFHEs plays a crucial role for the process design, since this type of heat exchanger allows a very high degree of versatility:

- the number of material streams managed in heat exchanger is quite unlimited,
- material streams can be withdrawn from the heat exchanger or introduced into the heat exchanger at any location/position.

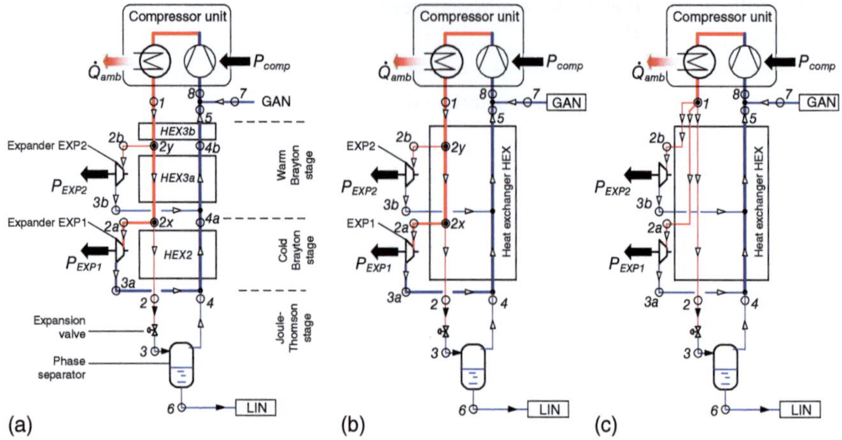

Figure 8.12 Modifications of the dual-expander nitrogen Claude liquefaction process.

Figure 8.12 illustrates how this versatility may be applied for process modifications:

- Figure 8.12a shows the dual-expander Claude process described in Section 8.1.3.2.1. It includes three separate heat exchangers (HEX2, HEX3a, and HEX3b), each with two channels for treating two material streams.
- Figure 8.12b illustrates a similar process, whereas the three heat exchangers (HEX2, HEX3a, and HEX3b) are combined into a single heat exchanger unit by using a brazed aluminum PFHE. Basically, this heat exchanger consists of two large channels going from top to bottom, for the high-pressure and low-pressure material streams. The internal mechanical design of the heat exchanger allows the extraction of both expander streams from the high-pressure channel as well as injection of the material stream 3b (after the expansion in the warm-expander) into the low-pressure channel at the corresponding positions. This "single heat exchanger" design offers advantages in terms of compactness and reduced piping efforts: both expanders are connected to the corresponding nozzles of the heat exchanger during installation.
- Another heat exchanger configuration is shown in Figure 8.12c. It consists of four channels: three for high-pressure material streams and one for low-pressure stream. At the compressor unit outlet, the compressed gas is separated into three material streams – the Joule–Thomson stream, the cold-expander stream, and the warm-expander stream. These streams are directed to the corresponding expansion devices through separate channels of the heat exchanger.

These three process modifications have no impact on the main process parameters: the temperatures, pressure, and flowrates remain the same.

At first glance, such process/hardware modifications may appear to be of minor significance. But sometimes even small changes yield valuable results and can help to achieve the best possible process performance. The changes in process morphology can provide input for new and fruitful ideas.

8.2.3.5 Impact of Expander Power on Process Design

In the dual-expander nitrogen Claude cycle, the power generated in expanders is relatively high, in comparison to the power required for recompression of the expander stream to the initial high-pressure level. This thesis can be proven by means of the following procedure:

According to Eq. (5.10), the power produced in an expander is defined as

$$P^{EXP} = \eta_s \cdot \dot{m}^{EXP} \frac{R}{M_r} \cdot T_{in}^{EXP} \cdot \frac{\kappa}{\kappa - 1} \left[1 - (\pi)^{\frac{1-\kappa}{\kappa}}\right],$$

where $\pi = p_{HP}/p_{LP}$ – pressure ratio

The power required for compression of the same flow back to high pressure is defined by Eq. (5.35):

$$P^{COMP} = \frac{1}{\eta_T} \cdot \dot{m}^{EXP} \frac{R}{M_r} \cdot T_{in}^{COMP} \cdot ln\,(\pi)$$

The next equation describes the relationship between these two powers:

$$\frac{P^{EXP}}{P^{COMP}} = \eta_s \cdot \eta_T \cdot \frac{T_{in}^{EXP}}{T_{in}^{COMP}} \cdot \frac{\kappa}{\kappa - 1} \frac{[1 - (\pi)^{(1-\kappa)/\kappa}]}{ln(\pi)} \tag{8.17}$$

Assuming

- $\kappa = 1.4$ for nitrogen
- $\pi = 5 \cdots 10$
- temperature at compressor inlet = 300 K
- isentropic efficiency of expander = 85%
- isothermal efficiency of compressor = 70%

we can estimate

$$\frac{P^{EXP}}{P^{COMP}} = (0.44 \ldots 0.48) \cdot \frac{T_{in}^{EXP}}{300\,K} \tag{8.18}$$

In the dual-expander nitrogen Claude cycle, the temperature at the expander inlets is relatively high. For the cold-expander, it is 160 ± 20 K. For the warm-expander, it is 270 ± 20 K. At these values, according to Eq. (8.18), the power generated in the cold-expander is 23–25% of the power required for recompression of this flow, and the power generated in the warm-expander is 39–42% of the power required for recompression. Both of these values are quite high.

Learning 8.1 For an efficient nitrogen liquefaction process, the expander power must be utilized or integrated into the overall process.

8.2.4 Process Design for Nitrogen Liquefaction

A large number of liquefaction processes and process modifications were developed over the past 100 years by inspired engineers. It is not necessary to consider all these processes individually; the more efficient method is to develop a general understanding of the underlying principles and to employ this knowledge for analysis and synthesis of processes.

The nitrogen liquefaction process is normally based on the dual-expander Claude process. Numerous process modifications of this process relate mainly to:

- number and connection sequence of expander–booster units,
- position of the warm Brayton stage[s] and the cold Brayton stage[s] to each other (they can be placed with an overlap, without overlap, or even with a distance) in terms of temperature range.
- minimizing the complexity of heat exchangers and piping,
- optimizing the pressure level in the process by targeting the recycle compressor with reduced number of compression stages.

8.2.4.1 Guidelines/Principles of Process Design for Nitrogen Liquefaction

Despite the broad diversity of these processes, several common rules or guidelines can be derived from the analysis. For example, as learned in the Section 8.2.3.5,

1. the expander power is commonly utilized and integrated into the process by means of expander–booster units,
2. frequently, the process with three separate warm streams – Joule–Thomson stream, warm-expander stream, and cold-expander stream as shown in Figure 8.12c – is a starting point for process design.

Further principles include the following:

1. **Compression:**
   Since at least three compression devices – recycle compressor, booster stage driven by cold-expander, booster stage driven by warm-expander – are used in the process, there are several degrees of freedom available for the compression procedure: the process engineer is free to decide, which stream [or streams] is compressed in which order in which device. Theoretically, at the end of this step, all three of these streams can be compressed to three different pressure levels. The usual approach is that all three warm streams are first compressed to a medium-pressure level by means of the common recycle compressor. In the next step, these streams are pressurized further by means of both boosters driven by expanders. The most important goal is to compress the Joule–Thomson stream to the highest possible pressure.
2. **Expansion:**
   Although the pressure at the warm-expander inlet may differ from the pressure at the inlet of the cold-expander inlet, both expander streams are often expanded to the same low-pressure level. The delicate part is to avoid limitations (described in Section 6.3.1), mainly regarding the pressure ratio. Another aspect relates to volumetric flows of the expander stream and the stream compressed in the booster stage. The optimum situation would be if they were comparable. Under such conditions, from the fluid mechanics perspective, it is simpler to build a machine with the best possible performance/high efficiency, since the

optimal rotational speed and optimal geometry for expander impeller are close to the speed and geometry of the booster impeller.

3. **Free Process Parameters:**
   The most important process parameters are:
   - pressure at the outlet of the recycle compressor,
   - temperatures at the expander inlets,
   - split between the Joule–Thomson stream, cold-expander stream, and warm-expander stream.

8.2.4.2 Core Process for Nitrogen Liquefaction

An industrial nitrogen liquefier is a relatively complex system that usually consists of a core process and customized process extensions. The core process is discussed in this section.

One version of a dual-expander Claude process is shown in Figure 8.13. It contains all the typical features and the main elements used in industrial processes for nitrogen liquefaction:

(1) recycle compressor,
(2) two expander–booster units,

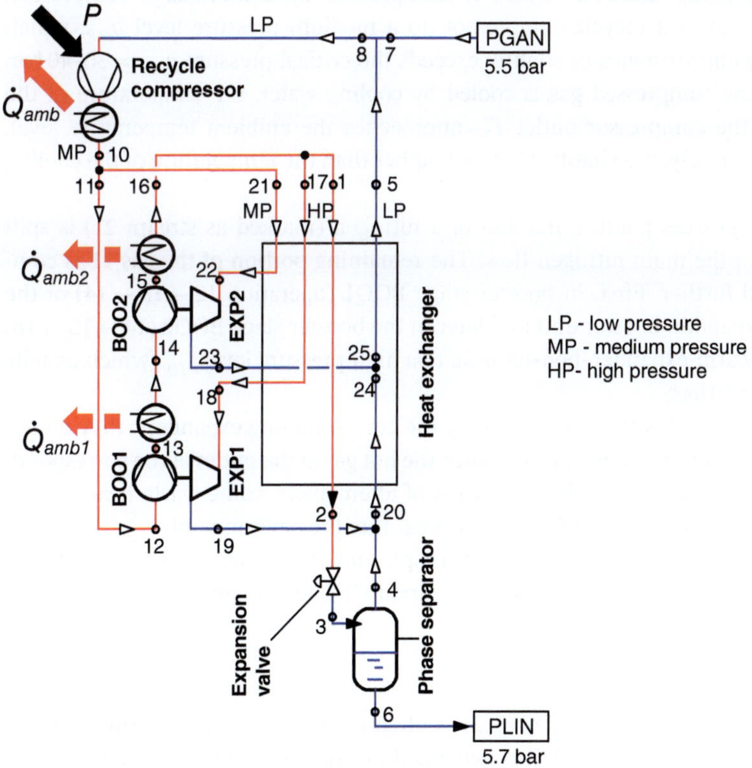

Figure 8.13 Core process for nitrogen liquefaction.

(3) heat exchanger with passages for three warm material streams and one cold material stream,
(4) Joule–Thomson valve,
(5) phase separator,

besides piping, instrumentation, thermal insulation, etc.

Three pressure levels are used in this process (LP – low-pressure level, MP – middle-pressure level, and HP – high-pressure level). The Joule–Thomson expansion is processed from the highest to the lowest pressure level.

The process functions as follows:

Feeding:

(a) The feed gas – pressurized gaseous nitrogen PGAN (state 7) – is supplied by an ASU. The feed gas is mixed with the low-pressure stream (state 5), leaving the heat exchanger at the warm end. The pressure level at this point corresponds to the PGAN pressure level, which is defined by the ASU process design and current operating conditions. It typically lies within the range of 5–6 bar. This pressure presets the low-pressure level of the entire liquefaction process p_{LP};

Compression:

(b) The resulting material stream is compressed by a multistage compressor, referred to as a recycle compressor, to a medium pressure level p_{MP}, which usually approximates or slightly exceeds the critical pressure p_{MP} = 30–40 bar. Since the compressed gas is cooled by cooling water, the temperature of the gas at the compressor outlet T_{10} approaches the ambient temperature level. More precisely, it is slightly (7–10 K) higher than the temperature of the cooling water.
(c) At this process point, a fraction of a nitrogen (marked as stream 21) is split off from the main nitrogen flow. The remaining portion of the gas 11 is compressed further. First, in booster stage BOO1 (operation 12 → 13 → 14) of the cold-expander–booster unit and later in the booster stage BOO2 (14 → 15 → 16) of the warm-expander–booster unit, to a high-pressure level p_{HP}, which usually exceeds 50 bar;
(d) Both of these boosters are driven by the corresponding expanders; they do not require external driving power. Since the hot gas at the outlet of the corresponding booster stage is cooled by means of aftercoolers, some waste heats ($\dot{Q}_{amb2}$ and $\dot{Q}_{amb1}$) are removed from the process into the environment.
(e) The high-pressure nitrogen stream is split into 2 streams: the material stream 1 (referred to as "Joule–Thomson stream") and stream 17 (referred to as "cold-expander stream").

Warm-expander Brayton Stage:

(f) The nitrogen stream marked as 21 is often referred to as the "warm-expander stream," since, at the end, it is expanded in expander EXP2, referred to as the "warm[4] expander," to low-pressure level p_{LP} (operation 22 → 23). However, before expansion, this gas is cooled in the heat exchanger (operation 21 → 22). The degree of precooling (and consequently, temperature T_{22}) depends on the process pressures (p_{LP} and p_{MP}). This is adjusted during process design to target a temperature level of 150–190 K after expansion.

(g) The expanded cold gas 23 is fed into the heat exchanger again, where it is injected into the cold low-pressure stream 24. The resulting gas 25 flows to the warm end of the heat exchanger providing cold for cooling of all warm material streams in the upper part of the heat exchanger.

Cold-expander Brayton Stage

(h) The cold-expander stream passes through the heat exchanger, where it is cooled to a temperature level of 150–190 K and then expanded in expander EXP1, referred to as the "cold-expander," to low-pressure level p_{LP} (operation 18 → 19). Due to expansion, the nitrogen is cooled down to approximately 100 K. This cold gas is provided through the entire heat exchanger from the cold end to the warm end and delivers the cold for cooling of all warmer high-pressure streams there.

Joule–Thomson Stage

(i) The Joule–Thomson stream (1) passes through the entire heat exchanger from the warm end to the cold end and is cooled there to approximately 100 K (operation 1 → 2). It is expanded in the Joule–Thomson valve to low-pressure level p_{LP} (operation 2 → 3) and partially liquefied. The flashgas (4) generated during the Joule–Thomson expansion is separated via a phase separator and mixed with the cold stream from the outlet of expander EXP1. The liquid nitrogen (6) can be withdrawn from the phase separator as a product.

The process is driven by the recycle compressor only; since both booster-compressors utilize power generated by expanders, they do not need any external power supply. The power of the recycle compressor depends on the pressure ratio $\pi = p_{MP}/p_{LP}$ and the flowrate of the nitrogen to be compressed there $\dot{m}^{COMP}$, which is the sum of flowrates of the Joule–Thomson-stream $\dot{m}^{JT}$, the warm-expander stream $\dot{m}^{WEXP}$ and the cold-expander stream $\dot{m}^{CEXP}$:

$$\dot{m}^{COMP} = \dot{m}^{JT} + \dot{m}^{WEXP} + \dot{m}^{CEXP}.$$

From the external perspective, gaseous nitrogen PGAN is fed into the apparatus and removed from it as liquid nitrogen PLIN. The pressure of PLIN is slightly higher than the pressure of PGAN due to pressure difference / pressure loss between the outlet phase separator (PLIN pressure) and the outlet heat exchanger (PGAN pressure) required for moving the cold gas through the heat exchanger on the low-pressure side.

4 Two expanders used in the process operate at different temperature levels: the "warm expander" EXP2 operates at higher temperature than "cold expander" EXP1.

8.2.4.3 Process Extension for Pressureless LIN As Product

For the medium-term bulk storage of liquid nitrogen, a specially designed cryogenic vessel – the flat bottom tank – is used. It is a very cost-efficient device: since it is designed for operation at ambient pressure, the wall thickness and hence the material cost are minimized (more information about flat bottom tanks can be found in Kerry [8].

The core process for nitrogen liquefaction produces liquid nitrogen PLIN at an elevated pressure level; this product cannot be filled into the pressureless flat bottom tanks without damaging the tank. The PLIN must be converted into the pressureless LIN before filling. For this purpose, the core liquefaction process is extended by an additional subprocess.

The idea for this subprocess is a very simple one: the basic operation is the isenthalpic expansion of the PLIN to the ambient pressure level. Since the expansion process becomes more efficient if the liquid is subcooled, a heat exchanger – known as the subcooler – is introduced before the expansion valve.

8.2.4.3.1 Simple Subcooling by Flashgas

Figure 8.14 shows a classic version of this process extension, whereby the cold flashgas (after expansion) is used for cooling of the PLIN: this subprocess consists of a subcooler, an expansion valve, and a phase separator. Morphologically, the process is identical to a Joule–Thomson process, and therefore often referred to as the "second Joule–Thomson stage.[5] The PLIN is precooled ($6 \rightarrow 26$) in the subcooler by cold flashgas ($29 \rightarrow 30$) and expanded to the ambient pressure level in the expansion valve 2 ($26 \rightarrow 27$). Most of the PLIN is liquefied and collected in the sump of the phase separator, and can then be removed as liquid nitrogen product LIN (28) at ambient pressure. The flashgas (29) is separated in the phase separator and first passed through the subcooler, then through the main heat exchanger. This quantity is small and can be blown off directly into the environment.

8.2.4.3.2 Subcooling by LIN Evaporation

Another version of the LIN extension is shown in Figure 8.15. It consists of a subcooler and two expansion valves (expansion valve 2 and expansion valve 3). The basic idea is the same – the PLIN is cooled in the subcooler before it is expanded into the flat bottom tank by means of expansion valve 2. The special feature of this process is that the PLIN is cooled in the subcooler so deeply that no flashgas (or only a very small fraction) occurs during this expansion.

For this purpose, a relatively large cooling capacity must be applied in the subcooler. This cooling capacity is generated by a partial stream (26b), which is split from the main liquid nitrogen stream 26 downstream of the subcooler. This slipstream is

- expanded in the expansion valve 3 to the ambient pressure level (operation $26b \rightarrow 29$),

5 Strictly speaking, in terms of thermodynamics, this expansion process cannot be considered as a Joule–Thomson expansion, since the fluid before the expansion valve 2 is not supercritical.

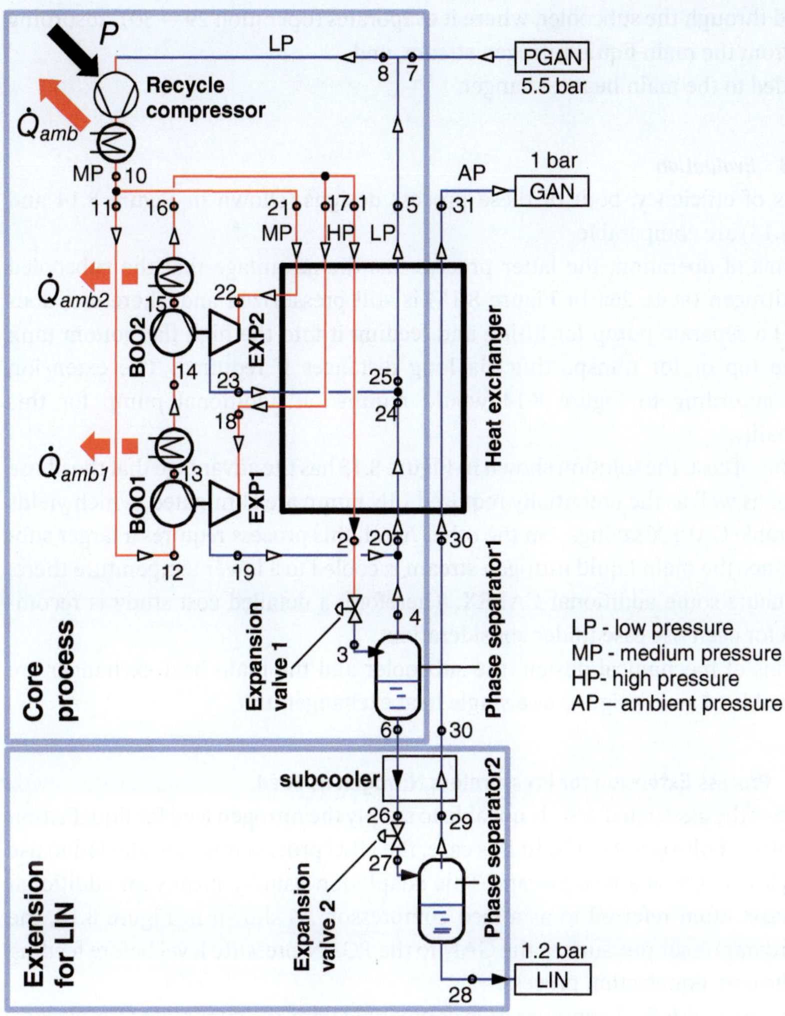

Figure 8.14 Expansion of the nitrogen liquefaction process, subcooling by flashgas.

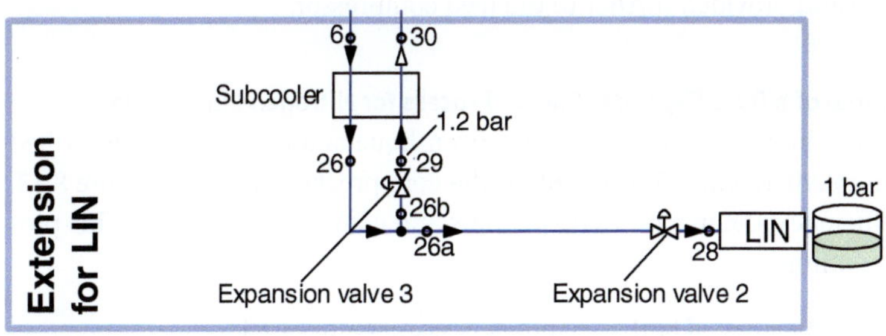

Figure 8.15 Process design for the LIN-extension.

- passed through the subcooler, where it evaporates (operation 29 → 30), adsorbing heat from the main liquid nitrogen stream, and
- provided to the main heat exchanger.

8.2.4.3.3 Evaluation

In terms of efficiency, both of these process designs (shown in Figure 8.14 and Figure 8.15) are comparable.

In terms of operation, the latter process has the advantage that the subcooled liquid nitrogen (state 26a in Figure 8.15) is still pressurized and, therefore, does not need a separate pump for lifting and feeding it into the high flat bottom tank from the top or for transporting via long distances if required. The extension scheme according to Figure 8.14 would require an additional pump for this functionality.

In terms of cost, the solution shown in Figure 8.15 has the advantage that the phase separator as well as the potentially required LIN pump are eliminated, which yields considerable CAPEX savings. On the other hand, this process requires a larger subcooler, since the main liquid nitrogen stream is cooled to a lower temperature there, which incurs some additional CAPEX. Therefore, a detailed cost study is recommended for each use case under consideration.

In terms of mechanical design, the subcooler and the main heat exchanger are often combined and designed as a single heat exchanger unit.

8.2.4.4 Process Extension for Pressureless Nitrogen as Feed

Sometimes the associated ASU is not able to supply the nitrogen feed for liquefaction as pressurized nitrogen PGAN. In this case, the NLU process must be adapted to use pressureless GAN as a feed stream. This adaptation usually means an additional compressor, often referred to as a feed compressor. As shown in Figure 8.16, the feed gas compressor pressurizes the GAN to the PGAN-pressure level before feeding it into the core liquefaction process.

The process with feed compressor is compatible with the subprocess for production of pressureless liquid nitrogen LIN, since it is able to utilize the small amount of GAN produced by the latter (stream 31 in Figure 8.14). This gas is added to the actual feed (32) and provided further to the feed compressor.

8.2.4.5 Example of a Dual-Expander Claude Process for Nitrogen Liquefaction

An example of a dual-expander Claude nitrogen liquefaction process is described in detail in Supplement 9.2.3. It is based on the core process shown in Figure 8.13. In terms of efficiency, the machines (expanders and compressors) used in this process are fairly common:

- the isentropic efficiency of both expanders is assumed to be 84%,
- the isothermal efficiency of the recycle compressor is specified as 70%,
- both boosters (driven by the expanders) have an isentropic efficiency of 72.5%.

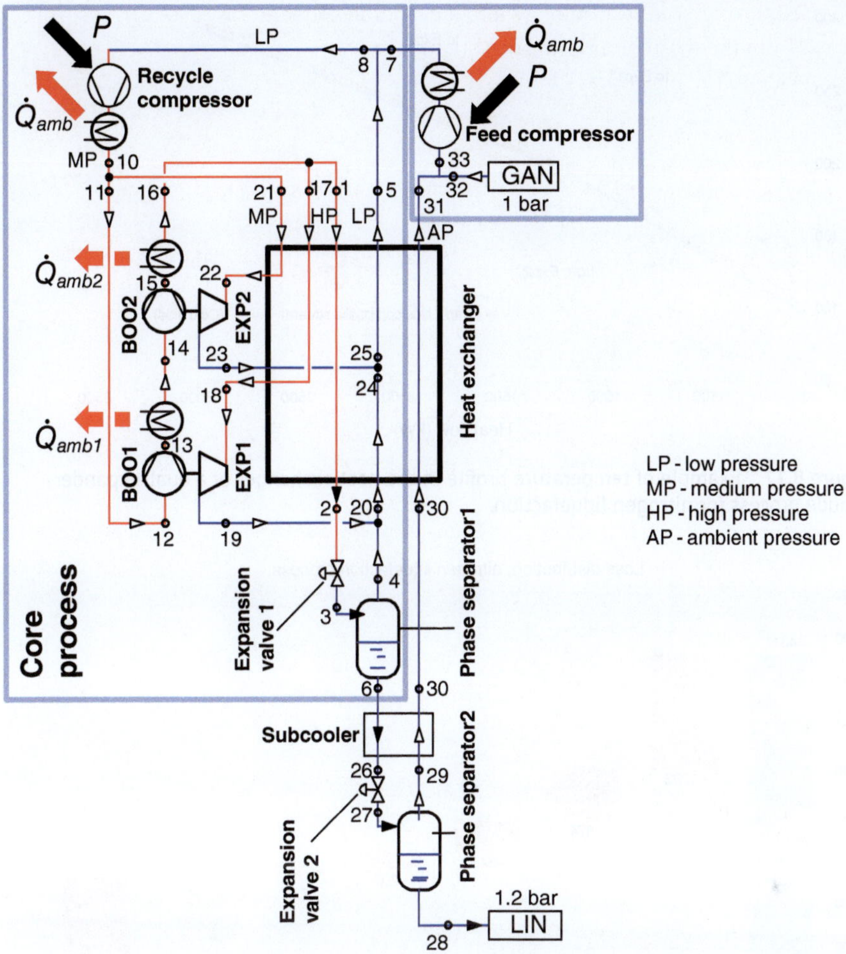

Figure 8.16 Integration of the feed compressor into the core liquefaction process.

The overall efficiency of the process is close to 42%, which is a remarkably high value. It means that the irreversibilities (and therefore the losses in this process) are low. As a result, the compressor power is converted into the liquefaction capacity in an efficient manner.

The temperature profile in the heat exchanger is shown in Figure 8.17. The temperature difference is persistently low, and the MTD in this heat exchanger amounts to less than 4 K. Additionally, the profile is favorable since the temperature difference at the cold end is minimized. Consequently, the loss in the heat exchanger is small, approximately 8% of the total loss. The low temperature-difference at the cold end of the heat exchanger means that the Joule–Thomson stream is cooled to a very low temperature. Consequently, the Joule–Thomson expansion causes little loss, approximately 3% of the total loss. The visible sign for such efficient expansion is the minor flash-gas fraction (in this case, less than 5%).

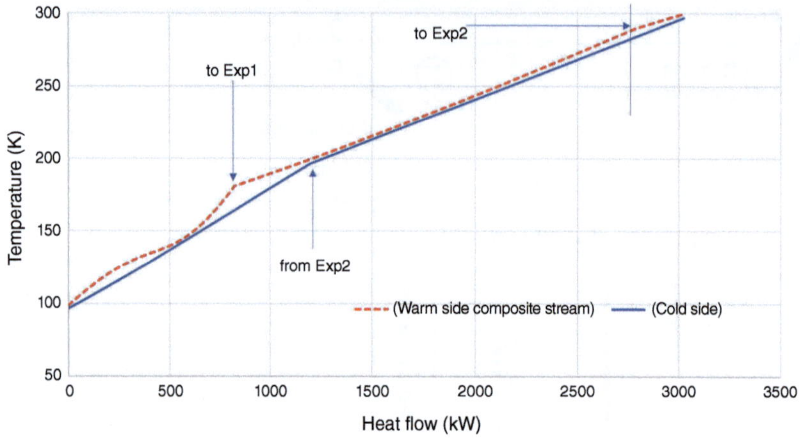

Figure 8.17 Example of temperature profile in the heat exchanger of a dual-expander Claude process for nitrogen liquefaction.

Figure 8.18 Example of loss distribution in a dual-expander Claude process for nitrogen liquefaction

The entire loss distribution in this process is shown in Figure 8.18. The large losses are associated with machines:

- 23% of losses are generated in the "cold" booster–expander unit,
- 14% are generated in the "warm" expander booster,
- 52% are generated in the recycle compressor.

These observations can be summarized as the following learning:

Learning 8.2 The efficiency of a liquefier is essentially determined by the efficiency of machines used. Therefore, the process design for an efficient liquefaction process is focused on development and searching for the best machine concept.

8.2.5 Example of an Industrial Plant

Figure 8.19 shows a stand-alone nitrogen liquefier manufactured by the US company Cosmodyne. The system includes a recycle compressor, two expander–booster combinations, and a heat exchanger placed in a cold box (thermally insulated box). The performance specification is given in Table 8.5. The design liquefaction capacity

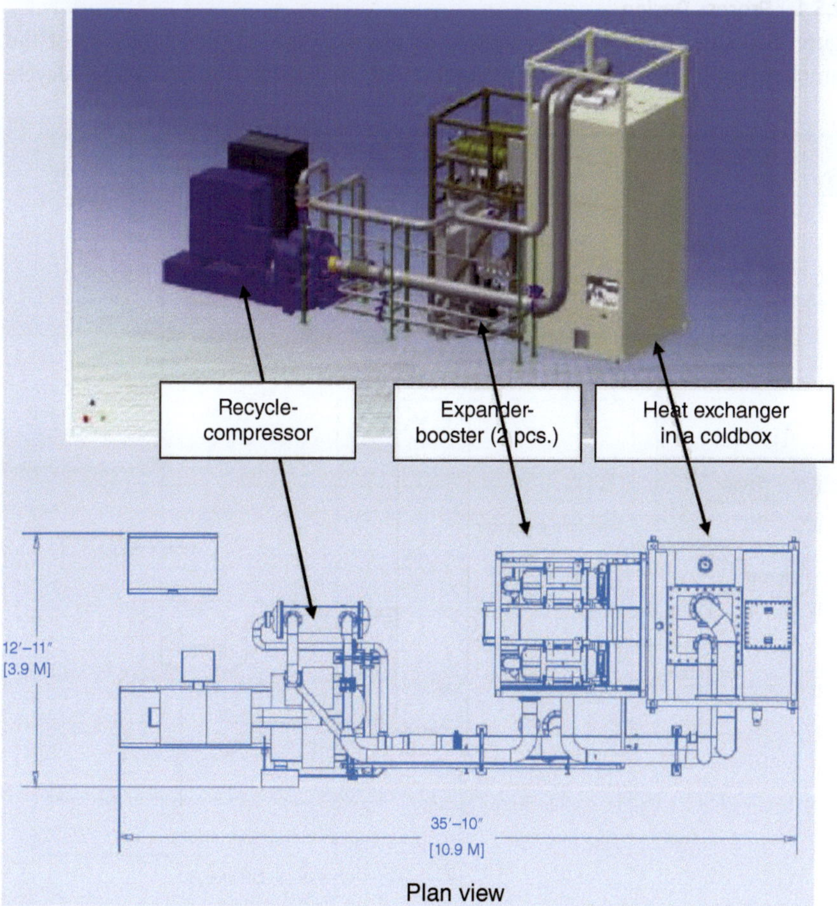

Figure 8.19 Nitrogen liquefier, Cosmodyne. Source: Cosmodyne, LLC.

Table 8.5 Performance specification, ELM 9000.

Liquefaction capacity	9300 nm^3/h LIN = 279 TPD (ton per day)
Product pressure	6.5 bar
Product temperature	82 K
Feed temperature	38 °C
Feed pressure	Min 1.03 bar abs.
Estimated power	5388 kW
Specific power	0.58 kWh/nm^3

is 9300 nm^3/h, which corresponds to approximately 280 tons per day [9]. This is a typical capacity for a middle-size nitrogen liquefier.

8.2.5.1 Process Design

Figure 8.20 shows the process diagram as given in the original brochure of the manufacturer. In terms of hardware components, it consists of a centrifugal recycle

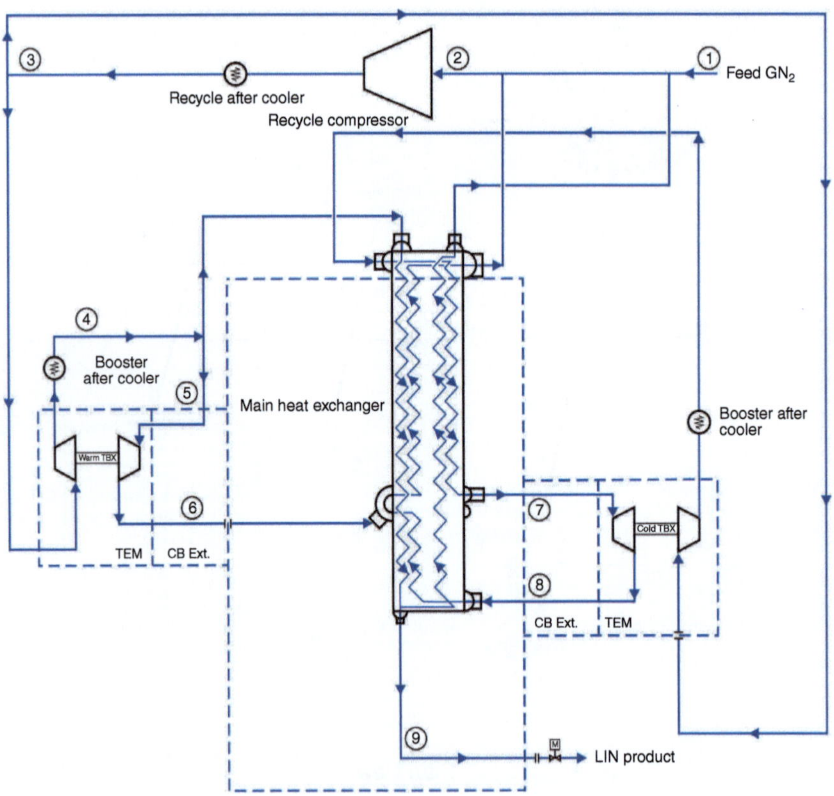

Figure 8.20 Process diagram of Cosmodyne process, as shown in manufacturer's brochure.

compressor, two expander–booster combinations, and aluminum PFHE. The plant is configured to use low-pressure GAN as a feed stream. Obviously, the process is a version of a classical dual-expander Claude process. It is very similar to the core process for nitrogen liquefaction described in Section 8.2.4.2:

Feeding:

(a) The feed nitrogen (1) is mixed with other nitrogen streams from the warm end of the heat exchanger.

Compression:

(b) The resulting nitrogen stream (2) is compressed isothermally in a multi-stage recycle compressor (the state at the outlet is 3).
(c) At this process point, a fraction of a nitrogen (cold-expander stream) is split from the main nitrogen flow.
(d) The remaining portion of the gas is compressed further in a booster stage of the warm-expander–booster unit (at left in the figure) and cooled in the aftercooler to ambient temperature level (4).
(e) This stream is divided again into two material streams: the Joule–Thomson stream and the warm-expander stream (5).
(f) The cold-expander stream is compressed further in a booster stage driven by the cold-expander (at right side in the figure) and cooled in the aftercooler to ambient temperature level.
(g) Both of these boosters are driven by their corresponding expanders. They do not need external driving power. Since the hot gas at the outlet of the corresponding booster stage is cooled by means of aftercoolers, some waste heat is removed from the process into the environment.
(h) Three pressurized material streams are generated:
(1) the Joule–Thomson stream,
(2) the warm-expander stream,
(3) the cold-expander stream,
whereby the Joule–Thomson stream and warm-expander stream have the same pressure.

Warm-expander Brayton Stage:

(i) The warm-expander stream (marked as 5) is expanded in the warm-expander to the low-pressure level. In contrast to the process described in Section 8.2.4.2, this gas is not provided through the heat exchanger at all; it is supplied to the warm-expander directly, where it is cooled by adiabatic expansion.
(j) The expanded cold gas (marked as 6) is fed into the heat exchanger (in the middle of heat exchanger). It flows to the warm end and provides cooling capacity for cooling of all pressurized material streams in the upper part of the heat exchanger.
(k) After leaving the heat exchanger, it is routed to the recycle compressor inlet.

Cold-expander Brayton Stage

(l) The cold-expander stream passes through the main heat exchanger, where it is cooled and then expanded in the cold-expander (at right in the figure) to the low-pressure level. Due to expansion, the nitrogen is cooled down to a temperature level close to the product temperature. This cold gas is provided through the entire heat exchanger from the cold end to the warm end and delivers the cooling capacity for cooling all warmer high-pressure streams there. It is routed to the recycle compressor inlet.

(m) **Joule–Thomson Stage:**

(n) The Joule–Thomson stream passes through the entire heat exchanger from the warm end to the cold end and cools down there. It is expanded in the Joule–Thomson valve to the specified product pressure level of 6.5 bar and the product temperature level of 82 K according to the specification (9).

(o) It looks like no flash-gas is generated during the Joule–Thomson expansion in this process. Probably due to extensive subcooling (first option in Table 8.3) by cold-expander stream. Additionally, in Figure 8.19, one can recognize a fraction of Joule–Thomson stream, which is split from the main Joule–Thomson stream at the cold end. This fraction is routed back through the entire heat exchanger from the cold end to the warm end and later mixed with the feed gas. This fraction possibly is used for subcooling of Joule–Thomson stream in similar way as shown in Figure 8.15.

Similar to the process described in Section 8.2.4.2, the entire process is driven solely by the recycle compressor. Since both booster stages are driven by expanders, they do not require any external power supply.

8.2.5.2 Mechanical Design

The mechanical design of the NLU is shown in Figure 8.19. On the left side of the figure, one can recognize a recycle compressor unit, which is connected by pipelines to the heat exchanger placed in a cold box (on the right side). Two expander–booster units are attached to the left side of the cold box. The footprint for the cryogenic part is roughly 12 × 4 m. The auxiliary units like utilities (power supply, cooling water, instrumentation air), the control unit, and the liquid nitrogen storage are not shown in this figure.

8.2.5.3 Brief Assessment of This Process

According to the process specification, the specific power amounts to 0.58 kWh/nm^3 = 464 kWh/ton. This corresponds to a total liquefier efficiency of η_{LIQ} = 46%, which is a relatively high value, but it is not optimal. It can be presumed that it is rather a cost-optimized than an energy-optimized process. One indication for this hypothesis is, for example, that the warm-expander stream is not conducted through the heat exchanger before expansion but provided to the expander directly. This process modification is probably not optimal in terms of efficiency. But it is very efficient in terms of cost, since: (i) the heat exchanger design becomes easier, as it contains less passages, (ii) the piping around the warm-expander becomes simpler, and the corresponding piping effort is reduced. Another indication of the "cost-efficient" process design is that the subcooler is integrated into the main heat exchanger, with corresponding savings in CAPEX.

8.2.6 Operation of Nitrogen Liquefier in Combination with ASU

The setup shown in Figure 8.11 demonstrates the principle of functional separation extremely clearly:

- By nature, the liquefier's main function is the efficient generation of the cooling capacity required for liquefaction purposes. The NLU's process is optimized for this job. In this combination, the NLU performs the cold generation for liquefaction and for the ASU in an optimum manner.
- The main function of an ASU is the separation of the air. This setup offers the advantage that the ASU does not need the equipment and components for an active internal cold generation. The process can be designed in a much simpler and efficient way, with focus on the best separation performance, such as increased recovery rate or high purity for all products.

8.2.6.1 Liquid Nitrogen as Cold Carrier for Other Liquid Products

From the process design perspective, the option of returning the liquid nitrogen to the ASU, shown as a dotted line in Figure 8.11, is an interesting one. Functionally, the liquid nitrogen returned to the ASU is a kind of cold carrier. Inside the ASU, this cooling capacity can be used for production of other liquid products, such as liquid oxygen (LOX) and liquid argon (LAR), as shown in Figure 8.21. From an external perspective, it appears that the PLIN is converted to LOX and LAR.

8.2.6.2 Flexible Operation for Optimization of Power Bill

The combination of an ASU and a separate nitrogen liquefier provides a high degree of operational flexibility in terms of load range and ramping rate. It means this process is able to switch quickly between

(a) normal operation (specified product quantity and corresponding energy consumption),
(b) turn-down operation (reduced product quantity and therefore reduced energy consumption) and

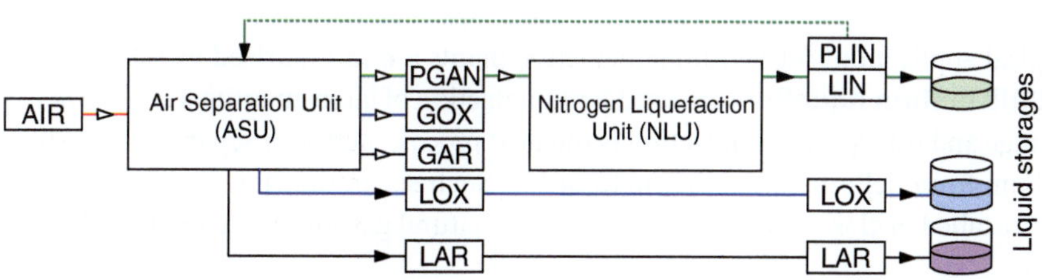

Figure 8.21 Combination of air separation unit and nitrogen liquefier for production of other liquid gases.

(c) max-load operation with increased production and consequently higher energy consumption.

This operational flexibility allows efficient operation of a cryogenic plant in electrical grids that are subject to wide fluctuations in power tariffs.[6] The electricity bill can be optimized through use of the following strategy:

(a) maximum load (maximum liquid production) whenever cheap energy is available,
(b) minimum load (minimum liquid production) whenever the power tariff becomes expensive.

Usually, the conventional ASU with integrated liquefier cannot follow such rapid fluctuations due to its complexity. Consequently, it cannot benefit much from the flexible power tariffs. However, a combination of ASU–NLU can do this since the NLU part, which is the largest consumer of the power, can be switched on and off, independently from the ASU part. Due to the fact that liquefaction requires a lot of power, this switch-on/switch-off operation of the NLU provides considerable savings in the power bill.

The most salient disadvantage of the ASU–NLU combination are the increased capital expenditures (in comparison to the integrated solution), since two separate cold boxes and correspondingly higher effort for engineering and manufacturing are required for this system.

8.2.7 Status Quo Nitrogen Liquefaction

Nitrogen liquefaction systems are increasingly becoming a commodity. In terms of process design, the technology is well understood. The know-how is focused on the design of hardware components – centrifugal compressors, expanders, and efficient heat exchangers.

Further improvements can be achieved by further development of hardware components in both directions: increasing efficiency and reducing costs. Since the rotating machinery (compressor and expander) are primary contributors to OPEX and CAPEX, further development of these components must be pursued.

8.3 Helium Liquefaction

Typical applications for helium liquefaction plants are summarized in Table 8.6.

Bulk helium liquefiers are used for liquefaction of large quantities of helium for storage and transport. Liquefaction is the next process step after separation of crude helium from helium-containing natural gas. Such separation/liquefaction systems are installed in close proximity to appropriate natural gas sources or fields. A facility

6 This can be the case e.g. for electrical grids with a high share of renewable power sources like solar and wind. These power sources are characterized by strong natural fluctuations, resulting in changing power tariffs.

Table 8.6 Typical applications for helium plants.

Application	Cooling temperature (K)	Liquefaction capacity	Process
Bulk helium liquefier	4.3	1000–4000 l/h	Claude
Laboratory helium liquefier	4.3–4.6	20–200 l/h	Claude
Liquefier/refrigeration plants for high-energy physics	1.8 & 4.4 & 40–80	1–20 kW	Claude

for separation and liquefaction of helium is a major investment. This investment can be justified by correspondingly large volumes of recoverable helium. The economics usually work well for large natural gas fields with relatively high helium content of more than 0.1%. However, natural gas often does not contain helium. Large natural gas sources with a suitably high helium fraction are rather a rarity; they are not often explored and developed. Therefore, the demand for this kind of cryogenic plant is not large.

Helium plants for applications in high energy physics are the most sophisticated helium systems, due to challenging requirements and complex functionality that can be provided by such systems. In terms of **functionality**, the typical tasks to be performed are to deliver:

(a) liquefaction capacity (for backup, local LHe-based infrastructure, etc.) as well as
(b) large refrigeration capacity.

The first hard requirement is that the required cooling capacity must be provided at **multiple temperature levels**: for example, at 1.8 K and 4.4 K for superconducting devices (magnets, cavities, and other instruments and components), and at 40–80 K for shielding and precooling purposes.

Another challenging requirement is that the plant must provide high **flexibility** in terms of operation modes. It must be able to operate in a pure refrigeration mode or in a mixed refrigeration–liquefaction mode, in turndown operation, or at maximum cooling capacity.

The most challenging **requirements** are:

- requirements for an extremely low cooling temperature of 1.8 K, since refrigeration at that temperature level, requires sophisticated processes and additional customized/made-to-order hardware components,
- high requirements for plant efficiency. Even in an ideal process, according to the Carnot equation, the generation of cooling capacity at temperatures below 5 K and especially below 2 K requires gigantic amounts of power. In real systems, due to losses, the power demand is even higher, roughly by an order of magnitude. This leads to huge power consumption. Therefore, every improvement in efficiency – even by a small amount – can lead to appreciable reduction in power costs.
- requirements for cost efficiency, which is a more or less self-explaining aspect.

Laboratory helium liquefiers: Demand for **laboratory helium liquefiers** is relatively high, since the number of customers (such as universities and government laboratories) and the corresponding sales are high. It leads to an interesting effect: due to high demand, implementation of new ideas by original equipment manufacturers (OEMs) is executed quickly, in parallel with continuous routine engineering jobs like standardization of hardware components, application of pre-engineered process units, etc. Due to these activities as well as due to current progress in automatization and digitalization, the commercial product becomes better and better – more efficient, more cost-efficient, and more customer-friendly. The attractive product stimulates new sales. This effectively results in a kind of self-reinforcing process: high demand stimulates continuous development, which leads to a more attractive product, and so on. In recent years, laboratory helium liquefiers have essentially become a sort of commodity since almost every university owns such a system.

8.3.1 Overall Helium Liquefier Setup, Feed Gas, and Product

The helium liquefier is usually a part of a superordinate helium liquefaction system, which includes at least:

- storage capacity for liquid helium,
- storage capacity for pure and unpurified gaseous helium at different pressure levels and corresponding compression systems,
- units for purification of unpurified helium,
- devices for filling of liquid helium vessels.

Often, this infrastructure includes a liquid nitrogen storage vessel as well as utilities such as power supply, cooling water, and instrumentation air. Physically, these units are connected by corresponding piping and equipped with the required instruments. An example of such a laboratory installation is shown in Figure 8.22. Figure 8.23 visualizes the main connections.

The liquid helium is supplied to customers/users in transportable cryogenic dewar vessels (100–400 l) shown as item 7 in Figure 8.22. Later, these LHe vessels are returned to the liquefaction facility. In the optimal case, they contain a minimal amount of liquid helium and, correspondingly, they are still cold. In less-than-optimal cases, the vessel is completely emptied and returned warm. The main drawback is that such dewars must be cooled before filling with LHe. For these purposes, an additional amount of liquid helium (and corresponding effort for liquefaction) is required. But the worst case of all is if the vessel is contaminated with other gases or substances. The maintenance of contaminated containers requires extensive effort for cleaning and/or purging, as well as for cooling after such maintenance.

At the customer site, liquid helium is evaporated, and the boil-off gas is partially collected (and partially lost), then compressed to high pressure (100–200 bar) by means of an on-site recycle compressor unit and returned to the liquefier in gas

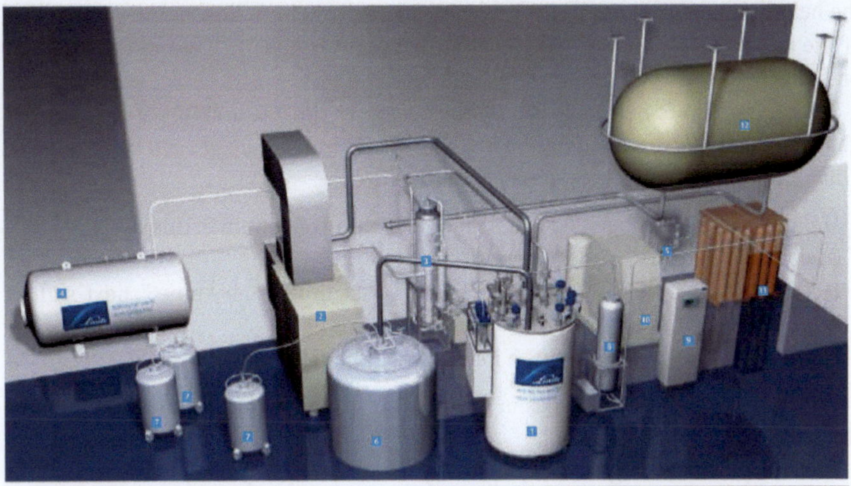

1 - Liquefier,
2 - Recycle compressor,
3 - Oil removal system,
4 - Medium pressure helium storage,
5 - Manifold for connection of storages,
6 - Liquid hydrogen storage vessel,
7 - Dewars for distribution of liquid helium,
8 - Dryer,
9 - Control and instrumentation panel,
10 - Compressor for helium storage system,
11 - High pressure storage tank,
12 - Bag for low pressure gaseous helium

Figure 8.22 Laboratory liquefier with some infrastructure, copyright Linde plc.

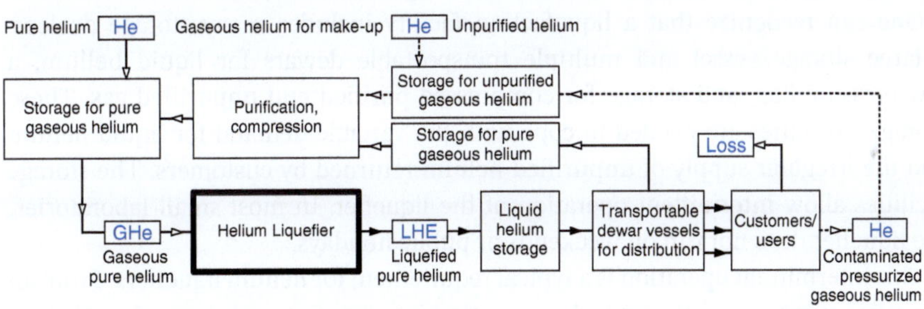

Figure 8.23 Overall setup for a helium liquefier.

cylinders. This helium is usually contaminated by humidity, atmospheric gases, and potentially by compressor oil if an oil-lubricated compressor was used for pressurization. The level of contamination depends on the customer's qualification and the equipment available on the customer's site. In the worst case, it can contain more than 20% impurities. This helium is considered to be unpurified helium. In the helium liquefaction facility, this gas is processed through a number of purification steps (dehumidification, pre-purification, and final purification) before it is stored in a corresponding gaseous storage vessel (marked as 11 in Figure 8.22) at medium or high pressure. This purified compressed helium is used as feed gas for the helium liquefier. Commonly, it is mixed into the helium flow in the suction line of the compressor. The pressure level at this point is usually close to the ambient pressure.

Since a fraction of helium is lost on the customer site, the amount of disappeared helium is usually purchased on the free market. This make-up gas can be sourced in different qualities or purities; typical qualities are:

- Helium 4.6, the number 4.6 means that the fraction of helium in this gas is 99.996% (four times sign "nine," next sign is "six").
- Helium 6.0, this quality means 99.99990% (six times sign "nine," next sign is "zero").
- Helium 2.5, this quality means a helium fraction of 99.5%.

Depending on quality, this gas is provided to "pure" or "unpurified" helium gas storage.

Obviously, the feed gas is liquefied by means of a liquefier (shown as 1 in Figure 8.22). The product – liquified helium (LHe or LHE) – is produced at low pressure, close to ambient pressure. The liquid is collected in a large liquid helium vessel (5000–10 000 l or even more), and placed close to the liquefier (6).

The transportable cryogenic dewar vessels for distribution to customers are filled from this large vessel. During the storage phase, some amount of liquid helium inside these vessels evaporates. Usually, a well-managed liquefaction facility includes a reservoir for collection of boil-off helium gas. In the simplest form, this is a huge bag made of an appropriate helium-tight material (12).

One can recognize that a liquefaction facility includes many storage devices: a large storage vessel and multiple transportable dewars for liquid helium, a low-pressure bag, and storage for compressed purified and unpurified gas. These storage capacities are needed to cope with the variable demand for liquid helium and the irregular supply of unpurified helium returned by customers. The storage facilities allow intermittent operation of the liquefier. In most small laboratories, the liquefier does not run on weekends or public holidays.

This intermittent operation is a typical requirement for helium liquefiers. From an operation perspective, the typical expectation is a very fast cool-down time (hours) from pushing the "ON" button until the total liquefaction capability is achieved. This requirement is new; it is not typical for nitrogen liquefiers or natural gas liquefiers since those plants run continuously for years at a time.

8.3.2 Performance Considerations

8.3.2.1 Power Requirements

A very rough estimation of power requirements can be made using the concept of minimum liquefaction work/power introduced in Section 8.1.2 and Eq. (8.7) ($\eta_{LIQ} = P_{rev}/P$). The minimum liquefaction work for helium amounts to

$$\boldsymbol{p}_{\boldsymbol{rev}} = 6.844\ \text{kJ/g} = 27.4\ \text{kJ/mol} = 855.5\ \text{kJ/(liter of liquid helium)}$$

$$= 0.238\ \text{kWh/(liter of liquid helium)}$$

Assuming an efficiency of $\eta_{LIQ} = 10\%$, the following values for required specific power can be derived.

$$\boldsymbol{p} = 68.44\,\text{kJ/g} = 274\,\text{kJ/mol} = 8555\,\text{kJ/(liter of liquid helium)}$$
$$= 2.38\,\text{kWh/(liter of liquid helium)}.$$

Similar values can be found in the literature of original equipment manufacturers (see Table 8.7 and Table 8.8).

This results in a power cost of

0.12 $/l if a rate of 0.05 $/kWh is assumed or
0.48 $/l for a rate of 0.2 $/kWh.

Table 8.7 and Table 8.8 provide data for power consumption for state-of-the-art helium liquefiers from two different manufacturers. They are on the same order of magnitude.

The specific power for liquefaction can be reduced by 50% if liquid nitrogen precooling is used; this is a great improvement. The corresponding nitrogen demand is about 0.5 l of liquid nitrogen per 1 l of liquid helium. In large industrial liquefaction plants, specific power of 0.5 kW/(l/h) can be achieved by further process optimization and application of more efficient machines [10].

Table 8.7 Performance of a laboratory helium liquefier L70 at temperature <4.4 K, Linde Kryotechnik.

Liquefaction capacity without LIN precooling (l/h)	Liquefaction capacity with LIN precooling (l/h)	Compressor power (kW)	Specific power without LIN precooling (kWh/L)	Specific power with LIN precooling (kWh/L)
20	40	45	2.25	1.13
27	55	55	2.04	1.00
35	70	75	2.14	1.07

Source: Adapted from Ref. [2].

Table 8.8 Performance of helium liquefier Helial, Air Liquide (https://advancedtech.airliquide.com/helial-helium-liquefiers).

	Liquefaction capacity without liquid nitrogen (L/h)	Liquefaction capacity with liquid nitrogen (L/h)	Compressor power (kW)	Expected specific power at maximum efficiency (kWh/L)
HELIAL SL	15–25	30–50	45–55	1
HELIAL ML	35–70	75–150	75–132	0.9
HELIAL LL	110–145	215–330	160–250	0.8

Source: Adapted from Ref. [3].

8.3.2.2 Definition of Liquefaction Capacity for Helium Liquefiers

In contrast to nitrogen liquefaction, the liquefaction capacity of a helium liquefier can be defined in different ways. In the past, the flowrate of the feed gas $\dot{m}^{FEEDGAS}$ was used for characterization. However, recently sales engineers use to provide the data for the flowrate of liquid helium entering the storage vessel $\dot{m}^{LIQ\_VESSEL}$, whereby the last number is usually given for a special type of measurement based on the increase of liquid level in the liquid helium storage vessel (the phase separator downstream from the Joule–Thomson valve) during operation without liquid withdrawal from this vessel. The first reason for this specification is that this measurement in fact corresponds to the normal operation of the helium liquefier. The second reason is that $\dot{m}^{LIQ\_VESSEL}$ is impressively higher than $\dot{m}^{FEEDGAS}$, and therefore this number simply looks better.[7]

The explanation for this phenomenon ($\dot{m}^{LIQ\_VESSEL} > \dot{m}^{FEEDGAS}$) is as follows: in the operation mode without liquid withdrawal, the product – LHe – displaces the cold vaporous helium from the top of the liquid helium storage vessel; the displaced cold vapor returns back in the liquefaction process. In effect, it means the entire liquefaction process is fed from two sources:

(a) from the liquid helium storage vessel by displaced cold vaporous helium $\dot{m}^{COLD\_VAP}$ and
(b) by conventional warm gaseous feed gas $\dot{m}^{FEEDGAS}$.

In this operation mode the flowrate of liquid helium going into the storage vessel $\dot{m}^{LIQ\_VESSEL}$ represents the sum of both feeds:

$$\dot{m}^{LIQ\_VESSEL} = \dot{m}^{COLD\_VAP} + \dot{m}^{FEEDGAS}.$$

This is inherently higher than the flowrate of the warm gaseous feed gas $\dot{m}^{FEEDGAS}$ alone, which actually characterizes the correct liquefaction capacity of a given system.

The difference between the actual liquefaction capacity $\dot{m}^{FEEDGAS}$ and the liquefaction capacity measured in the closed liquid helium $\dot{m}^{LIQ\_VESSEL}$ is roughly 14%, since the density of displaced vaporous cold helium at the dew point $\rho^{VAP} = 16.89\,\text{kg/m}^3$ is a relatively high number in comparison to the density of liquid helium $\rho^{LHE} = 125\,\text{kg/m}^3$. For comparison, the density of liquefied nitrogen LIN at 1 bar $\rho^{LIN} = 806.6\,\text{kg/m}^3$, and the density of vaporous cold nitrogen at 1 bar (dew point) is $\rho^{VAP\_GAN} = 4.6\,\text{kg/m}^3$, which is only 0.6% of LIN-density. For a nitrogen liquefier, the deviation between the actual liquefaction capacity and capacity measured in LIN storage vessel $\dot{m}^{LIQ\_VESSEL}$ is very minor and commonly not considered. But in helium liquefaction, this deviation is significant and must be considered anyway.

8.3.2.3 Performance Comparison

Principally, a fair performance comparison of two different systems/processes is not a complex task – the general approach is well described by a popular adage

7 From the customer's perspective, this is sometimes confusing.

about comparing apples with apples. However, the implementation of this principle in cryogenics requires a high degree of accuracy: the "apple-with-apple" principle presumes that the key performance indicators of two systems being compared are given for:

- the same environmental conditions (cooling water, ambient temperature, and ambient pressure)
- the same feed gas conditions,
- the same product quality (pressure, saturated or subcooled liquid, etc.),
- the same liquefaction capacity definition,

since all these parameters have a significant impact on the power consumption of the compressor.

One problem that often occurs is that the required data are only available for different conditions. In this case, the comparison procedure becomes challenging and requires higher effort and intelligence. The following five-step approach may be useful:

1. In the first step, all available information concerning process efficiency and operating conditions is collected.
2. In the second step, the process simulation is performed for the given operating conditions. The efficiencies of key hardware components – compressors, expanders, and heat exchangers – are tuned to reach the claimed process performance.
3. In the third step, the process simulation is performed again, this time for a different set of operating conditions referred to as a reference case (apples), assuming similar efficiencies[8] for comparable key hardware components. The process performance for the reference-case conditions is quantified in the form of key performance indicators such as specific power and Carnot efficiency.
4. The procedure is repeated for the second process/system.
5. Finally, the key performance indicators are compared for reference-case conditions.

8.3.3 Main Elements and Impact on Process Design

The main elements of a helium liquefier are the compressor, heat exchanger, and expander.

8.3.3.1 Heat Exchanger

Brazed aluminum PFHEs (described in Section 6.2.1) are used for helium liquefaction. In terms of size, these apparatuses are smaller than heat exchangers used for nitrogen liquefaction.

8 An appropriate method is used, if possible, for recalculation of efficiencies of key hardware components under reference-case conditions.

8.3.3.2 Compressor

There are not many hardware components designed and developed specifically for helium liquefiers. The conventional engineering approach is to consider the components available on the market that can satisfy the requirements resulting from the preliminary process design. In the next iteration, the process design is adapted to the performance of available hardware components. This working sequence is absolutely standard for helium liquefaction, especially for compressors. The machines used for helium compression are of screw or reciprocating types (see Sections 6.5.3 and 6.5.4). Originally, these were developed for conventional air conditioning and refrigeration purposes or for common pressurized air applications – e.g. for driving pneumatic devices or instruments. These machines do not match the requirements of a perfect helium liquefaction process. Therefore, the process design is adapted and optimized to the performance and features of this available hardware component.

The operation and application ranges for three of the most popular compressor types are shown in Figure 6.3.4. Unfortunately, the centrifugal compressor is not applicable due to low molecular weight of helium, as described in Section 6.5.2.4. In the past, reciprocating compressors were used for helium liquefaction, but recently, the screw compressor has become established as the primary standard compressor for helium liquefaction. Reciprocating machines are used mainly for auxiliary functions, e.g. for compression of helium to pressures above 20 bar for storage purposes.

An important criterion for application of air screw compressors is that they are originally designed for a suction pressure of 1 bar (ambient pressure) and discharge pressure of 8–10 bar. This discharge pressure is higher than the critical pressure of helium gas. That is an important enabler, since it allows the liquefaction process to be designed based on the Claude process, which is a supercritical process.

Conventional oil-flooded screw compressors do not exactly meet the requirements for helium liquefaction machines. For compression of helium, these machines must be "tuned." This tuning concerns mainly

- tightness of the compressor housing: the target is to avoid leakage of helium into environment, since the cost for makeup helium increases the operational expenditures,
- selection of the compressor oil: an appropriate oil is required in order to guarantee tribology, viscosity, and thermal stability even at the relatively high operating temperatures typical for helium compression.

Since the compressed fluid at the outlet of the compressor is contaminated with compressor oil, efficient oil removal system (ORS) is required, which includes at least the following components:

(a) oil separator
(b) rough filter-coalescer for filtering of large oil droplets
(c) fine filter-coalescer for filtering of fine oil droplets
(d) adsorber for removal of vaporous oil components

In some cases, an additional super-fine coalescer is integrated for filtering out the finest aerosol-like oil droplets. The ORS is a reliable process unit if sized and maintained properly. In normal operation, it does not require external power.

In terms of process design, the ORS is considered during process design as a pressure drop and potentially a minor parasitic bypass gas flow through the small pipeline for the return of separated oil to the compressor inlet. The cooling water system has an impact on the temperature at the outlet of the compressor unit.

8.3.3.3 Helium Expander

For helium expanders, the conventional engineering approach of applying the hardware components available on the market does not work, since no appropriate expanders are available. Therefore, original equipment manufacturers are forced to develop their own corresponding expansion machines. To this end, access to helium expander technology acts as a sort of pre-condition to entering the helium liquefier market.

The first commercial reciprocating devices were developed in 1940s and 1950s. The first turboexpanders were introduced in the 60s and 70s and met with great success. More recently, centripetal machines have dominated the market. This is a unique technology: a considerable part of the know-how for helium liquefaction relates to the expansion turbine since other hardware components (compressors and heat exchangers) are more or less available on the free market.

In helium liquefaction, small-size expanders with gas bearings (see Section 6.3.4) are applied, and the power is usually less than 0.1 MW. This number is by one order of magnitude lower than the power of typical expanders for nitrogen liquefaction.

The thermodynamics of expanders as process units as well as typical restrictions and limitations have already been described in Section 5.2.

8.3.3.4 Impact of Expander Power on Process Design

In helium liquefaction, the so-called gas brake is used for utilization of mechanical power generated during expansion (see Section 5.2.5). This means the generated mechanical power is converted into heat, which is dissipated into the environment or, in other words, simply destroyed. The reasons for this technical solution are not obvious. The approach described in Section 8.2.3.4 for nitrogen liquefiers, together with Eq. (8.17), aids in understanding the underlying logic.

For example, the equation is as follows for the process shown in Figure 8.24, where two serial expander driven by the same compressor are applied:

$$\frac{P^{EXP1}+P^{EXP2}}{P^{COMP}}=\eta_s\cdot\eta_T\cdot\frac{T_{in}^{EXP1}+T_{in}^{EXP2}}{T_{in}^{COMP}}\cdot\frac{\kappa}{\kappa-1}\,\frac{[1-(\pi^{EXP})^{(1-\kappa)/\kappa}]}{ln(\pi^{COMP})} \tag{8.19}$$

Assuming

- $\kappa = 1.67$ for helium
- pressure ratio in each expander $\pi^{EXP} = 2.5\ldots4$
- pressure ratio in compressor $\pi^{COMP} = (\pi^{EXP})^2 = 6.2\ldots16$
- temperature at the compressor inlet $\boldsymbol{T^{COMP}} = 300$ K
- isentropic efficiency of expanders: $\boldsymbol{\eta_s} = 75\%$

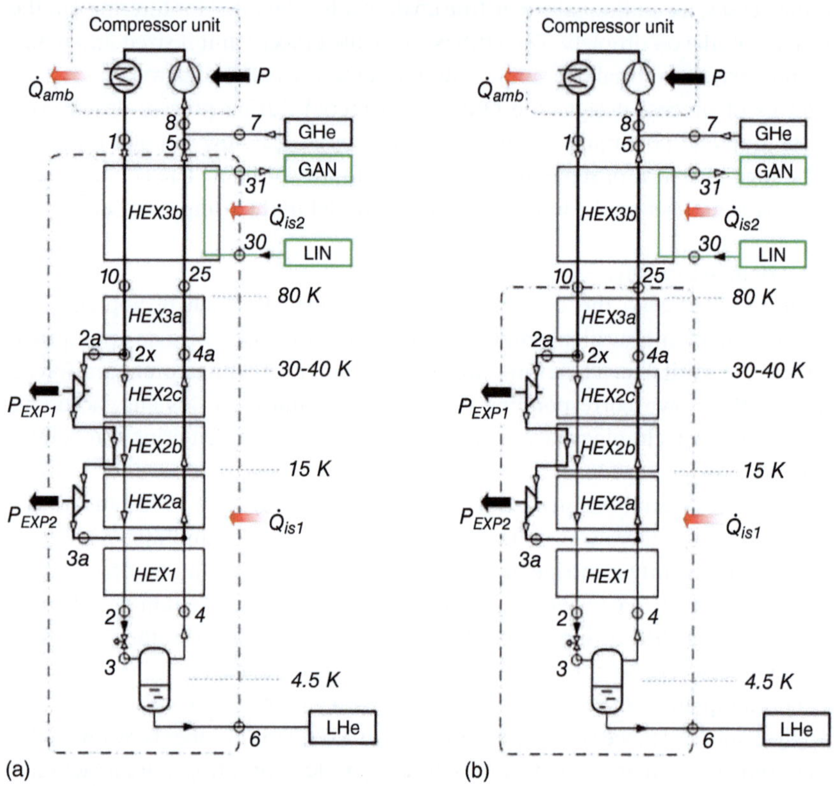

Figure 8.24 Core process for helium liquefier.

- isothermal efficiency of the compressor $\eta_T = 55\%$

we can estimate

$$\frac{P^{EXP1} + P^{EXP2}}{P^{COMP}} = (0.16 \ldots 0.17) \cdot \frac{T_{in}^{EXP1} + T_{in}^{EXP2}}{300\,K} \tag{8.20}$$

In a helium plant, the temperature at the expander inlets is relatively low. For the cold-expander EXP2, it is 15 ± 5 K, and for the warm-expander EXP1, it is 30 ± 5 K. For these values, according to Eq. (8.20), the power generated in both expanders together is roughly 2.5% of the power required for re-compression of this flow. This exercise shows that recovery of expander power has only a minor effect on the total power consumption of a helium plant.

In helium liquefaction, criteria such as simple and reliable design, low maintenance efforts, and cost efficiency are considered to be extremely important, at least in comparison to power savings in 2% order of magnitude. Therefore, application of a gas brake (a simple and reliable device) is better suited to this approach:

Learning 8.3 In helium liquefaction, integration of expander power has a minor effect on the overall power consumption of the plant.

8.3.4 Process Design

Similar to the case of nitrogen liquefaction, a number of helium liquefaction processes and process modifications were developed by engineers over the past 80 years, but it is not necessary to consider all these processes individually; a better way is to develop a general understanding of the underlying principles.

8.3.4.1 Typical Features of Helium Liquefaction Process

The core helium liquefaction process is normally based on the Claude process. Numerous process modifications of this process may be found in published literature. These often relate to development of customized solutions based on helium liquefiers, using several special process units (cold compressors, vacuum pumps, injectors) for customized processes.

In terms of process design, some features common to helium liquefaction processes can be derived from the analysis of processes. For example, as learned in Section 8.3.3.4, expander power is normally not utilized and not integrated into the main process. Other observations of note are:

(a) The number of integrated Brayton stages can vary from 1 to 4.
(b) The serial expansion in two or even three single-stage expanders is often used in a single Brayton stage, since the compressor pressure ratio is higher than can be managed by a single-stage helium centripetal expander (the thermodynamics for this limitation are described in Section 6.3.2.1).
(c) The number of expander streams corresponds to the number of Brayton stages, but the number of expanders is usually higher.

Other features include the following:

Compression:

The number of compressors may vary from a single compressor for small-scale plants to several machines for large-scale systems. In the latter case, they can be switched

- in parallel for higher volumetric flowrate,
- in series either for lowering the suction pressure or for increasing the discharge pressure, or
- as a combination of serial and parallel machines.

Usually, only a single high-pressure stream leaves the compression process unit, even if several machines are used for compression. This high-pressure helium stream is provided through a single ORS and is forwarded to the cold part of the process. Later, this stream is splited into a Joule–Thomson stream and an expander stream[s].

In the case of a single compressor (or parallel compressors), two pressure levels are available:

- high pressure (discharge last compressor),
- low pressure (suction first compressor[s]).

In the case of serial compressors, for process design, three pressure levels are available:

- high pressure (discharge last compressor),
- low pressure (suction first compressor[s])
- intermediate pressure (between two serial compressors).

This means more degrees of freedom are available for process design considerations.

Expansion:

Commonly, for every integrated Brayton stage, the pressure on the "warm" side corresponds to the discharge pressure of the compression unit. The pressure at the "cold" side is either low-pressure or intermediate pressure.

If serial expansion (in two single expanders) is used, intermediate cooling of the expander stream in a heat exchanger between two expansion steps is a typical process feature.

Generally, since the availability and performance of expanders is an issue, the process (expander streams) is often "tuned" to the expanders, i.e. the process parameters are varied to achieve volumetric flows and pressures that are manageable by available machines.

In large helium plants, instead of a pure expansion valve, an expander can be introduced in the Joule–Thomson stage as the so-called the Joule–Thomson expander (see Section 6.3.5).

Free process parameters:

The obvious degrees of freedom for process design are:

- discharge pressure at the compression unit outlet,
- temperatures at the expander inlets,
- pressure ratio[s] in expander[s],
- split between the Joule–Thomson stream and expander stream[s]

LIN-precooling

LIN precooling is the common option for process design.

8.3.4.2 Core Process for Helium Liquefaction

Similar to the nitrogen liquefier, the helium liquefier is a complex system, consisting of a core process and several process extensions.

The core process is discussed in this section. Figure 8.24 shows a Claude process with two expanders and liquid nitrogen precooling, which is used often as a starting point for further process design. The process is derived from the helium Claude process for refrigeration purposes described in Section 7.4.5. It utilizes the following components:

- recycle compressor,
- six heat exchangers,
- two expanders,
- Joule–Thomson valve,
- phase separator, which simultaneously performs the function of liquid helium storage, since it is sized as a large cryogenic vessel.

8.3.4.2.1 Process Description

The sequence of the helium Claude process is as follows:

(a) The low-pressure helium stream is compressed from the low-pressure level p_{LP} to the high-pressure level p_{HP} by means of a compressor and cooled to approximately ambient temperature.
(b) This compressed helium is cooled
   - in the countercurrent heat exchanger HEX3b to a temperature of approximately 80 K by returning cold low-pressure helium and evaporating liquid nitrogen, and further
   - in the countercurrent heat exchanger HEX3a to a temperature of 30–40 K by the cold low-pressure stream.
(c) At this point, the refrigerant flow is divided into two sub-streams, the Joule–Thomson stream and the expander stream. The flowrate of the expander stream is higher than the flowrate of the Joule–Thomson stream.
(d) The expander stream is:
   - expanded in expander EXP1 from the high-pressure level to an intermediate pressure level;
   - routed through heat exchanger HEX2b, where it is further cooled to approximately 15 K
   - before it is expanded in expander EXP2 from intermediate pressure to low pressure. The temperature of this stream is reduced correspondingly. This stream is gaseous. This cold gas passes through heat exchangers HEX2a, HEX2b, and HEX2c. The cold of the cold-expander stream is applied there for cooling of the Joule–Thomson stream. Simultaneously, the expander stream is warmed up.
   - routed further through heat exchangers HEX3a and HEX3b to the compressor inlet, where it is warmed up to nearly ambient temperature.
(e) The Joule–Thomson stream is:
   - routed through heat exchangers HEX2c, HEX2b, and HEX2a, where it is cooled by the cold-expander stream in countercurrent direction and enters the Joule–Thomson stage, where it is
   - cooled in the heat exchanger and HEX1 to a temperature of 6–8 K (the state at this point corresponds to the supercritical liquid) and

- expanded in the expansion valve to the low-pressure level p_{LP}. This process is a typical Joule–Thomson expansion, whereby the Joule–Thomson stream partially liquifies and a part of the refrigerant evaporates, generating flashgas.
- this two-phase helium stream is routed into a phase separator, which also serves as liquid helium storage. The LHe is collected in this vessel, whereas the flashgas and the vapor displaced by LHe are returned to the Joule–Thomson heat exchanger HEX1; thermodynamically, the state of this gas corresponds to the dew point condition.
- this cold vapor is provided through heat exchanger HEX1, cooling the Joule–Thomson stream.
- at the outlet of the heat exchanger HEX1, the Joule–Thomson stream is mixed with the expander stream, and it flows through all heat exchangers to the compressor inlet.

(f) The feed gas is injected either into the low-pressure helium stream or upstream of the compressor unit.
(g) The product stream – liquefied helium – is collected in liquid helium storage vessel. Usually, its withdrawal occurs discontinuously according to demand.

8.3.4.2.2 Process Considerations for Expander

Requirements on expander capacity can be estimated through the application of energy balance for the part of the process below the pre-cooling stage. The corresponding system boundary is shown in Figure 8.24b: high-pressure material stream 10 with corresponding enthalpy $\dot{H}_{10}$ and parasitic heat flow through the thermal insulation $\dot{Q}_{is1}$ are entering this system. All other streams are leaving the system; they are as follows: material streams 25 and 6 with corresponding enthalpies $\dot{H}_{25}$ and $\dot{H}_6$ as well as both powers P_{EXP1} and P_{EXP2} generated by expanders:

$$\dot{H}_{10} + \dot{Q}_{is1} = H_6 + H_{25} + P_{EXP1} + P_{EXP2} \quad \text{or}$$
$$\dot{m}_{10}h_{10} + \dot{Q}_{is1} = \dot{m}_6 h_6 + \dot{m}_{25}h_{25} + P_{EXP1} + P_{EXP2} \tag{8.21}$$

taking into account mass balance equation ($\dot{m}_{10} = \dot{m}_6 = \dot{m}_{25}$), this equation can be transformed to

$$\dot{m}_6\left(h_{10} - h_6\right) = \dot{m}_{25}\left(h_{25} - h_{10}\right) + P_{EXP1} + P_{EXP2} - \dot{Q}_{is1} \tag{8.22}$$

For a first estimation, it can be assumed that the pre-cooling system works properly. It means the temperature difference at the warm end of the system (T_{10} – T_{25}) is quite small (less than 5 K). Consequently, the corresponding enthalpy difference ($h_{25} - h_{10}$) is negligibly small (see Figure 7.28 for verification): $(h_{25} - h_{10}) \rightarrow 0$. In this case, nearly all the cooling capacity required for liquefaction is produced by expanders:

$$\dot{m}_6\left(h_{10} - h_6\right) + \dot{Q}_{is1} \approx P_{EXP1} + P_{EXP2} \tag{8.23}$$

Example Calculation For a very rough estimation, we can assume that heat flow through insulation is negligibly small $\dot{Q}_{is1} \rightarrow 0$. For a typical supercritical pressure level of 10–15 bar, the enthalpy difference $(h_{10} - h_6) \approx 420\,\text{J/g}$. Therefore, for a liquefaction capacity (feed gas flowrate) of 100 liter per hour (≈3.4 g/s), the expanders must generate at least $P_{EXP1} + P_{EXP2} = 0.42\ \text{kJ/g} \times 3.4\ \text{g/s} = 1.4\ \text{kW}$ of power. This translates to roughly 700 W per expander if two expanders of the same size are used in this process. This number (700 W) was obtained based on rather optimistic assumptions (no heat leakage, pre-cooling stage is well designed). Therefore, this number can be considered as a minimum value only; the actual values are usually higher.

8.3.4.2.3 Process Extension, Pre-cooling Stage

As described in Section 8.3.4.2.1, in a helium liquefier, the compressed helium is cooled first from the ambient temperature T_{amb} to liquid nitrogen temperature level (roughly 80 K), since the pre-cooling operation is often realized by means of liquid nitrogen. One interesting fact about helium pre-cooling is that the pre-cooling process is independent of the process design below pre-cooling stage; it can be considered separately. The process design for pre-cooling stage, once properly developed, can be combined with every helium liquefaction process.

The functionality is shown in Figure 8.24, where the liquid nitrogen (material stream 30) is introduced in the pre-cooling stage, where it is evaporated and warmed up to the ambient temperature level (operation 30 → 31). The warm high-pressure helium stream 1 is cooled (operation 1 → 10) not only by returning cold low-pressure helium (25 → 5) but also by cold nitrogen.

This figure (Figure 8.24) visualizes the functionality only. In terms of real process design/real mechanical design, this part of the system can be designed in different ways. A process example is given in Figure 8.25. Two main elements are a small phase separator fed by liquid nitrogen and an additional heat exchanger HEX4. The liquid nitrogen from the phase separator is used as coolant for warm, high-pressure helium stream provided through this heat exchanger. The liquid

Figure 8.25 Example of a process with pre-cooling by means of liquid nitrogen.

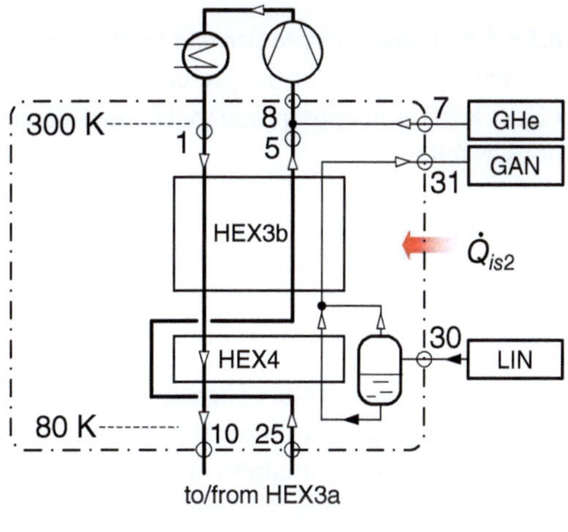

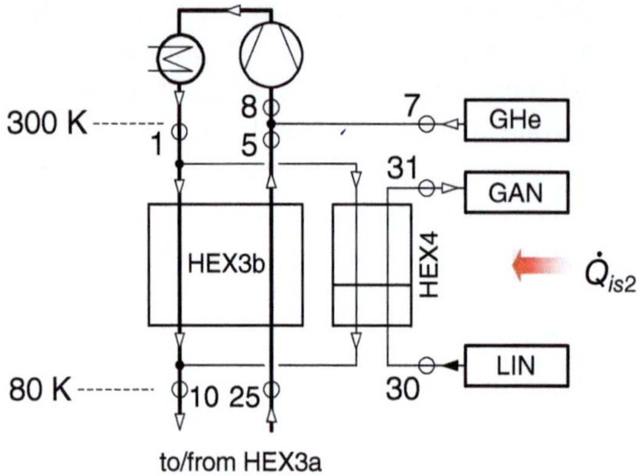

Figure 8.26 Example of a process with pre-cooling by means of liquid nitrogen.

nitrogen evaporates within the heat exchanger. The vaporous nitrogen from HEX4 is merged with gaseous nitrogen from the phase separator and directed further to the heat exchanger of the main cooling process HEX3b, supporting the main cooling process. For the heat exchanger HEX4, at least theoretically, all typical heat exchanger configurations (counter-current flow, co-current flow, and cross-counter flow) are suitable since the liquid nitrogen boils at constant temperature.

Another process example is given in Figure 8.26. In this arrangement, the warm, high-pressure helium stream is split into two sub-streams. The largest one is directed through the heat exchanger HEX3b; it is cooled there in conventional way by returning cold, low-pressure helium. The second partial stream is directed through an additional heat exchanger HEX4, where it is cooled by liquid nitrogen. Both high-pressure helium streams are merged again (downstream of heat exchangers HEX3b and HEX4) and provided further to the heat exchanger HEX3a. The heat exchanger HEX4 can be designed as "once-through" heat exchanger, as shown in Figure 8.26, or in the way shown already in Figure 8.25 as a combination of an evaporator (where the liquid nitrogen boils) and a gas–gas heat exchanger in the upper warmer part.

8.3.4.2.4 Liquid Nitrogen Demand for Pre-cooling Purpose

The energy balance for the pre-cooling stage (with system boundaries as shown in Figure 8.25) is an appropriate tool for estimation of liquid nitrogen demand. It functions as follows:

$$\dot{m}_1\, h_1 + \dot{m}_7\, h_7 + \dot{m}_{25}\, h_{25} + \dot{m}_{LIN}\, h_{30} + \dot{Q}_{is1} = \dot{m}_8\, h_8 + \dot{m}_{10}\, h_{10} + \dot{m}_{LIN}\, h_{31} \tag{8.24}$$

where

$\dot{m}_{LIN}$ – mass flowrate of liquid nitrogen
$\dot{Q}_{is2}$ – parasitic heat flow through thermal insulation
h_i – specific enthalpies of corresponding material streams
$\dot{m}_i$ – mass flowrates of corresponding material streams

Taking into account mass balance equation ($\dot{m}_{10} = \dot{m}_7 = \dot{m}_{25}$) as well as the fact that $\dot{m}_1 = \dot{m}_8$, this equation can be transformed to

$$\dot{m}_1 \left(h_1 - h_8\right) + \dot{m}_7 \left(h_7 - h_{10}\right) + \dot{m}_{25} \left(h_{25} - h_{10}\right) + \dot{m}_{LIN} \left(h_{30} - h_{31}\right) + \dot{Q}_{is2} = 0 \tag{8.25}$$

For well-designed processes, for a very rough estimation, the following assumptions can be made:

1) the temperature difference at the warm end of the system ($T_1 - T_8$) is small (less than 5 K). Consequently, the corresponding enthalpy difference ($h_1 - h_8$) is negligibly small: $(h_1 - h_8) \rightarrow 0$,
2) the temperature difference at the cold end of the pre-cooling stage ($T_{10} - T_{25}$) is small. Consequently, the corresponding enthalpy difference ($h_{10} - h_{25}$) is negligibly small: $(h_{10} - h_{25}) \rightarrow 0$,

For these conditions, Eq. (8.25) can be simplified as follows:

$$\dot{m}_{LIN} \left(h_{31} - h_{30}\right) = \dot{m}_7 \left(h_7 - h_{10}\right) + \dot{Q}_{is2} \quad \text{or}$$
$$\dot{m}_{LIN} = \left[\dot{m}_7 \left(h_7 - h_{10}\right) + \dot{Q}_{is2}\right] / \left(h_{31} - h_{30}\right) \tag{8.26}$$

This equation expresses that the liquid nitrogen is required mainly for (i) cooling of the feed helium stream from ambient temperature level to temperature level of liquid nitrogen and (ii) removing the heat inleak through thermal insulation.

Example Calculation For a very rough estimation, we can assume that heat flow through insulation is negligibly small $\dot{Q}_{is2} \rightarrow 0$.

Suppose other parameters as:

- $T_7 = 300$ K,
- $T_{10} = 80$ K,
- $T_{31} = 295$ K,
- pressure level of 12–15 bar for pressurized helium stream,
- saturated liquid nitrogen at pressure close to ambient pressure.

The enthalpy differences can be estimated as $h_7 - h_{10} \approx 1140$ kJ/kg and $h_{31} - h_{30} \approx 420$ kJ/kg.

With these numbers, the outcome of the calculation is as follows:

- the flowrate of liquid nitrogen required for pre-cooling purposes is 1140/420 = 2.7 times higher than the flowrate of the helium feed: 27 kg of liquid nitrogen for 10 kg of liquid helium.
- In terms of volumetric flowrates, the coefficient is roughly 0.4. It means more than 40 l of liquid nitrogen for 100 l of liquid helium. That is due to higher liquid nitrogen density ($\approx$800 kg/m^3) in comparison to relative low liquid helium density of $\approx$120 kg/m^3.

The coefficient of 2.7 (respectively 0.4 for volumetric flowrate) was obtained based on rather optimistic assumptions (no heat leakage, external cooling process designed for high efficiency). Therefore, this number can be considered as a minimum value only. The actual values are a bit higher, they approach 0.5.

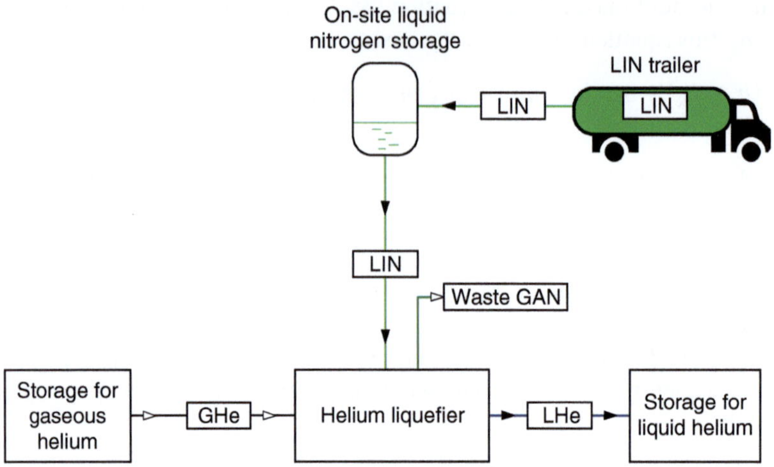

Figure 8.27 Supply of liquid nitrogen to helium liquefier by means of LIN trailer.

8.3.4.2.5 Supply of Liquid Nitrogen for Pre-cooling Purposes

For small-size and medium-size helium liquefiers, liquid nitrogen can be supplied by existing liquid nitrogen infrastructure, as shown in Figure 8.27, stored in an appropriate storage tank adjacent to the helium liquefier, and used for pre-cooling purposes in an appropriate manner.

8.3.4.3 Process Extension, Helium Purification System

A typical process extension is an integrated purifier unit, which is applied for final purification of contaminated helium before liquefaction.

Generally, before the final purification, the unpurified helium (from customers) is processed through a number of purification steps[9] (drying, pre-purification) in order to remove the main contaminants and problematic substances, such as humidity. The resulting feed gas for the purifier is not so heavily contaminated, containing only traces of low-boiling substances like nitrogen, argon, oxygen, and hydrogen. The purpose of final purification is to remove these remaining contaminants.

Figure 8.28 shows an example of such a process. The integrated purifier consists of

- a set of heat exchangers (HEX11 and HEX12),
- a phase-separator,
- a cold adsorber.

Thermal integration into the main liquefaction process is realized by means of two connections. A fraction of cold helium gas (for example, a fraction of the Joule–Thomson stream) is provided to the purifier for cooling purposes. Inside the purifier, this stream is warmed up and returned into the main process upstream of the compressor unit.

9 These process steps are not considered in current section.

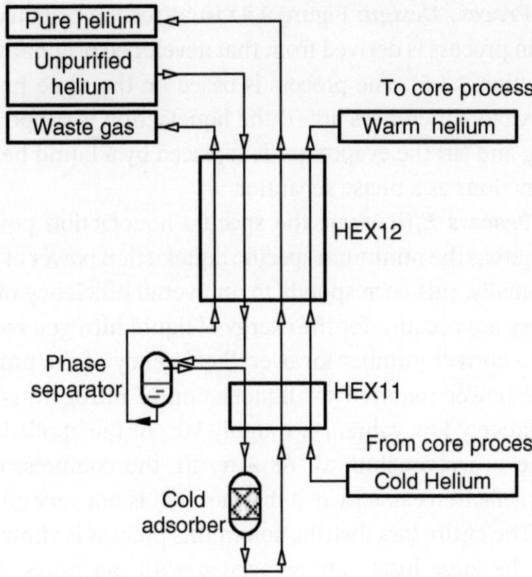

Figure 8.28 Purifier for final purification of helium gas, copyright Linde Kryotechnik.

Inside the purifier unit,

(a) The compressed feed gas is cooled first to liquid nitrogen temperature level by means of heat exchanger HEX12.
(b) At this temperature, the air gases will condense. The condensate is removed by means of a phase separator. The removed liquid is passed back through the heat exchangers, where it is warmed up.
(c) The vaporous phase from the phase separator is cooled further by means of heat exchanger HEX11 to a very low temperature, with the aim of freezing out all remaining substances. These solidified contaminants are filtered by means of a cold adsorber.
(d) The purified cold helium flows through all heat exchangers (HEX11 and HEX12) and is stored in appropriate storage for pure helium.
(e) Over time, the cold adsorber may become saturated with impurities. Periodic regeneration helps to maintain the functionality of this unit. During the regeneration phase, the adsorber is heated (usually by means of an electric heater) until the contaminations disappear, then cooled again to the normal operating temperature.

The purifier is commonly operated discontinuously. The frequency and duration depend mainly on the supply of unpurified helium and the level of contaminants in this gas.

8.3.4.4 Example of a Claude Process For Helium Liquefaction

An example of a helium liquefaction process is described in detail in Supplement 9.2.4. The following text section presents only an excerpt of the process design.

Process Design: Figure 9.9 visualizes the primary process concept: the liquefaction process is derived from that developed originally for refrigeration purposes (see Section 7.4.6). The process is based on the same hardware components. The most obvious differences are (i) the liquefaction is supported by liquid nitrogen precooling and (ii) the evaporator is replaced by a liquid helium storage vessel, which also functions as a phase separator.

Process Efficiency: the specific liquefaction power amounts to 1.86 kW/(l/h), whereas the minimum specific liquefaction power at given pressure is 0.22 kW/(l/h). Visually, this corresponds to an overall efficiency of 11.8%. However, this number does not account for the exergy of liquid nitrogen required for precooling purposes. The correct number for overall efficiency of the process, which takes into account the power required for liquefaction of nitrogen, is approximately 11%. That is a relatively low value, i.e. roughly 90% of the applied power is dissipated in process due to irreversibilities. As a result, the compressor power is converted into the liquefaction capacity in a manner that is not very efficient.

The entire loss distribution in this process is shown in **Figure 9.12**.

The large losses are associated with machines: 49% of losses are generated in the compressor, 23% are generated in expansion units (expanders and expansion valves), whereby the expansion valve contributes only 5.8% only, and the largest share – 17% – is associated with expanders. This is mainly due to moderate efficiency of these components: the isentropic efficiency of both expanders is assumed to be 70%, and the isothermal efficiency of the recycle compressor is specified as 55%. For a screw compressor, the considered machine is quite good. But in comparison with a typical centrifugal air/nitrogen compressor, the efficiency of this device is rather low. The efficiency of the liquefaction process can be considerably improved through use of machines with better efficiencies. Unfortunately, the development of highly efficient machines requires a considerable investment and consumes much time.

The second-largest block of losses – 28% – is associated with heat exchangers. **Figure 9.13** shows the distribution of losses between heat exchangers. The largest losses occur in heat exchangers HEX3b, HEX3a, and HEX2b, since these units are the largest. In given case, the loss in heat exchanger HEX1 is quite high and potentially can be reduced. The overall temperature profile is favorable, but the overall level of temperature difference can be reduced, especially at the cold end. Potentially, the losses from heat exchangers can be reduced through use of heat exchangers with larger heat transfer surfaces, even in state-of-the-art technology.

These observations can be summarized as the following learning:

Learning 8.4 The efficiency of a helium liquefier depends on the efficiency of all hardware components used in the process: compressors, expanders, and heat exchangers. For a power-efficient process, the best machines must be used in combination with large heat exchangers. For a cost-efficient design, the right balance must be found between available machines and heat exchanger size.

8.3.5 Example of a Commercial Helium Liquefier

Figure 8.23 shows the overall setup of a standard laboratory helium liquefier. The corresponding performance specification is given in Table 8.7. This section describes the process design and two subsystems directly related to the main process:

1. the compressor system, which is usually spatially separated,
2. the cold box, which contains all cryogenic hardware components except for the liquid helium storage vessel, which is usually designed as a separate system located next to the cold box.

8.3.5.1 Process Design

Main Elements: Figure 8.29 shows a section of an original process flow diagram of this system. The main elements are the recycle compressor, six heat exchangers, two expanders, the Joule–Thomson valve, and a liquid helium storage vessel. The process is quite similar to the Claude process previously discussed in Section 8.3.4.2 (see Figure 8.24), especially If

- the heat exchanger E3130 is considered as HEX1,
- E3120 as HEX2a,
- E3110B as HEX2b,
- E3110A as HEX2c,
- E3100B as HEX3a,
- E3100A as HEX3b,
- expansion turbine T1 as expander EXP1,
- expansion turbine T2 as expander EXP2,

LIN Precooling: LIN precooling is integrated into heat exchanger E3100A.

Integrated Purifier: The process contains an integrated purifier and is similar to the process discussed in Section 8.3.4.2.2 (see Figure 8.28) if

- heat exchanger E3470 is considered as HEX11,
- heat exchangers E3460, E3450, and E3470 as HEX12.

Additional Elements: The process in Figure 8.29 contains several additional hardware components:

- filter F3110, which is required for protection of expansion turbines from mechanical contamination
- valve upstream from filter F3110 for control of (a) flowrate through expanders and (b) pressure at the expander inlet
- ORS after compressor
- adsorber A3125 for separation of further potential impurities such as traces of air gases
- numerous control valves, for example, one upstream of the compressor unit, the next one downstream of the compressor unit, and the third one in the pipeline connecting the liquid helium storage outlet and the inlet of heat exchanger E3130.

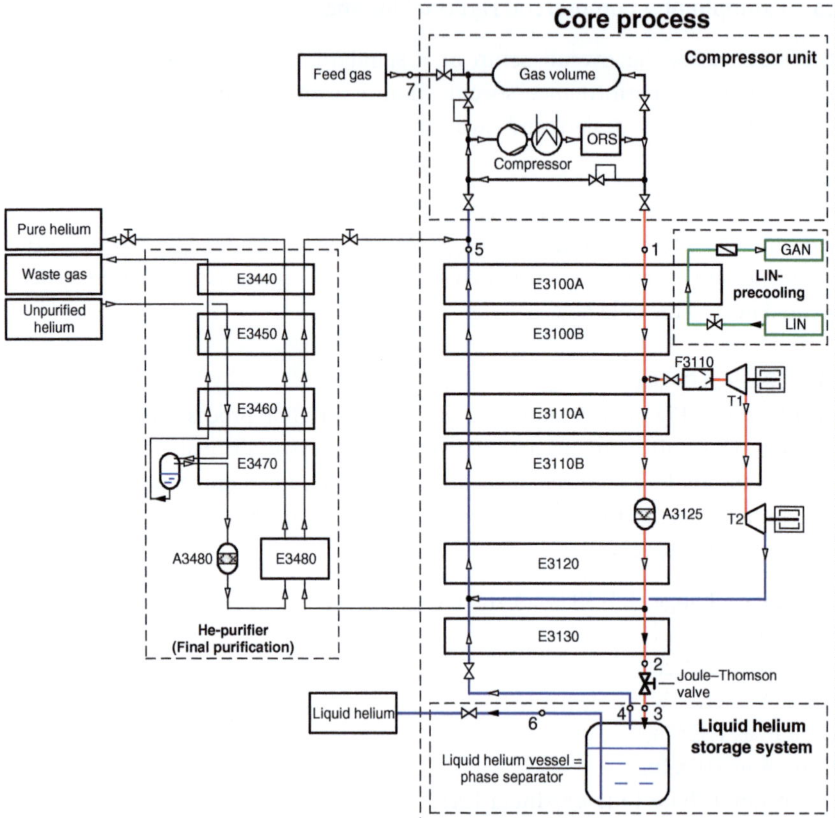

Figure 8.29 Core process section of a laboratory helium liquefier, copyright Linde Kryotechnik.

In terms of process simulation, at least the pressure drop caused by these hardware components must be considered.

8.3.5.2 Compressor System

In process design, the compressor is often considered as a single process unit, where the fluid is pressurized from a lower pressure (1–1.5 bar) to a higher pressure (8–16 bar) and cooled to ambient temperature level. In a real system, the process includes at least one step more, namely the removal of compressor oil from the compressed helium stream, since the applied compressor is an oil-flooded machine and the working fluid is contaminated with compressor oil as described in Section 6.5.3.

Figure 8.29 shows some key elements, such as the gas buffer, bypass valve, and several other valves, which are integral to gas management. The purpose of this subsystem is to ensure that operation pressures remain stable and are not disturbed by changes in feed gas flowrate, removal of liquid, load changes, and similar situations.

The entire compression unit is even more complex. It includes numerous subsystems required for regular operation, such as

- ORS
- pressure management system
- flowrate control and further controls
- cooling water (or cooling air) system
- gas purity control
- oil management system (compressor oil)
- power supply and
- automatization, digital documentation.

The compressor system, which can sometimes be very noisy during operation, is positioned as far as possible from the rest of the plant in order to reduce noise pollution and avoid mechanical vibrations.

8.3.5.3 Cold Box, Mechanical Design

Figure 8.30 discloses details of a cold box for a vertical cylindrical vessel. One can recognize some devices placed on the flat top flange, for example:

- two nozzles for connection to the liquid helium vessel (phase separator/storage): one for the Joule–Thomson stream to the liquid helium vessel and one for returned low-pressure vaporous helium,
- two expanders can be easily recognized as devices raised higher than others, with two connections each for cooling water,
- seven control valves, one of which is a Joule–Thomson valve,
- some feedthrough for instrumentation.

These are accessible from the top surface of this flange.

Mechanical design of hardware components under the flange is not recognizable in this picture. Obviously, all hardware components required for the main cryogenic process unit as well as components for process extensions (liquid nitrogen precooling, final purifier) are located there.

This flat flange is used as a cover for a cylindrical vessel so that the entire cryogenic part is enclosed by this vessel, as shown in Figure 8.31. The vessel is evacuated by means of a vacuum pump.

8.3.5.4 Cold Box, Thermal Insulation

The impact of heat leakage on the performance of cooling systems for the liquid helium temperature level was discussed briefly in Section 7.4.6: since this cold is extremely valuable in terms of cost and power, all additional generation of cold to make up for heat leakage bears a considerable impact on the power consumption of the refrigerator. In order to minimize the heat inflow from outside, the best available thermal insulation – multilayer vacuum insulation – is applied.

Figure 8.32 shows an example of a cold box (in horizontal design) during installation. The cylindrical shell is removed (visible at left), and one can see several

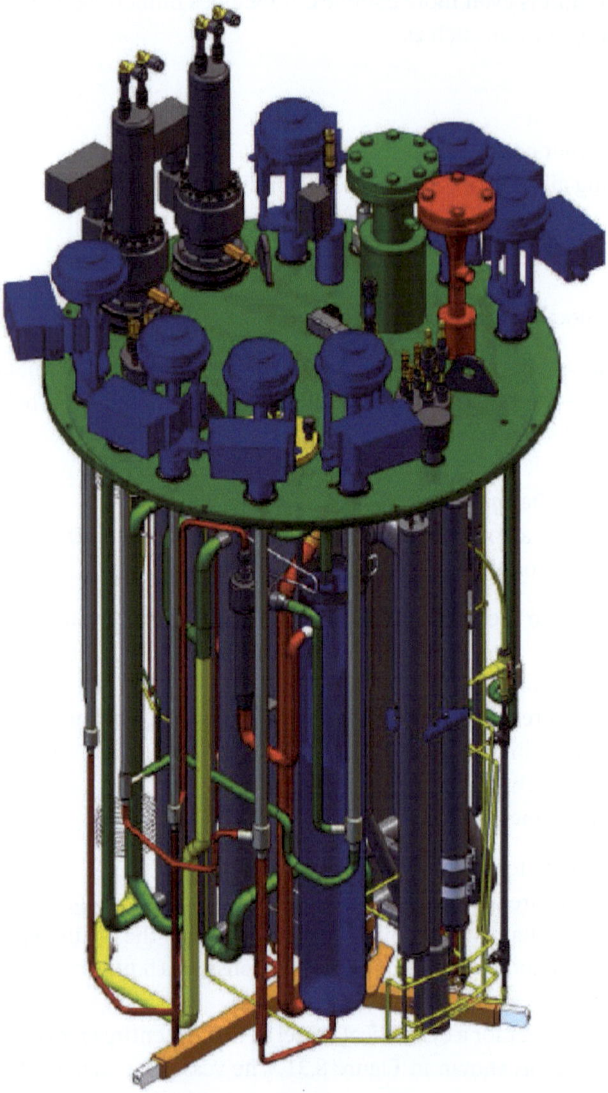

Figure 8.30 Cold box of a laboratory helium liquefier.

components (a heat exchanger and piping) wrapped in the multilayer insulation; the heat exchanger visible at the top is not insulated yet.

8.3.6 Status Quo, Helium Liquefaction

Helium liquefaction remains a fascinating and unique technology. There are only a few manufacturers who are able to supply this type of plant. Access to modern

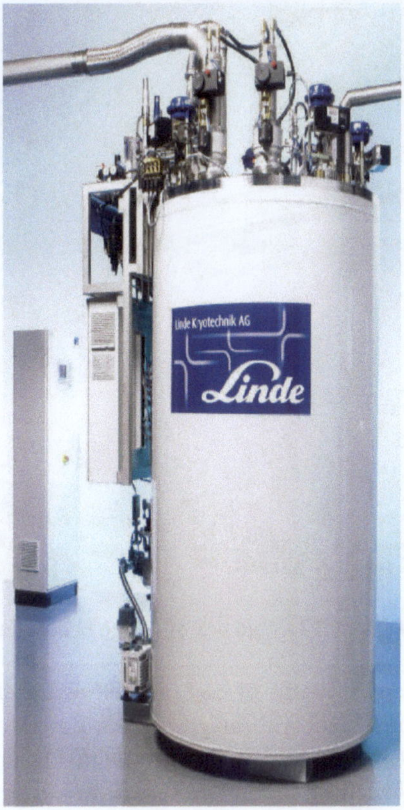

Figure 8.31 Cold box of a laboratory helium liquefier from external perspective. Source: Linde plc.

helium expander technology is a key factor to success, as all other necessary hardware components, such as compressors and heat exchangers, are readily available. The process design depends on helium expander performance; often, the process is adapted to the available expansion machines.

In terms of process design, the most challenging requirement is the temperature level of 1.8 K (superfluid helium), which requires a special technical solution for the low-pressure side of the process, since subatmospheric conditions are required to maintain this cooling temperature level.

Small-scale helium plants for laboratory applications are well-developed and standardized technology, and delivery time is relatively short. Large-scale helium plants are usually tailor-made systems, the customization of which requires extensive engineering effort. Therefore, projects of this kind may take years to realize and require close cooperation between supplier and customer.

Although this technology is well established, further continuous development is still required. Since the process design depends heavily on the hardware components used, further development of those components is a primary target. Engineers are constantly on the lookout for compressors that are more efficient, both technically

Figure 8.32 Cold box of a large helium liquefier, the vacuum vessel is removed.

and in terms of cost, as well as for efficient and reliable expanders. Potentially, machines with magnetic bearings may be introduced in the future. New kinds of compact heat exchangers with higher thermal efficiencies may provide further benefits for future processes.

8.4 Natural Gas Liquefaction

Natural gas liquefaction became a major industry in the 1960s. This is partly due to (i) the establishment of natural gas as a primary energy source and feed for chemistry at that time,[10] and (ii) progress and significant maturation of cryogenic technologies for liquefaction, storage, and transport of large amounts of liquified natural gas (LNG).

Geographically, natural gas sources and natural gas consumers are located far away from each other: there are countries where far more natural gas is produced than can be consumed, and countries where the reverse is true. In Western Europe, for example, more natural gas is consumed than produced. In contrast, several countries, such as Algeria, Qatar, Norway, and Russia, have rich sources of natural gas available, whereby natural gas production exceeds their own needs.

10 Previously, mainly liquid fuels (oil and its derivates) or solid fossil fuels (coal) were used as fuels.

The natural gas is liquefied for logistical reasons. LNG has a density that is 600 times higher (and therefore the volume to be transported is about 600 times smaller) than natural gas at normal pressure and room temperature. Therefore, the effort and cost of transporting natural gas over long distances (further than 2000–2500 km) in liquid form is lower than for gaseous conditions (via pipelines); that is the main driver for this business. The second most important one is flexibility in terms of transport routes and delivery points since many destinations in different countries can be approached and supplied by the same LNG carriers. This is a clear advantage in comparison to pipelines with fixed routes and fixed end points.

In the 2000s, a disruptive development appeared: a new cost-efficient technology for well stimulation and extraction of natural gas and oil from hydrocarbon-containing bedrock formations (commonly referred to as fracking[11]) was introduced and commercialized. This production method led to a rapid increase in natural gas production in the United States of America: after one decade, production capacity grew large enough to cover the United States internal demand and even to export an excess of natural gas to other markets. This growth led to a sharp drop in natural gas prices (in the United States of America and worldwide) due to availability of large quantities of cheap natural gas. Industrial companies are looking for ways to export the natural gas, e.g. as LNG. Therefore, the worldwide demand for natural gas liquefaction plants is rising.

8.4.1 Natural Gas Composition

Natural gas consists of valuable substances: methane (CH4), fractions of further hydrocarbons (ethane, propane, etc.), and sometimes helium as well. However, the crude gas contains numerous contaminants, some of which (e.g. nitrogen) are more or less harmless, but others, such as CO_2, H_2O, H_2S, and mercury, are problematic since they are either inherently dangerous or can impact the operation of liquefiers or even cause damage to equipment and machines. The composition of natural gas varies from source to source. Large natural gas plants are therefore practically adapted individually to the requirements of the respective location and its particular natural gas composition – an unlimited source of inspiration for creative engineers.

In Table 8.9 below, several natural gas compositions from different geographies are listed as examples to illustrate the range of possible compositions:

- Natural gas from the Netherlands has a very high nitrogen content and therefore a low heating value.
- Natural gas from Kansas also has a high nitrogen content, similar to that from Netherlands. However, the value of this gas is higher because of its relatively high helium content of 0.35%.

11 The fracking method is still controversially discussed in society, because of related environmental issues.

Table 8.9 Natural gas composition, in mol%.

	Miskar field, Tunisia14	Canada, Alberta14	Southwest Kansas14	Holland13 (Netherlands)	North Sea
Helium	0	0	0.35		
Nitrogen	16.93	3.20	14.64	10.28	0.87
Methane	63.90	77.10	72.89	84.11	90.73
Ethane	3.35	6.60	6.27	3.43	5.61
Propane	0.96	3.10	3.74	0.56	0.72
Butanes	0.54	2.00	1.38	0.10	0.09
Pentanes and higher	0.63	3.00	0.62	0.03	0.02
Carbon dioxide	13.59	1.70	0	1.32	1.80
H2S	0.09	3.30	0		

a) Source: Open Grid Europe (OGE), reference values 2015.
b) Source: Arthur and William [11].

- Natural gas from Kansas, Alberta, and the North Sea contains a large fraction of high-boiling hydrocarbons such as ethane and propane.
- In terms of contaminants, Tunisian natural gas, with over 13% CO_2 and some H_2S, is the most difficult to process, as it requires complex purification and removal of these problematic substances before liquefaction. In the past, development of such natural gas sources with similarly difficult compositions was avoided due to high purification effort and consequently high operating costs as well as high investment costs. However, recent developments in purification technology make the exploitation of these problematic sources lucrative. In the future, natural gas sources with similar gas compositions will play an increasingly important role.

8.4.2 LNG Infrastructure and Process Chain

LNG infrastructure is quite complex. Natural gas is usually

(a) transported from the source in the middle of a country through a pipeline to the coast,
(b) liquefied onshore by means of an LNG plant and stored in large-scale intermediate storage vessels, then
(c) later, the liquid is loaded onto LNG carriers (ships). This facility is often referred to as the "export LNG terminal."

In the recipient country, at the "import LNG terminal," the LNG is

(d) unloaded and stored in large-scale storage vessels, then
(e) the LNG is pressurized by means of high-pressure pumps, evaporated and warmed to ambient temperature level, then fed into the natural gas pipeline grid according to demand, which is manifested in the form of fluctuating pressure.

It is interesting to note that there are two links in this transport chain where cryogenic technology is required: (i) the natural gas liquefaction plant at the export terminal, and (ii) the system for evaporation of natural gas at the import terminal. At the first point, a large amount of cryogenic cold is generated, which requires a great amount of power. At the other end, this valuable cold is dissipated into the environment. From the perspective of energy utilization, this approach is not optimal. Therefore, engineers are always trying to utilize the LNG cold either directly for cooling purposes or indirectly by converting this cold into power or a similar utility. However, in truth, a generic solution with positive techno-economics has not yet been found.

8.4.2.1 Infrastructure Process Chain

The process chain behind this infrastructure (up to liquefaction) is shown in Figure 8.33. It starts directly at the well, where the crude natural gas is preconditioned. The main unit operations are removal of mechanical contaminants and dehydration. Removal of water is necessary to

(a) prevent formation of solid-phase hydrates,[12]
(b) avoid formation of liquid phase in the pipeline.

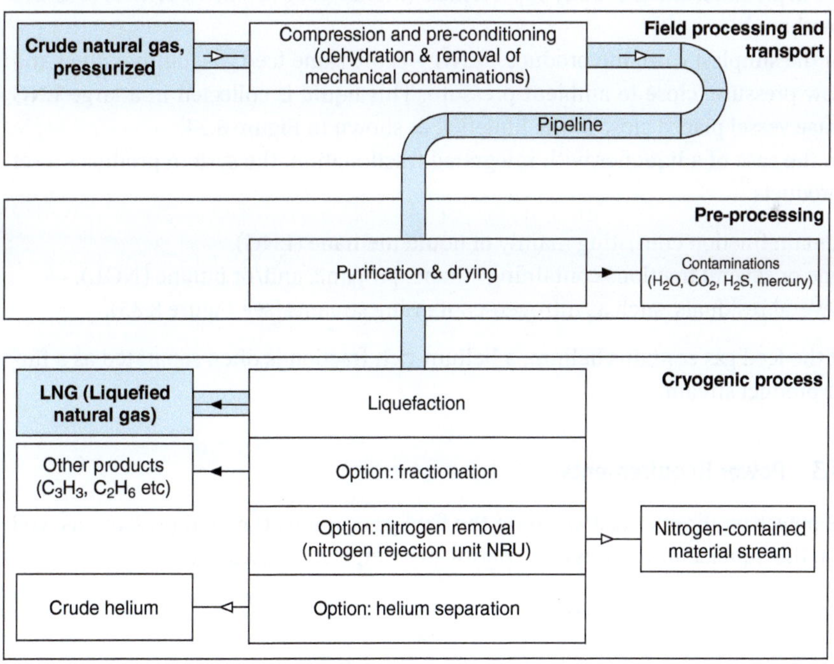

Figure 8.33 Natural gas liquefaction, overall process chain.

12 Gas hydrate is a solid compound, in which a large amount of methane is trapped within a crystal structure of water, forming a solid substance similar to conventional water ice.

This first stage of dehydration (on the well) is performed mainly by means of mechanical devices like phase separators, cyclones, and filters. The final drying, purification, and removal of critical contaminations (CO_2, H_2S, H_2O, mercury, etc.) is typically based on chemical technologies (adsorption, absorption (amine scrubber), membranes, cryogenics, or combinations of all of these). Commonly, these process units are located at the other end of the pipeline, upstream of the actual liquefaction system.

During the liquefaction process, the natural gas can also be fractionated, i.e. liquid fractions consisting of high-boiling components such as propane, butane, or ethane (these fractions are referred to as natural gas liquids (NGL)) as well as vaporous fractions containing gases like nitrogen or helium can be separated from the entire feed gas by means of the appropriate rectification processes.

The fractionation processes as well as the process operations for drying and purification are not considered in this book, which is focused solely on the actual liquefaction process. More information about these operations can be found in Ref. [12].

8.4.2.2 Feed Gas and Product

From the perspective of the liquefaction process, the feed gas is a dry, decontaminated, compressed natural gas. The pressure level is predefined by the pressure in the feed pipeline, and is usually supercritical (higher than 46 bar, the critical pressure of methane).

In the simplest case, the product – LNG – is the same feed gas, but liquefied and at low pressure, close to ambient pressure. This liquid is collected in a large LNG storage vessel placed close to the liquefier, as shown in Figure 8.34.

In the case of a liquefier with integrated fractionation, the system produces a set of products:

- a main fraction consisting mainly of liquid methane (LNG),
- one or several fractions containing ethane, propane, and/or butane (NGL),
- several residuals, such as nitrogen-containing streams (see Figure 8.35).

If the feed gas contains helium, a helium-rich fraction is often separated as a further product stream.

8.4.3 Power Requirements

The power requirements depend on the feed gas composition and pressure as well as on the liquefaction process and hardware components used, etc.

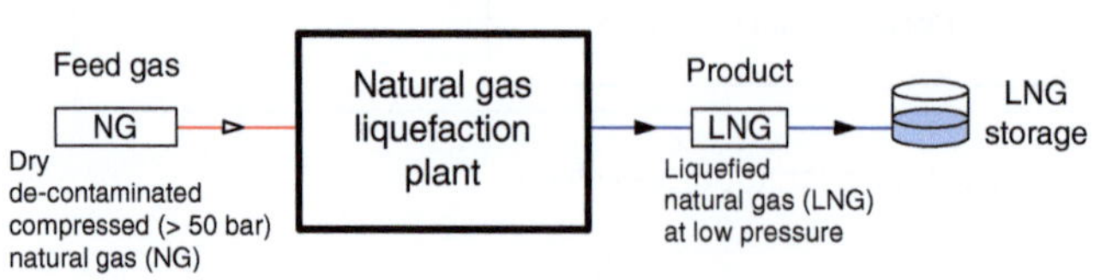

Figure 8.34 Natural gas liquefier, feed, and products.

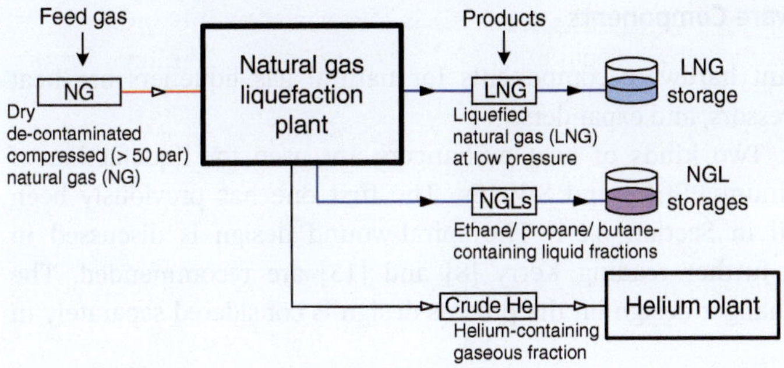

Figure 8.35 Natural gas liquefier with integrated fractionation, feed, and products.

Table 8.10 Minimum liquefaction work for CH_4, depending on feed gas pressure, T_{amb} = 300 K, p_{amb} = 1 bar abs.

Feed gas pressure bar	Minimum liquefaction work kJ/kg	kJ/mol	kJ/nm³	kWh/t
1	1095.4	17.53	782.4	304.3
20	634.5	10.15	453.2	176.3
30	574.1	9.19	410.1	159.5
40	531.9	8.51	379.9	147.8
50	499.8	8.00	357.0	138.8
60	473.9	7.58	338.5	131.6
70	452.3	7.24	323.1	125.7
80	434.0	6.94	310.0	120.6
90	418.0	6.69	298.6	116.1
100	403.9	6.46	288.5	112.2

The concept of minimum liquefaction work introduced in Section 8.1.2 and Eq. (8.7) ($\eta_{LIQ} = P_{min}/P_{real}$) may be used for very rough estimation. If the feed gas is considered as pure pressureless methane, the minimum liquefaction work amounts to 1095 kJ/kg (see Table 8.1). But, as considered in Section 8.4.2.2, the feed gas is usually pre-compressed, hence the liquefaction work is considerably lower (see Table 8.10). For example, for 50 bar pressure, the minimum liquefaction work amounts to c. 500 kJ/kg, which is less than half the value for pressureless methane.

Assuming

- liquefaction plant efficiency of η_{LIQ} = 50%,
- feed gas pressure 50 bar and
- pure methane (CH_4) as natural gas,

the value of 1000 kJ/kg = 278 kWh/t can be derived for specific power. This results in a power cost of 13.9 \$/t if a power tariff of 0.05 \$/kWh is assumed.

8.4.4 Key Hardware Components

The most important hardware components for natural gas liquefiers are heat exchangers, compressors, and expanders.

Heat Exchanger: Two kinds of heat exchangers are used for liquefaction of natural gas: aluminum PFHEs and SWHEs. The first one has previously been described in detail in Section 6.2.1. The spiral-wound design is discussed in Section 6.2.2. For further reading Kerry [8] and [13] are recommended. The impact of heat exchanger design on the process design is considered separately in Section 8.4.6.

Compressors: The mixed-fluid compressor is extraordinarily important for the overall performance and techno-economics of the plant. This is very similar to the case with nitrogen liquefaction. Modern industrial LNG plants are based on centrifugal machines, thanks to their capability to compress large gas flows in a single unit and generally better techno-economics. The thermodynamics of the compressor as a process unit have been described in Section 5.5. The specifics of centrifugal machines, as well as their typical restrictions and limitations, are described in Section 6.5.2.4. Mechanical design is discussed in Section 6.5.2.3.

Expanders: Expanders for natural gas liquefaction are a bit larger than those used for nitrogen liquefaction, but their mechanical design is very similar. Therefore, the basics illustrated in Section 6.3.3 for nitrogen expanders (for gas expanders as well as for Joule–Thomson expanders) are applicable to machines used for natural gas liquefaction.

8.4.5 Process Design for Natural Gas Liquefaction

The basic concept of the process for liquefaction of natural gas is not very sophisticated. In the simplest version shown in Figure 8.36a, the dry and pressurized (supercritical) feed gas is

(a) cooled down to a temperature of 120–130 K,
(b) expanded in an expansion device (e.g. Joule–Thomson expansion valve or Joule–Thomson expander) to low pressure, close to ambient pressure,
(c) after expansion the feed gas is already liquefied, at least partially. The liquid fraction is separated by a phase separator S1, as shown in Figure 8.36a, and routed to a storage vessel.

This is the entire process. Terminologically, this part of the process (the feed gas path) can be considered as a Joule–Thomson process, designed as an open cycle.

In detail, the process may be more complex, especially if

- a relevant amount of flashgas (which depends on feed gas composition, pressure, and temperature upstream of the Joule–Thomson valve) is generated after expansion. This flashgas must be implemented or integrated into the entire process in an optimal manner. Several potential options for this are discussed in Section 8.1.5.2.

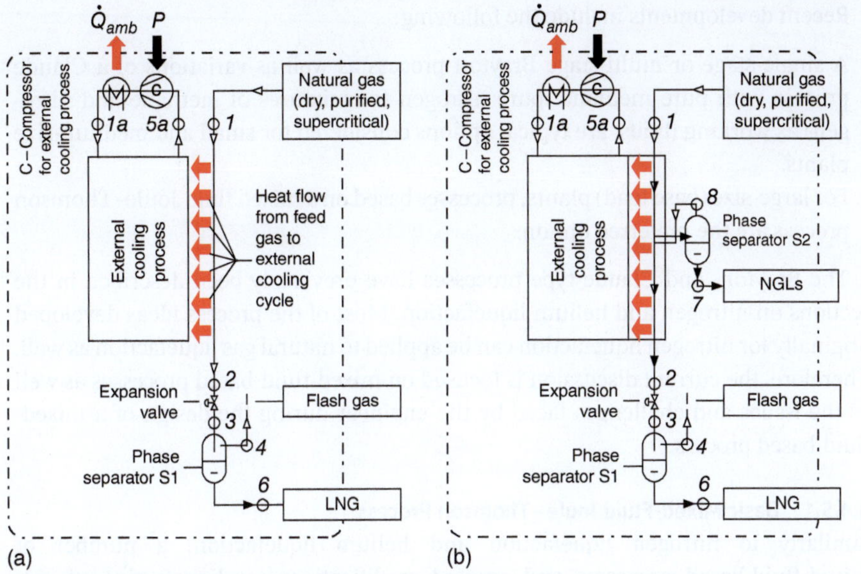

Figure 8.36 Principle of natural gas liquefaction by means of an external cooling process.

or

- fractionation of the feed gas is required. One example of a process with integrated fractionation is shown in Figure 8.36b. A phase separator S2 is integrated here into the feed gas path at an intermediate temperature level. The condensed components of the feed gas are separated here from the main feed gas and removed from the process as "natural gas liquids" (NGLs). The remaining gaseous feed stream is cooled down further.

In the case of natural gas liquefaction, the process know-how mainly concerns how to cool the feed gas to a temperature of 120–130 K. Through their architecture, processes with external cooling cycles are better suited to this task. Figure 8.36 shows the common principle: two fluids in two separate cycles are linked to each other: an open Joule–Thomson process for feed gas on the right side and an external cooling cycle on the left side, which delivers cooling capacity for cooling of feed gas to the required temperature level.

In principle, there are no limitations concerning the choice of a process for the external cooling cycle; any combination is conceivable.

Early natural gas liquefaction plants were based on a process referred to as the cascade process (see Isalski [14] for more information). This process consists of three linked Rankine-type refrigeration cycles, which are widely used in conventional refrigeration. Each subprocess uses its own refrigerant, whereby propane, ethane, and methane are typical working fluids. This technology is a sort of extension of conventional refrigeration technique to cryogenic temperature level. The thermodynamics of a Rankine cycle are already well described in the literature, for example in Teil et al. [15].

Therefore, the cascade process is not considered in this book.

Recent developments include the following:

- A single-stage or multi-stage Brayton process as well as variations of a Claude process with pure methane, pure nitrogen (or mixtures of methane and nitrogen) as working fluids, are typical options considered for small and medium-size plants.
- For large-size (base load) plants, processes based on a mixed fluid Joule–Thomson process are the preferred option.

The Brayton- and Claude-type processes have previously been described in the sections on nitrogen and helium liquefaction. Most of the process ideas developed originally for nitrogen liquefaction can be applied to natural gas liquefaction as well. Therefore, the current discussion is focused on mixed-fluid-based processes as well as the issues and challenges faced by the engineer during the design of a mixed-fluid-based process.

8.4.5.1 Basic Mixed-Fluid Joule–Thomson Process

Similarly to nitrogen liquefaction and helium liquefaction, a number of mixed-fluid-based processes and process modifications for liquefaction of natural gas were developed and realized by inspired engineers over the past 70 years. It is not necessary to consider all these processes individually; a more efficient approach is to develop a general understanding of the underlying principles and employ this knowledge for analysis and synthesis of processes.

8.4.5.1.1 Process Concept

The basic concept is shown schematically in Figure 8.37. Morphologically, this is a process with external cooling cycle, whereby

(a) feed gas is liquefied by means of a Joule–Thomson process described in Section 8.4.5, and
(b) a mixed fluid Joule–Thomson process (as described in Section 7.5) is applied for cooling the feed gas to low temperature.

An important element of this process is a common multi-stream heat exchanger (described in Section 5.3.3) for interlinking these two subprocesses.

In detail, the mixed-fluid part of the process functions as follows:

(a) Multicomponent refrigerant is compressed by an isothermal multistage compressor from low pressure to high pressure (process operation 5a → 1a). At the compressor outlet (state 1a), the temperature of the working fluid is close to ambient temperature.
(b) Compressed multicomponent refrigerant is routed through the heat exchanger, where it is cooled to a low temperature of 120–130 K (similar to the feed gas) by means of cold low-pressure mixed fluid on the other side of the heat exchanger (operation 1a → 2a). The process in the heat exchanger is described in detail in Section 7.5.3. During this cooling process, the fluid goes through several phase

Figure 8.37 Principle of natural gas liquefaction via external cooling process based on mixed-fluid Joule–Thomson process.

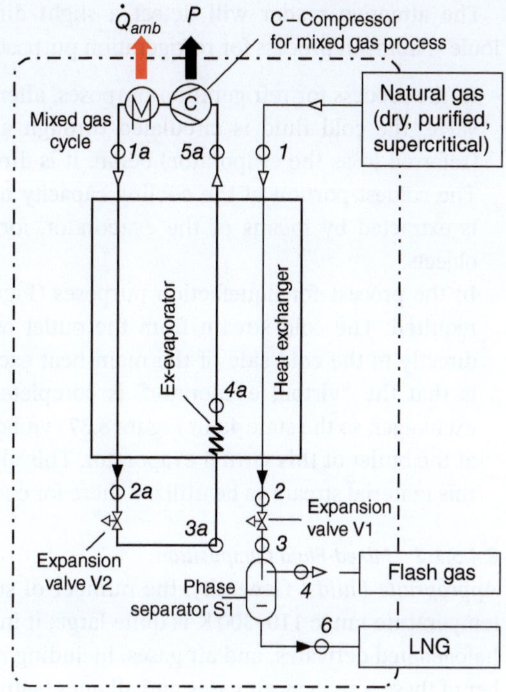

transitions: starting from vaporous or two-phase (vapor–liquid) conditions at the "warm end" of the heat exchanger, it is completely liquefied and even subcooled at the "cold end" of the heat exchanger.

(c) Cold compressed fluid is expanded through an expansion valve from high pressure to low pressure (operation 2a → 3a). This expansion causes a small temperature drop, with targeted end temperature usually 110–120 K, which corresponds to the LNG temperature. Another expansion effect is that the fluid is converted into liquid (not subcooled) and flashgas. The amount of flashgas depends on the feed gas composition, pressure, and temperature upstream of the expansion valve. In a well-designed process, it is usually not more than 10 mol%.

(d) This two-phase fluid is circulated through the common heat exchanger. It flows from the cold end to the warm end of the heat exchanger as a countercurrent to both of the warmer material streams – feed gas and high-pressure mixed fluid – and cools both these streams to the required temperature, whereby the cold low-pressure fluid itself is warmed up from LNG temperature to ambient temperature (operation 3a → 4a → 5a). This operation is described in detail in Section 7.5.3: the working fluid goes through several phase transitions, starting from two-phase (vapor–liquid) conditions at the "cold end" of the heat exchanger. At the "warm end" of the heat exchanger, it is completely evaporated and even superheated.

The attentive reader will detect a slight difference between the mixed-fluid Joule–Thomson process for refrigeration purposes and the process for liquefaction:

- In the process for refrigeration purposes, after expansion in the Joule–Thomson valve, the cold fluid is circulated through a separate heat-exchanging device (referred to as the evaporator) before it is directed to the main heat exchanger. The coldest portion of the cooling capacity of the returning low-pressure fluid is extracted by means of the evaporator, for use in cooling of corresponding objects.
- In the process for liquefaction purposes (Figure 8.37), this last function is not required. The cold stream from the outlet of the expansion valve is provided directly to the cold side of the main heat exchanger. A standard interpretation is that the "virtual evaporator" is completely integrated into the main heat exchanger, so the state 4a in Figure 8.37 symbolizes the state of the working fluid at the outlet of this virtual evaporator. This allows the entire cooling capacity of this material stream to be utilized there for cooling both of the warmer streams.

8.4.5.1.2 Mixed-Fluid Composition

Appropriate Fluids: Generally, the number of substances in a mixed fluid for the temperature range 110–300 K is quite large; it includes mainly hydrocarbons, their halogenated derivates, and air gases, including rare gases. But only a limited number of these – such as air gases and alkanes (saturated hydrocarbons) like methane, ethane, propane, etc. – are natural substances that are already available either in the natural gas itself or in ambient air. Other appropriate fluids, such as ethylene, propylene, and halogenated derivatives, are synthetic substances produced by means of special chemical plants. The synthetic substances require more expenditures in terms of production, transport, storage, and management at the LNG site. Therefore, from an economic standpoint, natural substances are the preferred fluids for natural gas liquefaction. One synthetic fluid – ethylene – presents a potentially appropriate option for LNG plants, since this low-cost substance is produced in large amounts all over the world and is readily available.

Thermodynamic properties and features of typical mixtures are described in Section 7.5.2.

Solidification Issue: One important requirement for failure-free operation of a system based on a mixed-fluid process is that the working fluid must not form a solid phase. The solid particles can cause clogging and disturb the normal circulation of the fluid in the system. In this sense, the most critical locations are at the cold end of the heat exchanger and in the expansion valve, since the lowest temperatures are achieved there and the risk of solidification is higher. The solidification issue was already considered in Section 7.5.5.3 for mixed-fluid refrigeration. These considerations hold valid for liquefaction process as well.

8.4.5.1.3 Example of Basic Mixed-Fluid Process for Natural Gas Liquefaction

An example of a mixed-fluid-based process for natural gas liquefaction is available in the supplement in section 9.2.5. In this process example, process parameters

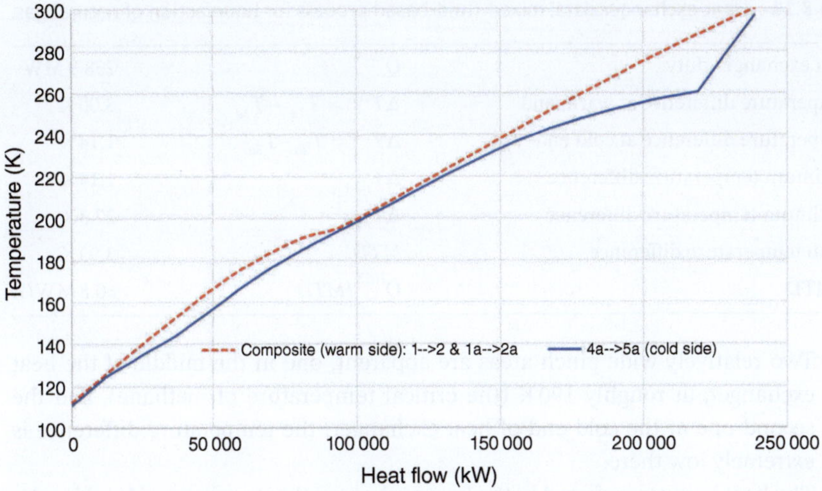

Figure 8.38 Temperature profile (T,Q-diagram) in heat exchanger.

(temperatures, pressure levels, and flowrates) as well as composition of mixed fluid are chosen more or less instinctively; the process design is not optimized by much. The main targets of this exercise are:

(a) to illustrate the process concept and process features,
(b) to establish a connection to the mixed-fluid Joule–Thomson process for refrigeration purposes described in Section 7.5 and
(c) to explain typical process-design challenges.

As shown in Section 7.2, for Joule–Thomson processes all important conclusions about process performance can easily be drawn by considering the T,Q-diagram. The T,Q-diagram for the given process example is shown in Figure 8.38. The solid line corresponds to the cold low-pressure mixed-fluid stream, which is heated here from 110.7 to 297 K. The dashed line corresponds to the warm composite stream, which is cooled from 300 to 112 K.

Heat exchanger data are summarized in Table 8.11. Noteworthy observations derived from consideration of Figure 8.38 are summarized as follows:

1. **Observations Related to Temperature Profile and Heat Exchanger Loss:**
   - In the upper fifth of the heat exchanger (in terms of temperature), a relatively great temperature difference of 27.3 K is reached locally, but since the temperature level is relatively high (above 250 K) there, the exergy loss in this part of the heat exchanger is moderate.
   - In the remaining four-fifths of the heat exchanger, the temperature difference is continuously low. At the end, the total mean temperature difference in the heat exchanger amounts to 3.9 K, which is a surprisingly low value.

Table 8.11 Heat exchanger data, mixed-fluid-based process for liquefaction of natural gas.

Heat exchanger duty	$\dot{Q}^{HEX}$	238.9 MW
Temperature difference at warm end	$\Delta T^{WE} = T_{1a} - T_{5a}$	3.00 K
Temperature difference at cold end	$\Delta T^{CE} = T_{2a} - T_{3a}$	1.14 K
Minimum temperature difference	ΔT_{min}	1.14 K
Maximum temperature difference	ΔT_{max}	27.3 K
Mean temperature difference	MTD	3.93 K
Q/MTD	$\dot{Q}^{HEX}/MTD$	60.8 MW/K

- Two relatively wide pinch areas are apparent, one in the middle of the heat exchanger, at roughly 190 K (the critical temperature of methane), and the second one at the cold end of heat exchanger; the temperature difference is extremely low there.
- The heat being transferred in the heat exchanger (heat exchanger duty) is relatively high. It is difficult to derive this conclusion directly from the T,Q-diagram with composite flow. In this case, consideration of contributions made by single flows can be useful. In the given case, it is evident that the minimal heat duty (required for liquefaction of feed gas only) is close to 60 MW, but the overall duty (with composite stream) of roughly 240 MW is four times higher than the minimal value. For an energy-optimized process, this value is a bit too high.
- Summarizing, the temperature profile is quite advantageous in terms of temperature difference (mean temperature difference is low, and temperature difference at the cold end is low). But since a large amount of heat is managed in the apparatus, the expected total loss is rather moderate than low.

2. **Observation Related to Expansion:**
   - In the process under consideration, the temperature difference at the cold end corresponds to the temperature drop during expansion. The observed small value of 1.14 K is a clear indication of low loss during expansion.
3. **Observation Related to Feed Gas Side:**
   - The feed gas is cooled to a very low temperature of 112 K; the compressed natural gas is strongly subcooled at this condition. Therefore, the expectation is that only little (if any) flash-gas will be generated during expansion. Expansion loss is minor.
4. **Observation Related to Phase Conditions of Working Fluid:**
   - High-pressure side: since no sharp kink is visible in the upper part of the heat exchanger, this indicates that the dew temperature for the given fluid is higher than 300 K, therefore it is not visible in the given diagram. It means that the working fluid at the inlet in heat exchanger (state 1a) is in two-phase conditions. This is a critical issue since homogeneous distribution of two-phase fluid is a challenging task; see Section 8.4.5.2 for further information.

Indeed, the exact evaluation of process performance and exergy losses (executed in Section 9.2.5.7) confirms the considerations summarized there. The overall process efficiency of 46% is quite moderate, whereby approximately 21% of the power is lost in the heat exchanger and 8.4% is lost in the two expansion valves. Therefore,

(1) a further improvement of the process is possible,
(2) a further improvement of the process is possible by minimization of heat exchanger loss.
(3) Since the total mean temperature difference in the heat exchanger with 3.9 K is already minimized so far, a further improvement of the process is possible by focusing more on minimization of heat exchanger duty. This can be achieved by increasing the high-pressure level in the mixed-fluid loop, with corresponding adaptation of the mixed-fluid composition, and by increasing the feed gas pressure.

Summary for Process Example This process example shows that mixed-fluid processes can easily be applied for liquefaction of natural gas. It requires only a limited number of hardware components, yet this process can achieve high efficiency in terms of power. Potentially, the first feature can be translated into competitive capital expenditures (CAPEX), while the second feature means competitive operational expenditures (OPEX). These are the main reasons why the mixed-fluid-based process is attractive for natural gas liquefaction.

8.4.5.1.4 Liquefaction Capacity of a Mixed-Fluid-Based Process

In this section, the liquefaction capacity of a mixed-fluid-based process for natural gas liquefaction is considered.

Energy Balance for Entire Cold Part *First*, the energy balance for the entire cold part of the process, as shown in Figure 8.39, is considered.

For simplification purposes, it is assumed that

(a) heat leakage through thermal insulation is very small and can be neglected: $\dot{Q}_{is} \approx 0$,
(b) feed gas is completely liquefied after expansion: $\dot{m}_4 = 0$, therefore $\dot{H}_4 = 0$.

Consequently $\dot{H}_6 = \dot{H}_3 = \dot{H}_2$

This results in the following enthalpy balance:

$$\dot{H}_{1a} + \dot{H}_1 - \dot{H}_{5a} - \dot{H}_6 = 0 \quad \text{or}$$

$$\dot{H}_{1a} - \dot{H}_{5a} + \dot{H}_1 - \dot{H}_6 = 0 \tag{8.27}$$

Basically, this equation consists of two enthalpy differences:

$\Delta\dot{H}_{NG} = \dot{H}_1 - \dot{H}_6$ - enthalpy difference on the natural gas side

$\Delta\dot{H}_{mix}^{WE} = \dot{H}_{5a} - \dot{H}_{1a}$ - enthalpy difference of mixed fluid

Correspondingly, Eq. (8.27) can be rewritten as follows:

$$\Delta\dot{H}_{NG} = \Delta\dot{H}_{mix}^{WE} \tag{8.28}$$

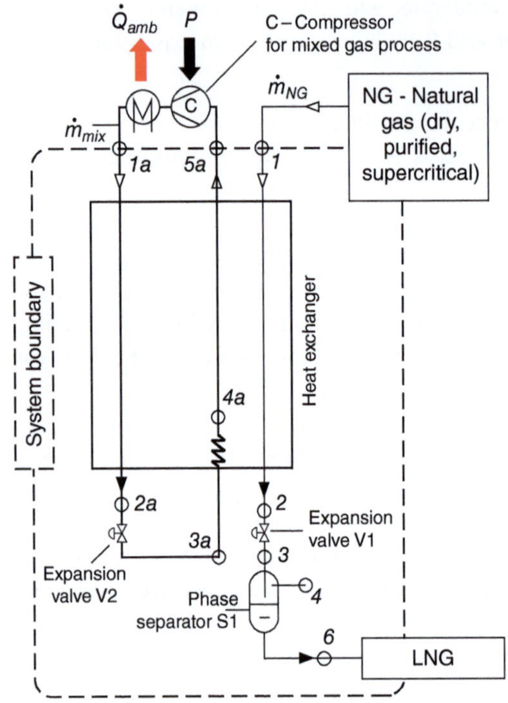

Figure 8.39 System boundary for the entire cold part of natural gas liquefaction process based on mixed-fluid Joule–Thomson process.

or in terms of specific values, as follows:

$$\dot{m}_{NG}\Delta h_{NG} = \dot{m}_{mix}\,\Delta h_{mix}^{WE} \quad \text{or}$$
$$\Delta\, h_{mix}^{WE} = A \cdot \Delta h_{NG} \quad \text{with}\, A = \frac{\dot{m}_{NG}}{\dot{m}_{mix}} \tag{8.29}$$

$\dot{m}_{NG}$ – mass flowrate of natural gas
$\dot{m}_{mix}$ – mass flowrate of mixed fluid
$\Delta h_{NG} = h_1 - h_6$ – specific enthalpy difference on the natural gas side
$\Delta h_{mix}^{WE} = h_{5a} - h_{1a}$ – specific enthalpy difference of mixed fluid

For quantitative calculations of liquefaction capacity, all equations above (Eqs. 8.28 and 8.29 are not well suited, since the temperature difference (and correspondingly the enthalpy difference) at the warm end, for mixed-fluid-based processes, is basically unknown. However, these equations contain one important message:

Learning 8.5 The total cooling capacity required for liquefaction of feed gas is generated by a mixed-fluid process; the liquefaction capacity of a mixed-fluid process is defined by the enthalpy difference of mixed fluid at the warm end of heat exchangers ΔH_{mix}^{WE}: the higher the enthalpy difference of mixed fluid at the warm end, the higher the liquefaction capacity of the process.

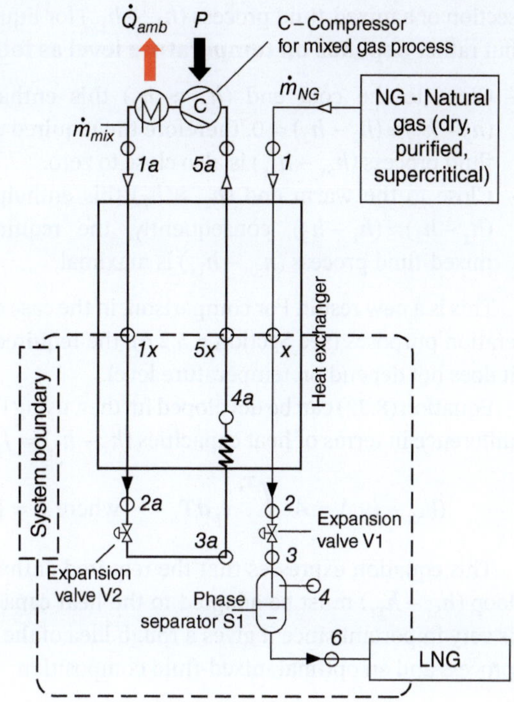

Figure 8.40 "Moving" system boundary for natural gas liquefaction process based on mixed-fluid Joule–Thomson process.

Extension of Energy Balance Equation Further learnings can be obtained by considering Eq. (8.30), which represents an extension of Eq. (8.27) for a "moving" system boundary as shown in Figure 8.40b. In this model, the heat exchanger is virtually cross-cut, and the cold part of the process below this crosscut is considered:

$$\dot{H}_{1x} + \dot{H}_x - \dot{H}_{5x} - \dot{H}_6 = 0 \tag{8.30}$$

Basically, it consists of two enthalpy differences:

$\Delta\dot{H}_{NG\_x} = \dot{H}_x - \dot{H}_6$ enthalpy difference on the natural gas side
$\Delta\dot{H}_{mix\_x} = \dot{H}_{5x} - \dot{H}_{1x}$ enthalpy difference of mixed fluid

Correspondingly, Eq. (8.27) can be rewritten as follows:

$$\Delta\dot{H}_{NG\_x} = \Delta\dot{H}_{mix\_x} \tag{8.31}$$

or in terms of specific values, as follows:

$$\begin{aligned}
&\dot{m}_{NG}(h_x - h_6) = \dot{m}_{mix}(h_{5x} - h_{1x}) \text{ or}\\
&\quad (h_{5x} - h_{1x}) = [\dot{m}_{NG}/\dot{m}_{mix}]\,(h_x - h_6) \text{ or}\\
&\quad (h_{5x} - h_{1x}) = [\dot{m}_{NG}/\dot{m}_{mix}]\,(h_x - h_2) \text{ or}\\
&\quad (h_{5x} - h_{1x}) = A\,(h_x - h_2), \text{ where } A = [\dot{m}_{NG}/\dot{m}_{mix}] = \text{const}
\end{aligned} \tag{8.32}$$

In this equation, h_x is a variable. It varies with temperature level (cross-cut position). Eq. (8.32) expresses that the **required enthalpy difference** in every cross section of a mixed-fluid process $(h_{5x} - h_{1x})$ for liquefaction purposes is not constant but rather **depends on temperature level as follows:**

- Close to the cold end $(h_x \approx h_2)$ this enthalpy difference is close to zero $(h_x - h_2) \approx (h_2 - h_2) = 0$, therefore the required enthalpy difference in the mixed-fluid process $(h_{5x} - h_{1x})$ is also close to zero.
- Close to the warm end $(h_x \approx h_1)$ this enthalpy difference is at its maximum $(h_x - h_2) \approx (h_1 - h_2)$, consequently, the required enthalpy difference in the mixed-fluid process $(h_{5x} - h_{1x})$ is maximal.

This is a new result. For comparison, in the case of a mixed-fluid process for refrigeration purposes (see Section 6.5.2.6), the required enthalpy difference is constant; it does not depend on temperature level.

Equation (8.32) can be developed further, using the classical definition of enthalpy difference in terms of heat capacities $(h_x - h_2) = \int_{T_2}^{T_x} c_p\, dT$ as follows:

$$(h_{5x} - h_{1x}) = A \int_{T_2}^{T_x} c_p \boldsymbol{dT}, \qquad \text{where } A = [\dot{m}_{NG}/\dot{m}_{mix}] = \text{const} \tag{8.33}$$

This equation expresses that the required enthalpy difference in the mixed-fluid loop $(h_{5x} - h_{1x})$ must be adapted to the heat capacity of the feed gas. This concept is very important since it gives a rough idea of the design of an optimal mixed-fluid process and an optimal mixed-fluid composition:

Learning 8.6 The optimal enthalpy difference $\Delta h_x = (h_{5x} - h_{1x})$ for an optimal mixed-fluid composition must fulfill the following criteria:

- **At the cold end (low-temperature level) it is minimal.**
- **In the heat exchanger, it increases from minimum to maximum value.**
- **The maximum is achieved at the warm end of the heat exchanger.**
- **The optimal profile $\Delta h_x = (h_{5x} - h_{1x})$ must be matched to the heat capacity of the feed gas stream.**

Fit to Thermodynamic Properties of Mixed Fluids An interesting observation can be made through further examination of Eq. (8.33). Assuming a constant heat capacity[13] of feed gas $c_p = \text{const}$, and therefore $(h_x - h_2) = c_p\,(T_x - T_2)$, the following equation can be derived:

$$(h_{5x} - h_{1x}) = A\, c_p\, T_x - B, \tag{8.34}$$

where A and B are constant values:

$$A = [\dot{m}_{NG}/\dot{m}_{mix}] = \text{const},$$

$$B = A\, c_p\, T_2 = \text{const}$$

This equation expresses that the best match will be achieved if the required enthalpy difference in the mixed-fluid loop $(h_{5x} - h_{1x})$ looks like a simple linear function.

13 This is a critical assumption, since the heat capacity of supercritical gas can be considered as constant only at very high pressures.

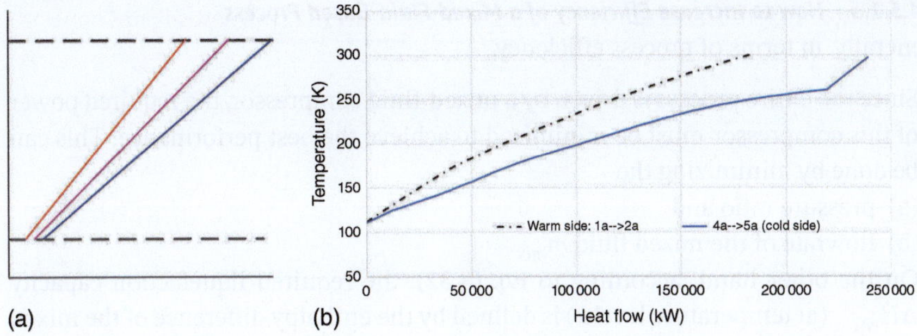

Figure 8.41 (a) typical isobar pattern for nitrogen–hydrocarbon mixture in two-phase area (greatly simplified), (b) T,Q-diagram for mixed-fluid process, process example from Section 9.2.5.

Surprisingly, the required enthalpy profile corresponds well to typical patterns for nitrogen-hydrocarbon mixtures in the two-phase area, as shown in Figure 8.41a: at low temperatures, the isobars are very close together, i.e. the distance between the lines is relatively small; at higher temperatures, the distance between the isobars becomes larger and larger, so that a typical fan-like pattern is formed, see Figure 8.41a (this unique feature of mixed fluid was discussed in Section 7.5.2.4 regarding the mixed-fluid Joule–Thomson process for refrigeration, see Figure 7.60). This behavior reflects the fact that in the two-phase area, the effective[14] **heat capacity of a low-pressure mixed-fluid stream is higher than the effective heat capacity of the same mixed-fluid stream at high pressure** by nature.[15]

This fact is an extremely useful one. It means, that some additional heat capacity, e.g. in the form of an additional material stream (for example feed gas) can be added to the existing mixed-fluid material stream on the warm side of heat exchanger to balance the excess heat capacity on the low-pressure side. Consequently, both these warm material streams can be cooled to required cooling temperature by single cold mixed-fluid material stream. Obviously, the cold mixed-fluid stream will be warmed (in the heat exchanger) to a higher temperature, which can be calculated by simple enthalpy balance for the heat exchanger.

Simultaneously, this consideration serves to explain why the mixed-fluid-based process for liquefaction is so efficient. This is based primarily on two facts:

1. By nature, the thermodynamic properties of nitrogen-hydrocarbons mixtures fit well with the temperature profile required for a typical liquefaction process.
2. Use of mixed fluids provides high flexibility and degrees of freedom for optimization: the mixed-fluid composition may be designed or customized in such a way that the temperature difference in the heat exchanger and the corresponding heat exchanger loss are minimized; for further information, see Section 8.4.5.1.5.

14 In the given case, the term $(dh/dT)_p$ is considered as effective heat capacity.

15 In the gas area, this behavior is always reversal: the heat capacity of a low-pressure gas is lower than the heat capacity of high-pressure gas.

8.4.5.1.5 How to Increase Efficiency of a Mixed-Fluid-Based Process

Generally, in terms of process efficiency,

- Since the entire process is driven by a mixed-fluid compressor, the required power of this compressor must be minimized to achieve the best performance. This can be done by minimizing the
  (a) pressure ratio and
  (b) flowrate of the mixed fluid $\dot{m}_{mix}$.
- On the other hand, according to Eq. (8.32), the required liquefaction capacity $\Delta\dot{H}_{NG\_x}$ (at temperature level x) is defined by the enthalpy difference of the mixed fluid cycle $\Delta\dot{H}_{mix\_x}$ (at this temperature level x), which depends on the
  (a) flowrate of mixed fluid $\dot{m}_{mix}$ and
  (b) specific enthalpy difference of the mixed-fluid cycle Δh_{mix_x} in the considered cross-section.

Therefore, the search for optimum efficiency is something like reaching a compromise between (a) minimal pressure ratio in compressor and (b) maximal enthalpy difference Δh_{mix_x}, since this enthalpy difference depends directly on the pressure ratio.

The following considerations in this section are focused more on maximization of enthalpy difference Δh_{mix_x}.

In compressed form, the argumentation chain looks as follows:

Maximization of enthalpy difference Δh_{mix_x} can be achieved by

- increase of high-pressure level and/or by
- optimization of mixture composition, usually by admixing of high-boiling components, such as propane, butane, and even higher hydrocarbons;
  Consequently,
- the liquid fraction in every cross-section of the heat exchanger will increase,
- the entire process moves in the two-phase area,
- the high-pressure mixed fluid on the warm end of heat exchanger (state 1a) becomes partially liquefied. The two-phase condition there [at the warm end of heat exchanger, high-pressure mixed fluid] is a distinguished indicator for a well-optimized process.

8.4.5.2 Impact of Maldistribution on Design of a Mixed-Fluid-Based Process

The maldistribution phenomenon refers to a non-uniform distribution of fluid between multiple channels of a heat exchanger. If this occurs, maldistribution leads to reduced heat exchanger performance in terms of both heat transfer as well as pressure drop. In most cases, the consequences are very dramatic and become visible and tangible in the form of diminished production capacity and low efficiency.

Uniform distribution of a two-phase flow is extremely challenging. Mainly due to the fact that the two-phase flow is generally highly non-homogeneous in terms of the following two criteria: (i) form (geometry): at a fixed moment, in every cross-section, the liquid and vapor phases are distributed nonuniformly and (ii) time: in every cross-section, the form, structure, and fraction of liquid and vapor phases change erratically and irregularly. Figure 8.42 shows some typical

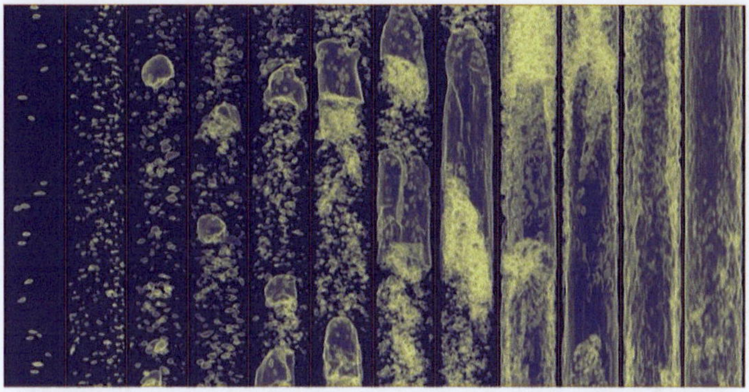

Figure 8.42 Flow-patterns in a two-phase flow, vertical channels (https://www.hzdr.de/db/Cms?pOid=24991&pNid=393&pLang=de); here the velocity changes from 0.004 m/s on the left side to 3.2 m/s on the right side. Source: 21, Helmholtz-Zentrum Dresden - Rossendorf e. V.

patterns of two-phase flow in vertical[16] channels as an impressive illustration of non-homogeneity of two-phase flow.

8.4.5.2.1 Connection to Mixed-Fluid-Based Process:

In the previous Section (8.4.5.1.5), we learned that the two-phase condition at the warm end of the heat exchanger, and generally within the heat exchanger, is an important attribute of an efficient mixed-fluid process. In particular, it is advantageous if the high-pressure working fluid is in two-phase condition at the warm end of heat exchanger, since that way the enthalpy difference Δh_{mix}^{WE} and, correspondingly, the liquefaction capacity of the process will be maximized. The process example considered in Section 8.4.5.1.3 is representative in this sense: the corresponding material stream (state 1a) is in two-phase condition: the vapor fraction at this state amounts to 0.32 mol/mol (vapor fraction is 0.68 mol/mol). This means the mixed-fluid-based process is currently facing the problem of maldistribution.

The mitigation of potential risks caused by maldistribution is an important consideration during the process design phase. The development target is always to develop a process design that is able to manage the maldistribution phenomena.

8.4.5.2.2 Approaches for Management of Maldistribution

Unfortunately, a universal technical answer to manage the problem of maldistribution is not available. Three different approaches for management of maldistribution in mixed-fluid processes are commonly used.

1. In the first approach, the non-homogeneity of two-phase flow is considered as the main issue. It is addressed by measures that target the high degree of homogenization of the fluid before distribution. This can be achieved by means of special mechanical devices such as baffles or nozzles for conversion of fluid into a homogeneous spray or mixing devices of any kind.

16 In a horizontal channel, the behavior of two-phase flow is even more complicated.

2. The main idea of the second approach is to convert the two-phase flow into two single-flow streams and distribute each of these streams separately. Since distribution of single-phase fluids is an easier task (by one order of magnitude) that usually can be managed successfully, the level of maldistribution risks is significantly minimized.
3. Another option for avoidance of maldistribution is to convert the two-phase fluid into a single-phase fluid by increasing the process pressure so greatly that the entire process becomes supercritical (simultaneously, the composition of mixed fluid will be adapted), which means a single-phase fluid.

The third option is more or less self-explanatory, it does not merit dedicated discussion here. Other approaches are pursued by engineers more or less simultaneously. In this section, only the second approach is considered, since it relates to process design, and falls within the scope of this book.

One example of an appropriate process-design-based solution is shown in Figure 8.43a. The process chain includes three elements, i.e. one phase separator and two distributors, for liquid and for vapor phase. The two-phase material stream is split into liquid and vapor fractions by means of a phase separator; each of these phases is distributed by means of corresponding distribution devices more or less homogeneously over the same heat exchanger channel. As a result, the two phases are merged back into one. This flow schematic is referred to as a "D-type" separator since the phase separator is applied for **D**istribution purposes.

Another example of this approach is a solution shown in Figure 8.43b. Principally, the same elements used in the previous approach are applied: a phase separator, a distribution device for liquid, and a distribution device for vapor. The two-phase material stream is converted into two single-phase material streams by means of phase separators. The only difference is that the liquid and vapor are provided to different channels of the heat exchanger or to different locations of the process, aiming toward a positive impact on process performance. This flow schematic is referred to as a "**P**-type" separator since the phase separator is applied for **P**rocess purposes. Although it may not appear much different from the approach shown in Figure 8.43a, this schematic introduces new degrees of freedom for process design and has triggered some remarkable developments.

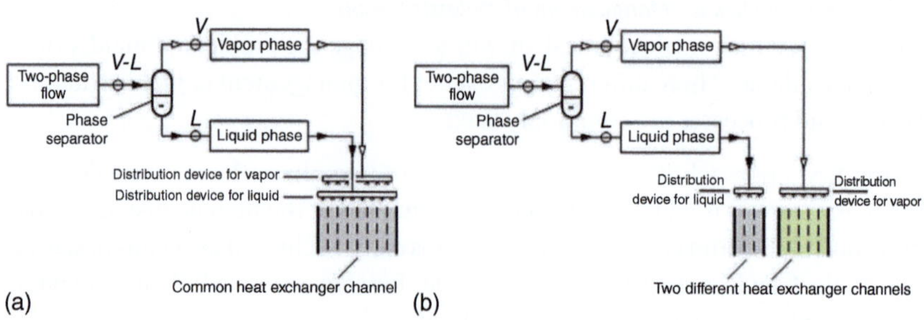

Figure 8.43 Application of phase separator for distribution of two-phase flow (a) D-type separator for Distribution purposes, (b) P-type separator for Process purposes.

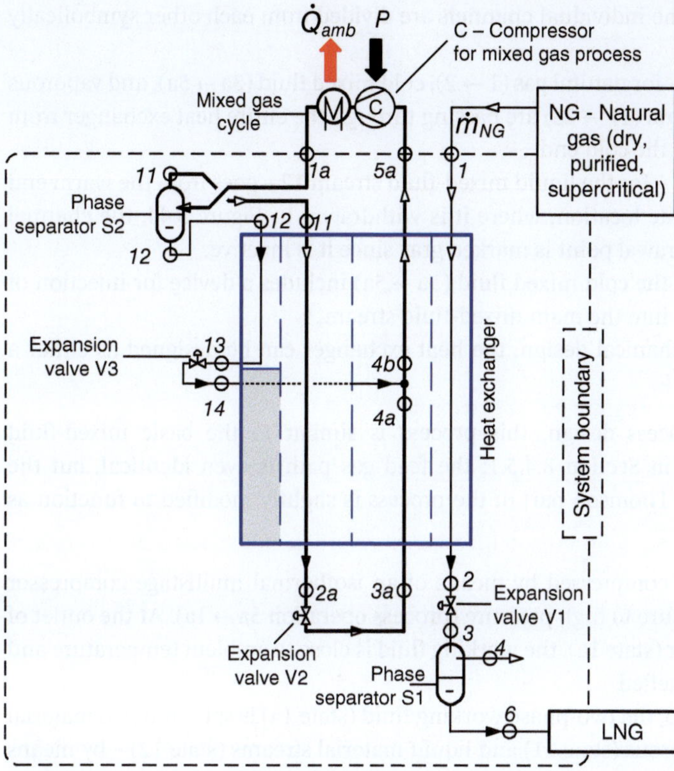

Figure 8.44 Mixed-fluid Joule–Thomson process with single phase-separator for natural gas liquefaction.

8.4.5.3 Mixed-Fluid Joule–Thomson Process with Single Phase-Separator

The mixed-fluid Joule–Thomson process with a single separator is considered in this section. The major advantage of this process is improved resilience against maldistribution and solidification.

8.4.5.3.1 Primary Concept

The process concept is shown in Figure 8.44. Visually, this process is quite similar to the basic mixed-fluid process discussed in Section 8.4.5.1. But it includes several new elements, including the following:

1. phase separator S2 is introduced upstream of the heat exchanger in the high-pressure mixed-fluid path; this is a P-type phase separator,
2. an additional expansion valve V3,
3. an additional channel in the main heat exchanger.

Since the required heat exchanger is quite complex, several comments are in order for clarification purposes:

- The heat exchanger is designed such that it is able to manage four material streams. For this purpose, the heat exchanger consists of four channels.

In Figure 8.44, the individual channels are divided from each other symbolically by dashed lines.

- Three channels – for natural gas (1 → 2), cold mixed fluid (3a → 5a), and vaporous mixed-fluid stream (11 → 2a) are passing through the entire heat exchanger from the warm end to the cold end.
- But one channel – for the liquid mixed-fluid stream 12 – goes from the warm end to an intermediate location, where it is withdrawn. In Figure 8.44, the channel below the withdrawal point is marked gray since it is inactive.
- The channel for the cold mixed fluid (3a → 5a) includes a device for injection of liquid stream 14 into the main mixed-fluid stream,
- In terms of mechanical design, the heat exchanger can be designed as either a PFHE or a SWHE.

In terms of process design, this process is similar to the basic mixed-fluid process discussed in Section 8.4.5.1: the feed gas path is even identical, but the mixed-fluid Joule–Thomson part of the process is slightly modified to function as follows:

(a) Mixed fluid is compressed by means of an isothermal multistage compressor from low pressure to high pressure (process operation 5a → 1a). At the outlet of the compressor (state 1a), the working fluid is close to ambient temperature and is partially liquefied.

(b) In the next step, the two-phase working fluid (state 1a) is split into two material streams – vaporous (state 11) and liquid material streams (state 12) – by means of phase separator S2.

(c) During further processing, the vaporous stream (11) is treated as the main mixed-fluid stream in the basic process described in Section 8.4.5.1. It is
  - provided to the heat exchanger (the risk of maldistribution is minimal, since this fluid is a single-phase fluid),
  - cooled/liquefied in the heat exchanger (operation 11 → 2a),
  - throttled from high pressure to low pressure via an expansion valve (operation 2a → 3a),
  - warmed back up/evaporated in the heat exchanger (operation 3a → 5a) and
  - returned to the compressor for pressurization in a more or less conventional manner.

(d) The new feature is a separate loop for processing of liquid material stream (12) from the separator:
  - This stream is provided through a separate channel of main heat exchanger and is cooled there to an intermediate temperature level (operation 12 → 13);
  - This subcooled liquid is withdrawn from the heat exchanger and expanded in expansion valve V3 from high pressure to low pressure (operation 13 → 14). In a well-designed process, the degree of subcooling is optimized such that after expansion, this liquid remains liquid and no flash-gas is generated. This is an important feature of the process: the material stream remains single-phase fluid. Consequently, further distribution of this material stream is not challenging.

- This liquid is routed to the appropriate location in the heat exchanger, distributed homogeneously by means of the corresponding distributor device, and merged with the cold low-pressure material stream (4a).
- The resulting material stream (4b) has the same flowrate and composition as the mixed fluids 5a and 1a. It is routed through the heat exchanger to the compressor inlet.

8.4.5.3.2 Cooling Capacity

If the system boundary as shown in Figure 8.44 is considered, Eq. (8.27) ($\dot{H}_{1a} + \dot{H}_1 - \dot{H}_{5a} - \dot{H}_6 = 0$) for the cooling capacity of this process still holds. In this sense, the considered process is very similar to the basic process described in Section 8.4.5.1.

Since $\dot{H}_{1a} = \dot{H}_{11} + \dot{H}_{12}$, the following equation is also valid:

$$\dot{H}_{11} + \dot{H}_{12} + \dot{H}_1 - \dot{H}_{5a} - \dot{H}_6 = 0 \quad \text{or}$$

$$\dot{H}_{5a} - (\dot{H}_{11} + \dot{H}_{12}) = \Delta\dot{H}_{NG} \tag{8.35}$$

Learning 8.5 is fully applicable to the process with a single phase-separator.

8.4.5.3.3 Mixed-Fluid Composition

Equation (8.35) can be interpreted in such a way that mixtures (mixed-fluid compositions) and process parameters developed and optimized originally for the basic process without a phase separator (described in Section 8.4.5.1) are principally applicable to the process with a single separator shown in Figure 8.44.

In reality, the best mixed-fluid composition for a process with a single phase-separator deviates from the best mixed-fluid composition developed for the basic process without a phase separator. Therefore, optimization is still required. The good news is that the best mixed-fluid composition developed for the basic process is always a good starting point for the process with a single phase-separator: it accelerates the optimization process considerably.

8.4.5.3.4 Example of Mixed-Fluid-Based Process with Single Phase-Separator

An example of a mixed-fluid-based process for natural gas liquefaction is available in the supplement, Section 9.2.6. For comparison purposes, the process parameters (temperatures, pressure levels, and flowrates) as well as the composition of mixed fluid are chosen close to the process example considered in Section 8.4.5.1, i.e. the mixed-fluid composition is slightly tuned. The main target of this exercise is to emphasize the benefits achieved by application of a P-type separator.

The process description is given in Section 8.4.5.3.1; the most important new parameter is the temperature of the liquid mixed-fluid stream before expansion, 180 K. The temperature of this stream after expansion is 178.6 K. The temperature after merging this fluid with the main low-pressure mixed fluid (16) is roughly 181 K.

Phase Separator S2 Since the phase separator S2 is a new element, it makes sense to analyze the operation of this device first. The compositions of fluids upwards and downwards of this device are shown in Table 8.12.

Table 8.12 Composition of fluids at inlet and outlets of phase separator.

	State 1a (mol/mol)	**State 11 (mol/mol)**	**State 12 (mol/mol)**
Nitrogen	0.055	0.078	0.006
Methane	0.315	0.425	0.082
Ethane	0.335	0.364	0.273
Propane	0.055	0.04	0.086
Butane	0.24	0.092	0.553

The main observations are the following:

- The feed fluid consists primarily of methane, ethane, and butane; the fractions of nitrogen (5.5 mol%) and propane (5.5 mol%) are relatively small.
- The composition of vaporous material stream 1V differs considerably from the composition of the feed: it consists of mainly low-boiling gases – more nitrogen, methane, and ethane, and less propane and butane.
- Similarly, the composition of the liquid material stream 12 differs considerably from the composition of the feed. It is rich in high-boiling substances: the fraction of butane is higher than 55 mol%, and the fraction of propane is higher than in the feed. This liquid consists of very little methane and practically no nitrogen (0.6 mol%).
- The medium-boiling fluid – ethane – is distributed between both phases.

The most important observation is that the fraction of components with relatively high freezing temperature (in this case it is butane) in the mixed-fluid stream (11) provided to the cold end is considerably reduced by means of a separator, in the given case from 24 to 9.2 mol%. It means that the risk of solidification at low temperatures is considerably reduced or even mitigated entirely.

Heat Exchanger As shown in Section 7.2, for Joule–Thomson processes, all important conclusions about process performance can easily be made by considering the T,Q-diagram. The T,Q-diagram for the given process example is shown in Figure 8.45. The solid line corresponds to the cold low-pressure mixed-fluid stream, which is heated here from 110.7 to 297 K. The dashed line corresponds to the warm composite stream, which is cooled from 300 to 112 K. This composite stream is a virtual (composite) material stream generated by summation of (enthalpies or heat capacities) three warm material streams – feed gas 1, high-pressure vaporous mixed fluid 1V, and high-pressure liquid 1L – to a single material stream. The heat exchanger data are summarized in Table 8.13.

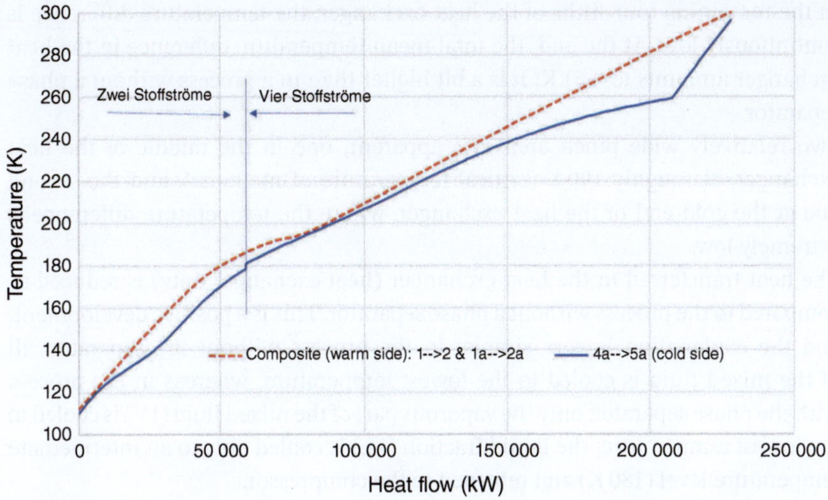

Figure 8.45 T,-Q-diagram for heat exchanger (mixed-fluid JT process with single separator).

Table 8.13 Heat exchanger data, mixed-fluid-based process for liquefaction of natural gas with single separator.

		Process with single separator	**Basic process (section 8.4.5.1)**
Heat exchanger duty	$\dot{Q}^{HEX}$	228 MW	238.9 MW
Temperature difference at warm end	$\Delta T^{WE} = T_{1a} - T_{5a}$	3.8 K	3.00 K
Temperature difference at cold end	$\Delta T^{CE} = T_{2a} - T_{3a}$	2.48 K	1.14 K
Minimum temperature difference	ΔT_{min}	1.04 K	1.14 K
Maximum temperature difference	ΔT_{max}	27.1 K	27.3 K
Mean temperature difference	MTD	4.63 K	3.93 K
Q/MTD	$\dot{Q}^{HEX}/MTD$	49.5 MW/K	60.8 MW/K

Noteworthy observations derived from consideration of Figure 8.45 can be summarized as follows:

- The T,Q-diagram for the given process is quite similar to the diagram for the process without a phase separator, see Section 8.4.5.1. In the temperature range of 180–300 K, the two diagrams are identical, and only below 180 K do any differences become apparent.
- In the upper fifth of the heat exchanger (in terms of temperature), a relatively high-temperature difference of 27.3 K is reached locally, but since the temperature there is relatively high (above 250 K), the exergy loss in this part of the heat exchanger is moderate.

- In the remaining four-fifths of the heat exchanger, the temperature difference is continuously low. At the end, the total mean temperature difference in the heat exchanger amounts to 4.61 K. It is a bit higher than in a process without a phase separator.
- Two relatively wide pinch areas are apparent, one in the middle of the heat exchanger, at roughly 190 K (critical temperature of methane), and the second one at the cold end of the heat exchanger, where the temperature difference is extremely low.
- The heat transferred in the heat exchanger (heat exchanger duty) is reduced as compared to the process without a phase separator. This is a positive development, and the explanation is very simple: in the process without the separator, all of the mixed fluid is cooled to the lowest temperature, whereas in the process with the phase separator, only the vaporous part of the mixed fluid (1V) is cooled to the lowest temperature; the liquid fraction (1L) is cooled only to an intermediate temperature level (180 K) and returned to the compressor.
- Since the required heat exchanger is smaller and the mean temperature difference is higher, the Q/MTD-value is smaller as compared to the process without the separator. This means the heat exchanger needs less heat transfer surface. In terms of mass and volume, it is smaller and potentially cheaper.
- In the considered process, the temperature difference at the cold end corresponds to the temperature drop during expansion. The observed small value is a clear indication of low loss during expansion.
- The feed gas is cooled to a very low temperature of 112 K; the compressed natural gas is deeply subcooled at this point. Therefore, it is to be expected that only a small amount of flash-gas (if any at all) will be generated during expansion. Expansion loss is minor.

In summary, the temperature profile is quite advantageous in terms of both heat exchanger duty and temperature difference: mean temperature difference is moderate, and the temperature difference at the cold end is low. Therefore, the expected losses in the process are moderate. Indeed, the exact evaluation of process performance and exergy losses (given in Section 9.2.6) confirms the consideration mentioned above. At 46%, the overall process efficiency is quite moderate.

8.4.5.3.5 Brief Assessment

The major advantage of the process with a single phase-separator is improved resilience against solidification and maldistribution:

- The risk of maldistribution is considerably reduced, since only single-phase fluids are managed/distributed in this process.
- The risk of solidification is reduced, since the problematic substances (butane and higher hydrocarbons), which may solidify at low temperatures, are removed from the main working fluid (1V) by means of phase separator S2.

A further advantage relates to tangible CAPEX savings achieved due to a smaller heat exchanger, since the heat transferred in the heat exchanger (heat exchanger duty) is reduced as compared to the basic process.

In terms of power efficiency, the process with a single phase-separator is considered as potentially more efficient than the basic process. This is mainly due to the same phenomenon – that the heat transferred in the heat exchanger is reduced. Principally, this means the exergy loss in the heat exchanger can be lowered correspondingly (see Eq. 5.22). Unfortunately, in the process example shown in Section 8.4.5.3.4, this advantage was not realized, since the mixed-fluid composition was not fully optimized there.

8.4.5.4 Mixed-Fluid Joule–Thomson Process with Two-/Multiple-Phase Separators

The process example considered in the previous Section (8.4.5.3) illustrates that a significant improvement can be achieved through relatively simple process modifications, such as implementation of a simple phase separator in the process. This is an important learning, and it is the motivation for process design engineers to invest more time and effort in the development of further processes. As a result, a great number of innovative and advantageous mixed-gas-based processes and process modifications for liquefaction of natural gas have been developed and realized over the past 70 years. Two processes derived from the process with a single phase-separator are described further, with the objective of demonstrating several patterns of thought that are typical among process design engineers:

a) process with multiple phase separators (an example of such a process is available in the Supplement, Section 9.2.7),
b) process with external pre-cooling stage, described in next section 8.4.5.5.

8.4.5.5 Mixed-Fluid Joule–Thomson Process with External Pre-Cooling Stage

The approach with multiple-phase separators works well, but it introduces a new level of complexity, as can be seen in the process flow diagram with numerous phase separators, control valves, knotty piping work, and a complex heat exchanger design, even if some of the integrated devices for withdrawal and for injection of different mixed-fluid streams are not shown there. The composition of the working fluid changes disruptively downstream of every phase separator, making the process challenging to control. This complexity is often undesirable since it increases technical risks. This issue of process complexity is addressed by the second approach described in this section. The primary objective of this process is to reduce the number of hardware components and simplify the heat exchanger design.

8.4.5.5.1 Primary Concept

This goal is achieved, at least partially, by splitting the mixed-fluid cycle into two separate sub-cycles:

– one with liquid refrigerant, referred to as the precooling cycle,
– one with gaseous mixed-fluid refrigerant (at high pressure and ambient temperature) as the main working fluid.

The principle of the process is shown in Figure 8.46. Visually, this process is quite similar to the basic mixed-fluid process discussed in Section 8.4.5.1. But it includes three new elements:

1. compressor C2,

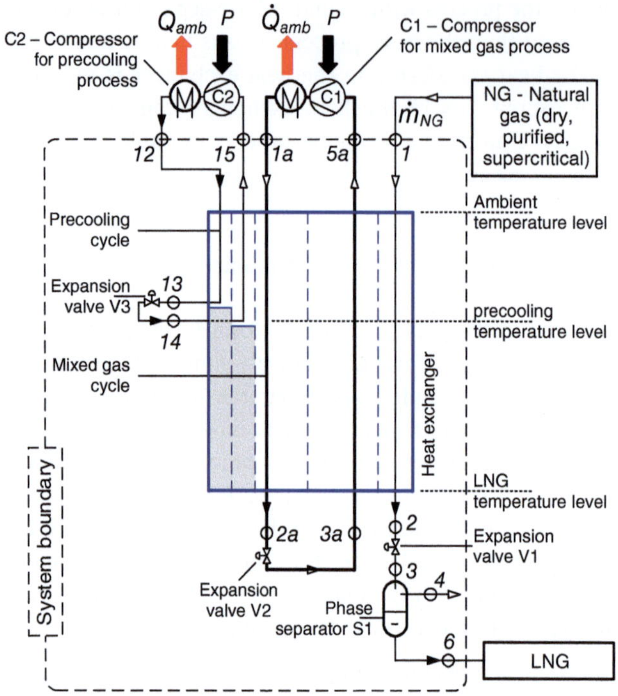

Figure 8.46 Mixed-fluid Joule–Thomson Process with precooling stage, new system boundary.

2. expansion valve V3,
3. the heat exchanger is designed such that it is able to manage five material streams.

The main mixed-fluid cycle (1a → 2a → 3a → 4a → 5a) is designed as a simple closed loop without any withdrawal or injection. At the inlet of the heat exchanger (state 1a), the mixed-fluid refrigerant is gaseous.

The external precooling cycle (12 → 13 → 14 → 15) is designed in a similar manner as a closed loop without any withdrawal or injection. At the inlet to the heat exchanger (state 12), the corresponding working fluid is liquid. This is an important feature of the process.

Formally, the process is characterized by three temperature levels:

1. ambient temperature at the warm end of the heat exchanger,
2. LNG temperature at the cold end of the heat exchanger
3. precooling temperature, i.e. the temperature after expansion (13 → 14) in the precooling cycle.

There are many ways to explain this process; the morphology can be derived from classical cascade process as well as from the nitrogen Joule–Thomson process with precooling. In this section, the connection between the mixed-fluid Joule–Thomson process and a single phase-separator is discussed. From a didactic standpoint, it is easier, since the process with a single separator is considered in Section 8.4.5.3.4 in this book in detail.

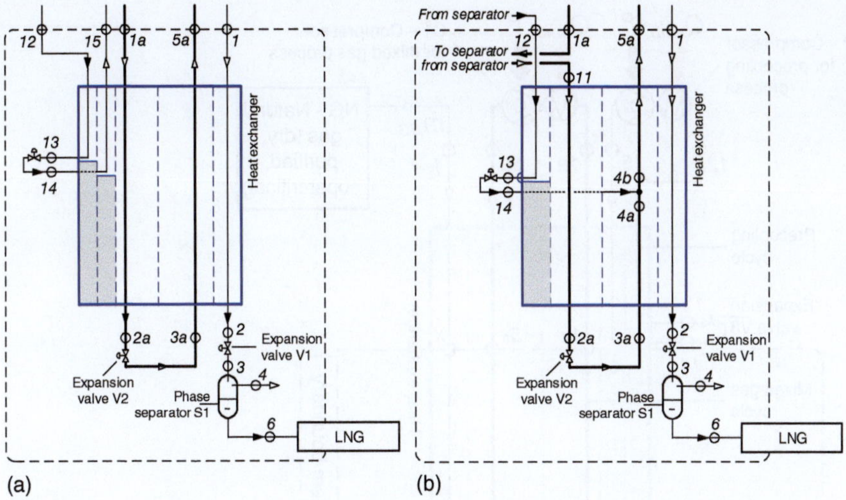

Figure 8.47 (a) upper part of mixed-fluid Joule–Thomson Process with precooling stage (b) upper part of mixed-fluid Joule–Thomson process with single phase-separator.

Morphologically, as well as in terms of phase states, both of these processes are very similar, especially if compressors are hidden, as shown in Figure 8.47. The main difference in the cold part is as follows:

Process with single phase-separator:

After subcooling (12 → 13) in the heat exchanger and expansion (13 → 14) in the expansion valve, the liquid stream (12) **is mixed** with the cold mixed-fluid stream and passed through the heat exchanger as a part of the total mixed fluid to the single-mixed-fluid compressor unit (C in Figure 8.44),

Process with external precooling:

After subcooling (12 → 13) in the heat exchanger and expansion (13 → 14) in the expansion valve, the liquid stream (12) **is not mixed** with any other material streams, but it is passed through the heat exchanger by means of a separate channel to a separate compressor unit (C2 in Figure 8.48) for compression. The mixed fluid is compressed by its own compressor unit (C1 in Figure 8.48).

8.4.5.5.2 Cooling Capacity

If the system boundary as shown in Figure 8.46 is considered, the following equation can be derived:

$$\dot{H}_{1a} - \dot{H}_{5a} + \dot{H}_{12} - \dot{H}_{15} + \dot{H}_1 - \dot{H}_6 = 0 \quad \text{or}$$
$$\Delta\dot{H}_{NG} = (\dot{H}_{5a} - \dot{H}_{1a}) + (\dot{H}_{15} - \dot{H}_{12}) \quad \text{or}$$
$$\Delta\dot{H}_{NG} = \Delta\dot{H}_{MIX}{}^{WE} + \Delta\dot{H}_{PRE}{}^{WE} \tag{8.36}$$

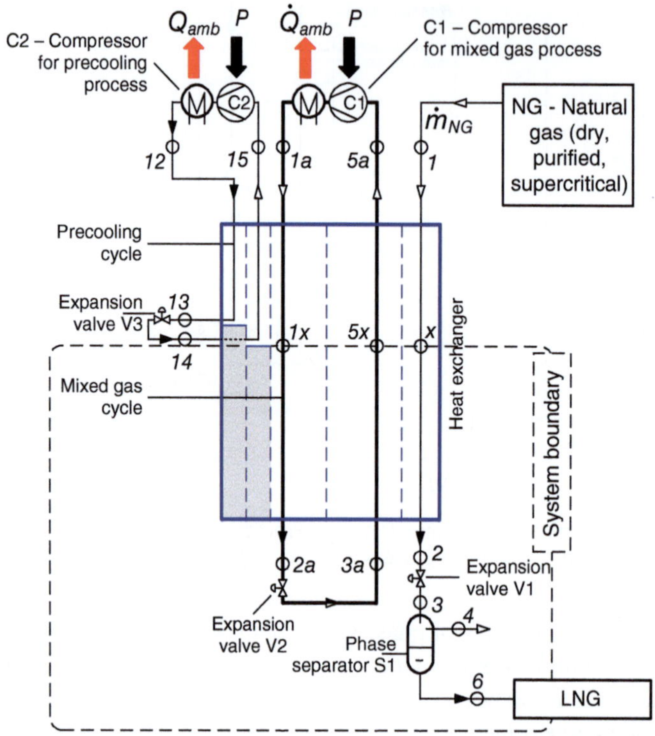

Figure 8.48 Mixed-fluid Joule–Thomson Process with precooling stage, new system boundary.

with

$\Delta\dot{H}_{NG} = \dot{H}_1 - \dot{H}_6$	– enthalpy difference on the natural gas (feed) side
$\Delta\dot{H}_{MIX}{}^{WE} = \dot{H}_{5a} - \dot{H}_{1a}$	– enthalpy difference of the mixed-fluid cycle at the warm end of the heat exchanger
$\Delta\dot{H}_{PRE}{}^{WE} = \dot{H}_{15} - \dot{H}_{12}$	– enthalpy difference of the precooling fluid cycle at the warm end of the heat exchanger

This equation expresses that the total cooling capacity required for liquefaction of feed gas is generated by both refrigeration loops: (i) mixed-fluid process and (ii) precooling process.

Precooling Cycle: Since the precooling cycle operates with refrigerant, which is liquid in state 12, and since enthalpy of a fluid in the liquid state is considerably lower than enthalpy of a fluid in the gaseous state, one can assume that the enthalpy difference of the precooling cycle at the warm end of the heat exchanger is larger than the enthalpy difference of the mixed-fluid cycle at the warm end of the heat exchanger: $\Delta\dot{H}_{PRE}{}^{WE} \gg \Delta\dot{H}_{MIX}{}^{WE}$. Therefore, the liquefaction capacity of the considered process is defined not only (but mainly) by the enthalpy difference of the precooling cycle at the warm end of the heat exchanger:

$$\Delta\dot{H}_{NG} \approx \Delta\dot{H}_{PRE}{}^{WE} \tag{8.37}$$

This is new as compared to the mixed-fluid processes previously considered (basic mixed-fluid process, process with single phase-separator, or process with multiple-phase separator).

Mixed-fluid cycle: If the system boundary as shown in Figure 8.48 is considered, whereby the part of the process below the precooling cycle is considered, the following equation can be derived:

$$\dot{H}_{1x} - \dot{H}_{5x} + \dot{H}_x - \dot{H}_6 = 0 \quad \text{or}$$
$$\dot{H}_x - \dot{H}_6 = \dot{H}_{5x} - \dot{H}_{1x} \tag{8.38}$$

This equation expresses that the cooling capacity required for cooling and liquefaction of feed gas at temperatures below precooling cycle is generated only by a mixed-fluid process.

Taken together, these last two equations (Eqs. 8.37 and 8.38) articulate the idea of this process as follows:

- The purpose of the precooling cycle is to cool the feed gas and the compressed mixed fluid to the precooling temperature level.
- The purpose of the mixed-fluid cycle is to cool and liquefy the feed gas downwards of the precooling temperature.

8.4.5.5.3 Working Fluids

The connection to the mixed-fluid process with a single phase-separator mentioned above has the advantage that it gives an idea of the fluids/refrigerants for both subprocesses:

(1) **Working Fluid for Precooling Cycle**
Resembles liquid fluid at state 12 in Table 8.12: this consists mainly of high-boiling substances like propane, butane, or even higher hydrocarbons. If a lower precooling temperature is targeted, it can consist of ethane or similar fluids. There are many degrees of freedom, so this refrigerant can be designed as a mixed fluid or as a pure refrigerant. In the latter case, propane is the preferred fluid.

(2) **Working Fluid for Mixed-Fluid Cycle**
Resembles gaseous fluid at state 11 in Table 8.12: this consists of mainly low-boiling substances like nitrogen, methane, and ethane; the fraction of propane and higher hydrocarbons (if any) is low.

8.4.5.5.4 Process Variations

The process with an external precooling stage is described here only in its simplest version, purely to explain the principle. Many variations of this process are known, differing mainly by the precooling temperature level, the fluid for the precooling cycle, and the design of the heat exchanger (or the heat exchanger system). An example of an industrial plant based on the process with an external precooling stage is described in Section 8.4.6.1.

8.4.5.5.5 Brief Assessment

The main disadvantages are as follows:

- Relative complex design of the upper part of the heat exchanger, where the precooling cycle is implemented.
- Rotating equipment concept based on two compressors. This can mean higher maintenance and service effort in operation, as well as higher CAPEX.

The CAPEX issue relates primarily to small- and medium-sized plants. For large plants, multiple machines are often required anyway for compression of large amounts of fluid. In that case, the process with an external precooling stage may even become advantageous, since it provides more degrees of freedom regarding the choice of compressors.

Benefits

- Simplicity in terms of process:
  - relatively simple process with only few hardware components,
  - simple piping and instrumentation,
  - easy control strategy,
- Simplicity in terms of mixed fluid: a simple mixed fluid containing few substances is used for the main mixed-fluid cycle. The design of the mixture as well as the management of the mixed-fluid composition in the process is less complex and requires less effort.
- Resilience against solidification:
  high-boiling substances such as butane and pentane, which can solidify at low temperatures, are completely avoided.
- Resilience against maldistribution:
  The mixed fluid is optimized such that it remains gaseous at ambient temperature. Therefore, the risk of maldistribution in the heat exchanger is reduced.
- High efficiency: potentially, the process with an external precooling stage can achieve high efficiency if designed properly.

8.4.6 Examples of Industrial Process for Natural Gas Liquefaction

The design of an LNG process depends considerably on the heat exchanger design. Both of the main cryogenic heat exchanger technologies – aluminum PFHEs as well as SWHEs – are applicable for LNG processing. The choice of heat exchanger type depends on the process design. And vice versa: the specifics of heat exchanger design impact the process design considerably. This thesis is confirmed by the following two process examples.

8.4.6.1 Process Example with Propane PreCooling

The C3MR process for natural gas liquefaction commercialized by the company Air Products is considered in this section. The abbreviation "C3" means propane since this gas is used as the refrigerant for the precooling cycle. The abbreviation MR means mixed refrigerant since a mixed fluid is used as the working fluid in

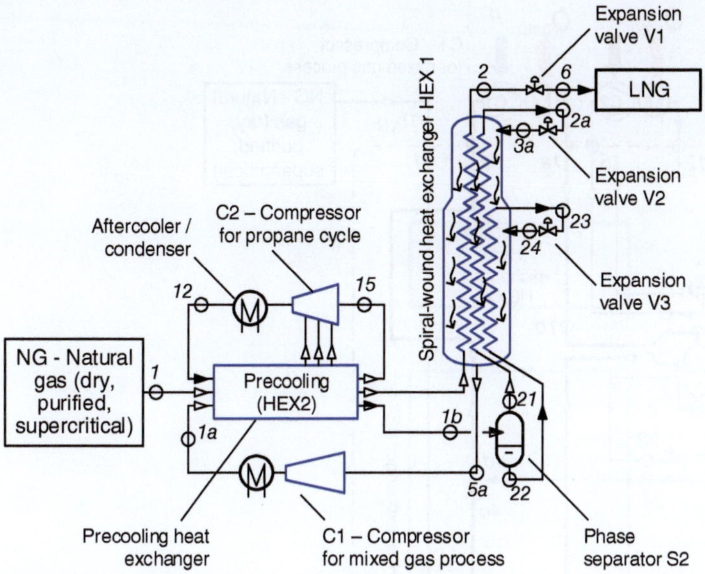

Figure 8.49 LNG Air Products C3MR process [16].

the main refrigeration cycle. Terminologically, the C3MR process is a mixed-fluid Joule–Thomson process with external precooling stage based on the propane cycle.

Figure 8.49 shows the C3MR process as it is often presented in brochures and catalogs. Figure 8.50 shows the same process in the more conventional manner used in this book. The representation of the process in Figure 8.49 is probably a bit better, since it shows more details of the heat exchanger design as well as the correct arrangement of the heat exchanger and the material streams inside it.

In terms of hardware, the single SWHE HEX1 is the most important element, since the process is designed around this heat exchanger. The unique feature of the SWHE is that this heat exchanger can be designed and built as a very large single unit, roughly two orders of magnitude larger than the largest PFHE. Therefore, it is possible to build a processing plant with gigantic liquefaction capacity, based on a single SWHE with all its corresponding advantages.

SWHE: In terms of process, the special attribute of SWHE (see Section 6.2.2) is that several different material streams can be processed on the tube side, but only one material stream is processed on the shell side there. Therefore, the C3MR process is designed in an appropriate manner and contains only one cold mixed-fluid flow in this heat exchanger.

The arrangement of the heat exchanger HEX1 within the plant is quite unconventional. Commonly, a cryogenic heat exchanger is implemented in a plant such that the warm end is on the top and the cold end is at the bottom of the cold box. However, in the C3MR plant, the heat exchanger is positioned in exactly the opposite way: the cold end is at the top. This configuration offers several advantages: it allows better management of the cold two-phase mixed fluid on the shell side: on the top, the two-phase flow is distributed homogeneously by

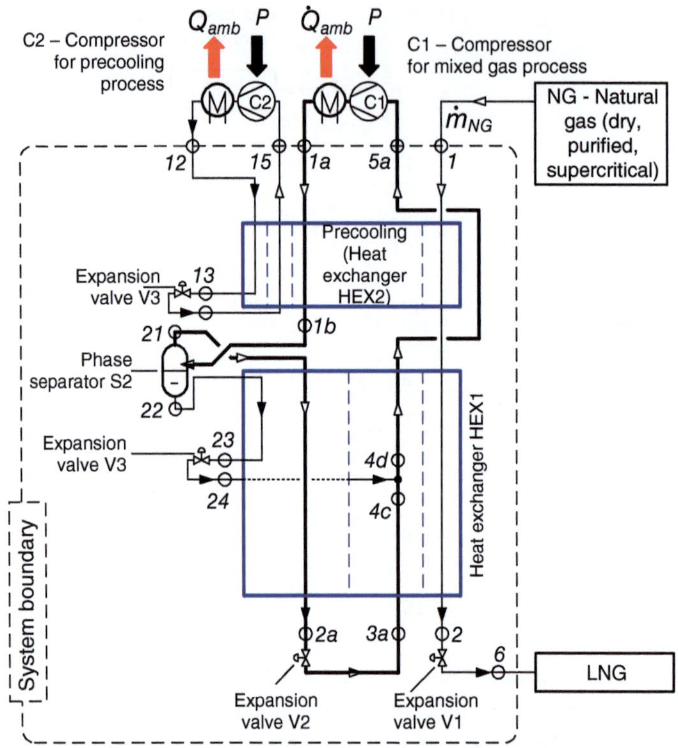

Figure 8.50 LNG Air Products C3MR process presented in conventional manner.

means of distributors (see Section 8.4.5.2) over the entire cross-section and flows downwards driven by pressure drop and gravity. The conventional arrangement (warm end on the bottom) would require some additional pressure for pushing the liquid phase against gravity and would make homogenous distribution even more challenging.

Precooling cycle: Besides the SWHE, the process utilizes another heat exchanger device referred to as the precooling/heat exchanger HEX2. The purpose of this heat exchanger is to perform cooling of two material streams – (i) the natural gas feed and (ii) high-pressure mixed fluid – to a precooling temperature of 235–240 K by means of the propane cycle.

The details of the propane cycle are not well represented by the figures. In reality, this propane cycle is quite complicated. It consists of up to 4 propane loops, switched in parallel and driven by the same compressor. Only one such loop is shown in Figure 8.50:

(a) The propane is compressed by means of multistage compressor C2 to high pressure,
(b) Then it is cooled to ambient temperature by an aftercooler. The pressure level is chosen such that the propane liquefies within this heat exchanger. Therefore, this aftercooler is often referred to as the condenser.

(c) The liquid propane is subcooled in HEX2 (12 → 13) and expanded through an expansion valve (13 → 14) to a low-temperature level (at pressure slightly above ambient level).
(d) After expansion, the liquid propane evaporates in heat exchanger HEX2 at constant temperature, and the propane vapor is routed to the inlet of compressor C2 (14 → 15).

In terms of mechanical design, every kind of heat exchanger, for example, PFHEs, may be applied for precooling purposes.

Mixed-fluid cycle: Further cooling of the natural gas feed is performed by means of a mixed-fluid cycle. This mixed-fluid process is a copy of a process with a single separator described in Section 8.4.5.3. The returning cold low-pressure mixed fluid from the SWHE is routed directly to the compressor unit C1 for compression.

The existing plants based on the C3MR liquefaction process have proven to be very energy efficient, thermodynamically stable, operator friendly, and operationally reliable. This process is used primarily for large-scale plants.

8.4.6.2 Process Example with Aluminum Plate-Fin Heat Exchanger

Another example is a process referred to as "Smartfin," developed and commercialized by the company Air Liquide. The key element of this process is an aluminum PFHE. This connection (the "fin" suffix) appears to be the reason for the name "Smartfin." The following features are unique to the PFHE:

(1) The number of material streams to be processed is practically unlimited.
(2) Material streams can be withdrawn or introduced at practically any location within the heat exchanger.

The Smartfin process consists of so many material streams, and these material streams are combined in such a sophisticated way, that this process can be realized only by means of a PFHE; other types of heat exchangers, including SWHEs, cannot meet the process requirements.

Figure 8.51 shows the process as presented in brochures and catalogs. Visually, it consists of 6 phase separators: five units are integrated in the mixed-fluid cycle and one additional unit (S6) in the natural gas path for removal of heavier hydrocarbons. The phase separators S4 and S5 are of type D since these phase separators are applied for homogenous **D**istribution of two-phase material streams. The other three units – S1, S2, and S3 – are of P-type since they are actually used for **P**rocess purposes (see Section 8.4.5.2 for definitions). Therefore, terminologically, the Smartfin process is a mixed-fluid Joule–Thomson process with three process phase separators.

Figure 8.52 shows the same process in the conventional manner used in this book. For simplification purposes, the phase separators S4 and S5 are not shown there. Other key hardware components are:

1. The multistage compressor C1, consisting of two sections C1.1 and C1.2 for compressing mixed-fluid,
2. A common aluminum PFHE with multiple channels for warm and cold material streams,
3. A number of expansion valves.

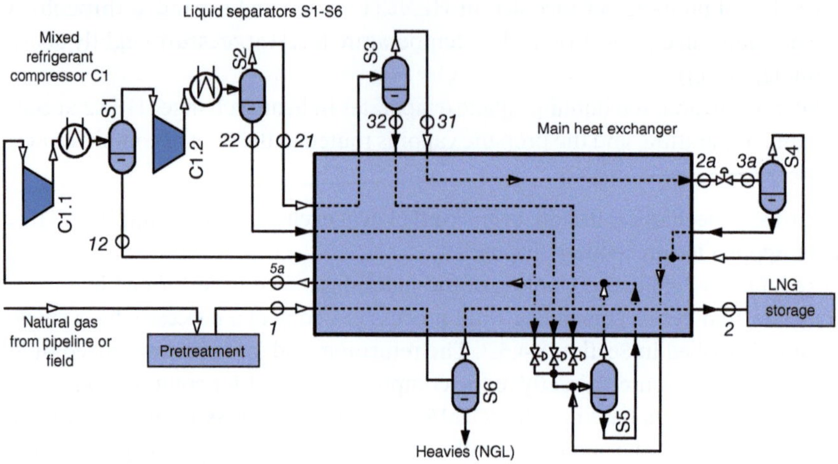

Figure 8.51 Smartfin© process. Source: Adapted from Ref. [17].

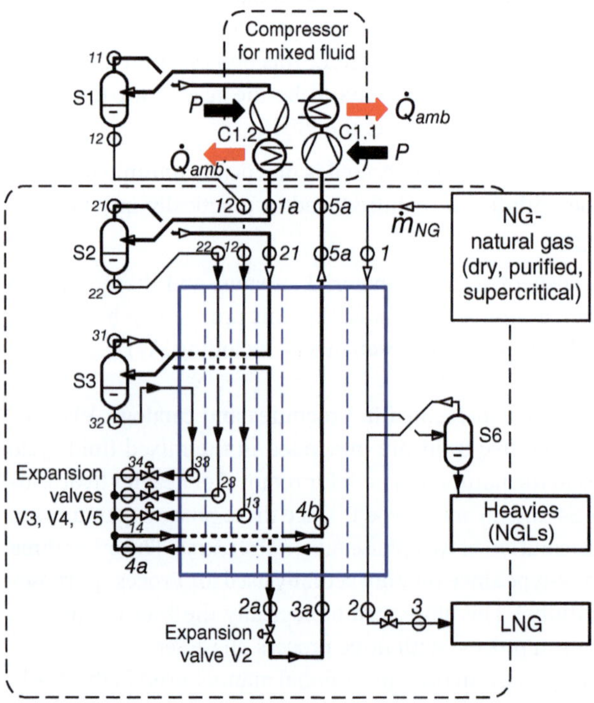

Figure 8.52 Smartfin© process, presented in conventional manner.

8.4.6.2.1 Process Description

The process is quite similar to the process with two-phase separators described in Section 8.4.5.4. The similarity is remarkable if you consider the part of the process marked in Figure 8.52 by a dashed line with the process shown in Figure 8.45. Therefore, this part of the process does not require a separate explanation.

In comparison to the process with two-phase separators, the novelty consists of the additional phase separator (S1) placed between the two compressor sections C1.1 and C1.2. The liquid phase separated there (material stream 12) is cooled in the main heat exchanger (12 → 13) in the same manner as the liquids from phase separators S2 and S3.

Other process features include the following:

- All liquid material streams (13, 23, and 33) are cooled to approximately the same temperature level, withdrawn from the heat exchanger at the same position, and expanded to the same low-pressure level. This is smart.
- The resulting material stream is injected into the main cold low-pressure mixed fluid flowing through the heat exchanger from the cold end to the warm end (3a → 4a → 5a). For this mixing operation, the cold two-phase mixed-fluid material stream (4a) is withdrawn from the heat exchanger. After mixing, the resulting two-phase material stream (4b) is returned to the heat exchanger.

8.4.6.2.2 Brief Assessment

(a) In comparison to SWHEs, PFHEs are more cost-efficient. Therefore, potentially, the LNG plant based on PFHE has better specific CAPEX.
(b) The process design with three-phase separators is quite advanced; in combination with an appropriate mixed-fluid composition, the process may become very competitive in terms of specific power.
(c) Due to "smart" process design, the piping and instrumentation are minimized, but the process itself remains rather sophisticated. It probably requires greater effort to control.
(d) Due to the existing limitation of PFHE's size, this process is better suited to small- and medium-size LNG plants.

8.4.7 Outlook, Natural Gas Liquefaction

As mentioned before, the worldwide demand for natural gas liquefaction plants is still rising. The expectation is that the LNG market will increase from today's 360 to 700 million tons by 2040. As a result, more LNG plants will be built and commissioned in next decade. Therefore, the first focus will be on the further scale and increase in capacity of single LNG trains. The economics of the LNG plant depend on power consumption. Therefore, the second focus will be on efficiency.

In terms of process design, the expectation is that mixed-fluid process will dominate the market of large plants. Since the increasing of high pressure contributes positively to both – CAPEX and OPEX (the higher high-pressure level the larger the enthalpy difference on the warm end of heat exchanger, which defines the specific cooling capacity of the process. Consequently, the mixed fluid flowrate may be reduced. This way, the heat to be transferred in heat exchangers, as well as compressor flowrate, will be required). Therefore, the second expectation is that the process development will go toward increasing of high-pressure level.

Further improvements can be achieved by further development of hardware components in both directions: increasing efficiency and reducing costs. Since both kinds of equipment – the rotating machinery and heat exchangers – are primary contributors to OPEX and CAPEX, further development of these components must be pursued.

References

1 Kolbe, L. (1920). *Flüssige Luft Sauerstoff, Stickstoff, Wasserstoff*. Leipzig: Verlag von Johann Ambrosius Barth.

2 Standard helium liquefier/refrigerator L70/LR70 Datasheet, https://www.linde-kryotechnik.ch/wp-content/uploads/2016/10/Datasheet_L70-LR70-E.pdf.

3 Helial liquefier brochure, Air Liquide, https://advancedtech.airliquide.com/sites/alat/files/2022-11/helial_liquefier_brochure_en_06.21_sd.pdf.

4 Ohlig, K. and Decker, L. (2014). The latest developments and outlook for hydrogen liquefaction technology. In: *AIP Conference Proceedings*, vol. 1573, 1311. https://doi.org/10.1063/1.4860858 Published Online: 17 February 2015.

5 J Kumar, SR Nair, RS Menon, M Goyal, NA Ansari, A Chakravarty and V Joemon (2017). Iop Conference Series: Materials Science and Engineering. **171** 012029. https://www.researchgate.net/journal/IOP-Conference-Series-Materials-Science-and-Engineering-1757-899X.

6 (2008). *Industrial Gases Processing*. Wiley-VCH Verlag GmbH & Co KGaA, ISBN: 978-3-527-31685-4.

7 Haselden, G.G. (ed.) (1971). *Cryogenic Fundamentals*. London, New York: Academic Press xiii, 757 p. illus. 24 cm. ISBN 0123305500.

8 Kerry, F.G. (2007). *Industrial Gas Handbook Gas Separation and Purification*, 158–159. CRC Press, ISBN 0-8493-9005-2.

9 ELM 1000-9000 Series nitrogen liquefiers. https://www.cosmodyne.com/wp-content/uploads/2015/08/ELM-Series-Liquefiers-0917.pdf.

10 Decker, L. and Ohlig, K.Industrial cryogenic plants, script for the VDI cryogenics course.

11 Arthur, J.K. and William, R. (2006). *Parrish: Fundamentals of Natural Gas Processing*. Taylor and Francis, ISBN -13: 978-0-8493-3406-1.

12 Kidnay, A.J. and Parrish, W.R. (2006). *Fundamentals of Natural Gas Processing*. Taylor and Francis ISBN-13: 978-0-8493-3406-1.

13 The standards of the brazed Aluminum plate-fin heat exchangers manufacturers' association ALPEMA. https://alpema.org/standards.html (accessed July 2023).

14 Isalski, W.H. (1989). *Separation of Gases*. Oxford: Clarendon Press, ISBN 0-19-854811-7.

15 Teil, I., Bosnjakovic, F., and Knoche, K.F. (1988). *Technische Thermodynamik*, 362–368. Darmstadt: Steinkopff Verlag, ISBN 3-7985-0759-7.

16 Pilarella M., Liu Y.N., Petrowski J, et al. The C3MR liquefaction cycle: versatility for a fast growing, ever changing LNG industry, LNG 15. Proceedings of the 15th International Conference and Exhibition of Liquefied Natural Gas2007.

17 LNG, Liquefied natural Gas technologies, brochure von Air Liquide: "Technology Handbook", Air Liquide, page 61: https://engineering.airliquide.com/sites/engineering/files/2022-09/technohandbook11oct.pdf.

9

Supplements

9.1 Supplements for Cryogenic Refrigeration

9.1.1 Example of Techno-economic Optimization

The process design is always based on the hardware concept (see Section 2.1), which is usually developed simultaneously with the process idea. Since the compressor is the most expensive hardware component required for the Joule–Thomson process, techno-economic optimization is often focused on finding the compressor with the best techno-economics for the overall system.

Basically, there are no compressors available that are designed and developed specifically for Joule–Thomson refrigerators. The conventional approach is to survey the machines available on the market, which may more or less meet the requirements; it is then necessary to narrow the selection to the most appropriate compressors and evaluate which of them is most suitable.

This procedure is quite complex. Usually, for every compressor considered, an optimization study (similar to that described in Section 7.2.4 but considering the compressor efficiency curve η_T) is required, since the performance of the Joule–Thomson JT process depends on the compressor efficiency η_T. In the second step, the optimized processes are evaluated and assessed in terms of cost expenditures, whereby both cost components – capital expenditures as well as operational expenditures – are considered.

Tables 9.1 and 9.2 show examples of typical cost considerations. In the given case, a laboratory requires a cryogenic cooling system with a cooling capacity of 1 kW. The laboratory will operate the refrigeration system more or less continuously for two years, with a target of 8200 operating hours per year. The electric power tariff is given as 0.25 €[1]/kWh. This system may be built based on compressor A or compressor B:

- The first machine is less expensive: the total costs of the entire cooling system (including engineering, procurement, construction/installation and commissioning) are €220 000, but the refrigerator based on this compressor is less efficient; the COP of the optimized process amounts to 3%.

1 The euro (symbol €) is the official currency of 20 of the 27 member states of the European Union used as currency in most European countries.

Process Design for Cryogenics, First Edition. Alexander Alekseev.

Table 9.1 Cost comparison for a Joule–Thomson system based on different compressors and two years of operation.

		System based on compressor A	System based on compressor B
Total system cost	€	220 000	300 000
Cooling capacity	kW	1	1
Project duration	years	2	2
Operation	h/year	8200	8200
Specific power cost	€/kWh	0.25	0.25
COP	%	3.0%	4.0%
Power consumption	kW	33.3	25.0
Total power bill	€	136 667	102 500
Total cost of ownership	€	356 667	402 500

Table 9.2 Cost comparison for a Joule–Thomson system based on different compressors and five years of operation.

		System based on compressor A	System based on compressor B
System cost	€	220 000	300 000
Cooling capacity	kW	1	1
Project duration	years	5	5
Operation	h/year	8200	8200
Specific power cost	€/kWh	0.25	0.25
COP	%	3.0%	4.0%
Power consumption	kW	33.3	25.0
Total power Bill	€	341 667	256 250
Total cost of ownership	€	561 667	556 250

- The alternative solution is based on compressor B, which is more expensive: the total system costs are €300 000, but the corresponding optimized process provides higher efficiency, with a COP of 4%.

The question is then which system offers better economics.

According to Table 9.1, for the system based on compressor A, the total expenses are roughly €360 000. At €136 666, the power bill over the considered duration is relatively high, since system A is less efficient. The power bill for the system based on compressor B is lower, amounting to only €102 000. But the total cost of ownership (roughly €400 k) is higher since the cooling system itself is more expensive. Therefore, in the end, the system based on compressor A is a better solution for given scenario.

Techno-economics depends on many parameters. For example, if the operation horizon for the same system is extended to five years, the system based on compressor B becomes more favorable (Table 9.2) due to power savings.

9.1.2 Process Design Approaches for Improvement of Joule–Thomson Process

Further process improvement can be achieved by a slight modification of the process morphology. Two corresponding examples are described in this section.

Both these approaches are based on the specifics of the Joule–Thomson coefficient, which can be easily derived by considering the T,-s-diagram for the given working fluid. There are two important findings:

1) The first observation is that the slop of the isenthalps (Joule–Thomson coefficient) depends on pressure. Commonly, the Joule–Thomson coefficient is relatively high in the pressure range $(0.25 \cdots 10) \cdot p_c$,[2] at lower pressures it is relatively low. Haselden in his book [1, pp. 39–40] comments the nitrogen diagram as follows: "it is apparent from the slop of isenthalps on the T,s-diagram that an isothermal compression of nitrogen up to 20 bar produces negligible enthalpy change, whereas the further compression to 200 bar generates a higher enthalpy increase. Thus, virtually the same cooling effect could be obtained by circulating the gas only between higher pressures (for example 20 and 200 bar instead of 1 and 200 bar) whilst the work of compression will be reduced in the approximate proportion log(200/20) to log(200/1), i.e. 1: 2.3."
2) The second is that the Joule–Thomson coefficient and the Joule–Thomson effect for the same pressure range depend on the temperature: the lower the temperature (or the closer to the two-phase area), the larger the cooling effect (see Section 5.1). This means that the cooling capacity of a classical Joule–Thomson process may be increased if the temperature T_1 (inlet heat exchanger, high pressure side) is reduced, even if all other parameters (pressure, refrigerant, etc.), remain the same.

9.1.2.1 Relocate the Process to a Higher Pressure Level

This approach is based on the first observation. The main idea is to modify the process in such a way that the isenthalpic expansion (and corresponding compression of the working fluid) occurs in the preferred pressure range of $(0.25–10) \cdot p_c$, at least partially.

A process implementing this idea is described in most textbooks (for example, see Ref. [2], pp. 80) as the so-called "Linde dual pressure process." The term "dual pressure" refers to the fact that the process includes two cold streams at two different pressure levels: (i) low-pressure level as in the classic nitrogen Joule–Thomson process; and (ii) intermediate sub-critical pressure level referred to as medium pressure level MP.

---

2 Where p_c = is the critical pressure.

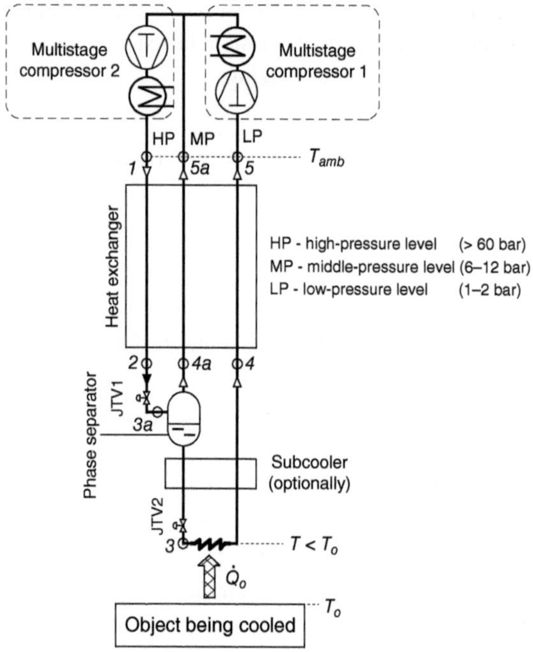

Figure 9.1 Joule–Thomson refrigerator with an intermediate pressure.

In a simplified version, the process is shown in Figure 9.1. It is based on the same main elements known from the classic nitrogen Joule–Thomson process: heat exchanger, expansion valves, evaporator, thermal insulation, and a multistage compressor unit. But it includes some new features:

- Phase separator.
- The heat exchanger consists of three channels for three material streams: (i) cold LP nitrogen, (ii) warm high-pressure nitrogen, and (iii) cold medium-pressure nitrogen.
- The process includes two expansion valves. In the first valve –JTV1 – the working fluid is expanded from high-pressure level HP to the medium pressure level MP, in the second valve -JTV2 – nitrogen is expanded further from the medium-pressure level MP to the low-pressure level LP.
- In terms of mechanical design, the compressor unit may consist of two separate multistage compressors, as shown in Figure 9.1a, or, alternatively, it can be designed as a single machine with three nozzles: (i) inlet nozzle for low-pressure gas, (ii) outlet nozzle for high-pressure gas, and (iii) additional nozzle for feeding a material stream in the compressor at an intermediate pressure (MP) level between compressor stages.

In terms of morphology, this process consists of two overlapping Joule–Thomson subprocesses:

1) The first Joule–Thomson subprocess, marked as 1 → 2 → 3a → 3 → 4 → 5 in Figure 9.1 is driven by both multistage compressors 1 and 2, corresponds to the classic nitrogen Joule–Thomson process described in Section 7.2.

2) The second Joule–Thomson subprocess, marked as 1 → 2 → 3a → 4a → 5a in Figure 9.1 is driven by multistage compressor 2 only, is a new feature.

The process functions as follows:

(a) After cooling in the heat exchanger (1 → 2), the high-pressure Joule–Thomson stream (state 2) is expanded in the first expansion valve JTV1 to the medium pressure level (2 → 3a). During this expansion process, the Joule–Thomson stream is partially liquefied and partially converted into flashgas.
(b) This flashgas is separated from the liquid by means of a phase separator; this cold gas is provided through the heat exchanger, where it is warmed up quite close to the ambient temperature (4a → 5a). It flows further to the compressor (inlet compressor 2), where it is compressed from medium pressure to high pressure and cooled to ambient temperature level.
(c) The liquid separated in the phase separator is expanded further to low pressure (LP) by means of the second expansion valve JTV2.
(d) The path of this material stream is already known from the classic nitrogen Joule–Thomson process: the cold two-phase stream is provided through the evaporator (3 → 4) and further through the heat exchanger (4 → 5) to the inlet of compressor 1, where this material stream is compressed from low pressure to medium pressure and mixed with the MP nitrogen stream.

Main observations:

- The flashgas fraction after the first expansion (2–3a) is high: this flash-gas is removed from the main stream by means of a phase separator at medium pressure and directed to the inlet of compressor 2. This means the largest fraction of the fluid is circulated in the loop connecting the medium-pressure and high-pressure levels (second Joule–Thomson subprocess). The mean Joule–Thomson coefficient within this pressure range is quite high; therefore, it is possible to generate high cooling capacity with a relatively low flowrate.
- The liquid remaining in the phase separator is already quite cold. Therefore, the fraction of flash-gas after expansion in the valve JTV2 is relatively low; the largest portion of the fluid is liquefied. Therefore, the flowrate of the LP stream (flowing through the evaporator and heat exchanger and compressed in multistage compressor 1) is reduced by an order of magnitude in comparison to the classic Joule–Thomson process.
- Consequently, the working fluid is processed mainly by the second Joule–Thomson subprocess at relatively high pressure levels (MP and HP). The flowrate of the nitrogen circulating in the first Joule–Thomson subprocess (between HP and LP) is relatively low. The power required for compressor 1 is minor, as the process is driven mainly by compressor 2.

All these process changes have positive side effects:

- The temperature profile in the heat exchanger becomes more favorable due to the fact that there is not much difference in heat capacity between the warm HP stream and the cold MP stream. Therefore, the loss in the heat exchanger is reduced. Consequently, the overall process becomes more efficient.

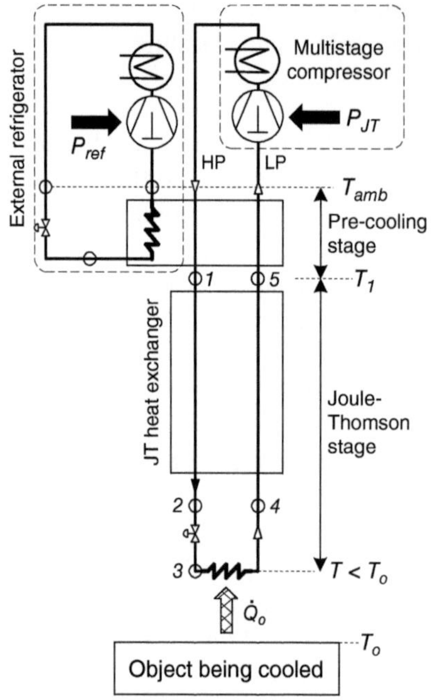

Figure 9.2 Joule–Thomson refrigerator with external precooling stage.

- In terms of size, compressor 1 remains compact since the flowrate of the gas undergoing compression is reduced considerably. This has a positive impact on the compressors cost.

9.1.2.2 Implementation of a Precooling Stage

Another approach is based on the additional observation that the Joule–Thomson coefficient becomes larger at low temperatures. The basic idea here is to reduce the temperature (T_1) at the inlet of the Joule–Thomson heat exchanger. Since the Joule–Thomson coefficient becomes larger at lower temperatures, the specific cooling capacity of the Joule–Thomson process increases. Consequently, the mass flow rate of refrigerant can be reduced, as can the power required for compression.

An active reduction of the temperature T_1 may be achieved through the implementation of an external refrigeration system to precool the Joule–Thomson system to the target temperature level, as shown in Figure 9.2. Since this external refrigeration system requires some driving power $\boldsymbol{P_{ref}}$, the power required for the overall system $\boldsymbol{P}$ will be the sum of $\boldsymbol{P_{ref}}$ and the power required for driving the Joule–Thomson system $\boldsymbol{P_{JT}} : \boldsymbol{P} = \boldsymbol{P_{ref}} + \boldsymbol{P_{JT}}$. Therefore, this approach makes sense only if the power savings in the Joule–Thomson stage $\boldsymbol{\Delta P_{JT}}$ are higher than the power $\boldsymbol{P_{ref}}$ required to drive the external refrigeration system.

Most textbooks on cryogenics refer to the process implementing this concept as a "precooled Linde-Hampson[3] system"; see [2, pp. 73].

3 The origin of this name is unknown and unfortunately misleading because neither Linde nor Hampson realized, described, or commercialized such a refrigeration system.

9.1.2.3 A Combination of These Approaches

Both of these approaches can be combined; a process implementing both of these concepts was indeed commercialized as the so-called "Linde process with high pressure recycle and precooling" for liquefaction purposes.

9.1.2.4 Approaches for Improvement of Joule–Thomson Process, Summary

The process modification targeting the relocation of the Joule–Thomson process or the Joule–Thomson stage of a process in the temperature/pressure area in order to maximize the Joule–Thomson effect is particularly worthwhile. It is often implemented in cryogenic processes. For example

- The intermediate-pressure approach is implemented wherever possible, especially in nitrogen liquefaction and air separation.
- The precooling concept is widely used in the liquefaction of natural gas (see Section 8.4.5.5) in the form of a propane precooling stage.

In a broader sense, the idea of the Claude process pursues and implements the precooling approach, whereby the Brayton stage assumes the role of precooling stage for the main Joule–Thomson process.

9.2 Supplements for Liquefaction

9.2.1 Original Claude Process

Figure 9.3 shows the original depiction of the Claude process, published first in Ref. [3]. Figure 9.4 shows the same process in the manner commonly used in this

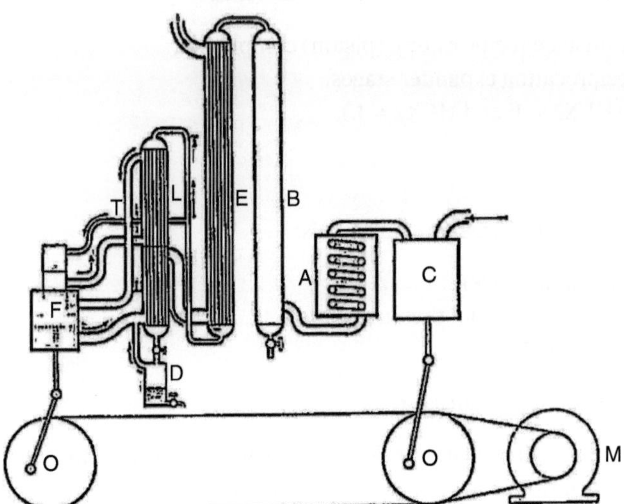

Figure 9.3 Original Claude process for air liquefaction. C – compressor, A – aftercooler, B – drying/purification unit, E – heat exchanger, L – heat exchanger, F - two-stage reciprocating expander, D – phase separator, M – motor.

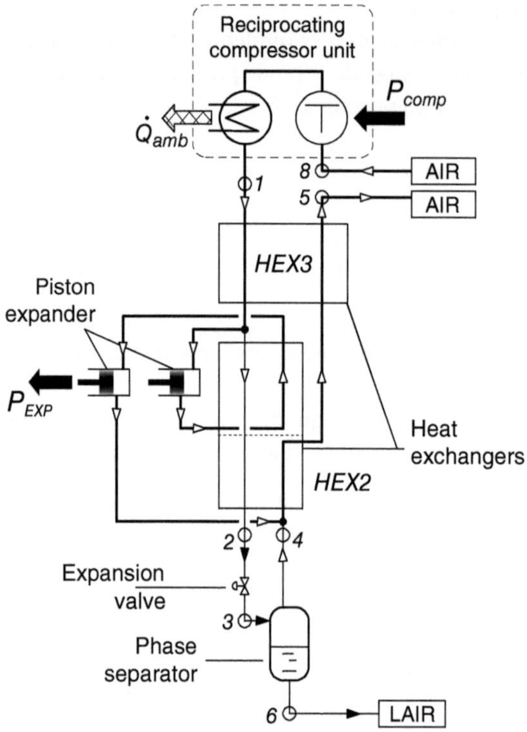

Figure 9.4 Original Claude process for air liquefaction AIR – ambient air, LAIR – liquefied air.

book, whereby several of the elements of Figure 9.3 (the motor and the mechanics that connect the motor, compressor, and expander) are not shown.

The original Claude process includes the following elements:

- compressor unit: a multistage reciprocating (piston) compressor,
- expander unit: two reciprocating expander stages,
- two heat exchangers (HEX3 = E and HEX2 = L),
- expansion valve,
- phase separator.

The process is characterized by three pressure levels:

1) low pressure LP (ambient pressure),
2) high pressure HP (compressor outlet pressure), and
3) intermediate pressure level MP.

The most salient process design details:

- Two expander stages are implemented in series: the expander stream is expanded in the first expander stage from high pressure HP to intermediate pressure MP, then directed through the heat exchanger, and expanded further in the second expander stage from intermediate pressure to low pressure LP.
- The upper heat exchanger, HEX3 is a conventional heat exchanger with two channels for two material streams. The design of the lower heat exchanger,

HEX2 is more complex. Principally, it includes two channels for two material streams, but it is able to manage three material streams: in the lower part, the Joule–Thomson stream is cooled by the cold stream from the outlet of the second expander, whereas in the upper part (in the middle of the process), the Joule–Thomson stream is cooled by the cold stream from the outlet of the first expander (intermediate pressure stream).

The process concept is easily recognizable:

a) Ambient air is compressed to high pressure by a compressor unit and cooled to ambient temperature (state 1).
b) This material stream (1) is cooled in the heat exchanger HEX3 and split into two material streams, the Joule–Thomson stream and the expander stream.
c) Joule–Thomson path: the Joule–Thomson stream is cooled further in the heat exchanger HEX2 before it is expanded in the expansion valve (2 → 3) and partially liquefied. The liquefied fraction of this stream is separated by a phase separator and removed from the system as product stream 6. The flash-gas (stream 4) is removed from the top of the phase separator, then directed through the heat exchangers (4 → 5) and exhausted into the ambient environment,
d) Brayton path: the expander stream is expanded in both expanders to low pressure LP as described above. After expansion, it is mixed with the flash-gas from the phase separator, directed through the heat exchangers (4 → 5) and exhausted into the ambient environment.

9.2.2 Kapitsa Process

This process was introduced by Peter Kapitsa in the Soviet Union in late 1930s. Generally, in 1930s and1940s, the majority of air separation plants were engineered for production of liquid oxygen, mainly for the merchant market, since the market for gaseous product was not yet developed. Therefore, Peter Kapitsa focused on the development of a process for oxygen liquefaction. The process was never used or applied for refrigeration purposes. This is an important aspect.

Figure 9.5 shows the original depiction of the Kapitsa process and Figure 9.6 shows the same process in the manner commonly used in this book, whereby some elements of Figure 9.5 (the filter, oil separator, switch valves, thermal mass, control valve, withdrawal valve, electrical heater, and brake device) are not shown.

Visually, In terms of the process flow diagram, the Kapitsa process appears very similar to a liquefier based on a single expander Claude process – see Figure 7.40 for comparison – but this similarity is illusive. The most important characteristic attribute of the Kapitsa process is that the maximal process pressure with approximately 6 bar is relatively low. Therefore, in terms of process classification, the Kapitsa process is definitely a subcritical (low-pressure) process. In this sense, the Kapitsa process is more closely related to the Brayton cycle rather than to the Claude process family, which is basically a supercritical (high-pressure) process.

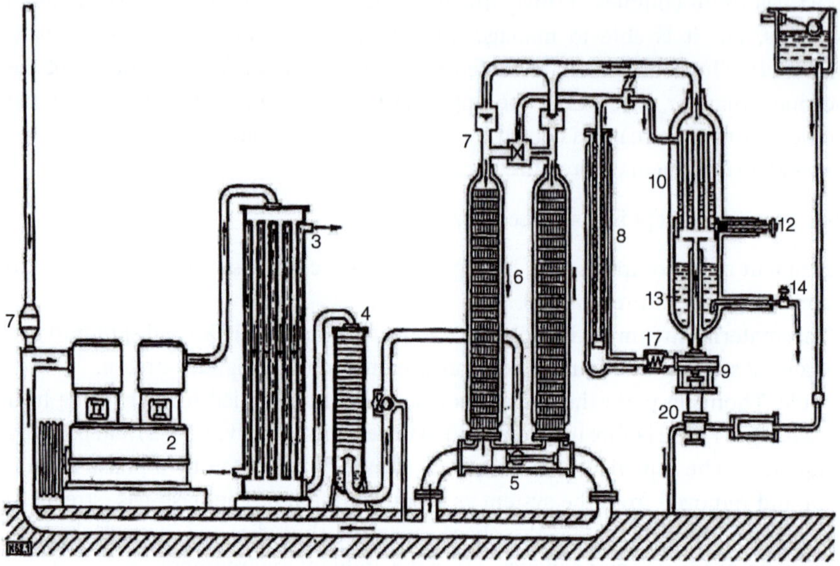

Figure 9.5 Kapitsa process for air liquefaction, original version according to Ref. [4]. 1 – filter, 2 – air compressor, 3 – compressor cooler, 4 – oil separator, 5&7 – switch valves, 6 – regenerators, 8 – thermal mass (for smoothing of temperature fluctuations), 9 – expander, 10 – condenser, 11 – control valve, 12 – expansion valve, 13 – phase separator, 14 – withdrawal valve, 17 – electrical heater, 20 – brake device. Elements 6, 8, 10, 12, 13, 17 are thermally insulated by means of vacuum insulation

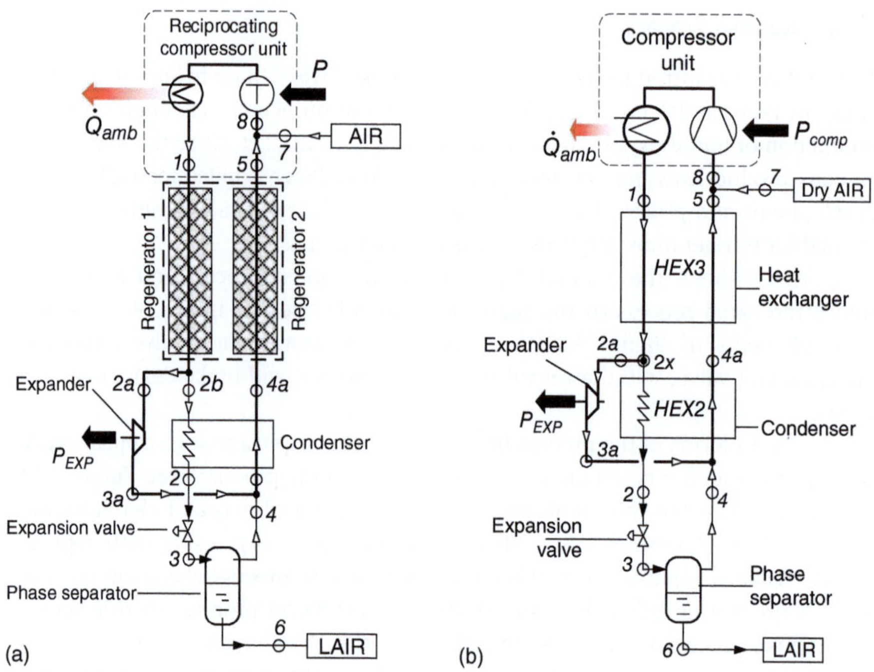

Figure 9.6 Kapitsa process for air liquefaction, AIR – ambient air, LAIR – liquefied air: a) original version with regenerators, b) virtual Kapitsa process based on conventional heat exchangers.

Hardware components:

- The key hardware component of the Kapitsa process was a (then) novel centrifugal expander, developed by the team around Peter Kapitsa in Moscow. Since the entire process takes place at low pressure, the mechanical design of such a LP turboexpander was/is easier and less challenging, at least in comparison to the high pressure turboexpander required for a Claude-type process. Additionally, the Kapitsa machine was considerably better in terms of efficiency and reliability due to the innovative geometry of the impeller, which was designed rather like a water turbine in contrast to the steam turbine-like design previously pursued by other engineers.
- The next key component is a set of regenerators[4] used for heat transfer purposes. The application of regenerators was not novel. At that time, it was a kind of state-of-the-art technology. A couple of regenerators with corresponding switching valves provide the functionality of a single heat exchanger.
- Another key element is a heat exchanger, referred to as a condenser. On the warm side, the air enters this heat exchanger as a gaseous substance, but it leaves the heat exchanger as a liquid. This is a new element. Typically, the Joule–Thomson, Brayton, or Claude process does not use a heat exchanger with phase transition; all material streams are designed as single-phase streams.

The process works as follows:

a) The working fluid (air-based) is compressed by means of a multistage reciprocating compressor unit (recycle compressor) to a pressure level of 6 bar and cooled to the ambient temperature level (state 1).
b) This material stream (1) is cooled in regenerator 1 and split into two material streams: The first stream (2a) is referred to as the expander stream, since it will be expanded later in the expander. The second material stream (2b) is often referred to as a Joule–Thomson-stream, but strictly speaking that is not entirely accurate, since this stream is not supercritical (see definition in Section 5.1.7). It is probably better to use the term "6bar air" for this material stream to avoid any misinterpretation.
c) Brayton path: The expander stream is considerably larger than the "6bar air." It is expanded in a centrifugal expander to a low pressure level LP (2a → 3a) of roughly 1.6 bar. After expansion, the cold expander stream is directed through a heat exchanger referred to as the condenser (3a → 4a) and later through the regenerator (4a → 5) to the inlet of the recycle compressor.

4 Regenerators were invented by Bavarian engineer Matthias Fraenkl and commercialized by the company Linde in the 1920s. An important point is that this device provides additional functionality, namely the removal of humidity, CO_2, and other similar solidifiable contaminations from the feed ambient air (these substances are solidified on the regenerator's surface, and the remaining dry and decontaminated air is provided further to the cold part of the process). This function is valuable for all cryogenic plants that use ambient air as feedstock, for example, air separation plants. For further information, see Ref. [5], pp. 169.

d) 6 bar air path: The 6 bar air is provided through the condenser (2b → 2). Due to the cooling capacity provided by the cold expander stream, the 6 bar air is liquefied there. At the outlet of the condenser (state 2), the 6 bar air is completely condensed. This liquid is expanded in the expansion valve (2 → 3) and partially evaporates. The liquefied fraction of this stream is separated by means of a phase separator and removed from the system as product stream 6. The flash-gas (stream 4) is removed from the top of the phase separator, mixed with the expander stream, and directed further through the condenser and regenerators (4 → 5) to the inlet of the recycle compressor.
e) The feed air is injected into the cycle at the inlet of the recycle compressor.

Two interesting facts:

(1) The composition of the liquid (LAIR) produced by the Kapitsa liquefier corresponds to the composition of the ambient air.
(2) The composition of the working fluid circulated in the Kapitsa cycle differs slightly from the composition of the ambient air; it contains less oxygen.

Historically, the first laboratory-sized plant based on the Kapitsa process was a small air liquefier, as shown in Figure 9.5. This system was designed mainly for the testing and qualification of the novel expander device. The prototype expander passed the testing phase in 1939, and the team was granted approval for commercialization. As a result of this development, two industrial systems for production of liquid oxygen were built in the 1940s: 0.2 t/h in 1942 and 1.6 t/h in 1944.

At that time, an efficient and reliable **high-pressure** centrifugal expander was not yet available.[5] The industrial liquefiers were designed based on high-pressure Claude-type processes using reciprocating machines for expansion. Principally, the Claude-type process is more suitable for liquefaction, but since the available reciprocation expanders were less efficient and required high effort for maintenance, the overall system was not optimal. The low-pressure Kapitsa expander was more efficient, more reliable, and more powerful. Therefore, at that time, Kapitsa's industrial plant for liquid oxygen was probably the best system available worldwide.

To tell the truth, in terms of process design, the Kapitsa process has an inherent weak point: the extremely high flowrate of the expander stream. The reason for this is that the specific cooling capacity of every expander in a low-pressure process is low, since expander power depends on the pressure ratio in the expander, which (in a low-pressure process) is low by definition. But every liquefaction process requires high cooling capacity by nature. In the Kapitsa process, the high cooling demand is met by the extreme mass flowrate through expander. This process feature – high flowrate of the expander stream – means a high flowrate of the air circulated in the cycle. This results in (i) higher CAPEX for larger hardware components (compressor as well as heat exchanger/regenerator) required for management of large material streams, and (ii) increased OPEX due to increased losses in all main elements

5 "the first cryogenic high-pressure expansion turbines…were produced in 1936 by Linde in Germany, but the mechanical design was not optimal; efficiencies were expected to be about 65%, but the value was only 53%," according to Ref. [1].

(compressor, heat exchanger/regenerator, and expander). These handicaps are the reason why the Kapitsa process was not really successful in the end; it was realized by Kapitsa's team twice, but it was never used or applied further, neither within nor outside of the Soviet Union.

Development after World War II: After World War II, the market for air separation plants moved toward large-scale on-site systems for the production of gaseous oxygen and nitrogen as the main products, driven mainly by oxygen demand for large steel mills and nitrogen demand for ammonia, first in the USA and later in Europe. Addressing this new market segment, a novel large-size air separation process was developed and commercialized in late 1940s in the USA.[6] This process is commonly referred to by experts as the "Lachmann process." Surprisingly, the LP Kapitsa-type turboexpander was a key element of this new air separation process. According to Ref. [1], (pp. 404), the corresponding commercial expander unit was "designed by the Elliot Corporation and manufactured by the Sharples Corporation" since 1942.

In comparison to the original Kapitsa process, the purpose of the expander in the Lachmann process is completely different: the expander capacity is used not for liquefaction but for refrigeration purposes, as the cooling power was required a) for balancing heat leakage through non-ideal thermal insulation and b) for balancing the temperature difference at the warm end of the heat exchanger (or regenerators). The cooling demand for these purposes is minor in comparison to the cooling capacity required for liquefaction. Since the required cooling capacity was reduced, the flowrate of the expander stream was reduced to 20–25% of the total air flowrate or even lower. As a result, the major disadvantages of the original Kapitsa process (high flowrate of expander stream, large heat exchanger surface, and large compressor) were eliminated.

The LP Lachmann process with a single LP compressor and a single LP Kapitsa-type expander was a great innovation; it was optimal in terms of both efficiency and cost. Therefore, this air separation process has been in common use for many decades.[7] It must be acknowledged that the advantages of the Kapitsa-type turbine (high efficiency, high reliability/low maintenance efforts) contributed much to the success of the Lachmann process. The Kapitsa-type expander is still widely used in air separation for gaseous products.

In summary:

- The Kapitsa process was a type of transitional process design solution applied in 1940s.
- The development of a LP centrifugal Kapitsa expander was a disruptive development that triggered considerable acceleration in air separation engineering.

---

6 Unfortunately, it could not be determined conclusively which company was the first on the market. It could be either the company "Air Products" or the company "Kellogg" or the company "Linde Air Products Company," later known as "Union Carbide," then as "Praxair," and later again as "Linde" following its merger with "Linde AG" in 2019.

7 Not until the 1990s, when the market for air separation plants turned once again to production of high-pressure gaseous products by means of internal compression, did the Lachmann process become rather obsolete. But even today, this process remains a solid benchmark for the production of gaseous LP oxygen.

9.2.3 Example of a Nitrogen Liquefaction Process

An example of a process for nitrogen liquefaction is presented in this section.

9.2.3.1 Task Definition

The goal is to develop a process design for a nitrogen liquefier based on a dual expander Claude process. The target is to determine:

1. process parameters (flowrate, temperature, pressure, phase state, composition, enthalpy, entropy, and exergy) at all relevant process points,
2. total power consumption, specific power consumption, efficiency as well as exergy loss in every process unit, and
3. figures required for the design of the heat exchanger and rotating machinery (compressors, expanders)

The process description is given in Section 8.2.4.2. Only the core part of the process (see Figure 8.13) is considered. Pressurized gaseous nitrogen (PGAN) is used as feed. The pressure of the PGAN is given as 5.5 bar, see Table 9.3. The product stream is known as pressurized liquid nitrogen (PLIN). Usually, the pressure of the liquid produced is a bit higher than the feed pressure. In the given process example, a product pressure of 5.7 bar is assumed; see Table 9.3.

The targeted liquefaction capacity of this system is defined as 30 tons per day which corresponds to 3476 g/s.

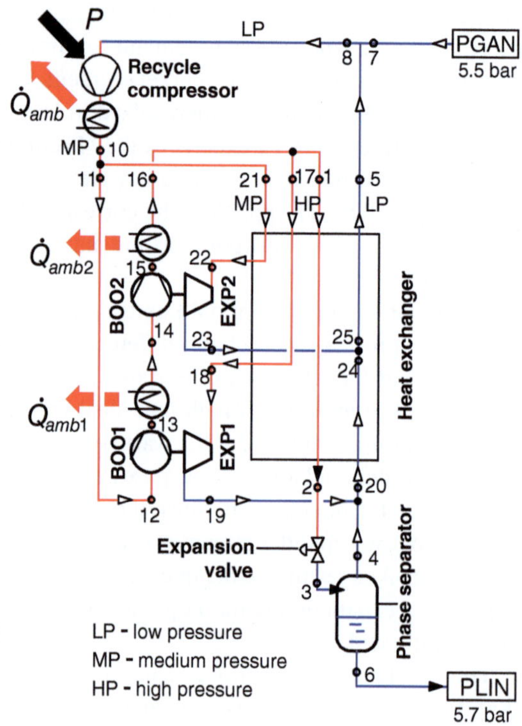

Ambient parameters are given in Table 7.5. It is assumed that the cooling water system is sufficient enough to cool every material stream to a temperature of $T = 300$ K.

9.2.3.2 Process and Hardware Concept

The process concept is based on application of the following key hardware components:

(a) an efficient aluminum brazed plate-fin heat exchanger;
(b) an oil-free multistage centrifugal compressor;
(c) both turboexpanders are designed as expander-booster combination;
(d) efficient multilayer insulation.

Table 9.3 Specifications for feed gas, product, and cooling water.

Feed gas	
Flow rate (feed gas)	3476 g/s
Feed gas pressure	5.5 bar
Feed gas temperature	300 K
Purity	99.9999 mol%
Product	
Product (PLIN) pressure	5.7 bar
Temperature specification	saturated liquid at product pressure
Purity	as in feed gas
Flowrate	as close as possible to feed gas flowrate
Cooling water	
Cooling water	cools the gas to 300 K

Further process design considerations are as follows:

1) Compressor:
   - Since a multistage centrifugal compressor is applied, the compression operation can be considered quasi-isothermal.
   - It will be assumed that the high-pressure level corresponds to the $p_{10} = 28$ bar.
   - data for compressor efficiency can be derived from previous experiences or discussions with equipment manufacturers. In the given example, it is assumed to be 70% (typical value for small centrifugal compressors).
   - an oil removal system is not required since the compressor used is oil-free.
   - temperature T_{10} is close to the ambient temperature ($T_{10} \approx T_{amb} = 300$ K) due to the efficient cooling after compression
   - considerations for the recycle compressor are summarized in Table 9.4.
2) Expander-booster combination means that a booster stage is used as a brake device; the generated power is utilized for compression of the corresponding material stream.
   - Isentropic efficiency of 84% is assumed for the expander.
   - Isentropic efficiency of 72.5% is assumed for the booster stage.
   - Considerations for the recycle compressor are summarized in Table 9.4.
3) Thermal Insulation: heat leakage through thermal insulation is assumed to be negligibly small due to efficient insulation.
4) Heat exchangers:
   - For the first process design, minor pressure loss on the high-pressure side $\Delta p_{1\to2} = 0.2$ bar as well as minor pressure loss on the LP side $\Delta p_{4\to5} = 0.2$ bar may be assumed.
   - For the temperature difference at the warm end of the heat exchanger the value of $\Delta T^{WE} = 3$ K is assumed.
   - minimal temperature difference in the heat exchanger is 2 K.

Table 9.4 Specifications for rotating machinery.

Recycle compressor	
Multistage centrifugal compressor quasi-isothermal compression	
Pressure at outlet of recycle compressor unit	$p_{10} = 28\,\text{bar}$
Multistage compressor with isothermal efficiency η_T	$\eta_T = 70\%$
Expander-booster combinations	
Expander stage, isentropic efficiency	$\eta_S = 84\%$
Booster stage, isentropic efficiency	$\eta_S = 72.5\%$
Thermal insulation	
Thermal insulation, heat leakage	$\dot{Q}_{is} = 0$

Table 9.5 Selected parameters for process simulation (assumed).

Temperature, inlet coldbox	$T_1 = 300\,\text{K}$
Pressure drop, heat exchanger, warm side	$\Delta p_{1\to2} = 0.2\,\text{bar}$
Pressure drop, heat exchanger, cold side	$\Delta p_{4\to5} = 0.2\,\text{bar}$
Pressure drop, heat exchanger, cold side, lower part	$\Delta p_{20\to24} = 0.1\,\text{bar}$
Temperature difference at warm end of heat exchanger	$T_1 - T_5 = 3\,\text{K}$
Approach for minimal temperature difference in heat exchanger	$\Delta T_{min} = 2\,\text{K}$

All these assumptions can be reviewed later, when the detailed design of the cold part is finished and the real values are available, but for the initial process design, these are acceptable assumptions. These considerations are summarized in Table 9.5.

9.2.3.3 Process Design Calculation, Cold Part of Process

The process simulation task usually has multiple correct solutions, as described in Section 2.1. Therefore, an optimization procedure usually must be carried out in order to determine the best solution. In this section, only the very first solution will be elaborated, which is required for initialization of the optimization procedure. The target is not to find out the best solution but, unpretentiously, any solution.

An appropriate process simulator (see Section 2.2) is usually required for the process design procedure.

The overall procedure can be considered a sort of trial-and-error method (some starting values for required input parameters are assumed and the calculation is performed based on these values). If the quality of the start values is not sufficient, the calculation/simulator may not be able to deliver usable results; in such cases, the procedure must be restarted using better starting values, especially in the initial phase. However, since the process is not terribly complex, a suitable solution can be reached very rapidly.

The most important ***free parameters*** are:

1) temperature at cold expander inlet
2) temperature at warm expander inlet
3) temperature at Joule–Thomson valve inlet
4) flowrate, cold expander stream
5) flowrate, warm expander stream
6) flowrate, Joule–Thomson stream.

The following considerations are useful for setting the ***Starting values***:

1) Starting value for temperature upstream of Joule–Thomson-valve: this is assumed to be $T_2 = 99$ K, based on following considerations:
   - It is known that in a well-designed process, the Joule–Thomson stream must be precooled as far as possible before expansion (to minimize flash-gas).
   - The lowest temperature in the given process is the temperature of the material streams 3 (saturated liquid at 5.7 bar) and 4 (saturated vapor at 5.7 bar), which can be easily calculated: $T_3 = T_4 = 95.6$ K,
   - The assumed value $T_2 = 99$ K is only slightly higher than T_3 and can be optimized later.
2) The starting value for the temperature upstream of the warm expander is assumed to be $T_{22} = 290$ K.
3) The starting value for the temperature upstream of the cold expander is assumed to be $T_{18} = 180$ K.
4) The starting value for the flowrate of the Joule–Thomson stream is 3650 g/s, which is roughly 5% higher than required for the product stream. The idea behind this is that flash-gas generation during Joule–Thomson expansion is roughly 5% due to deep precooling.
5) The starting value for the flowrate of the warm expander stream: $\dot{m}_{21} = 7300$ g/s, which is roughly double that of the Joule–Thomson stream.
6) The starting value for the flowrate of the cold expander stream: $\dot{m}_{17} = 11\,000$ g/s, which is roughly triple that of the Joule–Thomson stream.
7) The starting value for the pressure of the Joule–Thomson stream: $p_1 = 60$ bar; this value was used in Section 7.4.4 in the process calculation for a nitrogen Claude process.

Process design sequence:

8) **Cold expander:** the flowrate, pressure at the inlet and outlet, and temperature at the inlet are defined. Therefore, this process unit can be calculated immediately. As a result, the outlet temperature and generated power can be determined.
9) **Booster BOO1 with corresponding aftercooler:** the flowrate, pressure, and temperature at inlet, and power applied (which corresponds to the power generated in the cold expander) are defined. Therefore, this process unit can be calculated immediately. As a result, the outlet temperature and pressure can be determined.

10) **Warm expander:** the flowrate, the pressure at inlet and outlet, and the temperature at the inlet are defined. Therefore, this process unit can be calculated immediately. As a result, the outlet temperature and generated power can be determined.
11) **Booster BOO2 with corresponding aftercooler:** the flowrate, the pressure at inlet, the temperature at the inlet, and the power applied (which corresponds to the power generated in the warm expander) are defined. Therefore, this process unit can be calculated immediately. As a result, the outlet temperature and pressure can be determined.
12) **Joule–Thomson valve:** the flowrate, the pressure at inlet and outlet, and the temperature at inlet are defined. Therefore, this process unit can be calculated immediately. As a result, the state at the outlet (temperature, liquid fraction) can be determined immediately.
13) **Phase separator:** inlet conditions are known from the previous calculation step (Joule–Thomson expansion). Therefore, this process unit can be calculated directly after the calculation of the Joule–Thomson expansion. As a result, the state at the liquid outlet and at the vapor outlet can be determined.
14) **Heat exchanger:** since (i) all warm material streams (flowrates, pressures, inlet temperatures, and outlet temperatures) are known and (ii) all cold material streams (flowrates, pressures, inlet temperatures) are known, the heat exchanger can be calculated. As a result,
    - state 5 (outlet cold stream), and correspondingly, the temperature difference at the warm end of the heat exchanger can be found by means of simple enthalpy balance, and
    - the entire temperature profile and the minimal temperature difference in the heat exchanger can be determined.

First runs:

15) The first target is to achieve a suitable solution/process, which means a positive temperature difference in the entire heat exchanger (no crossing points). Usually,
    - the flowrate of the cold expander stream needs to be increased if a negative temperature difference is observed in the lower part of heat exchanger, and
    - the flowrate of the warm expander stream needs to be increased if a negative temperature difference is observed in the upper/warmer part of the heat exchanger.

Further simulation runs:

16) During further process simulation runs, free parameters are varied iteratively until specified values for the heat exchanger ($\Delta T_{min} = 2\,\text{K}$, $\Delta T^{WE} = 3\,\text{K}$) are reached.
17) Procedure for tuning flowrate of warm expander: the temperature difference at the warm end of the heat exchanger (and the temperature profile in the upper part of the heat exchanger) depends mainly on the flowrate of the warm expander. Correspondingly, the flowrate of this stream must be tuned to reach

the required value $\Delta T^{WE} = 3$ K. In a simulator, this procedure can usually be automated by introducing a dedicated controller.

18) Procedure for tuning temperature difference at the cold end of heat exchanger: the temperature at the inlet of the cold expander T_{18} has a direct impact on the temperature at the outlet of the cold expander. Also, the temperature at the outlet of the cold expander defines the temperature of the LP material stream (20) entering the heat exchanger at the cold end, since this material stream is larger than that of the flash-gas from the phase separator. This means that the temperature difference at the cold end of the heat exchanger depends strongly on the temperature at the inlet of the cold expander T_{18}. Therefore, this temperature is a rather efficient instrument for controlling the temperature difference at the cold end of the heat exchanger. In the given example, the targeted temperature at the outlet of cold expander was $T_{19} = 97$ K. The temperature at the inlet was tuned, resulting in the value $T_{18} = 181.28$ K.
19) Procedure for tuning the flowrate of the cold expander: the flowrate of the cold expander has a major impact on the temperature profile in the lower/colder part of the heat exchanger. It is the main instrument toward achieving a minimal temperature difference $\Delta T_{min} = 2$ K in the heat exchanger.

This status of process simulations is represented in Table 9.6. As previously mentioned, further optimization (systematic tuning of temperatures T_2, T_{22}, pressure at the outlet of recycle compressor p_{10} and other process parameters targeting the minimum power of the recycle compressor) is still required, but this simulation is already sufficient for process analysis and considerations.

Data package for fluid properties:

The Soave-Redlich-Kwong (SRK) equation of state is used for this process simulation.

9.2.3.4 Heat Exchanger

The heat exchanger data are given in Table 9.7. The temperature profile in the heat exchanger (T,Q diagram as well as ΔT,Q diagram) is shown in Figure 9.7, whereby a composite stream represents all three warm high-pressure material streams jointly. Two peaks can be recognized on the warm composite-stream side, at temperatures of 290 K and ≈180 K. They correspond to withdrawal of the warm expander stream (at 290 K) and cold expander stream(at ≈180 K) from the heat exchanger. The heat capacity of the material stream changes abruptly at both of these points due to the change in flowrate; this fact is visible in the form of a peak. On the cold side, similar behavior (peaking) can be recognized at a temperature of roughly 200 K. The background is exactly the same: a cold material stream (23) is fed into the heat exchanger at this temperature; the flowrate and correspondingly the heat capacity of the cold stream increases abruptly at this point.

Thermodynamically, the temperature profile looks quite good. The temperature difference is minimized; it reaches the minimal required value $\Delta T_{min} = 2$ K at the cold end as well as in the middle (at roughly 140 K kW) and in the upper part (above 200 K). Only in one segment, where the cold expander stream is removed,

Table 9.6 Process data for nitrogen liquefaction process based on dual expander Claude process.

	Temperature (K)	Pressure (bar)	Vapor fraction (mol/mol)	Specific enthalpy (J/g)	Specific entropy (J/g/K)	Flow rate (g/s)	J/g	Exergy (kW)
1	300.00	53.84	1	−9.0	4.077	3650	355.0	1295.7
2	99.00	53.64	0	−381.4	1.757	3650	678.6	2476.8
3	95.59	5.70	0.0476	−381.4	1.828	3650	657.1	2398.6
4	95.59	5.70	1	−217.4	3.544	174	306.5	53.3
5	297.00	5.50	1	−2.4	4.776	17 354	151.8	2634.3
6	95.59	5.70	0	−389.6	1.742	3476	674.7	2345.3
7	300.00	5.50	1	0.7	4.787	3476	151.8	527.6
8	297.50	5.50	1	−1.9	4.778	20 831	151.8	3161.8
10	300.00	28.00	1	−4.0	4.288	20 831	296.7	6179.8
12	300.00	28.00	1	−4.0	4.288	14 550	296.7	4316.5
13	343.00	39.67	1	41.2	4.325	14 550	330.8	4813.7
14	300.00	39.57	1	−6.3	4.178	14 550	327.5	4765.2
15	337.96	53.94	1	33.8	4.211	14 550	357.7	5203.9
16	300.00	53.84	1	−9.0	4.077	14 550	355.0	5165.2
17	300.00	53.84	1	−9.0	4.077	10 900	355.0	3869.5
18	181.28	53.74	1	−155.4	3.450	10 900	396.7	4323.9
19	96.99	5.70	1	−215.8	3.561	10 900	302.9	3301.7
20	96.97	5.70	1	−215.8	3.561	11 074	303.0	3355.0
21	300.00	28.00	1	−4.0	4.288	6281	296.7	1863.2
22	290.00	28.00	1	−14.9	4.251	6281	296.9	1864.4
23	196.48	5.60	1	−107.8	4.338	6281	177.9	1117.5
Reference state	300.00	1.00	1	1.7	5.296	28	0.0	0.0

the temperature difference does increase rapidly to roughly 17 K, but this segment is relatively short, so that in the end the mean temperature difference in this heat exchanger amounts to quite low value of 3.8 K (see Table 9.7). This means that the loss in the heat exchanger is low. Further optimization is, of course, possible, but this process already appears to be well designed in terms of power efficiency.

9.2.3.5 Compressor

The process data for the recycle compressor are given in Table 9.8. The compression can be performed by means of a 4-stage machine (for example, by means of an integrally geared compressor), which is more or less of conventional design. It requires neither special development nor additional costs for tuning and/or adaptation. In this sense, the recycle compressor is well designed.

Table 9.7 Heat exchanger data.

Heat transferred in the heat exchanger (heat exchanger duty)	$\dot{Q}^{HEX}$	kW	3024
Mean temperature difference	MTD	K	3.82
Minimal temperature difference	ΔT_{min}	K	2.01
Maximal temperature difference	ΔT_{max}	K	16.59
Q/MTD	Q/MTD = k F	kW/K	791.8

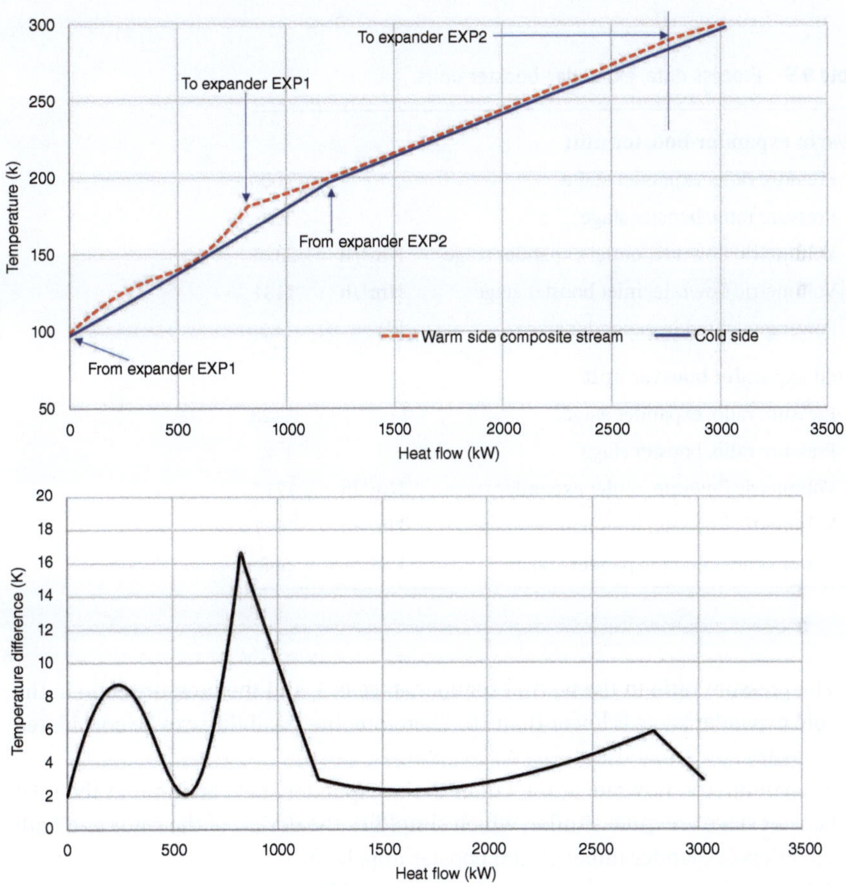

Figure 9.7 Temperature profile and temperature difference profile, both dependent on heat exchanger duty (heat to be transferred in the heat exchanger).

9.2.3.6 Expander-booster Units

The process data for expansion units is given in Table 9.9, separately for the "warm" machine and for the "cold" machine.

At first glance, in terms of fluid mechanics, the results conform with typical expander requirements related to pressure ratio and volumetric flows:

Table 9.8 Process data: recycle compressor.

Power required for reversible (isothermal) compression $\boldsymbol{P}_{rev}$	kW	3018
Compressor efficiency η_T	%	70
Real compressor power $\boldsymbol{P}$	kW	4311
Pressure ratio π	bar/bar	5.09
Potential compressor design number of stages x pressure ratio in a single stage		4×1.5
Volumetric flowrate at the inlet nozzle	m^3/h	12 040

Table 9.9 Process data, expander-booster units.

Warm expander-booster unit		
Pressure ratio, expander stage		5.000
Pressure ratio, booster stage		1.36
Volumetric flowrate, outlet expander stage	Nm^3/h	2325
Volumetric flowrate, inlet booster stage	Nm^3/h	1181
Power generated in expander stage	kW	583
Cold expander booster unit		
Pressure ratio, expander stage		9.428
Pressure ratio, booster stage		1.42
Volumetric flowrate, outlet expander stage	Nm^3/h	1717
Volumetric flowrate, inlet booster stage	Nm^3/h	1667
Power generated in expander stage	kW	658

1) The pressure ratio in the warm expander stage is 5, and the pressure ratio in the cold expander stage is lower than 10. Therefore, the feasibility can be considered as given;
2) The volumetric flowrate at the outlet of the expander stage and that at the inlet booster stage are quite similar, which simplifies the design of the rotor and both impellers (expander impeller and booster impeller);
3) A harmonious process design can be recognized by
   - similar power levels generated in the warm expander (583 kW) and the cold expander (658 kW) and
   - similar temperature reduction in the warm expander (93 K) and the cold expander (82 K).

At this point, it is recommended to query potential suppliers for feasibility, availability of appropriate machines, and performance maps of these machines. These data are useful for further processing considerations.

9.2.3.7 Overall Process

The efficiency of the process is calculated in Table 9.10. The overall efficiency is roughly 42.2%, which is not a perfect number; the process requires further optimization.

9.2.3.8 Losses

The losses are summarized in Table 9.11 and Figure 9.8. The largest source of losses is the recycle compressor, followed by the cold expander-booster unit and the warm

Table 9.10 Evaluation, overall process.

Cold capacity, required for liquefaction process	$\dot{Q}_o = (\dot{H}_7 - \dot{H}_6)$	kW	1357
Minimal (reversible) power required for liquefaction	$\dot{\boldsymbol{E}}_{\boldsymbol{Qo}} = \dot{\boldsymbol{E}}_{\boldsymbol{6}} - \dot{\boldsymbol{E}}_{\boldsymbol{7}}$	kW	1818
Power demand (compressor power) for liquefaction	$\boldsymbol{P}$	kW	4311
Specific power	$\boldsymbol{p_{spec}}$	kW/(nm^3/h)	43
Process efficiency	$\boldsymbol{\eta_L = \dot{E}_{Qo}/P}$	%	42.2%

Table 9.11 Losses.

Loss in recycle compressor	kW	1293.4
Loss in heat exchanger	kW	201.6
Loss in cold expander-booster unit (Exp1-Boo1)	kW	573.5
Loss in warm expander-booster unit (Exp2-Boo2)	kW	346.9
Loss in Joule–Thomson valve	kW	78.2
Further loss: mixing of material streams 4 and 19, and mixing of material streams 5 and 7	kW	0.06

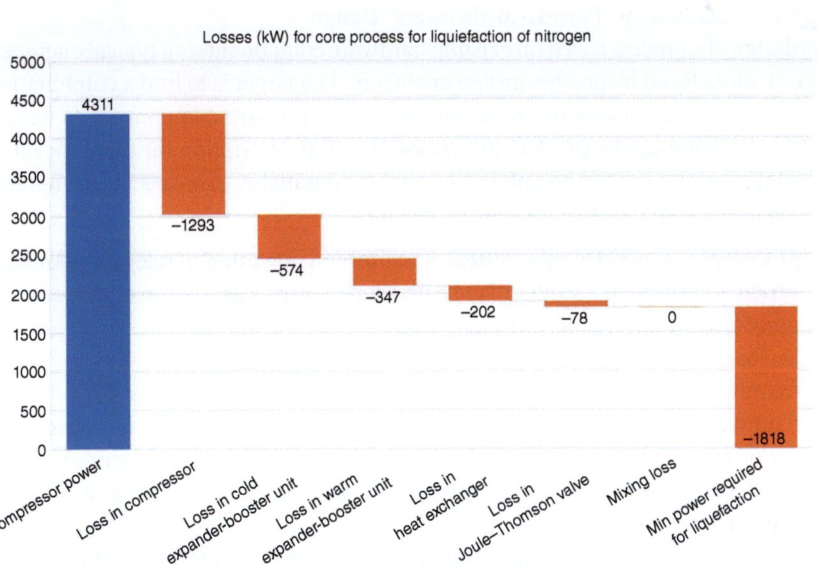

Figure 9.8 Losses in a process for nitrogen liquefaction.

expander-booster unit. The loss in the heat exchanger is moderate. The loss in the Joule–Thomson valve is extremely low, mainly due to deep precooling before expansion.

9.2.3.9 Evaluation

In terms of the process concept, the overall process is relatively simple; it consists of a minimal number of hardware components. Therefore, the process is easy to control.

In terms of efficiency, further process optimization is required. Specifically, more efficient rotating machinery (recycle compressor, booster stage) would help improve the efficiency of this process.

9.2.4 Helium Liquefaction Process with Integrated Liquid Nitrogen Precooling

This section considers an example of process design for a helium liquefier.

An example of a refrigeration process for liquid helium was described previously in Section 7.4.6. Here, this process is adapted for liquefaction purposes. A special feature of this process is the integrated precooling stage based on liquid nitrogen.

In all of the process design examples described in previous sections, the hardware specification follows the process design considerations: the process parameters are defined first by the process design engineer, and then the hardware components are sized based on these process parameters and purchased according to this specification. For the case to be considered here, the task definition is the converse: the hardware components are fixed (in stock), so the process design (and process parameters) must be adapted to suit them. This sort of task is often referred to as a "process-to-hardware" procedure.

9.2.4.1 Procedure for "Process-to-Hardware" Design"

The design of a process based on existing hardware components is a typical engineering task often faced by process design engineers. The target is to find a combination of parameters that allows the requirements of the primary components to be satisfied by existing hardware. A common step-by-step description for this procedure does not exist, and the method of finding a solution is highly case-specific. Generally, however, this is an iterative procedure, whereby

(a) process parameters (temperatures, pressure, and flowrates) at relevant points are varied according to a strategy (to be developed by process design engineer),
(b) process simulation is performed,
(c) key hardware components are dimensioned,
(d) examination is performed to ascertain the degree of similarity between the required hardware components and existing ones, (for example, for a compressor or expander, the efficiency for the required flowrate and pressure range is checked; for heat exchanging devices, the heat transfer surface and pressure drop are analyzed).
(e) If the deviation between the required and existing components is not excessive, the procedure is complete. Otherwise, a further iteration is executed.

The effort required for this procedure depends largely on the specified acceptable deviation between the required and existing components. The degree of deviation depends on the projects design phase. In some cases (preliminary design, simple process, one operation case), it can be concluded within a matter of hours. In other cases, however (for demanding customer requirements, final design, complex processes, guaranteed performance for several operation cases), it can take days, weeks or months.

9.2.4.2 Formal Task Definition

The objective is to adapt the process considered in Section 7.4.6 for liquefaction purposes. The following data are to be provided:

1) Process parameters (flowrate, temperature, pressure, phase state, composition, enthalpy, entropy, and exergy) at all relevant process points;
2) Maximum liquefaction capacity and efficiency as well as exergy loss in every process component;

Figure 9.9 shows the process for helium liquefaction with integrated precooling based on liquid nitrogen; all process-relevant points are marked and numbered there. The process concept is described in Section 8.3.4.2.

Figure 9.9 Helium liquefier with liquid nitrogen precooling.

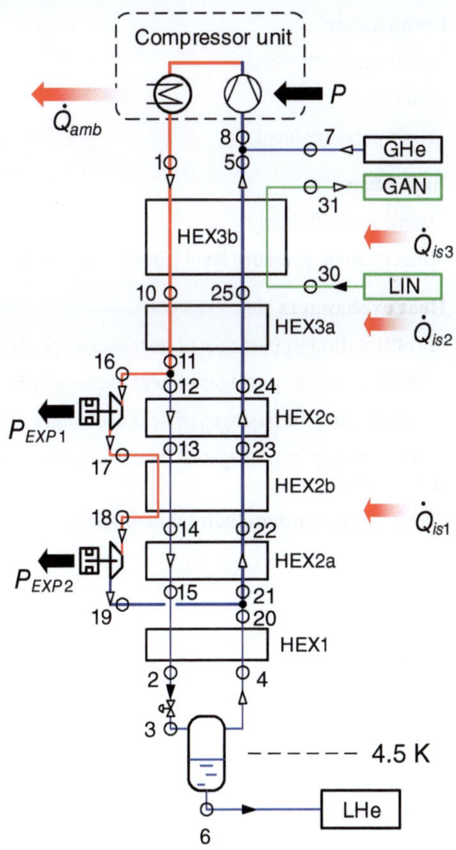

The ambient conditions are given in Table 7.5. The framework of limitations is specified by given hardware components as follows:

- **Compressor:** specifications for flowrate, pressure, and efficiency are given in Table 7.37.
- **Expanders:** specifications for pressure and efficiency are given in Table 7–38; flowrate and temperatures can be slightly varied.
- **Heat exchangers:** For the three largest heat exchangers- HEX3b, HEX3a, HEX2b – the required Q/MTDs as given in Table 7.36. For the three smallest

Table 9.12 Input data for process calculation: helium liquefaction process with LIN precooling.

Common process data	
Temperature T_1	$T_1 = 300\,K$
Product pressure (in liquid helium storage vessel)	$p_6 = p_3 = p_4 = 1.25\,bar$
Feed gas pressure	$p_7 > 1.1\,bar$
Feed gas temperature	$T_7 = 300\,K$
Saturated liquid nitrogen, supply pressure	1.4 bar
Compressor	
Isothermal efficiency	$\eta_T = 55\%$
Flowrate	45.7 g/s
High-pressure level	$p_1 = 12\,bar$
Expanders	
Isentropic efficiency	$\eta_{S\_EXP\_1} = \eta_{S\_EXP\_2} = 70\%$
Intermediate pressure level (for expander calculation)	$p_{17} = 4\,bar$
Heat exchangers	
Pressure drop in passages of heat exchanger HEX3b	0.120 bar
Pressure drop in passages of heat exchanger HEX3a	0.050 bar
Pressure drop in passages of heat exchanger HEX2b	0.030 bar
Pressure drop in passages of heat exchangers HEX2a, HEX2c, HEX1	0.0 bar
Heat inleak through thermal insulation	$\dot{Q}_{is} = 200\,W$
	$\dot{Q}_{is}$ (HEX3b) = 120 W
	$\dot{Q}_{is}$ (HEX3a) = 30 W
	$\dot{Q}_{is}$ (HEX2b) = 50 W
Q/MTD	$\dot{Q}^{HEX3b}/MTD = 8.5\,kW/K \pm 5\%$
	$\dot{Q}^{HEX3a}/MTD = 1.5\,kW/K \pm 5\%$
	$\dot{Q}^{HEX2b}/MTD = 2.4\,kW/K \pm 5\%$
	$\dot{Q}^{HEX1}/MTD < 0.33\,kW/K$
	$\dot{Q}^{HEX2a}/MTD < 0.18\,kW/K$
	$\dot{Q}^{HEX2c}/MTD < 0.23\,kW/K$

heat exchangers- HEX1, HEX2a, and HEX2c – the required Q/MTDs can be less than values given in Table 7.36 (it means that the heat exchanger is able to fulfill the required heat transfer job, but it is a bit oversized). For all heat exchangers: pressure drop is specified in Section 7.4.6; temperatures can be varied. The minimum temperature difference and the temperature difference at the warm end are not specified.

9.2.4.2.1 *Process Simulation*

With the given specification, the process is well described. Most of the parameters and variables can be taken over from the process simulation for helium refrigeration described in Section 7.4.6, but some parameters must be adapted to meet the Q/MTD values specified in Table 9.12. These variables are:

- split factor for flowrates of expander stream and Joule–Thomson stream,
- flowrate of liquid nitrogen,
- temperatures T_2, T_5, T_{10}, T_{11}, T_{13}, T_{14}, T_{15}, and T_{18}.

Thermodynamic properties are calculated using the fluid property software "miniREFPROP."

The results of iteration are summarized in Table 9.13. This is not yet the exact final design, but these results come quite close to the final version.

In comparison to the helium refrigeration process described in Section 7.4.6:

- the flowrate of the expander stream is higher,
- the flowrate of the Joule–Thomson stream is lower, and
- the temperatures at the inlets of expander EXP1, expander EXP2 and the expansion valve are higher.

Heat Exchangers

Table 9.14 summarizes the data related to heat exchangers. Figure 9.10 shows the temperature profile on conventional and in logarithmic scales. Figure 9.11 shows the corresponding temperature difference profile on different scales.

Generally, the temperature profile is very similar to the temperature profile of the helium refrigeration process described in Section 7.4.6. All comments given there are valid for the liquefaction process. However, one item is new: it relates to the temperature profile at roughly 80 K, where the cold-side line runs horizontally. The temperature difference profile (see Figure 9.11) at the same location is characterized by two remarkable spikes. The explanation for this is quite simple: that is the point where the liquid nitrogen is fed into the heat exchanger HEX3b. The liquid nitrogen evaporates there at a constant temperature of roughly 80 K, hence the horizontal section in Figure 9.10. Due to this additional cold, the temperature difference at the cold end of heat exchanger HEX3b is rapidly reduced (from 9.4 K to roughly 5 K) and creates the spike to the right in Figure 9.11. In the heat exchanger HEX3a, the temperature difference rises to 8.3 K due to differences in mass flowrates, hence the second spike.

9.2.4.3 Compressor

The compressor specification (Table 9.15) is similar to the compressor specification given in Table 7.37 for the refrigeration process.

Table 9.13 Output process data: helium liquefier.

	Vapor fraction (mol/mol)	Temperature (K)	Pressure (bar)	Mass flowrate (g/s)	Specific mass enthalpy (J/g)	Specific mass entropy (J/g/K)	Specific mass exergy (J/g)	Exergy (kW)
1	1	300.00	12.00	45.70	1566.76	22.84	1569.06	71.71
2	1	6.95	11.80	9.96	16.14	1.35	6468.32	64.44
3	0.78	4.50	1.30	9.96	16.14	3.54	5808.63	57.87
4	1	4.50	1.30	7.77	20.24	4.45	5539.53	43.04
5	1	296.29	1.10	43.51	1543.92	27.74	76.95	3.35
6	0	4.50	1.30	2.19	1.62	0.32	6762.04	14.83
7	1	300.00	1.10	2.19	1563.20	27.81	76.83	0.17
8	1	296.46	1.10	45.70	1544.85	27.75	76.94	3.52
10	1	82.00	11.88	45.70	433.21	16.12	2453.61	112.13
11	1	38.80	11.85	45.70	205.87	12.18	3407.57	155.73
12	1	38.80	11.85	9.96	205.87	12.18	3407.57	33.95
13	1	29.70	11.85	9.96	156.85	10.74	3790.65	37.76
14	1	13.50	11.80	9.96	63.43	6.15	5073.88	50.55
15	1	10.50	11.80	9.96	42.75	4.41	5574.40	55.54
16	1	38.80	11.85	35.74	205.87	12.18	3407.57	121.78
17	1	29.16	4.00	35.74	155.51	12.98	3118.09	111.43
18	1	13.50	3.95	35.74	70.92	8.83	4278.91	152.92
19	1	9.89	1.30	35.74	54.24	9.60	4029.78	144.01
20	1	9.91	1.30	7.77	54.36	9.61	4026.32	31.28
21	1	9.89	1.30	43.51	54.26	9.60	4029.16	175.30
22	1	10.76	1.30	43.51	58.99	10.06	3896.21	169.51
23	1	28.18	1.25	43.51	151.03	15.24	2433.44	105.87
24	1	30.33	1.25	43.51	162.25	15.63	2329.50	101.35
25	1	76.35	1.22	43.51	401.74	20.48	1112.07	48.38
Helium reference state	1	300.00	1.00	1.00	1563.35	28.06	0.00	0.00
GAN	1	294.00	1.40	5.33	−4.64	5.17	30.02	0.16
LIN	0	80.24	1.40	5.33	−423.11	1.37	753.53	4.02
Nitrogen reference state	1	300.00	1.00	1.00	1.70	5.30	0.00	0.00

Table 9.14 Heat exchanger data: helium Claude process.

		HEX3b	HEX3a	HEX2c	HEX2b	HEX2a	HEX1	Total
Duty Q^{HEX}	kW	51.80	10.39	0.49	3.95	0.21	0.27	67.11
MTD	K	6.09	6.85	4.03	1.66	1.38	1.52	5.22
Q/MTD = U A	kW/K	8.51	1.52	0.12	2.38	0.15	0.17	12.85
$\Delta T^{\min}$	K	3.71	5.65	1.52	1.11	0.61	0.59	0.59
$\Delta T^{\max}$	K	9.37	8.34	8.47	2.53	2.74	2.53	9.37

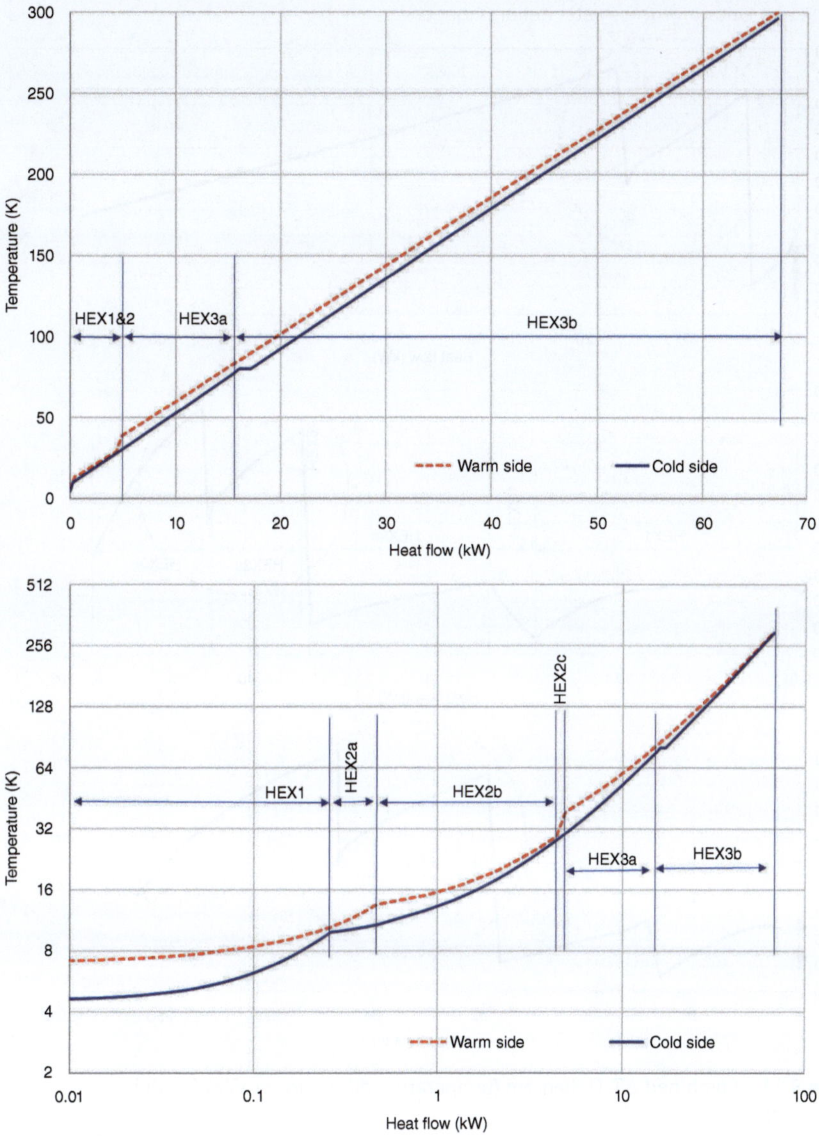

Figure 9.10 Combined T, Q diagram (temperature profile) for all heat exchangers.

9.2.4.4 Expanders

The expander specification (Table 9.16) differs from the specification in Table 7.38 for the refrigeration process: the flowrates and generated powers are approximately 20% higher. Since this deviation is quite high, it makes sense to check the isentropic efficiency of both expanders at these new conditions and to use corrected numbers for the next iteration.

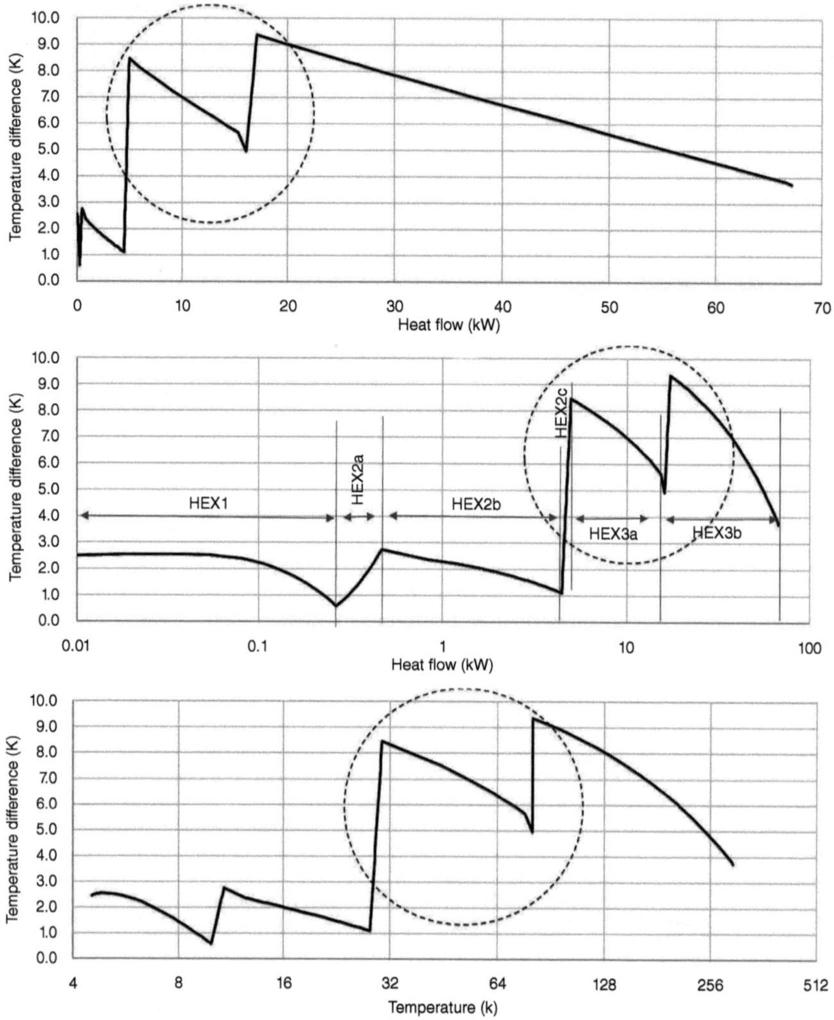

Figure 9.11 Combined ΔT, Q diagram (temperature difference profile) for all heat exchangers.

Table 9.15 Compressor specification: helium liquefier.

Compressor power	P^{COMP}	kW	124
Pressure, inlet	p_8	bar	1.1
Pressure, outlet	p_1	bar	12
Pressure ratio	π_{COMP}	bar/bar	10.9

Table 9.16 Expander specification: helium liquefier.

			Expander EXP1	Expander EXP2
Generated power	P^{EXP}	kW	1.8	0.6
Operating pressures (inlet/outlet)	p_{ein}/p_{aus}	bar/bar	11.85/4	3.95/1.3
Temperature at inlet	T	K	38.8	13.5
Volumetric flowrate, outlet nozzle	m^3/h	m^3/h	19.6	19.6
Pressure ratio	π_{EXP}	bar/bar	2.96	3.04

Table 9.17 Joule–Thomson expansion in helium Claude process.

Operating pressures (inlet/outlet)	bar/bar	11.8/1.3
Vapor fraction at the outlet	mol/mol	0.78
Inlet temperature	K	6.95

9.2.4.5 Expansion Valve

The expansion process in Joule–Thomson valves (Table 9.17) is not very efficient; only 22% of the Joule–Thomson stream is liquefied, and the largest fraction – 78% of the product – is flashgas. This is due to a relatively high inlet temperature of 6.95 K. One can see that the heat exchanger HEX1 is a bit undersized: Q/MTD = 0.17 kW/K in comparison to the 0.33 used in refrigeration process. Potentially, in the next iteration, the inlet temperature may be reduced, leading to a higher liquid fraction and better efficiency of the process.

9.2.4.6 Liquid Nitrogen

The helium liquefaction process utilizes some liquid nitrogen for precooling purposes. The power required for liquefaction can be estimated using the approach described in Section 8.1.2. The calculation is summarized in Table 9.18.

9.2.4.7 Overall Performance and Losses

Table 9.19 summarizes key numbers for performance estimation. The minimum liquefaction work for this amount of helium is given as the energy difference between feed stream and product stream. The definition of power required for the overall process is tricky since not only the compressor power P^{COMP} must be taken into account, but the power P^{LIN} required for liquefaction of liquid nitrogen used for precooling purposes must also be considered. The procedure for calculating this last parameter is shown in Table 9.18.

Table 9.18 Power demand for nitrogen liquefaction.

Minimum power required for LIN liquefaction $\dot{E}_{LIN} - \dot{E}_{GAN}$	kW	3.86
Assumed efficiency of nitrogen liquefier	%	40
Estimated power demand for liquefaction P^{LIN}	kW	9.64

Table 9.19 Performance estimation for helium liquefaction process.

Liquefaction capacity	$\dot{m}_6$	g/s	2.19
Cooling capacity required for helium liquefaction	$\dot{Q}_o = (\dot{H}_7 - \dot{H}_6)$	kW	3.42
Minimum liquefaction power for helium liquefaction	$\dot{\boldsymbol{E}}_{\mathbf{Qo}} = \dot{\boldsymbol{E}}_6 - \dot{\boldsymbol{E}}_7$	kW	14.66
Compressor power	$\boldsymbol{P}^{\mathbf{COMP}}$	kW	124
Power for liquid nitrogen production	$\boldsymbol{P}^{\mathbf{LIN}}$	kW	9.64
Total power requirement for process	$\boldsymbol{P} = \boldsymbol{P}^{\mathbf{LIN}} + \boldsymbol{P}^{\mathbf{COMP}}$	kW	133.62
Overall process efficiency	$\eta_L = \dot{\boldsymbol{E}}_{\mathbf{Qo}}/\boldsymbol{P}$		14.66/133.62 = 10.97%
Specific power		kW/(l/h)	124 kW/66.69 l/h = 1.86

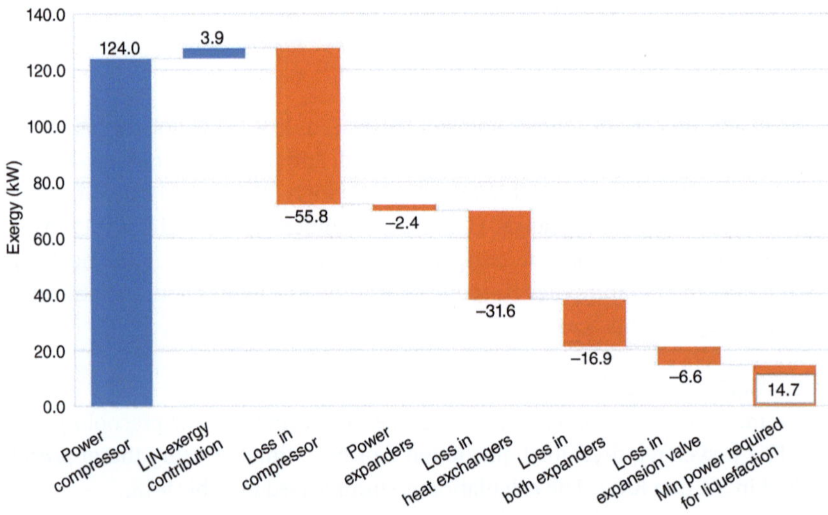

Figure 9.12 Distribution of losses in helium liquefaction process.

The overall process efficiency amounts to approximately 11%. The specific power is 1.86 kW/(l/h), which is considerably higher than state-of-the-art performance, as shown in Tables 8.7 and 8.8. This indicates that the process is still not optimized and can be improved further.

A brief process evaluation has been described in Section 8.3.4.4. The entire loss distribution of this process is shown in Figure 9.12, Table 9.20, and Figure 9.13.

9.2.5 Basic Mixed-Fluid Joule–Thomson Process for Natural Gas Liquefaction

An example of a process for liquefaction of natural gas is considered in this section. This process is a virtual process, since its realization is problematic due to potential

Table 9.20 Losses in helium liquefaction process.

Compressor loss	kW	55.79	49.30%
Loss HEX3b	kW	8.47	7.48%
Loss HEX3a	kW	9.37	8.28%
Loss HEX2c	kW	0.71	0.62%
Loss HEX2b	kW	9.37	8.28%
Loss HEX2a	kW	0.80	0.70%
Loss HEX1	kW	2.85	2.52%
Loss expander 1	kW	8.55	7.55%
Loss expander 2	kW	8.31	7.34%
Power expander 1	kW	1.80	1.59%
Power expander 2	kW	0.60	0.53%
Loss expansion valve	kW	6.57	5.81%
Loss evaporator	kW	0.00	0.00%
Total loss	kW	113.18	100.00%

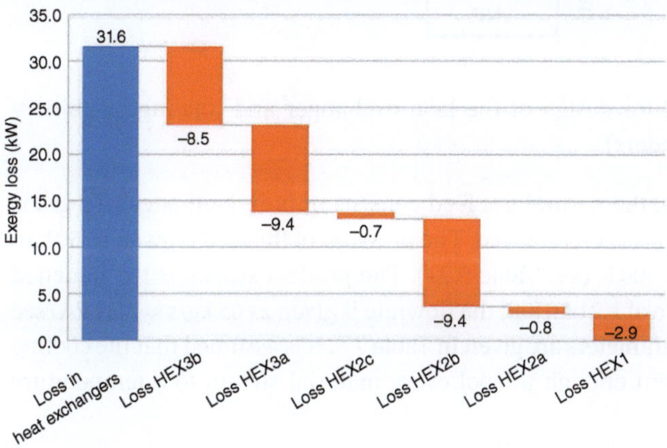

Figure 9.13 Distribution of losses in heat exchangers.

maldistribution on the warm end of the heat exchanger (see Section 8.4.5.2). The process description is given in Section 8.4.5.1.

9.2.5.1 Task Definition

The goal is to develop a process design based on the basic mixed fluid Joule–Thomson process as shown in Figure 9.14. The target is to determine:

1) the process parameters (flowrate, temperature, pressure, phase state, composition, enthalpy, entropy, and exergy) at all relevant process points,
2) the total power consumption, specific power consumption, efficiency, and exergy loss in every process unit and

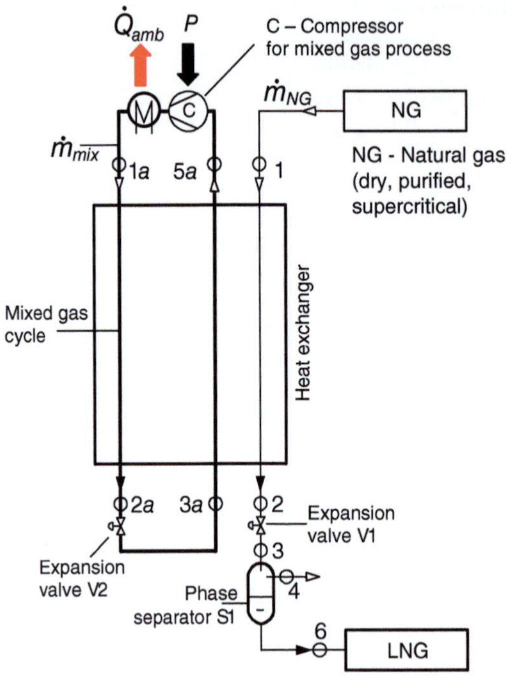

Figure 9.14 Basic mixed fluid-Joule–Thomson process for natural gas liquefaction.

3) the figures required for design of the heat exchanger and rotating machinery (compressors, expanders).

In the given example, the natural gas feed consists of pure methane. This gas is dry; all contamination has been removed. The pressure of the feed is given as 50 bar, and the temperature is 300 K (see Table 9.21). The product stream is the liquefied feed stream at a pressure of 1.2 bar, and the flowrate is given as 65 kg/s ≈ 2MTPA (see Table 9.21). Ambient parameters are given in Table 7.5. It is assumed that the cooling water system is sufficient enough to cool every material stream to a temperature $T = 300$ K.

9.2.5.2 Process and Hardware Concept

The process concept is based on application of the following key hardware components:

(a) an oil-free multistage centrifugal compressor
(b) an efficient aluminum brazed plate-fin heat exchanger
(c) efficient multilayer insulation.

Further process design considerations include the following:

1) Compressor:
   - Since a multistage centrifugal compressor is applied, the compression operation can be considered quasi-isothermal compression.

Table 9.21 Specifications for feed gas, product, and cooling water.

Feed gas	
Flow rate	65 kg/s
Composition	Pure methane
Feed gas pressure	50 bar
Feed gas temperature	300 K
Product	
Product (LNG) pressure	1.2 bar
Temperature specification	Saturated liquid at product pressure
Purity	As in feed gas
Flowrate	As close as possible to feed gas flowrate
Cooling water	
Cooling water	Cools the gas to 300 K

- Data for compressor efficiency can be derived from previous experience or discussions with equipment manufacturers. In the given example, it is assumed to be 75%.
- An oil-removal system is not required since the compressor used is oil-free.
- Temperature T_1 is close to the ambient temperature ($T_1 \approx T_{amb} = 300$ K) due to efficient cooling after compression.
- Considerations for the recycle compressor are summarized in Table 9.22.

2) Thermal insulation:
   Heat leakage through thermal insulation is assumed to be negligibly small ($\dot{Q}_{is} = 0$) due to efficient insulation.
3) Heat exchangers:
   - For the first process design, minor pressure losses on the low-pressure side $\Delta p_{3a \to 5a} = 0.2$ bar are assumed
   - Moderate pressure loss on the high-pressure side $\Delta p_{1a \to 2a} = 1$ bar;
   - For the temperature difference at the warm end of the heat exchanger the value of $\Delta T^{WE} = 3$ K or higher is assumed;
   - minimal temperature difference in the heat exchanger is 1 K.

All these assumptions can be reviewed later, when the detailed design of the cold part is finished and the real values are available; however, for the first process design, these are acceptable assumptions. These considerations are summarized in Table 9.22.

For simplification purposes, some input data are fixed:

- The mixed fluid composition is predefined (see Table 9.23).
- The pressures in the mixed fluid cycle are as specified in Table 9.22.

Table 9.22 Process design data, mixed fluid cycle.

Compressor	
Compressor outlet pressure (outlet aftercooler)	$p_1 = 30$ bar
Compressor inlet pressure	$p_1 = 2.3$ bar
Compressor efficiency	$\eta_T = 75\%$
Heat exchanger	
Temperature warm end, inlets	$T_1 = 300$ K
Minimal approach	$\Delta T_{\text{min}} = 1$ K
Minimal temperature difference at the warm end	>3 K
Pressure loss, high-pressure side	$\Delta p_{1a \to 2a} = 1$ bar
Pressure loss, low-pressure side	$\Delta p_{3a \to 5a} = 0.2$ bar

Table 9.23 Composition of mixed fluid.

Composition	Molar fraction (mol/mol)
Nitrogen	0.066
Methane	0.340
Ethane	0.285
Propane	0.055
n-Butane	0.254
Molecular weight	33.06

Table 9.24 Binary interaction parameters.

	Nitrogen	Methane	Ethane	Propane	*n*-butane
Nitrogen	0	0.03120	0.0405	0.059	0.0905
Methane	0.0312	0	0.00652	0.01519	0.0095
Ethane	0.0405	0.00652	0	0.00651	0.0123
Propane	0.059	0.01519	0.00651	0	0.0046
n-butane	0.0905	0.0095	0.0123	0.0046	0

9.2.5.3 Thermodynamic Properties

The SRK equation of state is used; the binary interaction parameters are given in Table 9.24.

9.2.5.4 Simulation Results

The simulation procedure is quite simple: the flowrate of the mixed fluid must be adjusted until predefined criteria for the temperature difference in the heat exchanger are fulfilled. The result is summarized in Table 9.25.

Table 9.25 Simulation results: process data/parameters.

	Temperature (K)	Pressure (bar)	Vapor fraction (mol/mol)	Specific enthalpy (J/g)	Specific entropy (J/g/K)	Flowrate (g/s)	Specific exergy (J/g)	Exergy (kW)
1	300	50	0	−4716	9.302	65 000	597.4	38 831.0
2	112	49	1	−5581	4.652	65 000	1127.4	73 281.0
3	113.77	1.2	1	−5581	4.751	65 000	1097.7	71 350.5
Reference state for natural gas	300	1		−4666	11.46	1	0.0	0.0
1a	300	30	0.679	−2828	4.09	314 000	279.1	87 637.4
2a	112	29	1	−3410	1.26	314 000	546.1	171 475.4
3a	110.65	2.5	0.0464	−3410	1.303	314 000	533.2	167 424.8
5a	296.97	2.3	0	−2649	5.251	314 000	109.8	34 477.2
Reference state for mixed fluid	300	1		−2667	5.557	1	0.0	0.0

9.2.5.5 Heat Exchanger

Three material streams are processed in the heat exchanger: two warm high-pressure streams – feed gas (1 → 2) and high-pressure mixed fluid (1a → 2a) – and a single cold material stream, which is low-pressure mixed fluid (3a → 5a). An approach for calculation of multistream heat exchangers is given in Section 5.3.4. Figure 9.17 visualizes the principle of creating a "composite stream" by compounding two streams: the warm, high-pressure mixed fluid stream and the natural gas stream.

Table 9.26 summarizes the heat exchanger data. Figures 9.15 and 9.16 show the overall temperature profile of the heat exchanger. Generally, the difference is small, with a mean value of less than 4 K. At the warm end of the heat exchanger, the temperature difference is roughly 3 K; at the cold end and in the middle of the heat exchanger, the temperature difference achieves values close to 1 K; but in one sector,

Table 9.26 Heat exchanger data for mixed-fluid Joule–Thomson process for natural gas liquefaction.

Heat exchanger duty	Q	MW	238.923
Temperature difference at the warm end	$\Delta T^{WE} = T_1 - T_5$	K	3
Temperature difference at the cold end	$\Delta T^{CE} = T_2 - T_4$	K	1.14
Minimal temperature difference	ΔT_{min}	K	1.14
Maximal temperature difference	ΔT_{max}	K	27.3
Mean temperature difference MTD	MTD	K	3.93
Q/MTD = KF	Q/MTD	MW/K	60.8

Table 9.27 Compressor-specification, basic mixed fluid Joule–Thomson process for natural gas liquefaction.

Compressor power	P_{COMP}	MW	70.88
Operating pressures (inlet/outlet)	p_{1a}/p_{5a}	bar/bar	30/2.3
Volumetric flow at inlet		m^3/h	360 000
Pressure ratio	π_{COMP}	bar/bar	$30/2.3 = 13$
Proposal compressor concept: number of compression stages x mean pressure ratio in a single stage			4×1.9

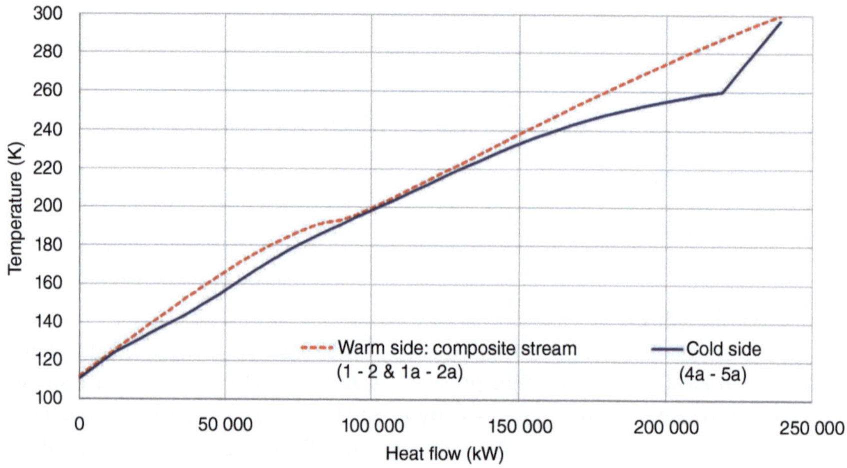

Figure 9.15 Overall temperature profile (T,Q diagram) in heat exchanger.

at temperatures above 250 K, the temperature difference is quite high and attains the maximum value of 27.3 K.

Figure 9.18 shows the temperature profile for the mixed fluid process separately in the form of a T,Q- diagram for two – warm and cold – mixed fluid streams. At the cold end, the temperature difference is small, but at the warm end the temperature difference is rather high.

Some salient observations are the following:

- In the mixed fluid process, the flowrate of the natural gas (65 kg/s in the given example) is typically smaller than the flowrate of the circulated mixed fluid (314 kg/s in the given example). The cold mixed gas fluid primarily cools the warm mixed fluid stream; only 20–25% of the available cooling capacity is used to cool the natural gas stream.
- In the mixed fluid process, the slope of the isobars is quite complex due to phase changes (on both sides) and changes in the composition of the vapor phase (and liquid phase) of the mixed gas fluid. Additionally, the heat capacity of the supercritical natural gas exhibits a complex curve (see Figure 9.17). The resulting

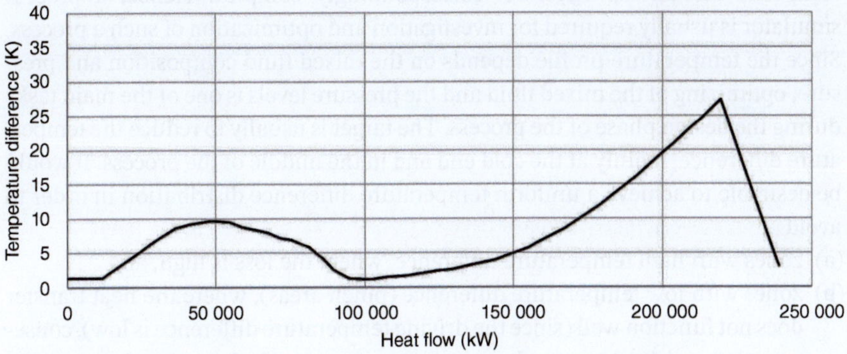

Figure 9.16 Temperature difference in heat exchanger (ΔT, Q diagram).

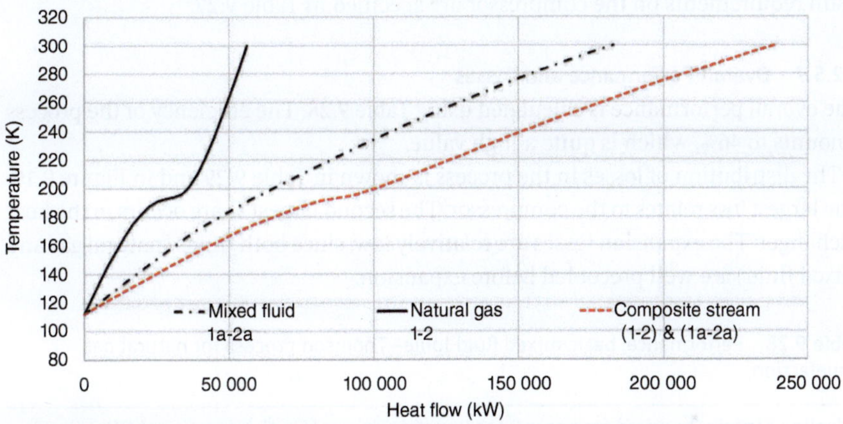

Figure 9.17 Warm composite stream as a composition of mixed fluid stream 1a → 2a and natural gas feed stream 1 → 2.

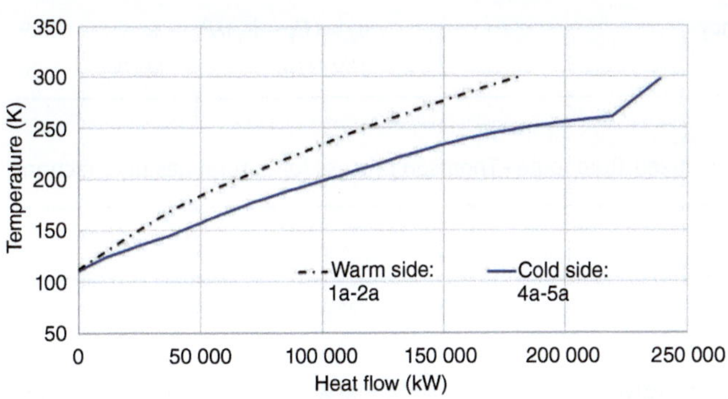

Figure 9.18 Temperature profile (T,Q diagram) for mixed-fluid cycle (without natural gas material stream).

composite stream is, therefore, correspondingly complex. Hence, a process simulator is usually required for investigation and optimization of such a process.

- Since the temperature profile depends on the mixed fluid composition and pressure, optimizing of the mixed fluid and the pressure levels is one of the main tasks during the design phase of the process. The target is usually to reduce the temperature difference, mainly at the cold end and in the middle of the process. It would be desirable to achieve a uniform temperature difference distribution in order to avoid
  (a) zones with high temperature difference, where the loss is high, and
  (b) zones with low temperature difference (pinch areas), where the heat transfer does not function well (since the driving temperature difference is low), consequently requiring large (and expensive) surface area for heat transfer purposes.

9.2.5.6 Compressor

Main requirements on the compressor are specified in Table 9.27.

9.2.5.7 Overall Performance and Losses

The overall performance is calculated using Table 9.28. The efficiency of the process amounts to 46%, which is quite a high value.

The distribution of losses in the process is shown in Table 9.29 and in Figure 9.19. The largest loss relates to the compressor. The second largest share occurs in the heat exchanger. The expansion losses are relatively low, since both fluids (natural gas and mixed fluid) are well precooled before expansion.

Table 9.28 Performance, basic mixed fluid Joule–Thomson process for natural gas liquefaction.

Cooling capacity required for natural gas liquefaction	$(\dot{H}_1 - \dot{H}_3)$	MW	56
Compressor power	P^{COMP}	MW	70.9
Liquefaction capacity	$\dot{m}_3$	kg/s	65
Minimum liquefaction power required	$(\dot{E}_3 - \dot{E}_1)$	MW	32.52
Overall process efficiency	$\eta_e = (\dot{E}_3 - \dot{E}_1)/P$	%	45.9
Specific power	$P^{COMP}/\dot{m}_3$	MJ/kg	1.09

Table 9.29 Losses, basic mixed-fluid Joule–Thomson process for natural gas liquefaction.

Components		
Compressor	MW	17.72
Heat exchanger	MW	14.66
Natural gas Joule–Thomson valve	MW	1.93
Mixed fluid Joule–Thomson valve	MW	4.05
Total loss	MW	38.4

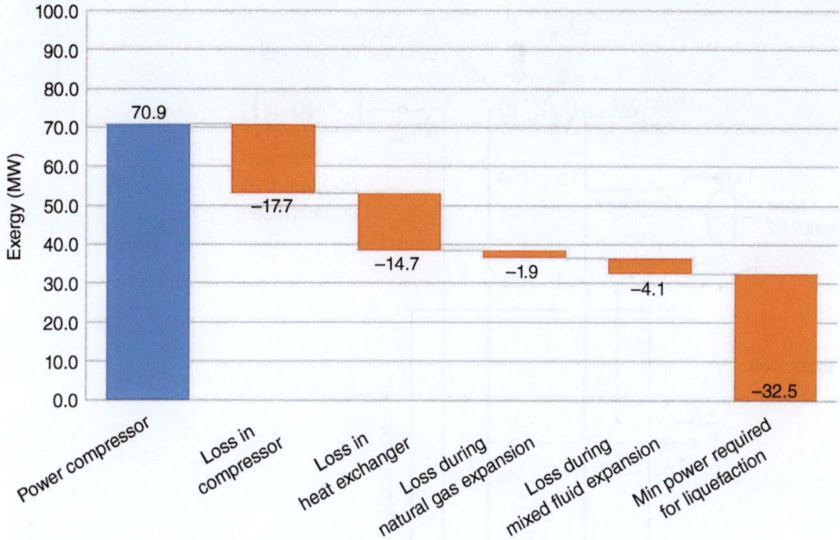

Figure 9.19 Loss distribution, mixed fluid Joule–Thomson process for natural gas liquefaction.

9.2.6 Mixed Fluid Joule–Thomson Process with Single-phase Separator

An example of the process for liquefaction of natural gas is considered in this section. This process design is a further development of the basic mixed fluid process described in Section 9.2.5. The main improvement is that this process is no longer virtual, so it can be realized without any limitations.

The process description is given in Section 8.4.5.3. Figure 9.20 shows the simplified process flow diagram.

9.2.6.1 Task Definition

The task definition is identical to that of the basic mixed fluid process described in Sections 9.2.5.1–9.2.5.3, and Tables 9.21, 9.22, and 9.24 remain valid. But the mixed fluid composition is slightly different, as shown in Table 9.30.

9.2.6.2 Simulation Results

Simulation results are summarized in Tables 9.31 and 9.32.

9.2.6.3 Heat Exchanger

In comparison to the basic mixed fluid process, the design of the heat exchanger for a mixed fluid process with a single phase separator is more complex, both in terms of process design as well as in terms of mechanical design. The first complication is a relatively high number of material streams – four – being managed in the upper part of this apparatus: three warm streams – natural gas 1 → 2, high-pressure liquid mixed fluid and 12 → 13 and high-pressure vaporous mixed fluid 11 → 2a, and a single cold material stream – LP mixed fluid 3a → 5a. The second difficulty relates

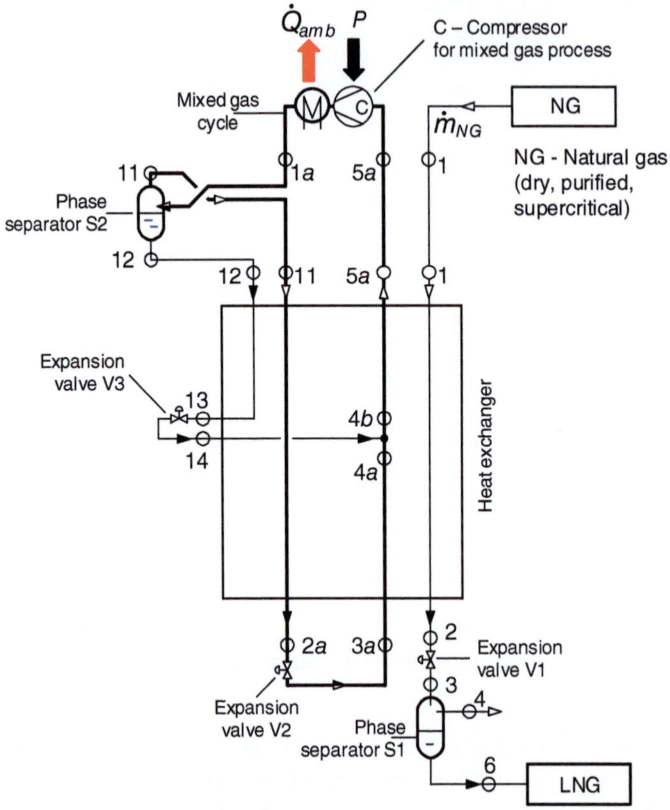

Figure 9.20 Mixed fluid process with single phase-separator.

Table 9.30 Composition of mixed fluid for processing with a single-phase separator.

Composition	Molar fraction (mol/mol)
Nitrogen	0.055
Methane	0.315
Ethane	0.335
Propane	0.055
n-butane	0.24
Molecular weight	33.04

to management of the liquid mixed fluid stream: in the middle of the heat exchanger this stream (13) is extracted from the heat exchanger, expanded in a valve (13 → 14) and reintroduced into the heat exchanger; it is injected into the cold mixed-fluid stream 4a to comprise the material stream 4b.

Table 9.31 Simulation results: process data/parameters.

	Temperature (K)	Pressure (bar)	Vapor fraction (mol/mol)	Specific enthalpy (J/g)	Specific entropy (J/g/K)	Flowrate (g/s)	Specific exergy (J/g)	Exergy (MW)
1	300.00	50	1.000	−4716.1	9.302	65 000	596.0	38.7
2	112.00	49	0.000	−5581.1	4.652	65 000	1126.2	73.2
3	113.77	1.2	0.000	−5581.1	4.751	65 000	1096.4	71.3
Reference state for natural gas	300.00	1	1.000	−4665.6	11.457	1	0.0	0.0
1a	**300.00**	**30**	**0.679**	**−2846.0**	**4.147**	**320 000**	**231.6**	**74.1**
5a	296.19	2.3	1.000	−2670.3	5.298	320 000	62.2	19.9
4b	180.93	2.4	0.415	−3198.3	2.977	320 000	230.2	73.7
Reference state for mixed fluid	300.00	1	1.000	−2661.1	5.535	1	0.0	0.0
12	**300.00**	**30**	**0.000**	**−2723.3**	**1.998**	**141 777**	**118.8**	**16.8**
13	180.00	30	0.000	−3000.0	0.838	141 777	190.1	26.9
14	178.56	2.4	0.040	−3000.0	0.866	141 777	181.6	25.7
Reference state for liquid fraction of mixed fluid	300.00	1	1.000	−2365.9	3.585	1	0.0	0.0
11	**300.00**	**30**	**1.000**	**−2943.6**	**5.857**	**178 223**	**299.6**	**53.4**
2a	112.00	29	0.000	−3687.5	2.265	178 223	633.2	112.8
3a	109.82	2.5	0.050	−3687.5	2.312	178 223	619.3	110.4
4a	178.56	2.4	0.606	−3356.1	4.639	178 223	252.4	45.0
Reference state for vapor fraction of mixed fluid	300.00	1	1.000	−2896.2	7.014	1	0.0	0.0

One approach for calculating multistream heat exchangers is described in Section 5.3.4. Figure 9.21 shows the resulting temperature profile, and Figure 9.22 shows the resulting temperature difference. Table 9.33 summarizes the most important characteristics of the heat exchanger.

In the upper part (between 180 and 300 K), the temperature profile is very similar to the profile generated in the basic mixed fluid process (see Section 9.2.5, Figure 9.15). But in the lower part, the temperature difference is a bit higher. For the overall heat exchanger, in comparison to the process considered in Section 9.2.5, the mean temperature difference is higher, but the heat exchanger duty is a bit reduced.

Table 9.32 Composition of fluids.

	Nitrogen (mol/mol)	Methane (mol/mol)	Ethane (mol/mol)	Propane (mol/mol)	*n*-butane (mol/mol)
1, 2, and 3	0	1	0	0	0
1a, 5a, and 4b	0.055	0.315	0.335	0.055	0.240
12, 13, 14	0.006	0.082	0.273	0.086	0.553
11, 2a, 3a, and 4a	0.078	0.425	0.364	0.040	0.092

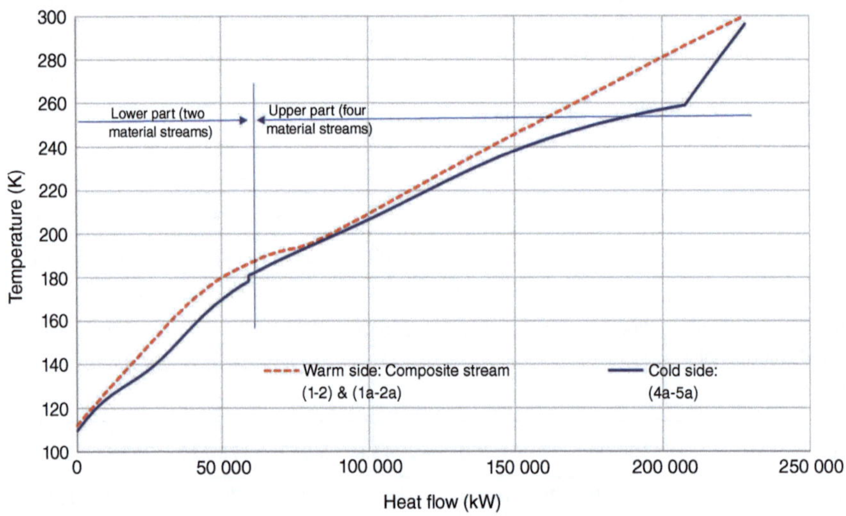

Figure 9.21 Temperature profile (T, Q diagram) in heat exchanger.

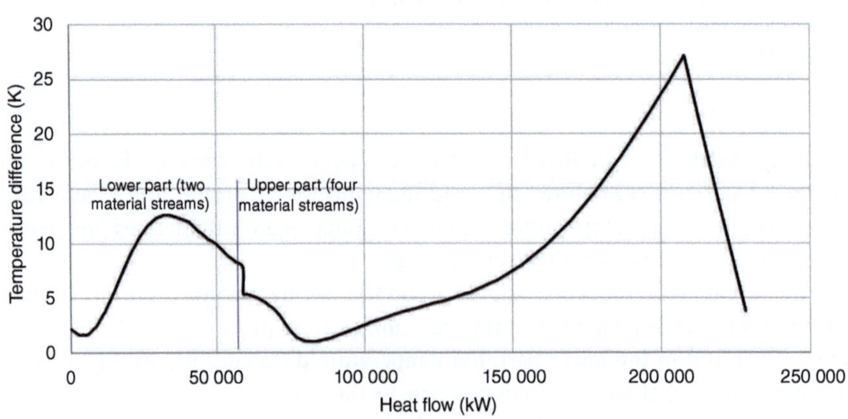

Figure 9.22 Temperature difference in heat exchangers (ΔT,Q diagram).

Table 9.33 Heat exchanger data: mixed fluid process with single-phase separator.

Heat exchanger duty	$\dot{Q}^{HEX}$	MW	228
Minimal temperature difference	ΔT_{min}	K	1.044
Maximal temperature difference	ΔT_{max}	K	27.12
Mean temperature difference	MTD	K	4.61
Q/MTD = k F	$\dot{Q}^{HEX}$/MTD	MW/K	49.47

Table 9.34 Compressor specification: mixed fluid process with single-phase separator.

Compressor power	P^{COMP}	MW	72.29
Operating pressures (inlet/outlet)	p_{5a}/p_{1a}	bar/bar	30 bar/2.3
Volumetric flow at the inlet		m^3/h	366 000
Pressure ratio	π^{COMP}	bar/bar	30/2.3 = 13
Proposal compressor concept: number of compressor stages × mean pressure ratio in a single stage			4 × 1.9

Table 9.35 Process performance: mixed-fluid process with single-phase separator.

Cooling capacity required for natural gas liquefaction	$(\dot{H}_1 - \dot{H}_3)$	MW	56.23
Compressor power	P^{COMP}	MW	72.29
Liquefaction capacity	$\dot{m}_3$	kg/s	65
Minimum Required liquefaction power	$(\dot{E}_3 - \dot{E}_1)$	MW	32.52
Overall process efficiency	$\eta_e = (\dot{E}_3 - \dot{E}_1)/P$	%	45
Specific power	$P^{COMP}/\dot{m}_3$	MJ/kg	1.11

9.2.6.4 Compressor

Main requirements on the compressor are specified in Table 9.34.

9.2.6.5 Overall Performance and Losses

The overall performance is calculated using Table 9.35. The efficiency of the process amounts to 45%, which is quite a high value.

The distribution of losses is shown in Table 9.36 and Figure 9.23. The largest loss relates to the compressor, and the second largest share is generated in the heat exchanger. All other losses are minor.

9.2.7 Mixed Fluid Joule–Thomson Process with Two or More Phase Separators

The concept of the first approach is quite straightforward: **if it works well with a single-phase separator, then it will probably work even better with**

Table 9.36 Losses in mixed-fluid process with single-phase separator.

Compressor	MW	18.07
Heat exchanger	MW	16.08
Natural gas expansion valve	MW	1.94
Mixed fluid Joule–Thomson valve 1	MW	2.47
Mixed fluid Joule–Thomson valve 2	MW	1.20
Total	MW	39.76

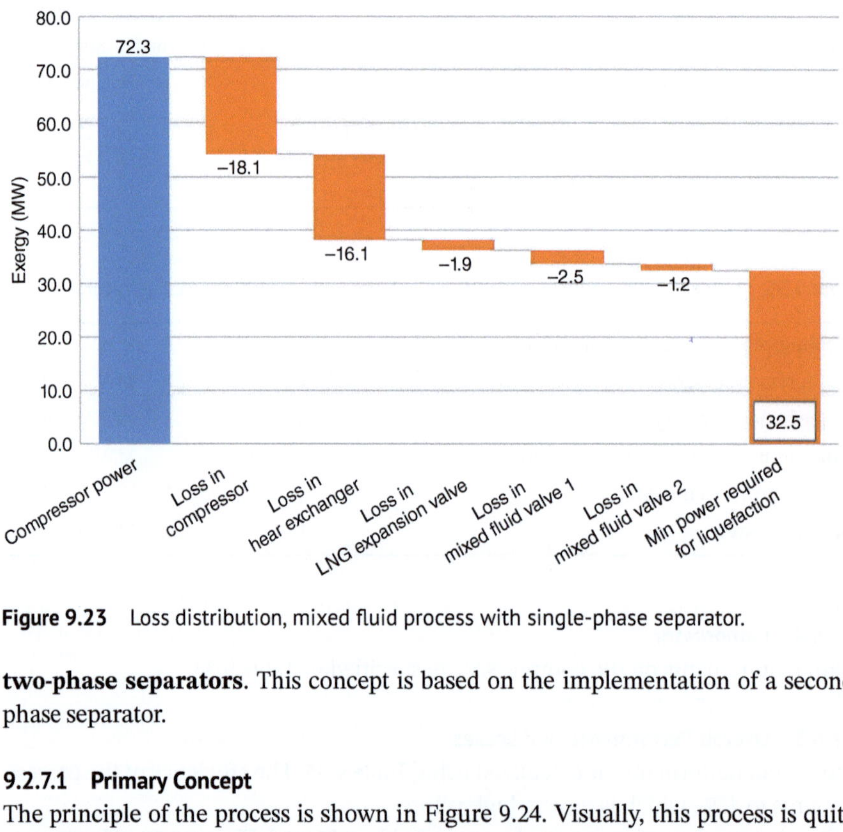

Figure 9.23 Loss distribution, mixed fluid process with single-phase separator.

two-phase separators. This concept is based on the implementation of a second phase separator.

9.2.7.1 Primary Concept

The principle of the process is shown in Figure 9.24. Visually, this process is quite similar to that with a single-phase separator. It includes three new elements:

1) phase separator S3 introduced upstream of the heat exchanger in the high-pressure mixed fluid path,
2) an additional expansion valve V4, and
3) an additional channel in the main heat exchanger. The heat exchanger is designed such that it is able to manage five material streams.

These elements are integrated into the mixed fluid part of the process in the same manner as phase separator S2 and expansion valve V3, but at lower temperature.

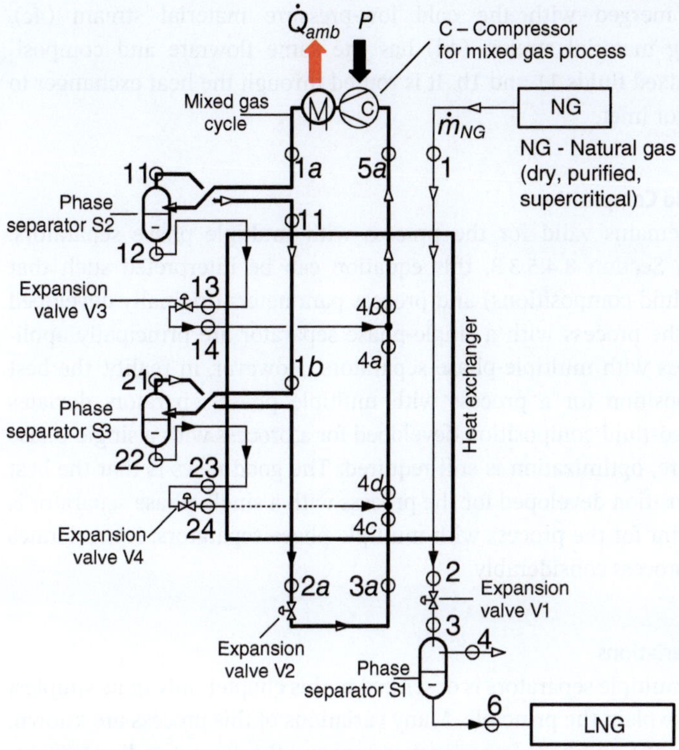

Figure 9.24 Principle of mixed-fluid Joule–Thomson process with two-phase separators.

At this temperature, the main mixed-fluid material stream introduced into the heat exchanger as vaporous stream 11 is already partially liquefied:

a) This two-phase material stream (state 1b) is withdrawn from the heat exchanger and routed to the phase separator S3, where the liquid fraction (state 22) is removed.
b) The remaining vaporous stream (state 21) is returned to the heat exchanger for further cooling, where it is treated as the main mixed fluid stream.
c) The liquid phase from phase separator S3 passes through a separate channel of the main heat exchanger, where it is cooled further (operation 22 → 23).
   - This sub-cooled liquid is withdrawn again from the heat exchanger, then expanded from high pressure to low pressure in expansion valve V4 (operation 23 → 24). In a well-designed process, the degree of subcooling (during the previous operation) is optimized such that after expansion, this fluid remains liquid and no flash-gas is generated. This is an important feature of the process: the material stream remains a single-phase fluid. Consequently, further distribution of this material stream does not present any challenges.
   - The sub-cooled liquid is routed to an appropriate location in the heat exchangers, then distributed homogeneously by means of a corresponding distribution

device and merged with the cold low-pressure material stream (4c). The resulting material stream (4d) has the same flowrate and composition as the mixed fluids 11 and 1b. It is routed through the heat exchanger to the compressor inlet.

9.2.7.2 Mixed-fluid Composition

Equation (8.30) remains valid for the process with multiple phase separators. Similar to that in Section 8.4.5.3.3, this equation can be interpreted such that mixtures (mixed-fluid compositions) and process parameters originally developed or optimized for the process with a single-phase separator are principally applicable to the process with multiple-phase separators. However, in reality, the best mixed-fluid composition for a process with multiple phase-separators deviates from the best mixed-fluid composition developed for a process with a single-phase separator. Therefore, optimization is still required. The good news is that the best mixed-fluid composition developed for the process with a single-phase separator is a good starting point for the process with multiple phase-separators; it accelerates the optimization process considerably.

9.2.7.3 Process Variations

The process with multiple separators is described in this chapter only in its simplest version, purely to explain the principle. Many variations of this process are known, differing mainly by the number of phase separators and the corresponding temperature levels of the phase separators.

An example of an industrial plant based on a process with multiple integrated phase separators is described In Section 8.4.6.2.

9.2.7.4 Brief Assessment

Gains: As the number of separators and multiple streams to be withdrawn from heat exchangers and returned again increases, the appropriate heat exchanger, piping, and instrumentation become increasingly complex. Therefore, the development of a process with multiple separators is driven by finding the right balance between complexity and power efficiency benefits.

Advantages:

- Even with multiple-phase separators and complex heat exchanger, piping, and instrumentation, the mixed-fluid process is still driven by a single compressor. The heat exchanger is a bit smaller in size, since the heat transferred in the heat exchanger is further reduced.
- In terms of power efficiency, the process with multiple phase separators is potentially more efficient than that with a single separator due to reduced loss in the heat exchanger.
- Further advantages include improved resilience against solidification and maldistribution.

References

1 Haselden, G.G. (ed.) (1971). *Cryogenic fundamentals*, vol. xiii, 757. London, New York: Academic Press. ISBN 0123305500.

2 Barron, R.F. (1985). *Cryogenic systems*. Oxford University Press. ISBN: 0–19–503 567-4.

3 Claude, G. (1920). *Air liquid, oxygene, azote*, in German Translation referred to as “Fluessige Luft,” (trans. L. Kolbe, Leipzig, Verlag von Johann Ambrosius Barth), page XXX

4 Hausen, H. (1941). Aussichten des Luftverflüssigungsverfahrens nach Kapitza. *Zeitschrift für die ges. Kälte-Industrie* 48 (2): 24–28.

5 Hausen, H. and Linde, H. (1985). *Tieftemperaturtechnik*. Berlin: Springer Verlag. ISBN: 3–540–13 972-9.

Index

a

aerospace 7, 24
air 19–22
  conditioning 326
  separation 6–7
air separation unit (ASU) 97, 235, 355–356, 362, 366, 373–374
aluminum plate-fin heat exchanger 102, 435–437
  mechanical design 102–104
  restrictions and limitations 107
  thermal and hydraulic design 104–105
argon 5, 7, 16, 19, 22–26, 99, 122, 181–182, 192, 207, 216, 279, 373, 392

b

basis of design 12, 194
bath evaporator 194–195, 245
bearing system 118–121, 124, 129–131, 133
booster 455
Brayton process 169, 331
  assessment 235–236
  cold gas heat exchanger 224–226
  compressor and expander 226–227
  control of Brayton refrigerator 233
  cooling capacity 217–219
  cooling temperature 217
  definition 220
  expander loss 229
  expander specification 226
  hardware components
    applications 235
    compressors 234
    expanders 234
    heat exchanger 233
  heat exchanger 224
  loss in cold gas heat exchanger 227–228
  loss in compressor 227
  overall performance 227
  phase state conditions 216
  process analysis 227–229
  process and hardware concepts 220–222
  process description 215–216
  process optimization 230–232
  process simulations 223–224
  properties data 220
  reversible Brayton process 219–220
  typical process pressure levels 217
  working fluids 216
Brayton stage 135, 238–240, 257, 261, 276–277, 343–345, 346, 351–352, 360, 362–363, 371–372, 385, 445
bulk helium liquefiers 374–375
bypassing 147

c

capital expenditures 439
carbon monoxide 19
Carnot efficiency 181, 202, 381
Carnot engine 40–41, 46
Carnot equation 42, 46, 47, 177–181, 202, 219, 335–336, 375

Process Design for Cryogenics, First Edition. Alexander Alekseev.

centrifugal compressor (booster compressor) 69–70, 234
control of 146–149
electric motor 149
gas turbine 149
impeller design 141–142
mechanical design 139–141
multi-staging 142–144
pressure ratio for light gases 145–146
pressure ratio in single stage 144–145
principle of compression 139–141
steam turbine 149
thermodynamics of 143–146
classical Joule Thomson process 107, 206–208, 213, 279, 280, 441
variables 208–209
Claude liquefier
liquefaction capacity of 341–342
Claude process 169, 331, 340, 342, 445
air gases 241
compressor 253–254
cooling capacity of 241–242
evaporator 252–253
expander and expansion valve 253–254
extended definition of 238
heat exchangers 249–252
helium 241
helium Claude process 257–276
high-pressure level and pressure dilemma 240–241
loss in compressor 254
loss in evaporator 257
loss in expander 256
loss in expansion valve 256–257
loss in heat exchanger 254–256
low-pressure level 241
main idea of 238–239
morphology 237–238
nitrogen Claude process 243–245
non-ideal gas, infinitely large heat exchanger 242
overall performance 254
process analysis 254
process and hardware concept 245–247
process description 239–240
process simulations 247–249
process units 249–252
real gas, real heat exchanger 242–243
simulation sequence 248–249
special feature 238
coefficient of performance (COP) 180–181, 194, 208, 209, 220, 231, 232, 245, 261, 309, 439, 440
coldbox
mechanical design 397
thermal insulation 397–398
cold compressed fluid 215, 409
cold expander Brayton stage 363, 372
cold-expander stream 345, 358, 360–363, 372, 387
cold gas heat exchanger 214–215, 217, 219, 221–230, 238, 344
cold surface 99, 172–175, 183
compressed multicomponent refrigerant 408–409
compressed refrigerant 212, 213, 215, 239, 243, 294, 344
compressor 453, 458, 464, 465, 472
energy transformation chain 137–138
housing, tightness of 152
lubrication 138
oil, selection of 152
wheel 125
constraints 15, 110, 122, 230, 276
conventional refrigeration 2, 38, 152, 154, 326–327, 407
cooling capacity, Joule–Thomson process 5, 173, 191
cooling temperature, refrigerator 172
counter-current heat exchanger 182, 215, 239, 243, 259, 294, 344, 387
cryogenic engineering 1, 4
cryogenic refrigeration

Brayton process 331
Carnot efficiency 181
classification of cryogenic processes 170–171
Claude process 331
coefficient of performance (COP) 180–181
definition 172–173
energy conservation 177
energy conservation law for 173–176
Joule Thomson process 330–331
minimum power requirement, for typical temperature levels 179–180
mixed fluid Joule–Thomson process 331
parasitic heat flow 176–177
reversible and Carnot equations 177–179
second law of thermodynamics 179
specific power 181
standard cryogenic processes 169
three interfaces 174, 175
cryogenics
applications 5–9
history 3–5
cryotechnology 1, 2

d

data package, for fluid properties 457
degrees of freedom 109–110, 153, 208, 230, 238, 323, 352, 355, 360, 386, 417, 420, 431
Dewar vessel 99–101, 376, 378
distribution fins 103, 104
double-tube (multitube) heat exchanger 72–74, 183
driving power 16, 42, 50, 83, 84, 89, 91, 173, 174, 177, 178, 180, 229, 313, 362, 371, 444
dry air 19, 21
D-type separator 420
dual-expander Claude process 344, 346, 351, 352, 358, 360, 361, 366–369, 371
for nitrogen liquefaction 366–369
dynamic gas-lubricated bearing 129, 131, 133

e

early liquefaction processes 338, 342–343
electric generator 40, 69, 70, 136, 220, 234, 238, 246, 278
electric motor 40, 45, 142, 148, 149, 160, 161, 174, 234
electric power tariff 439
Elliot Corporation 451
energy balance 14, 177, 187, 188, 217, 218, 241, 413–415
equation 415–416
energy conservation law 42, 52, 63–64, 68, 78, 79, 82–83, 85, 92–93, 173–177, 335
energy transformation chain 137–138, 160
enthalpy balance 223, 339, 341, 346, 413, 417, 456
entropy 42
to Boltzmann 44
to Rudolf Clausius 42–43
for thermodynamic analysis 44
evaporator, cold surface 183
exergy 45, 336, 337
applications 49–50
balance sheet 57
of cold/cooling power 47
of heat/heat flow 46
of material/material stream 47
of mechanical and electrical energy 45
for multicomponent materials/material flows 49
expander-booster units 360, 361, 372, 459–460
expanders 453, 464, 467
bearing system 118–121

expanders (*contd.*)
  expansion process 116–117
  helium 127–133
  isentropic efficiency 118, 119
  large turboexpanders with oil bearings 123–127
  limited pressure ratio 121–122
  no phase state change 122–123
  wheel 125
expansion device design 52, 60, 183
expansion valve, mechanical design 136
external cooling
  flash-gas management 353–354
  vs. internal 354–355
  liquefaction process with 351–352
  principle of 348–351
external pre-cooling stage
  cooling capacity 429–431
  primary concept 427–429
  process variations 431
  working fluids 431
external refrigeration process 351, 353

f
feed gas
  and product 404
  specifications for 453
flash-gas 450
  fraction 404
  management 353–354
  subcooling by 364
flow arrangement 110–111
fluids 19, 21
  lubricated bearings 131
  properties 15
fluorine 19

g
gap seal 125
gases
  brake 383
  data properties 19, 20
  industry 2
gas-lubricated bearings 120, 129, 133

h
hardware components, development of 11, 14
heat exchanger 297, 357, 442, 452, 453, 456, 457, 464, 465, 473, 475, 478
  aluminum plate-fin 102–107
  configuration 357
  double tube 72
  duty 426
  efficiency 81
  energy loss 78–81
  logarithmic temperature difference 74, 75
  loss 411–412
  mean logarithmic temperature difference 74–75
  multi-stream heat exchanger 77–78
  natural gas liquefaction 406
  Q/MTD and UA 75–76
  spiral-wound 108, 115
heat flow
  through heat bridges 96
  through thermal insulation 97
heat leak, simulation procedure 275
helium 31–33
  plants 7
  purification system 392–393
helium Claude process
  evaporator 269
  expander 269–270
  expansion valve 270
  heat exchangers 264–269
  loss by heat leak, thermal insulation 274
  loss in compressor 271
  loss in evaporator 274
  loss in expander 273–274
  loss in expansion valve 274
  loss in heat exchanger 271–273
  mechanical design 258, 260
  overall performance 270
  process analysis 270
  process and hardware concept 261–264
  process simulation 264

simulation procedure 275
staging 260, 261
helium expander
bearing system 129–131
control and instrumentation 133
mechanical design 128–129
utilization of expander power 132–133
helium liquefaction process
Claude process 393–394
compressor 382–383, 396–397, 465
core process for 386–388
expanders 467
expansion valve 469
features of 385–386
feed gas and product 376–378
formal task definition 463–465
heat exchanger 381, 465
helium expanders 383
liquid nitrogen 469
mechanical design 397
overall performance and losses 469–470
performance comparison 380–381
power requirements 378–379
process design 395
process extension 392–393
process simulation 465, 466
process-to-hardware-design 462–463
status quo 398–400
thermal insulation 397–398
high-energy physics 7
high-pressure processes 171, 183, 240, 242, 249, 277, 447
high-temperature superconductors (HTSC) 2, 8, 9, 26
hydrogen 27–31
based economy 8
liquefaction 235

i

impeller design 124, 141–142
industrial plant
mechanical design 372
process design 370–371
process specification 372
infrastructure process chain 403–404
inlet guide vanes (IGV) 126, 127, 140, 147–148
inlet nozzle assembly 116
integrally geared compressor 142–143, 458
integrated purifier 392, 395
intermediate-pressure approach 445
internal cooling 348, 353–355
irreversibility 39–40, 42, 44, 50, 80, 83, 239, 257, 274, 314
isenthalpic expansion
expansion valve 53
flash gas 61
Joule–Thomson coefficient 53–54
Joule Thomson effect 51–52
Joule–Thomson expansion 60–61
loss of exergy 57, 60
in temperature enthalpy diagram (nitrogen) 54, 56, 57
isentropic efficiency, impact of 84, 118, 119
isentropic exponent 67, 84–85, 145, 152, 160

j

Joule–Thomson coefficient 53–57, 60, 92, 190–192, 207, 340, 441, 444
Joule–Thomson effect 51–52, 54, 57, 216, 242, 340, 441, 445
Joule–Thomson expander 386
capacity 135
mechanical design 136
techno-economics 135
Joule–Thomson expansion 60–61, 133, 136, 195, 201, 204, 214, 245, 253, 259, 270, 274, 294, 296, 309, 342, 345, 353, 362, 363, 367, 372, 387, 406, 455, 456, 469
Joule–Thomson heat exchanger 268, 277, 343, 388, 443
Joule–Thomson liquefier
liquefaction capacity of 339–340

Joule–Thomson process 169, 330, 338, 340, 439
- compressor 183
- control 211–213
- cool-down procedure 185–187
- cooling capacity 187–188, 192
- evaporator 183–184
- expansion valve 183
- heat exchanger 182–183
- high-pressure level 192
- with infinitely large heat transfer exchanger 189–190
- low-pressure level 192–193
- with real heat exchanger 190–191
- sub-atmospheric (p<1 bar) conditions 193

Joule–Thomson refrigerators 182, 183, 213, 303, 321, 323, 327, 439, 442
Joule–Thomson system 325, 327, 440
Joule–Thomson valve 68, 182, 194, 211, 233, 256, 257, 274, 280, 314, 325, 338, 345, 362, 363, 372, 380, 387, 395, 397, 406, 410, 455, 456, 469

k

Kapitsa process 243, 447–451
Kapitsa, Peter 5, 243, 447–451
Kapitsa-type expander 451

l

laboratory helium liquefiers 376
Lachmann process 451
lambda line 32, 33
large turboexpanders with oil bearings
- control and instrumentation 126–127
- labyrinth sealing 125–126
- mechanical design 124–125
- nitrogen liquefiers 125–126
- seal gas 125–126

Linde dual pressure process 441
Linde process 339, 340, 348, 445
liquefaction
- capacity 335
- connection to refrigeration processes 335
- helium liquefaction process with integrated liquid nitrogen precooling 462–470
- Kapitsa process 447–451
- liquefier 334–335
- mixed fluid Joule–Thomson process 470–478
- nitrogen liquefaction process 452–462
- original Claude process 445–447

liquefied natural gas (LNG) 6
- infrastructure and process chain 402–404

liquid air 3, 22, 26, 279, 339
liquid hydrogen 1, 7, 8, 28–31, 99, 179
liquid nitrogen (LIN)
- evaporation 364–366
- helium liquefaction process 469
- precooling 395, 464

logarithmic temperature difference 74, 75
low-pressure processes 171, 207, 217, 234, 240, 447
lubrication 82, 125, 138, 151–153, 161, 162, 221, 305, 327

m

magnetic bearings 120, 121, 129, 130, 331, 400
magnetic levitation 121
main air compressor 86–88, 235
maldistribution 111, 112, 327, 418–421, 422, 426, 432, 471, 486
- management of 419–421

material stream, Kapitsa process 449
mean logarithmic temperature difference 74, 75
medical applications 8
metastability 321–323
methane (CH_4), cryogenic gases 25
middle pressure process 171
minimum liquefaction power 335
- via Carnot equation 335–336
- efficiency of 338
- via exergy 336–337

specific liquefaction power 338
typical cryogenic gases 337
miniREFPROP 15, 196, 220, 247, 264, 465
mixed fluid composition
ballast gas addition 319–320
initial mixed fluid composition 316, 318
optimization 316
solidification 320–323
mixed fluid Joule–Thomson process 169, 323, 331
challenging issues 330
comparison mixed fluid *vs.* nitrogen 291–294
components 281
compressor power 310
control of 323–324
cooling capacity 301–303
elements 280
evaporator 310
exchangers 327
hardware components 324–330
heat exchanger 297–304, 308–309
high efficiency 314
infinitely large heat exchanger 302–303
isobaric cooling/heating 289–291
location of pinch point 300
loss in compressor 313–314
loss in evaporator 314
loss in expansion valve 314
loss in heat exchanger 312–313
for or natural gas liquefaction
compressor 476, 477
definition 471–472
heat exchanger 475–478
overall performance and losses 478–479
process and hardware concept 472–474
simulation result 475
and nitrogen Joule Thomson process 298
overall performance 311
phase state conditions 294–297
pressure levels 303–304
process analysis 311–315
process and hardware concept 305–307
process feature 294–304
process simulation procedure 301
process simulations 307–308
process unit 308–311
properties data 305
with single phase separator
compressor 482, 483
heat exchanger 479–480
overall performance and losses 483
simulation results 479, 482
task definition 479
T,h diagram 286–289
T,s diagram 282–286
task definition 305
temperature difference at cold end 315
temperature difference at warm end 315
thermodynamic fundamentals 281–282
with two/multiple phase separators
brief assessment 486
mixed-fluid composition 486
primary concepts 484–486
process variations 486
mixed fluids
compositions 423
refrigerator 325–327
thermodynamic properties of 416–417
multi-component mixture over 114
multicomponent refrigerant 408
multi-phase flow 114
multi-stage compressors
compression processes, adiabatic and isothermal 94
energy conservation law 92
energy loss, isothermal compression 93–94
main air compressor 86–88

multi-stage compressors (*contd.*)
  non-reversible isothermal compression 89–91
  reversible isothermal compression 88–89
  rule of thumb for isothermal compression 91
  second law of multistage compressor 92–93

n

natural gas liquefaction 6
  capacity 413–417
  composition 401–402
  compressors 406
  efficiency 418
  heat exchanger 406
  Joule–Thomson process 408–418
  LNG infrastructure and process chain 402–404
  maldistribution 418–421
  mixed fluid composition 410
  mixed fluid process 410–413
  power requirements 404–405
neon 7, 19, 23–27, 67, 99, 216, 230, 235, 279, 280, 319, 337
  isotopes 27
nitrogen, cryogenic gases 22–23
nitrogen Joule–Thomson process 192
  compressor specification 202
  exergy losses in process units of 203
  loss in compressor 204–205
  loss in evaporator 206–207
  loss in expansion valve 203–204
  loss in heat exchanger 205–206
  performance of design 202
  process analysis 202–203
nitrogen liquefaction process
  compressor 458–459
  core process for 361–363
  dual expander Claude process for 366–369
  evaluation 462
  expander-booster units 459–460
  feed gas and products 355–356
  guidelines/principles of 360–361
  heat exchanger 357, 457–458
  large expander 357
  losses 461–462
  overall process 461
  power requirements 356
  process and hardware concept 452–454
  process design calculation, cold part of the process 454–457
  process extension for pressureless LIN 364–366
  process extension for pressureless nitrogen 366
  recycle compressor 356
nitrogen liquefier operation
  as cold carrier 373
  power bill optimization 373–374
  status quo 374

o

objective functions 15, 16
oil-brake/gas-brake 70
oil-lubricated bearings 120, 125, 129–131, 133
oil lubrication 82, 151, 153, 162, 221, 305
oil removal system (ORS) 138, 153, 221, 246, 261, 305, 306, 327, 453, 473
on-site cryogenic plant 5
optimization, process 15–17
overlapping Joule–Thomson subprocesses 442
oxygen
  cryogenic gases 23–24
  liquefaction 3, 447

p

phase separator, process design 456
phase separator S2 407, 421–424, 426, 484
plate fin heat exchanger (PFHE) 77, 96, 102–108, 114, 115, 183, 220, 233, 246, 261, 277, 305, 312, 357, 371, 422, 433–437, 452, 473

power consumption, total and specific 452
precooling cycle 427–432, 434
pressure drop, heat exchangers 52, 114, 275
pressurized gaseous nitrogen (PGAN) 355, 356, 362, 363, 366, 452
pressurized liquid nitrogen (PLIN) 355, 356, 363, 364, 373, 452
process calculation procedure
  compressor power 201
  evaporator 200–201
  heat exchanger 199–200
  output process data 199
  process and hard concept 194–196
  process simulation 196–199
process concept 11–13, 95, 220, 245, 261, 305, 394, 408–410, 421, 447, 452, 462–463, 472
process design
  Brayton stage 343–346
  expander power on 359
  for nitrogen liquefaction 359–363
  heat exchanger design on 357–359
  liquid nitrogen pre-cooling 346–348
process design sequence 15, 455–456
process optimization
  applications 279
  assessments 279
  Claude refrigerator 277
  constraints 276
  hardware components 277–278
  variables 276
process parameters, Claude process 452
process simulation, selected parameters 454
product, and cooling water 453, 473
propane pre-cooling 432–435

r

Rankine-type refrigerator 343, 349
reciprocating compressor
  control 165
  design and working principle 154–159
  impact of the isentropic exponent 159
  isentropic efficiency 159
  load range 162–165
  mechanical design 161–162
  mechanical efficiency 160
  multistage compressor 160–161
  pulsation and pulsation control 165–166
  short assessment of 166–167
  single-stage compressor 160–161
recycle compressor, compression device 356, 458
refrigeration system, control 17
refrigeration technology 1, 2
refrigerator/liquefier 173
remote cryogenic plant 5
reversibility 39–50, 80, 83, 239, 257, 274, 314
  concept 39–50
reversible adiabatic compression 83–85
reversible expansion process 64–66
reversible isothermal compression 88–89, 93
reversible processes 40, 42, 44, 178
reversible refrigerators 179, 335

s

screw compressors
  assessment of 153–154
  compressor control 153
  design and working principle 150–152
  oil lubrication 151–152
  oil removal system ORS 153
  process parameters 152–153
  tuning of 152
  working principle 150–151
second Joule–Thomson subprocess 443
shaft power 138
Sharples Corporation 451
shroud 108
simulation 15
  process simulations 247–249
  sequence 248–249
  step 13
single adiabatic compression stage
  energy conservation law 82

single adiabatic compression stage (*contd.*)
  energy loss, adiabatic compressor stage 85–86
  non-reversible process 83–85
  reversible adiabatic compression stage 83
single phase separator
  assessment 426–427
  cooling capacity 423
  example of 423–426
  mixed fluid composition 423
  primary concept 421–423
single-shaft compressor 143
Soave–Redlich–Kwong (SRK) equation of state (EOS) 282, 305, 457
solidification 24, 281, 320–323, 330, 410, 421, 424, 426, 432, 486
solid-liquid equilibria 321–322
spiral-wound heat exchangers (SWHE) 433, 434
  mechanical design 108–113
  restrictions and limitations 114
  thermal and hydraulic design 110–114
static gas-lubricated bearing 129, 131, 133
steam turbine 124, 149, 161, 449
subcooler 364, 366, 373
subcritical processes 171, 304
subprocess 238, 257, 277, 348–352, 364, 366, 407, 408, 431, 442–443
supercritical processes 171
superfluidity 3, 5, 32–33
supplements
  Joule Thomson process 441–445
  for liquefaction 445–157
  techno-economics 440–441
surge phenomenon 148–149

t

techno-economical optimization 439–441
temperature-enthalpy diagram 19, 34–36, 54, 55, 319
temperature-entropy diagram 19, 36–38, 54–57, 60, 64, 68, 71, 86, 189, 204
thermal insulation/cold box 453, 473
  Dewar vessel 99–101
  high-performance insulation 97–99
  low-performance insulation 97
thermal radiation 96, 101
thermodynamic analysis 44, 49–50
thermodynamic rule 3
thermomechanical stress 107
thermos flask 97, 100
thrust bearings 119
total plant cost 13, 14
total power consumption 13, 14, 194, 220, 245, 261, 305, 384, 452, 471
tube bundle 108–111, 114
turbine wheel or impeller 116
Turbo–Brayton refrigerators 235
turboexpanders 122, 383
  energy conservation 63–64
  energy loss 68–69
  generated power, loading devices 69–70
  isenthalpic expansion and expansion 70, 71
  non-reversible expansion 66–68
  reversible expansion process 64–66
two/multiple phase separators
  assessment 486
  mixed-fluid composition 486
  primary concept 484–486
  process variations 486
typical design procedure 12

v

vacuum furnace 105–107
variable frequency driver 148
virtual evaporator 410

w

warm expander Brayton stage 362–363, 371
warm-expander stream 344, 358, 360, 362–363, 371–372
working fluid, air-based 449